MATH
DICTIONARY
WITH SOLUTIONS

A Math Review 2nd

Chris Kornegay

SAGE Publications
International Educational and Professional Publisher
Thousand Oaks London New Delhi

For information:

SAGE Publications, Inc.
2455 Teller Road
Thousand Oaks, California 91320
E-mail: order@sagepub.com

SAGE Publications Ltd.
6 Bonhill Street
London EC2A 4PU
United Kingdom

SAGE Publications India Pvt. Ltd.
M-32 Market
Greater Kailash I
New Delhi 110 048 India

Printed in the United States of America

Library of Congress Cataloging-in-Publication Data

Kornegay, Chris.
    Math dictionary with solutions, 2nd: A math review / by Chris Kornegay.
        p.   c.m.
    Includes bibliographical references and index.
    ISBN 0-7619-1784-5 (cloth. — ISBN 0-7619-1785-3 (pbk.)
    1. Mathematics—Dictionaries.   2. Mathematics—Problems, exercises, etc.   I. Title.
    QA5.K692   1999
    510′.3—dc21   98-55316

99 00 01 02 03 04 05 7 6 5 4 3 2 1

Acquiring Editor: C. Deborah Laughton
Editorial Assistant: Eileen Carr
Production Editor: Diana E. Axelsen
Editorial Assistant: Nevair Kabakain
Book Designer: Ravi Balasuriya
Typesetter   Technical Typesetting, Inc.
Cover Designer: Candice Harman

Contents

Preface

Math Dictionary With Solutions is a compilation of as many major math topics as can be presented in a single volume. The articles vary from elementary math through beginning calculus and have been written in a style commensurate with the topics. In this manner, the readability of each article matches the level of difficulty of the concept being covered. A person interested in learning fractions can read the book just as easily as someone reading an article on derivatives.

The topics have been arranged alphabetically, each giving one or more of the following: a definition, an explanation, and examples with solutions. Within many of the articles are cross references. These can be used to clarify subtopics within the article. This allows the reader a means to backtrack to easier topics if the current one is too difficult or if the topic of discussion is not clear. At the back of the book is a natural logarithm table, a common logarithm table, a table of values of the trigonometric functions, a table of conversions, and an index.

Enjoy the success this book can bring. The personal growth of the reader has been my goal.

Dedicated to the
person who made me whole:
My wife, Meg

➤ **Abscissa**   The **abscissa** is the *x*-coordinate of a point. The **ordinate** is the *y*-coordinate of a point. Together they form an **ordered pair**. See **Coordinate Plane**.

➤ **Absolute Value**   The **absolute value** of a number is the distance the number is from the origin. Whether the number is on the right side or left side of the origin, the absolute value of the number is how far away the number is from 0.

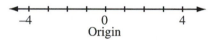

In the diagram, both 4 and −4 are 4 units away from the origin, 0. Hence, we say that the absolute value of each is 4, and we write $|4| = 4$ or $|-4| = 4$ where the symbol $|\ \ |$ means absolute value. As examples, we see that

$$|7| = 7, \qquad |-7| = 7, \qquad \left|2\tfrac{1}{3}\right| = 2\tfrac{1}{3}, \qquad \left|-2\tfrac{1}{3}\right| = 2\tfrac{1}{3},$$

$$\left|\tfrac{1}{2}\right| = \tfrac{1}{2}, \qquad \left|-\tfrac{1}{2}\right| = \tfrac{1}{2}$$

A simple definition, which works for numbers only, is to say that the absolute value of a positive number is positive and the absolute value of a negative number is also positive. However, this approach does not necessarily hold when it comes to algebra.

To apply the concept of absolute value to algebra, we can use the fact that the absolute value of −4 is $+4 = -(-4)$. Therefore, if we consider a number *x* on either side of the origin, we have two situations:

1. If *x* is on the right side of the origin, then the absolute value of *x* is just *x*, since *x* is a positive number.

2. If *x* is on the left side of the origin, then the absolute value of *x* must be −*x*, since *x* is a negative number.

A

It is the second situation that causes the most confusion. To clarify it, think of any number on the left side of the origin, say -4. Since its absolute value must be $+4$, we see that this is obtained by $-(-4)$. Hence, if x is a negative number, its absolute value is $-(x)$ or just $-x$.

Combining the two situations, we can formally write the algebraic definition of absolute value:

$$|x| = \begin{cases} x & \text{if } x \geqslant 0 \\ -x & \text{if } x < 0 \end{cases}$$

The first situation, $|x| = x$ if $x \geqslant 0$, means that if x is equal to 0 or greater than 0 (in other words if x is 0 or any number on the right side of 0), then the absolute value is x. The second situation, $|x| = -x$ if $x < 0$, means that if x is less than 0 (in other words if x is any number on the left side of 0), then the absolute value is $-x$.

For the absolute value of complex numbers, see **Complex Number, absolute value**. For the absolute value used in distance, see **Distance on a Number Line**.

Examples

1. Find the absolute value of the following: $|3|$, $|-3|$, $\left|\frac{4}{5}\right|$, and $|-\sqrt{2}|$.

   Since the absolute value of any number is positive, we have

   a. $|3| = 3$
   b. $|-3| = 3$
   c. $\left|-\dfrac{4}{5}\right| = \dfrac{4}{5}$
   d. $|-\sqrt{2}| = \sqrt{2}$

2. Simplify $|8 - 10|$.

   In order to do a problem like this, we first need to simplify the numbers inside the absolute value signs. After that we can evaluate the absolute value.

   $$|8 - 10| = |-2| = 2 \qquad \text{See } \textbf{Positive and Negative Numbers}.$$

3. Simplify $|-12 - 2| - |3|$.

   $$\begin{aligned} |-12 - 2| - |-3| &= |-14| - 3 \\ &= 14 - 3 \qquad \text{See } \textbf{Positive and Negative Numbers}. \\ &= 11 \end{aligned}$$

4. Simplify $\left|\frac{-22-3}{-6+14}\right|$.

   $$\left|\frac{-22 - 3}{-6 + 14}\right| = \left|\frac{-25}{8}\right| = \frac{25}{8}$$

5. If $a = -3$, $b = 2$, and $c = 5$, find $|a - c|^2$ and $\frac{|a-b|}{|c|-|a|}$.

   Substituting in each, we have

   $$|a - c|^2 = |-3 - 5|^2 = |-8|^2 = 8^2 = 64$$

   $$\frac{|a - b|}{|c| - |a|} = \frac{|-3 - 2|}{|5| - |-3|} = \frac{|-5|}{5 - 3} = \frac{5}{2}$$

6. Find $|x - 4|$ if $x \geqslant 4$.

   To deal with an algebraic expression, it helps to think of substitution numbers. Since $x \geqslant 4$ means that the x's we can use are either 4 or any number greater than 4, say 5, we see that $x - 4$ at its least is $4 - 4 = 0$, or $5 - 4 = 1$, or any number larger than this. If x is 6 then $6 - 4 = 2$. If x is 7 then $7 - 4 = 3$, etc. However, don't forget, that x is all of the numbers greater than 4, fractions, decimals, and irrational numbers as well as whole numbers.

   We see then that the expression $x - 4$ is either 0 or any number greater than 0 (positive). Since the absolute value of a positive number is the number itself, we have

   $$|x - 4| = x - 4$$

7. Find $|x - 4|$ if $x < 4$.

   Since $x < 4$ means x is always less than 4, the expression $x - 4$ is always negative. To see this, think of any number less than 4, say 3. When x is 3, the value of $x - 4$ is $3 - 4 = -1$, which is a negative number.

   This is the case for any number less than 4. Therefore, we see that whenever x is less than 4, the value of $x - 4$ is negative. Since the absolute value of a negative number is found by multiplying by a $-$, we have

   $$|x - 4| = -(x - 4) = -x + 4$$

   Combining examples 6 and 7, we have

   $$|x - 4| = \begin{cases} x - 4 & \text{if } x - 4 \geqslant 0 \quad \text{where } x - 4 \geqslant 0 \text{ is } x \geqslant 4 \\ -(x - 4) & \text{if } x - 4 < 0 \quad \text{where } x - 4 < 0 \text{ is } x < 4 \end{cases}$$

   This follows directly from the definition of absolute value. The absolute value of an algebraic expression is either itself, so to speak, if it is positive or $-$ itself (the negative of itself), if it is negative.

8. Find $|x + 2|$ if $x < -2$.

   If we choose some x-values less than -2, say -3 or -4, we see that the value of $x + 2$ is negative. Therefore, since the absolute value of a negative expression is $-$ itself (see example 7), we have

   $$|x + 2| = -(x + 2) = -x - 2$$

A

The other way of solving this problem is strictly by the definition of absolute value:

$$|x+2| = \begin{cases} x+2 & \text{if } x+2 \geqslant 0 \\ -(x+2) & \text{if } x+2 < 0 \end{cases} \quad \begin{array}{l} \text{where } x+2 \geqslant 0 \text{ is } x \geqslant -2 \\ \text{where } x+2 < 0 \text{ is } x < -2 \end{array}$$

Choosing the second situation, since $x+2 < 0$ (is negative), we have

$$|x+2| = -(x+2) = -x-2$$

9. Find $|x-2| + |x+3|$ if $x < 2$.

If $x < 2$ then $x - 2$ will always be negative (try substituting in some values for x). From this we know that

$$|x-2| = -(x-2) = -x+2$$

by the second part of the definition of absolute value. Similarly, we see that $x - 3$ is always negative or

$$|x-3| = -(x-3) = -x+3$$

Therefore, combining the two expressions, we have

$$|x-2| + |x-3| = -(x-2) + (-)(x-3) = -x+2-x+3 = -2x+5$$

10. Find $|x-2| + |x+3|$ if $2 < x < 3$.

If x is between 2 and 3, but not equal to 2 and 3, then $x - 2$ will be positive and $x - 3$ will be negative (substitute 2½ for x). Therefore,

$$|x-2| + |x-3| = x-2 + (-)(x-3) = x-2-x+3 = 1$$

► **Absolute Value Equations**   **Absolute value equations** are equations that have a variable inside the absolute value sign. To solve an absolute value equation, the absolute value must first be isolated, then the definition of an absolute value can be applied. By definition, an absolute value is composed of two situations:

$$|x| = \begin{cases} x & \text{if } x \geqslant 0 \\ -x & \text{if } x < 0 \end{cases}$$

As a result, we need to rewrite the absolute value equation as two equations, one which deals with the positive situation and the other which deals with the negative situation. Then each equation is solved. There are usually two answers to an absolute value equation.

Examples

1. Solve $-2 - 3|2x - 5| = -29$ for x.

First, we need to isolate the absolute value. We do this as though we were solving the linear equation $-2 - 3a = -29$ for a. See **Linear Equations, solutions of**.

$$\begin{aligned}-2-3a&=-29\\-3a&=-29+2\\-3a&=-27\\\frac{-3a}{-3}&=\frac{-27}{-3}\\a&=9\end{aligned} \qquad \text{or} \qquad \begin{aligned}-2-3|2x-5|&=-29\\-3|2x-5|&=-29+2\\-3|2x-5|&=-27\\\frac{-3|2x-5|}{-3}&=\frac{-27}{-3}\\|2x-5|&=9\end{aligned}$$

Next, we need to apply the definition of an absolute value:

$$|2x-5|=\begin{cases}2x-5 & \text{if } 2x-5\geqslant 0\\-(2x-5) & \text{if } 2x-5<0\end{cases}$$

The $|2x-5|=2x-5$ when $2x-5\geqslant 0$ or the $|2x-5|=-(2x-5)$ when $2x-5<0$. Each of these situations will give us a solution. Applying the definition and splitting the absolute value into two parts, we have

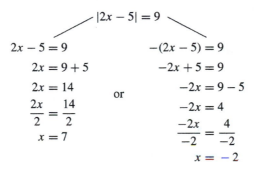

$$|2x-5|=9$$

$$\begin{aligned}2x-5&=9\\2x&=9+5\\2x&=14\\\frac{2x}{2}&=\frac{14}{2}\\x&=7\end{aligned} \qquad \text{or} \qquad \begin{aligned}-(2x-5)&=9\\-2x+5&=9\\-2x&=9-5\\-2x&=4\\\frac{-2x}{-2}&=\frac{4}{-2}\\x&=-2\end{aligned}$$

Check the answers in the original equation. They each yield -29.

Another way of solving $|2x-5|=9$ is to think of what could be inside the absolute value sign, $|?|=9$. Only a 9 or a -9 could be in there, $|9|=9$ or $|-9|=9$. This means that the expression $2x-5$ is either a 9 or a -9. Therefore, we have

$$|2x-5|=9$$

$$\begin{aligned}2x-5&=9\\2x&=9+5\\2x&=14\\\frac{2x}{2}&=\frac{14}{2}\\x&=7\end{aligned} \qquad \text{or} \qquad \begin{aligned}2x-5&=-9\\2x&=-9+5\\2x&=-4\\\frac{2x}{2}&=\frac{-4}{2}\\x&=-2\\x&=-2\end{aligned} \qquad \text{Set } 2x-5 \text{ equal to 9 and } -9.$$

2. Solve $-8+|2x+5|=0$ for x.

Solving by both of the methods described in example 1, we have

A

Solving by using the absolute value definition

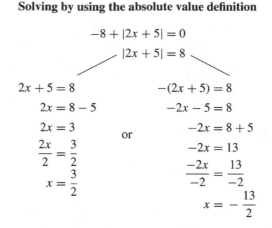

$$-8 + |2x + 5| = 0$$
$$|2x + 5| = 8$$

$2x + 5 = 8$ or $-(2x + 5) = 8$

$2x = 8 - 5$ $-2x - 5 = 8$

$2x = 3$ $-2x = 8 + 5$

$\dfrac{2x}{2} = \dfrac{3}{2}$ $-2x = 13$

$x = \dfrac{3}{2}$ $\dfrac{-2x}{-2} = \dfrac{13}{-2}$

                          $x = -\dfrac{13}{2}$

**Solving by thinking of what could be
in the absolute value**

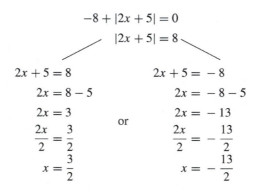

$$-8 + |2x + 5| = 0$$
$$|2x + 5| = 8$$

$2x + 5 = 8$ or $2x + 5 = -8$

$2x = 8 - 5$ $2x = -8 - 5$

$2x = 3$ $2x = -13$

$\dfrac{2x}{2} = \dfrac{3}{2}$ $\dfrac{2x}{2} = -\dfrac{13}{2}$

$x = \dfrac{3}{2}$ $x = -\dfrac{13}{2}$

Check the answers in the original equation.

3. Solve $3|6x| - 5 = 4$ for x.

Using the second method described in example 1, we have

$$3|6x| - 5 = 4$$
$$3|6x| = 4 + 5$$
$$3|6x| = 9$$
$$\dfrac{3|6x|}{3} = \dfrac{9}{3}$$
$$|6x| = 3$$

$6x = 3$ or $6x = -3$

$x = \dfrac{1}{2}$ $x = -\dfrac{1}{2}$

Check the answers in the original equation.

A

4. Solve $|3x + 14| - 2 = 5x$ for x.

$$|3x + 14| - 2 = 5x$$
$$|3x + 14| = 5x + 2$$

$3x + 14 = 5x + 2$	or	$-(3x + 14) = 5x + 2$
$3x - 5x = 2 - 14$		$-3x - 14 = 5x + 2$
$-2x = -12$		$-3x - 5x = 2 + 14$
$\dfrac{-2x}{-2} = \dfrac{-12}{-2}$		$-8x = 16$
$x = 6$		$\dfrac{-8x}{-8} = \dfrac{16}{-8}$
		$x = -2$

The solution $x = 6$ is the only answer. It satisfies the original equation when substituted. The solution $x = -2$ is not an answer since -2 substituted into the original equation yields $6 = -10$. This is why it is important to check your solutions. See **Extraneous Solutions (roots)**.

5. Solve $8 - |7x - 5| = 12$ for x.

$$8 - |7x - 5| = 12$$
$$-|7x - 5| = 12 - 8$$
$$\frac{-|7x - 5|}{-1} = \frac{4}{-1}$$
$$|7x - 5| = -4$$

Since absolute value cannot be negative, $|?| = -4$, there is no solution to this equation. The solution set is the empty set, $\varnothing$.

► **Absolute Value Inequalities**   **Absolute value inequalities** are inequalities that have absolute values with variables in them. To solve an absolute value inequality, we first need to isolate the absolute value. See **Absolute Value Equations**. After the absolute value has been isolated, we then do one of two things:

1. If the inequality sign is less than ($<$), solve the problem as a compound inequality. See **Inequalities, compound**.

2. If the inequality sign is greater than ($>$), apply the definition of absolute value and solve each inequality.

After the solution is obtained, write the answer in either set notation or interval notation. Sometimes a graph of the solution set is required.

Before beginning the examples, a familiarity with the article **Absolute Value Equations** is suggested. Also see **Inequality; Inequalities, graph on a number line;** or **Inequalities, compound**.

A

1. Solve the inequality $|x - 4| - 10 < -8$.

   We first need to isolate the absolute value:

$$|x - 4| - 10 < -8$$
$$|x - 4| < -8 + 10$$
$$|x - 4| < 2$$

Since the inequality sign is less than ($<$), we need to continue the problem as a compound inequality. Writing the inequality in compound form, we have

$$-2 < x - 4 < 2$$
$$-2 + 4 < x < 2 + 4 \qquad \text{Add 4 to all three parts.}$$
$$2 < x < 6$$

The solution implies that x is any number between 2 and 6. In set notation, the answer is written $\{x \mid 2 < x < 6\}$. See **Set Notation**. In interval notation, the answer is written $(2, 6)$.

See **Interval Notation**. The graph of the solution set is

See **Inequalities, graph on a number line**.

If solving by applying a compound inequality seems confusing, applying the definition of an absolute value will always work. Whether the inequality is $<$ or $>$, the method is the same. The only thing to keep in mind is that when the inequality is $<$ the solutions are combined by $\cap$ (and), and when the inequality is $>$ the solutions are combined by $\cup$ (or). See **Union and Intersection**.

To solve by applying the definition of an absolute value, we isolate the absolute value, apply the definition of absolute value, split the problem into two inequalities, and then solve each inequality. See **Inequalities, linear solutions of**.

$$|x - 4| - 10 < -8$$
$$|x - 4| < -8 + 10$$
$$|x - 4| < 2$$

$x - 4 < 2$		$-(x - 4) < 2$
$x < 2 + 4$	and	$-x + 4 < 2$
$x < 6$		$-x < 2 - 4$

Use an "and"

when the inequality

is $<$.

$$\frac{-x}{-1} > \frac{-2}{-1}$$
$$x > 2$$

When we divide by a "$-$," we turn the direction of the inequality sign around.

A

If we write the solution set at this point, we have $\{x \mid x > 2\} \cap \{x \mid x < 6\}$, or, in interval notation, we have $(2, \infty) \cap (-\infty, 6)$. However, if we notice that $x > 2$ is the same as $2 < x$, we can combine the inequalities and get the compound inequality $2 < x < 6$, which we've already seen.

The solutions in either case are acceptable although the compound inequality is preferable. So, if you solve a $<$ problem by applying the definition of an absolute value, just remember to write the solution as a compound inequality.

2. Solve the inequality $\left|\frac{x}{2} - 3\right| - 4 > -3$.

Since the inequality is $>$, we apply the definition of an absolute value:

$$\left|\frac{x}{2} - 3\right| - 4 > -3$$

$$\left|\frac{x}{2} - 3\right| > -3 + 4$$

$$\left|\frac{x}{2} - 3\right| > 1$$

$\dfrac{x}{2} - 3 > 1$		$-\left(\dfrac{x}{2} - 3\right) > 1$	Apply the definition
$\dfrac{x}{2} > 1 + 3$		$\dfrac{-x}{2} + 3 > 1$	of an absolute value.
$\dfrac{x}{2} > 4$	or	$-\dfrac{x}{2} > 1 - 3$	
$2 \bullet \dfrac{x}{2} > 2 \bullet 4$		$-\dfrac{x}{2} > -2$	
$x > 8$		$2 \bullet \left(-\dfrac{x}{2}\right) = 2 \bullet (-2)$	
Use an "or"		$-x > -4$	Divide by a "–," turn
when the inequality		$\dfrac{-x}{-1} < \dfrac{-4}{-1}$	the sign around.
is $>$.		$x < 4$	

In set notation, the answer is $\{x \mid x < 4\} \cup \{x \mid x > 8\}$. In interval notation, the answer is $(-\infty, 4) \cup (8, \infty)$. The graph of the solution set is

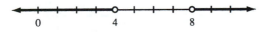

3. Solve the inequality $|2x + 4| - 6 \leqslant 2$.

Since the inequality is $\leqslant$, we can solve either by using a compound inequality or by applying the definition of an absolute value:

A

Compound Inequality

$$|2x + 4| - 6 \leqslant 2$$
$$|2x + 4| \leqslant 2 + 6$$
$$|2x + 4| \leqslant 8$$
$$-8 \leqslant 2x + 4 \leqslant 8$$
$$-8 - 4 \leqslant 2x \leqslant 8 - 4$$
$$-12 \leqslant 2x \leqslant 4$$
$$\frac{-12}{2} \leqslant \frac{2x}{2} \leqslant \frac{4}{2}$$
$$-6 \leqslant x \leqslant 2$$

Absolute Value

$$|2x + 4| - 6 \leqslant 2$$
$$|2x + 4| \leqslant 2 + 6$$
$$|2x + 4| \leqslant 8$$

$2x + 4 \leqslant 8$	$-(2x + 4) \leqslant 8$
$2x \leqslant 8 - 4$	$-2x - 4 \leqslant 8$
$\dfrac{2x}{2} \leqslant \dfrac{4}{2}$ and	$-2x \leqslant 8 + 4$
$x \leqslant 2$	$-2x \leqslant 12$
	$\dfrac{-2x}{-2} \geqslant \dfrac{12}{-2}$ Divide by a "−," turn
	$x \geqslant -6$ the sign around.

Since $x \geqslant -6$ is the same as $-6 \leqslant x$,
we have the compound inequality $-6 \leqslant x \leqslant 2$.

The solution set is $\{x \mid -6 \leqslant x \leqslant 2\}$. In interval notation, the solution is $[-6, 2]$. The graph of the solution set is

4. Solve the inequality $-2 + |4x + 2| \geqslant 12$.

Since the inequality is $\geqslant$, we apply the definition of an absolute value:

A

$$-2 + |4x + 2| \geqslant 12$$
$$|4x + 2| \geqslant 12 + 2$$
$$|4x + 2| \geqslant 14$$

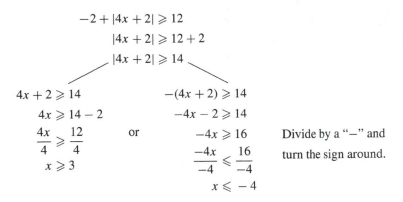

$4x + 2 \geqslant 14$		$-(4x + 2) \geqslant 14$	
$4x \geqslant 14 - 2$		$-4x - 2 \geqslant 14$	
$\dfrac{4x}{4} \geqslant \dfrac{12}{4}$	or	$-4x \geqslant 16$	Divide by a "−" and
$x \geqslant 3$		$\dfrac{-4x}{-4} \leqslant \dfrac{16}{-4}$	turn the sign around.
		$x \leqslant -4$	

The solution set is $\{x \mid x \leqslant -4\} \cup \{x \mid x \geqslant 3\}$. In interval notation, the solution is $(-\infty, -4] \cup [3, \infty)$. The graph of the solution set is

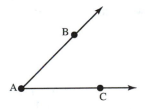

➤ **Absolute Value of a Complex Number**   See **Complex Number, absolute value**.

➤ **Acute Angle**   An **acute angle** is an angle with a measure greater than 0 and less than 90. In the figure, $\angle BAC$ (read "angle BAC"), is an acute angle.

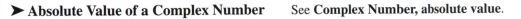

➤ **Acute Triangle**   If all the angles of a triangle are acute, the triangle is called an **acute triangle**.

➤ **Addition Formulas**   See **Sum and Difference Formulas**.

➤ **Addition of Decimal Numbers**   See **Decimal Numbers, addition of**.

➤ **Addition of Fractions**   See **Fractions, addition of**.

➤ **Addition of Whole Numbers**   **Addition of whole numbers** is a process that determines the combination of one number and another, or several numbers.

$$\begin{array}{r} \text{addend} \\ \text{addend} \\ + \text{addend} \\ \hline \text{sum} \end{array}$$

Each of the numbers to be combined is called an **addend** and the total combination is called the **sum**.

A

When adding two or more numbers, the units digit of each number must be aligned. Then each column can be added. If the sum of any column is greater than 9, write the units digit of the sum in the answer and carry the remaining figures to the next column. Then add the next column and so on. Columns are arranged from right to left.

Examples

1. Add 321 and 457.

   Since addition is commutative ($a + b = b + a$; see **Properties of the Real Number System**), we can add the numbers either way:

$$
\begin{array}{r} 321 \\ +\,457 \\ \hline 778 \end{array}
\quad \text{or} \quad
\begin{array}{r} 457 \\ +\,321 \\ \hline 778 \end{array}
$$

   The sum of 321 and 457 is 778.

2. Add 473 and 59.

   Aligning the units digits, i.e., the 3 and the 9, we have

$$
\begin{array}{r} 4^{1}73 \\ +\quad 59 \\ \hline 2 \end{array}
\qquad
$$
   Add 3 and 9 to get 12. Write the units digit (2) of the sum in the answer and carry the remaining figure (1) to the second column.

   Then add the second column:

$$
\begin{array}{r} {}^{1}4^{1}73 \\ +\quad 59 \\ \hline 32 \end{array}
\qquad
$$
   Add 1, 7, and 5 to get 13. Write the 3 in the answer and carry the 1 to the third column.

   Then add the third column:

$$
\begin{array}{r} {}^{1}4^{1}73 \\ +\quad 59 \\ \hline 532 \end{array}
\qquad
$$
   Add 1 and 4 to get 5.

   The sum of 473 and 59 is 532.

3. Add 5,964 and 352.

   Aligning the units digits, we have

$$
\begin{array}{r} {}^{1}5,{}^{1}964 \\ +\qquad 352 \\ \hline 6,316 \end{array}
$$

   After every third digit put in a comma.

4. Add 45, 631, 9 and 147.

$$\begin{array}{r} {}^2 45 \\ {}^1 631 \\ 9 \\ +\ 147 \\ \hline 832 \end{array}$$

5. Add 9,604; 476; 8; 4,876; 59; and 607.

$$\begin{array}{r} {}^2 9,{}^2 6{}^4 04 \\ 476 \\ 8 \\ 4,876 \\ 59 \\ +\quad 607 \\ \hline 15,630 \end{array}$$

➤ **Addition-Subtraction Method**   The **addition-subtraction method** is one of the methods used in solving systems of two linear equations in two unknowns. See **Systems of Linear Equations**.

➤ **Additive Identity**   The **additive identity** is one of the properties of the real number system. It states that for any real number a there is a unique number 0 such that $a + 0 = a$ and $0 + a = a$. The number 0 is called the **additive identity**. The equation $x + 6 = 10$ is solved by applying the additive identity.

$$\left. \begin{array}{r} x + 6 = 10 \\ x + 6 - 6 = 10 - 6 \\ x + 0 = 4 \\ x = 4 \end{array} \right\} \quad \text{The additive identity states that } x + 0 \text{ is } x.$$

See **Properties of the Real Number System**.

➤ **Additive Inverse**   The **additive inverse** is one of the properties of the real number system. It states that for any real number a there is a real number $-a$ such that $a + (-a) = 0$ and $(-a) + a = 0$. The equation $x + 6 = 10$ is solved by applying the additive inverse.

$$\left. \begin{array}{r} x + 6 = 10 \\ x + 6 - 6 = 10 - 6 \\ x + 0 = 4 \\ x = 4 \end{array} \right\} \quad \text{The additive inverse states that } 6 - 6 = 6 + (-6) = 0.$$

See **Properties of the Real Number System**.

➤ **Adjacent Angles**   **Adjacent angles** are two angles that share a common side, have the same vertex, but have no interior points in common. The angles must be in the same plane and combined may not be greater than $180°$.

A

In the figures, ∠1 and ∠2 are adjacent angles. And so are ∠*a* and ∠*b*.

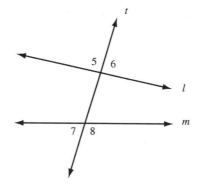

➤ **Algebra**   **Algebra** is a generalized form of arithmetic. It uses letters, positive and negative numbers, and various symbols to analyze concepts in terms of equations or formulas.

➤ **Algebraic Fractions**   See the various topics under **Rational Expression**.

➤ **Alternate Exterior Angles**   If two lines, not necessarily parallel, are intersected by a transversal, then the angles formed on the outside of the lines, and opposite each other, are called **alternate exterior angles**. In the figure, ∠5 and ∠8, and ∠6 and ∠7 are alternate exterior angles. If the lines *l* and *m* were parallel, then ∠5 ≅ ∠8 and ∠6 ≅ ∠7. See **Congruent Angles** or **Parallel Lines (geometric)**.

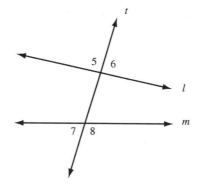

➤ **Alternate Interior Angles**   If two lines, not necessarily parallel, are intersected by a transversal, then the angles formed on the inside of the lines, and opposite each other, are called **alternate interior angles**. In the figure, ∠1 and ∠4, and ∠2 and ∠3 are alternate interior angles. If the lines *l* and *m* were parallel, then ∠1 ≅ ∠4 and ∠2 ≅ ∠3. See **Congruent Angles** or **Parallel Lines (geometric)**.

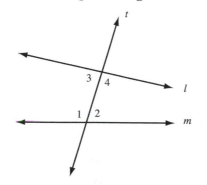

➤ **Altitude of a Triangle**   The **altitude of a triangle** is a segment drawn from a vertex and perpendicular to the opposite side or its extension. In the figure, the segment, $\overline{BD}$, is the altitude. Since there are three vertices, there are three altitudes to each triangle, one from each vertex.

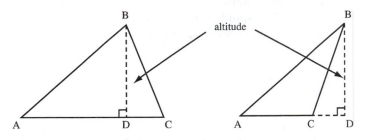

➤ **Amplitude**   Generally, the term **amplitude** pertains to trigonometry. In the equation $y = A \sin Bx$, the coefficient A determines the amplitude where the

$$\text{Amplitude} = |A|$$

The range of the function is $-A \leqslant y \leqslant A$. In reference to a graph, the amplitude represents how "high" or "low" the graph is drawn. For more, see **Trigonometry, graphs of the six functions**.

➤ **Analytic Geometry**   **Analytic geometry** is a branch of mathematics that uses algebraic techniques as a means to solve problems that are geometric in nature. Such an approach simplifies various topics in geometry and gives solutions to others that would otherwise be inaccessible.

➤ **Angle Bisector**   See **Bisector of an Angle**.

➤ **Angle, in geometry**   There are two articles on angles. This one considers angle in reference to geometry.

In geometry, an **angle** is the union of two noncollinear rays with the same endpoint. The rays are called the **sides** of the angle and their common endpoint, point A, is called the **vertex** of the angle. We refer to the angle in the diagram as $\angle BAC$ (read "angle BAC") with sides $\overrightarrow{AB}$ and $\overrightarrow{AC}$ with vertex A. We can also refer to the angle as $\angle 1$.

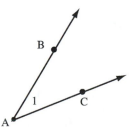

For more, see **Acute Angle, Obtuse Angle, Right Angle**, or **Congruent Angles**.

➤ **Angle, in trigonometry**   There are two articles on angles. This one considers angle in reference to trigonometry.

In trigonometry, an **angle** is determined by a different method than that of geometry. It is determined by rotating a ray around its endpoint.

A

In a coordinate system, the endpoint of the ray coincides with the origin and is called the **vertex** of the angle. The ray coinciding with the positive *x*-axis is called the **initial side**. The ray is then rotated around the origin in either a counterclockwise (positive) direction or a clockwise (negative) direction. The new position is called the **terminal side** of the angle. The angle formed between the initial side and the terminal side is called the **standard position** of the angle.

The following are examples of angles in standard position:

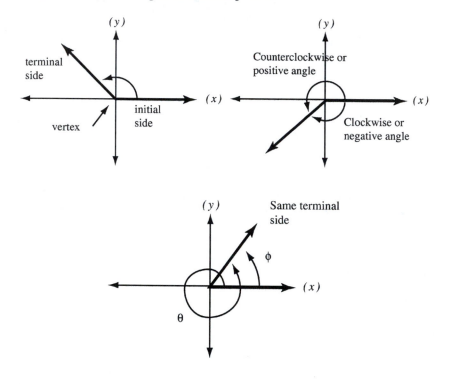

Examples

1. Sketch an angle of 120° or $\frac{2\pi}{3}$, where $\pi = 180°$.

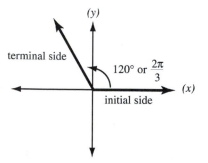

2. Sketch an angle of 225° or $\frac{5\pi}{4}$.

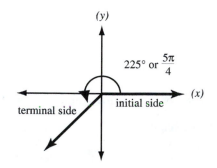

3. Sketch an angle of $-150°$ or $\frac{-5\pi}{6}$.

4. Sketch an angle of $-30°$ or $\frac{-\pi}{6}$.

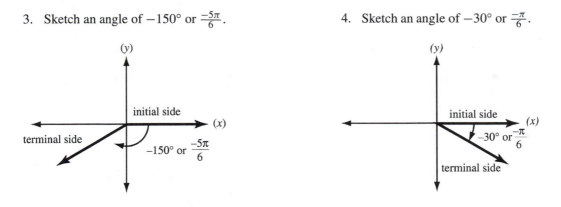

For related articles, see **Coterminal Angles**; **Trigonometry, reference angle**; **Radian**; **Converting Degrees to Radians**; or **Converting Radians to Degrees**.

➤ **Angle, negative**   See **Angle, in trigonometry**.

➤ **Angle of Depression**   The **angle of depression** is the angle formed from the horizontal, downward to the line of sight from an observer to an object.

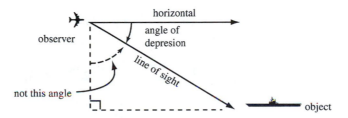

➤ **Angle of Elevation**   The **angle of elevation** is the angle formed from the horizontal, upward to the line of sight from an observed to an object.

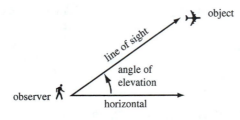

➤ **Angle of Rotation**   See **Angle, in trigonometry**.

➤ **Angle Sum**   See **Sum of the Measures of the Angles of a Triangle**.

➤ **Antilogarithm**   An **antilogarithm** is the number that corresponds to a given logarithm. It is the inverse of a logarithm. Consider the number 3. The logarithm of 3 is .4771 and the antilogarithm of .4771 is 3. See **Logarithm Table (chart), how to use**.

A

a. $\log_{10} 3 = .4771$ where .4771 is the logarithm of 3.

b. antilog $.4771 = 3$ where 3 is the antilogarithm of .4771.

It is easier to recognize the logarithm (exponent) and the antilogarithm (number) if $\log_{10} 3 = .4771$ is written in exponential form:

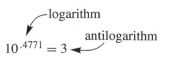

$$10^{.4771} = 3$$

If 10 is raised to an exponent (logarithm) of .4771, the result is the number 3 (antilogarithm). See **Logarithmic Form to Exponential Form, converting from**.

Since we already know that $\log_{10} 3 = .4771$, suppose we are given the equation $\log_{10} N = .4771$ and we are asked to find N. To find N, we need to take the antilog of both sides of the equation:

$$\log_{10} N = .4771$$
$$\text{antilog}(\log_{10} N) = \text{antilog} .4771$$
$$N = 3$$

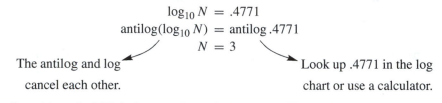

The antilog and log cancel each other.

Look up .4771 in the log chart or use a calculator.

The antilogarithm of .4771 is 3, as we knew it should be. The antilogarithm 3 is the number that corresponds to the given logarithm .4771.

Examples

1. Find N if $\log N = .7348$.

   To find N, we need to take the antilog of both sides:

   $$\log N = .7348$$
   $$\text{antilog}(\log N) = \text{antilog} .7348 \qquad \text{Look up .7348 in the log chart.}$$
   $$N = 5.43$$

2. Find N if $\log N = 2.7348$.

   $$\log N = 2.7348$$
   $$\text{antilog}(\log N) = \text{antilog} 2.7348$$
   $$N = 5.43 \qquad \text{with a characteristic of 2}$$
   $$N = 543$$

We look up the antilogarithm of the mantissa which is 5.43. We then move the decimal point two places because the characteristic is 2. The result is the number $N = 543$.

A

3. Find N if $\log N = 6.7348 - 10$.

$$\log N = 6.7348 - 10$$
$$\text{antilog}(\log N) = \text{antilog } 6.7348 - 10$$
$$N = 5.43 \qquad \text{with a characteristic of } -4$$
$$N = .000543$$

For more, see **Common Logarithms; Logarithm Table (chart), how to use; Logarithmic Form to Exponential Form, converting from**; or the various articles under logarithms.

➤ **Apothem** An **apothem** is the perpendicular segment connecting the center of a regular polygon to any one of its sides. It is perpendicular to the polygon at the point of tangency with its inscribed circle and bisects each of its segments. See **Polygon, Perpendicular (geometric)**, and **Inscribed (polygon)**. If a circle is inscribed in a regular polygon, then the segments $\overline{OA}, \overline{OB}, \overline{OC}, \overline{OD}$, and $\overline{OE}$, are apothems.

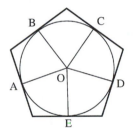

➤ **Arc** Let A, B, and C be three distinct points lying on a circle O. Then the **arc** determined by A, B, and C is the set of all points of the circle lying between A and C, through B, and including A and C. The arc from A to C, through B, is denoted $\overset{\frown}{ABC}$. If we consider another point D of the circle, the arc from A to C, through D, is denoted $\overset{\frown}{ADC}$.

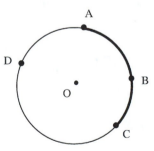

➤ **Arc Length (in geometry)** The length of an arc, or **arc length**, is that part of a circumference that is determined by the arc. It can be found by either of the following equations:

$$s = r\theta$$

where s is the arc length, r is the radius, and θ is the central angle subtended by the arc, measured in radians

or $\quad s = \dfrac{m}{180}\pi r \quad$ where m is the measure of the are measured in degrees.

A

Example

1. On a circle with a radius of 5 inches, find the length of the arc subtended by a central angle of 150°.

If we find the arc length by the first equation, we first need to change 150° to radians. See **Converting Degrees to Radians**. Using $1° = \frac{\pi}{180}$, we have

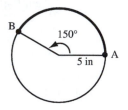

$$1° = \frac{\pi}{180}$$

$$(150)1° = \frac{\pi}{180}(150)$$

$$150° = \frac{150\pi}{180} = \frac{15\pi}{18}$$

$$150° = \frac{5\pi}{6}$$

Substituting into $s = r\theta$, we have

$$s = r\theta = 5\left(\frac{5\pi}{6}\right) = \frac{25\pi}{6} \approx \frac{25(3.14)}{6} \approx 13.08 \text{ inches}$$

If we find the arc length by the second equation, we have

$$s = \frac{m}{180}\pi r = \frac{150}{180}\pi(5) = \frac{25\pi}{6}$$

▶ **Arc Length (in calculus)**   If the function $y = f(x)$ is continuously differentiable on the interval $[a, b]$, then the **arc length** of $f(x)$, between a and b, is given by the formula

$$s = \int_a^b \sqrt{1 + [f'(x)]^2}\, dx$$

Or, if $x = g(y)$ is continuously differentiable on the interval $[c, d]$, then the **arc length** of $g(y)$, between c and d, is given by the formula

$$s = \int_c^d \sqrt{1 + [g'(y)]^2}\, dy$$

This can be seen by approximating the graph of $f(x)$ by n line segments whose endpoints are determined by the partition

$$a = x_0 < x_1 < x_2 < \cdots < x_{i-1} < x_i < \cdots < x_{n-1} < x_n = b \qquad \text{See } \textbf{Partition.}$$

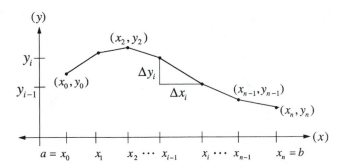

If we let $\Delta x_i = x_i - x_{i-1}$ and $\Delta y_i = y_i - y_{i-1}$, then, using the distance formula, the approximate length of the arc from a to b is

$$s \approx \sqrt{(x_1 - x_0)^2 + (y_1 - y_0)^2} + \sqrt{(x_2 - x_1)^2 + (y_2 - y_1)^2} + \cdots$$
$$+ \sqrt{(x_n - x_{n-1})^2 + (y_n - y_{n-1})^2}$$
$$\approx \sum_{i=1}^{n} \sqrt{(x_i - x_{i-1})^2 + (y_i - y_{i-1})^2} \approx \sum_{i=1}^{n} \sqrt{(\Delta x_i)^2 + (\Delta y_i)^2}$$

Now, if we take the limit as $n \to \infty$, we have

$$s = \lim_{n \to \infty} \sum_{i=1}^{n} \sqrt{(\Delta x_i)^2 + (\Delta y_i)^2} = \lim_{n \to \infty} \sum_{i=1}^{n} \sqrt{1 + \left(\frac{\Delta y_i}{\Delta x_i}\right)^2} \, \Delta x_i$$

where, by the mean value theorem (see **Mean Value Theorem**), there is a c_i, between x_i and x_{i-1}, such that

$$f(x_i) - f(x_{i-1}) = f'(c_i)(x_i - x_{i-1}) \qquad \text{and} \qquad f'(c_i) = \frac{\Delta y_i}{\Delta x_i}$$

Therefore, since $f'(x)$ is continuous on $[a, b]$ and $\sqrt{1 + [f'(x)]^2}$ is continuous (hence integrable) on $[a, b]$, we have

$$s = \lim_{n \to \infty} \sum_{i=1}^{n} \sqrt{1 + [f'(c_i)]^2} \, \Delta x_i = \int_{a}^{b} \sqrt{1 + [f'(x)]^2} \, dx$$

Examples

1. Find the distance between $(2, 3)$ and $(8, 11)$ by using the distance formula and then by using the formula for arc length.

By the Distance Formula

$$s = \sqrt{(x_2 - x_1)^2 + (y_2 - y_1)^2} = \sqrt{(8 - 2)^2 + (11 - 3)^2} = \sqrt{6^2 + 8^2} = 10$$

A

By the Formula for Arc Length

First, we need to find $f(x)$, where the slope $m = \frac{y_2 - y_1}{x_2 - x_1} = \frac{11-3}{8-2} = \frac{4}{3}$:

$$y - 3 = \frac{4}{3}(x - 2)$$

See **Linear Equation, finding the equation of a line**.

$$y = \frac{4}{3}x - \frac{8}{3} + 3$$

$$f(x) = y = \frac{4}{3}x + \frac{1}{3}$$

Next, using $f'(x) = \frac{4}{3}$, we have the distance

$$s = \int_a^b \sqrt{1 + [f'(x)]^2}\, dx = \int_2^8 \sqrt{1 + \left(\frac{4}{3}\right)^2}\, dx = \int_2^8 \sqrt{2\frac{7}{9}}\, dx$$

$$= \sqrt{2\frac{7}{9}} \int_2^8 dx = \sqrt{2\frac{7}{9}}\,(x)\Big|_2^8 = \sqrt{2\frac{7}{9}}\,(8 - 2) = 10$$

2. Find the arc length of $y = f(x) = \frac{x^5}{10} + \frac{1}{6x^3}$ on the interval $[1, 2]$.

First, we need to find $f'(x)$:

$$f'(x) = \frac{1}{10} \bullet 5x^4 + \frac{(-1)(18x^2)}{6x^3} = \frac{x^4}{2} - \frac{1}{2x^4}$$

Substituting $f'(x)$ into the formula for arc length, we have

$$s = \int_a^b \sqrt{1 + [f'(x)]^2}\, dx = \int_1^2 \sqrt{1 + \left(\frac{x^4}{2} - \frac{1}{2x^4}\right)^2}\, dx = \int_1^2 \sqrt{\frac{1}{2} + \frac{x^8}{4} + \frac{1}{4x^8}}\, dx$$

$$= \int_1^2 \sqrt{\frac{2x^8 + x^{16} + 1}{4x^8}}\, dx = \int_1^2 \frac{\sqrt{(x^8 + 1)^2}}{2x^4}\, dx = \int_1^2 \left(\frac{x^8}{2x^4} + \frac{1}{2x^4}\right) dx$$

$$= \frac{1}{2} \int_1^2 x^4\, dx + \frac{1}{2} \int_1^2 x^{-4}\, dx = \frac{1}{2}\left(\frac{x^5}{5}\right)\Big|_1^2 + \frac{1}{2}\left(\frac{x^{-3}}{-3}\right)\Big|_1^2$$

$$= \left(\frac{2^5}{10} - \frac{1^5}{10}\right) - \left(\frac{1}{6 \bullet 2^3} - \frac{1}{6 \bullet 1^3}\right) = \frac{779}{240}$$

➤ **Arc Measure**   The measure of a minor arc, or **arc measure**, is the same as the measure of its central angle. See **Central Angle of a Circle**. The measure of a **major arc** is equal to 360° minus the measure of the central angle of its minor arc. See **Major and Minor Arc**.

➤ **Arc of a Chord**   The phrase **arc of a chord**, unless otherwise specified, refers to the minor arc determined by the chord. See **Major and Minor Arc**.

A

➤ **Arccosine Function**   See **Trigonometry, inverse (arc) functions.**

➤ **Arcsine Function**   See **Trigonometry, inverse (arc) functions.**

➤ **Arctangent Function**   See **Trigonometry, inverse (arc) functions.**

➤ **Area**   **Area** represents the amount of square units that can be placed in a closed figure.

Examples

1. If a rectangle has a length of 4 and a height of 3, then it has 12 squares in it. The 12 squares represent the area of the rectangle. They can be measured in meters, inches, or any other unit of length.

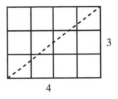

2. Draw a diagonal (dotted line) through the rectangle in example 1. Since the diagonal divides the rectangle into two equal halves, the two triangles formed each have a total of 6 squares per triangle. They each have an area of 6 square units.

➤ **Area Formula**   See **Trigonometry, area formula.**

➤ **Area (surface) of a Cylinder**   See **Area (lateral) of a Right Circular Cylinder (and total area).**

➤ **Area of a Regular Polygon**   The **area of a regular polygon** can be found from the formula

$$A = \frac{1}{2}aP$$

where A is the area of the polygon, a is the length of the apothem, and P is the perimeter of the polygon.

Example

1. Find the area of a regular hexagon with 4-inch sides.

$$A = \frac{1}{2}aP = \frac{1}{2}(2\sqrt{3})(24)$$
$$= 24\sqrt{3} \text{ in}^2$$

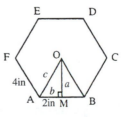

Reasons for the Solution

a. The **perimeter** of the polygon is

$$P = (6)(4) = 24 \text{ in}$$ See **Perimeter of a Polygon.**

A

b. To find the **apothem**, we first need to know two things:

1) The **central angle**, $\angle AOB$, has a measure of 60. This follows from the formula

$$\angle AOB = \frac{360°}{R} = \frac{360°}{6} = 60° \qquad \text{See Central Angle of a Regular Polygom.}$$

2) The $\triangle AOB$ is **equilateral** as well as **isosceles**. Since $OA = OB$,

$$m\angle OAB = m\angle OBA \qquad \text{See Isosceles Triangle.}$$

Also,

$$m\angle OAB + m\angle OBA + m\angle AOB = 180°$$
$$2m\angle OAB + 60° = 180° \qquad \text{Since } m\angle OAB = m\angle OBA.$$
$$2m\angle OAB = 120° \qquad \text{So does } m\angle OBA = 60°.$$
$$m\angle OAB = 60°$$

See **Sum of the Measures of the Angles of a Triangle**.

Therefore, since all of the angles of $\triangle AOB$ have a measure of 60, $\triangle AOB$ is equilateral.

See **Equilateral Triangle**.

c. The **apothem** can be found by either of two methods.

Since $\triangle AOB$ is equilateral, $AB = OA = c = 4$ in. Therefore, by the Pythagorean theorem,

$$2^2 + a^2 = 4^2$$
$$a^2 = 16 - 4$$
$$a^2 = 12$$
$$a = \sqrt{12} = \sqrt{4 \bullet 3} = 2\sqrt{3} \qquad \text{See Pythagorean Theorem.}$$

Or, since $\overline{OM}$ is perpendicular ($\perp$) to $\overline{AB}$, $\triangle OAM$ is a 30-60-90 degree triangle.

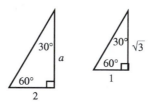

Therefore, by proportion of the triangles

$$\frac{a}{\sqrt{3}} = \frac{2}{1}$$
$$a = 2\sqrt{3} \qquad \text{See Thirty-Sixty-Ninety Degree Triangle.}$$

➤ **Area (lateral) of a Right Circular Cone (and total area)**   The **area (lateral) of a right circular cone** is the area of the outside of the cone excluding the area of the base. It is represented by the equation $L = \pi r s$ where L is the lateral area, s is the slant height, and r is the radius.

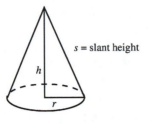

s = slant height

Example

1. Find the lateral area of a cone with a radius of 4 meters and a height of 12 meters.

   First, we use the Pythagorean theorem to find the slant height. See **Pythagorean Theorem**. Letting $h = 12$ and $r = 4$, we have

   $$s^2 = r^2 + h^2$$
   $$s^2 = 4^2 + 12^2 = 16 + 144 = 160$$
   $$\sqrt{s^2} = \sqrt{160}$$
   $$s = \sqrt{16 \bullet 10} = \sqrt{16} \bullet \sqrt{10} \quad \text{See } \textbf{Radicals, simplifying (reducing)}.$$
   $$s = 4\sqrt{10}$$

   Next, we find the lateral area with $r = 4$ and $s = 4\sqrt{10}$:

   $$L = \pi r s = \pi(4)\left(4\sqrt{10}\right) = 16\pi\sqrt{10}$$

   To find the total area, we add in the area of the base:

   $$\pi r^2 = \pi(4)^2 = 16\pi \quad \text{See } \textbf{Circle, area of}.$$
   $$\text{Total area} = \text{Lateral area} + \text{Area of base}$$
   $$= 16\pi\sqrt{10} + 16\pi$$
   $$= 16\pi\left(\sqrt{10} + 1\right) \quad \text{See } \textbf{Factoring Common Monomials}.$$

➤ **Area (lateral) of a Right Circular Cylinder (and total area)**
The **area (lateral) of a right circular cylinder (and total area)** is the outside area of a cylinder, not including the area of the bottom and the top. It is represented by the equation

$$L = 2\pi r h$$

where L is the lateral area, r is the radius, and h is the height.

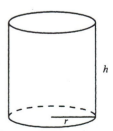

A

1. Find the lateral area of a right circular cylinder with a radius of 5 feet and height of 9 feet.

$$L = 2\pi rh = 2\pi(5)(9) = 290\pi \text{ ft}^2$$

To find the total area, add in the area of the top and the bottom:

$$\text{Area of top} = \text{Area of bottom} = \pi r^2 = \pi(5)^2 = 25 \text{ ft}^2 \qquad \text{See } \textbf{Circle, area of.}$$
$$\text{Total area} = \text{Lateral area} + \text{Area of top} + \text{Area of bottom}$$
$$= 90\pi + 25\pi + 25\pi = 140\pi \text{ ft}^2$$

The total area can also be found from the formula

$$A = 2\pi r(h + r)$$

where r is the radius of the cylinder and h is the height. Substituting $r = 5$ and $h = 9$ into the formula, we have

$$A = 2\pi r(h + r) = 2\pi(5)(9 + 5) = 2\pi(5)(14) = 140\pi \text{ ft}^2$$

➤ **Area of the Lateral Surface of a Regular Pyramid (and total area)**   The **lateral surface area of a pyramid** is the area comprising the sum of the area of the lateral faces. The equation for finding the lateral surface area is

$$L = \frac{1}{2}sP$$

where L is the lateral surface area, s is the slant height, and P is the perimeter of the base.

1. Find the lateral surface area of a regular hexagonal pyramid with a base edge of 12 inches and a slant height of 16 inches.

Since the perimeter of the base is

$$P = 6(12) = 72$$

we have

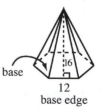

base

16

12
base edge

$$L = \frac{1}{2}sP = \frac{1}{2}(16)(72) = 576 \text{ in}$$

To find the total area, add in the area of the base:

$$A_{\text{base}} = \frac{1}{2}AP = \frac{1}{2}(6\sqrt{3})(72) = 216\sqrt{3} \text{ in}^2 \qquad \text{See } \textbf{Area of a Regular Polygon.}$$
$$\text{Total area} = \text{lateral surface area} + \text{Area of base}$$
$$= (576 + 216\sqrt{3}) \text{ in}^2$$

See **Pyramid** or **Regular Pyramid**.

➤ **Area of the Lateral Surface of a Right Prism (and total area)**   The **area of the lateral surface of a prism** is the area represented by the sum of the outside faces of the prism, not including the area of the top and bottom. The lateral surface area has an equation of

$$L = Pe$$

where L is the lateral surface area, P is the perimeter of the base, and e is the lateral edge or height.

Example

1. Find the lateral surface area of a hexagonal prism with a base that is 4 inches on a side and a lateral edge of 20 inches.

   First, we find the perimeter of the base

   $$P = (6)(4) = 24 \text{ in}$$

   Next, we find the lateral surface area

   $$L = Pe = 24(20) = 480 \text{ in}^2$$

   To find the total area, add in the area of the top and bottom:

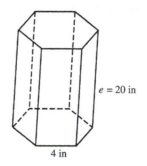

$e = 20$ in

4 in

$$\text{Area of either} = \frac{1}{2}aP = \frac{1}{2}2\sqrt{3}\,24 = 24\sqrt{3} \text{ in}^2 \qquad \text{See } \textbf{Area of a Regular Polygon}.$$
$$\text{Total area} = \text{Lateral surface area} + \text{Area of base} + \text{Area of top}$$
$$= 480 + 24\sqrt{3} + 24\sqrt{3} = 480 + 48\sqrt{3} \text{ in}^2$$

See **Prism, Regular Prism,** or **Radicals, combining**.

➤ **Arithmetic Means**   In an arithmetic progression, the terms between any two other terms of the progression are called **arithmetic means**. In the arithmetic progression 3, 7, 11, 15, 19 the terms 7, 11, and 15 are **arithmetic means**.

Examples

1. Find the four arithmetic means between −2 and 18.

   Since the numbers form an arithmetic progression, all we have to do is find the difference between each of them. This can be found from the equation

   $$l = a + (n - 1)d$$

A

where $l = 18$, $a = -2$, $n = 6$ (the total number of terms in the progression), and $d =$ the difference between each term. See **Arithmetic Progression**. Substituting, we have

$$18 = -2 + (6 - 1)d$$
$$18 + 2 = 5d$$
$$20 = 5d$$
$$4 = d$$
$$d = 4 \qquad \text{the difference between each term.}$$

Adding 4 to each term, we have the arithmetic progression

$$-2, \; 2, \; 6, \; 10, \; 14, \; 18$$

where 2, 6, 10, and 14 are the arithmetic means.

2. Find the three arithmetic means between 20 and 10.

Letting $l = 20$, $a = 10$, and $n = 5$, we have

$$10 = 20 + (5 - 1)d$$
$$-10 = 4d$$
$$\frac{-10}{4} = d$$
$$d = -2.5$$

Hence, the arithmetic progression is 20, 17.5, 15, 12.5, 10 where 17.5, 15, and 12.5 are the arithmetic means.

➤ **Arithmetic Progression** An **arithmetic progression** is a sequence of numbers in which one term subtracted from the next yields a common difference. The sequence 4, 7, 10, 13, 16, ... is an arithmetic progression where each term differs from the next by 3. The equation for finding any term in the progression is

$$l = a + (n - 1)d$$

where l is the last term of the sequence, a is the first term, and n is the total number of terms in the sequence.

Examples

1. Find the 10th term in the arithmetic progression 4, 7, 10, 13, 16, ...
   From the equation $l = a + (n - 1)d$ with $l =$ last term, $a = 4$, $n = 10$, and $d = 3$, we have

$$l = a + (n - 1)d = 4 + (10 - 1)3 = 4 + 27 = 31$$

which is the 10th term of the arithmetic progression. We can verify this by expanding the arithmetic progression to 10 terms:

$$4, 7, 10, 13, 16, 19, 22, 25, 28, \mathbf{31}, \ldots.$$

2. Find the eighth term in the arithmetic progression $7, 2, -3, \ldots$.

With $a = 7, n = 8$, and $d = -5$, we have

$$l = a + (n - 1)d = 7 + (8 - 1)(-5)$$
$$= 7 - 35 = -28 \quad \text{which is the eighth term.}$$

Verifying by expanding, we have $7, 2, -3, -8, -13, -18, -23, \mathbf{-28}$.

➤ **Arithmetic Sequence**   See **Arithmetic Progression**.

➤ **Arithmetic Series**   An **arithmetic series** is a sum of a given number of terms of an arithmetic progression. The sum of the first n terms of the series is given by

$$s = \frac{n}{2}(a + l)$$

where s is the sum, n is the number of terms in the series, a is the first term, and l is the last term of the series.

Example

1. Find the sum of the first 10 terms of the series $4 + 7 + 10 + \cdots$.

The number of terms $n = 10$, the first term $a = 4$, and the last term l can be found from the arithmetic progression formula (see **Arithmetic Progression**) with $d = 3$:

$$l = a + (n - 1)d = 4 + (10 - 1)(3) = 4 + 27 = 31$$

Substituting, we have

$$s = \frac{n}{2}(a + l) = \frac{10}{2}(4 + 31) = 5(35) = 175$$

We can verify this by adding the first 10 terms:

$$4 + 7 + 10 + 13 + 16 + 19 + 22 + 25 + 28 + 31 = 175$$

➤ **Ascending and Descending Order**   Polynomials are usually written in **descending order** of the exponents, i.e., $x^4 + 2x^3 - 5x^2 - x - 10$.

1. Polynomials in one variable are arranged so that the exponents either decrease in **descending order** or increase in **ascending order**.

2. Polynomials in several variables are arranged so that the exponents either **descend** or **ascend** according to a chosen variable.

A

Examples

1. One Variable:

   a. Arrange the polynomial $-3x + x^3 - 4 + 5x^8 - 4x^5$ in descending order.

   $$5x^8 - 4x^5 + x^3 - 3x - 4 \qquad \text{constants are placed at the end.}$$

   b. Arrange the polynomial $8x^9 - 5 - 7x + 12x^3$ in ascending order.

   $$-5 + 7x - 12x^3 + 8x^9 \qquad \text{constants are placed in the front.}$$

2. Several Variables:

   a. Arrange the polynomial $4 + y^5 - 6x^3 + 4x^3y + 9xy + x^2$ in descending powers of x.

   $$4x^3y - 6x^3 + x^2 + 9xy + y^5 + 4$$

   b. Arrange the polynomial $5xy^3 - 12x^3y^2 + 18xy - 7$ in ascending powers of y.

   $$-7 + 18xy - 12x^3y^2 + 5xy^3$$

➤ **Associative Property of Addition** The **associative property of addition** is one of the properties of the real number system. It states that, for any real numbers a, b, and c,

$$a + (b + c) = (a + b) + c$$

In general, this property allows for regrouping when adding.

Examples

1. The expressions $2 + (3 + 4)$ and $(2 + 3) + 4$ both simplify to the same number, 9.

2. Use the associative property of addition to simplify $8 + (9 + 4)$.

$$
\begin{aligned}
8 + (9 + 4) &= (8 + 9) + 4 \qquad \text{Regrouping the numbers.} \\
&= 17 + 4 \\
&= 21
\end{aligned}
$$

3. The associative property of addition is used to solve the equation $6x + (2x + 4) = 20$.

$$
\begin{aligned}
6x + (2x + 4) &= 20 \qquad \text{Regroup using the associative property of addition.} \\
(6x + 2x) + 4 &= 20 \\
8x + 4 &= 20 \\
8x &= 16 \\
x &= 2
\end{aligned}
$$

See **Properties of the Real Number System**.

➤ **Associative Property of Multiplication**   The **associative property of multiplication** is one of the properties of the real number system. It states that, for any real numbers a, b, and c,

$$a(bc) = (ab)c$$

In general, this property allows for regrouping when multiplying.

Examples

1. The expressions $3(4 \times 5)$ and $(3 \times 4)5$ both simplify to the same number, 60.

2. Use the associative property of multiplication to simplify $5(3 \times 6)$.

$$5(3 \times 6) = (5 \times 3)6 = 15 \times 6 = 90$$

See **Properties of the Real Number System**.

➤ **Asymptote**   An **asymptote** is a line that a graph may approach but never intersects.

Example

1. The graph of $f(x) = \frac{1}{x}$ is asymptotic to both the x- and y-axes (plural of axis). Each branch of the hyperbola will get closer and closer to each axis but will never touch or cross the axis. Hence, each axis is called an **asymptote**.

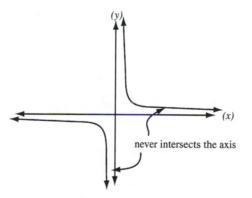

never intersects the axis

➤ **Augmented Matrix Solution**   See **Matrix, augmented (matrix) solution**.

➤ **Average**   See **Mean (average)**.

➤ **Axis**   In reference to a coordinate plane, the term **axis** refers respectively to the horizontal and vertical lines that determine the plane. See **Coordinate Plane**.

➤ **Axis of Symmetry**   In general, the **axis of symmetry** is used in reference to a parabola. Every parabola is symmetric with respect to this axis.
     For the parabola $y = ax^2 + bx + c$, the axis of symmetry is the vertical line passing through the vertex. The equation of the axis of symmetry is $x = -b/2a$.
     In conic sections, the axis of symmetry is referred to as the **principal axis**. See **Parabola**.

Bb

B

➤ **Base (of an exponent)** In reference to an exponent, the **base** is the term that is raised to a given exponent. When we write a^n, we refer to a as the base and n as the exponent. For application, see the articles under **Exponent** or **Exponents, rules of**.

➤ **Base Angles of an Isosceles Triangle**

In an isosceles triangle, the angles opposite the equal sides are referred to as the **base angles**.

In the figure, $\angle A$ and $\angle C$ are the **base angles**. For more, see **Isosceles Triangle**.

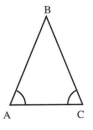

➤ **Base, change of** See **Change of Base Formula**.

➤ **Base, in reference to logarithms** See **Logarithm, base of**.

➤ **Base of an Isosceles Triangle**

In an isosceles triangle, the side opposite the vertex angle is referred to as the **base**.

In the figure, the segment $\overline{AC}$ is the **base**. For more, see **Isosceles Triangle**.

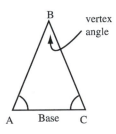

➤ **Base of Logarithms** See **Logarithm, base of**.

➤ **Base of a Trapezoid**

Each of the parallel sides of a trapezoid is referred to as a **base**.

In the trapezoid $ABCD$, the segments $\overline{BC}$ and $\overline{AD}$ are bases.

For more, see **Trapezoid**.

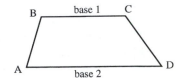

➤ **Bearing**   The term **bearing** is usually used in surveying and navigation. It measures the acute angle made between a north-south line and a path, or line, of sight. Measured from the north, the bearing in the figure is N45°E.

This means, an angle that is 45° east of north.

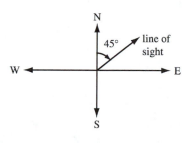

B

Examples

1. Draw an angle that describes a bearing of S30°E.

   First, we need to draw a north-south line, then from the south we need to draw a 30° angle to the east.

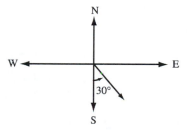

2. Describe the bearings in each of the figures below.

   In figure (a), measuring from the north, the bearing is N60°W. The 30° angle is not used. We would not say W30°N. Bearing is described in relation to a north-south line, not west-east.

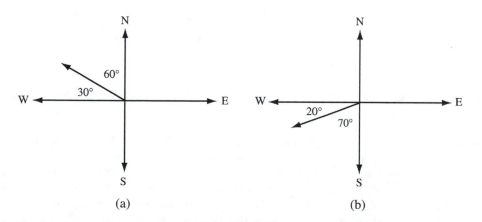

(a)   (b)

In figure (b), measuring from the south, the bearing is S70°W. The 20° angle is not used.

B

➤ **Betweenness of Points**   For three distinct points A, B, and C, B is said to be **between** A and C if the following are true:

1. A, B, and C are collinear.

2. $AB + BC = AC$.

This concept is referred to as **betweenness of points**. It is generally used in geometric proofs that relate to segments. See **Proof in Geometry**, example 2.

➤ **Biconditional (equivalent) Statement**   A statement that can be written in **if and only if** form, denoted by $p \rightarrow q$, is called a **biconditional** or **equivalent statement** (read "If p then q, and if q then p" or "p, if and only if q" or "p is equivalent to q"). A biconditional statement is a single statement composed of a conditional statement $p \rightarrow q$, see **Conditional Statement**, and its converse statement $q \rightarrow p$, see **Converse**. In a biconditional statement, either the conditional statement and its converse are both true, or both are false.

Examples

1. Write the following conditional and converse statements as a biconditional statement:

   Conditional: If two angles are congruent, then they have equal measures.

   Converse: If two angles have equal measures, then they are congruent.

   To write the biconditional statement, we need to combine the conditional and converse statements into an "if and only if" form. Hence, we have the biconditional statement:

   Two angles are congruent if and only if they have equal measures.

2. Separate the following biconditional statement into its conditional and converse statements: Two angles are complementary if and only if the sum of their measures is 90.

   The conditional statement is $p \rightarrow q$ and the converse statement is $q \rightarrow p$. Hence, we have

   Conditional: If two angles are complementary, then the sum of their measures is 90°.

   Converse: If the sum of the measures of two angles is 90, then they are complementary.

   For more, see **Logic**.

➤ **Bigger and Smaller**   For bigger and smaller decimals, see **Decimal Numbers, comparing**. For fractions, see **Fractions, comparing**.

➤ **Binomial**   A **binomial** is a polynomial with two terms. The following are binomials: $3x + 2$, $7x^2 - 12y^4$, and $8x^4y^3 - 7x^2y^2z^2$.

   Each of the two terms of a binomial is separated by a positive $(+)$ or negative $(-)$ sign, which belongs to the second term. It is this factor that determines a binomial, not the number of variables in each term. In the binomial $8x^4y^3 - 7x^2y^2z^2$, there are two terms separated by a $-$ sign. The term $8x^4y^3$ is composed of a constant 8 and two variables x and y while the term $-7x^2y^2z^2$ is composed of a constant -7 and three variables x, y, and z, each of which is squared. For more, see **Polynomial**.

➤ **Binomial Coefficient**   The coefficients in front of each of the terms in the expansion of $(a + b)^n$ are called **binomial coefficients**. For an expansion of $(a + b)^n$, see **Binomial Theorem**.

B

Examples

---

1. Use Pascal's triangle to find the binomial coefficients of $(a + b)^5$.

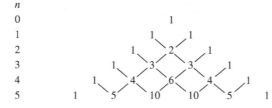

n

0   1

1   1   1

2   1   2   1

3   1   3   3   1

4   1   4   6   4   1

5   1   5   10   10   5   1

From the last row, with $n = 5$, the binomial coefficients in the expansion of $(a + b)^5$ are 1, 5, 10, 10, 5, and 1.

2. Use $\binom{n}{k}$ (read "*n choose k*" in relation to permutations and combinations) to find the **binomial coefficients** in the expansion of $(a + b)^4$.

The binomial coefficient

$$\binom{n}{k} = \frac{n!}{k!(n - k)!}$$

where n and k are nonnegative integers with $k \leqslant n$.

For an explanation of $n!$ (read "*n factorial*"), see **Factorial**. With $n = 4$ and k ranging from 0 to 4, we have

$$\binom{4}{0} = \frac{4!}{0!(4 - 0)!} = \frac{4!}{1(4!)} = 1$$

$$\binom{4}{1} = \frac{4!}{1!(4 - 1)!} = \frac{4 \cdot 3 \cdot 2 \cdot 1}{1(3 \cdot 2 \cdot 1)} = 4$$

$$\binom{4}{2} = \frac{4!}{2!(4 - 2)!} = \frac{4 \cdot 3 \cdot 2 \cdot 1}{2 \cdot 1(2 \cdot 1)} = 6$$

$$\binom{4}{3} = \frac{4!}{3!(4 - 3)!} = \frac{4 \cdot 3 \cdot 2 \cdot 1}{3 \cdot 2 \cdot 1(1)} = 4$$

$$\binom{4}{4} = \frac{4!}{4!(4 - 4)!} = \frac{4 \cdot 3 \cdot 2 \cdot 1}{4 \cdot 3 \cdot 2 \cdot 1(0!)} = 1 \qquad \text{where 0! is 1}$$

The **binomial coefficients** in the expansion of $(a + b)^4$ are 1, 4, 6, 4, and 1. Compare this with the row that corresponds to $n = 4$ in the Pascal triangle.

➤ **Binomial Theorem**   If n is a positive integer, then

$$(a + b)^n = a^n + \frac{n}{1}a^{n-1}b + \frac{n(n - 1)}{1 \cdot 2}a^{n-2}b^2 + \frac{n(n - 1)(n - 2)}{1 \cdot 2 \cdot 3}a^{n-3}b^3 + \cdots + b^n$$

Using the binomial coefficient $\binom{n}{k}$, this can also be written as

$$(a+b)^n = \binom{n}{0}a^n + \binom{n}{1}a^{n-1}b + \binom{n}{2}a^{n-2}b^2 + \binom{n}{3}a^{n-3}b^3 + \cdots + \binom{n}{n}b^n$$

For an explanation of $\binom{n}{k}$, see **Binomial Coefficient**, example 2.

B

Examples

1. Write the expansion of $(x - 3y)^4$.

    First, expand $(a+b)^4$ by either method described above. This process can be made easier by writing the a's in descending order, the b's in ascending order, and filling in the coefficients using the Pascal triangle with $n = 4$. See **Pascal's Triangle**.

    $$a^4 + 4a^3b + 6a^2b^2 + 4ab^3 + b^4$$

    Now let $a = x$, $b = -3y$, and substitute these values into the expansion of $(a + b)^4$:

    $$\begin{aligned}
    (a + b)^4 &= a^4 + 4a^3b + 6a^2b^2 + 4ab^3 + b^4 \\
    \left(x + (-3y)\right)^4 &= x^4 + 4x^3(-3y) + 6x^2(-3y)^2 + 4x(-3y)^3 + (-3y)^4 \\
    (x - 3y)^4 &= x^4 - 12x^3y + 54x^2y^2 - 108xy^3 + 81y^4
    \end{aligned}$$

2. Expand $(x - 3y)^4$ using the binomial coefficient $\binom{n}{k}$ only. In the expression $(a + b)^n$, let $a = x$, $b = -3y$, and $n = 4$, then

    $$\begin{aligned}
    (a + b)^n &= (x - 3y)^4 \\
    &= \underset{1}{\binom{4}{0}} x^4 + \underset{4}{\binom{4}{1}} x^{4-1}(-3y) + \underset{6}{\binom{4}{2}} x^{4-2}(-3y)^2 + \underset{4}{\binom{4}{3}} x^{4-3}(-3y)^3 \\
    &\quad + \underset{1}{\binom{4}{4}} x^{4-4}(-3y)^4 \\
    &= x^4 + 4x^3(-3y) + 6x^2 9y^2 + 4x\left(-27y^3\right) + 81y^4 \\
    &= x^4 - 12x^3y + 54x^2y^2 - 108xy^3 + 81y^4
    \end{aligned}$$

3. Find the third term of $(x - 3y)^4$.

    To find any rth term in the expression of $(a + b)^n$, use the formula

    $$\frac{n(n - 1)(n - 2) \cdots (n - r + 2)}{(r - 1)!} a^{n-r+1}b^{r-1}$$

    In our example, $a = x$, $b = -3y$, and $r = 3$ (the third term)

    $$\begin{aligned}
    &= \frac{4 \cdot 3}{(3 - 1)!} x^{4-3+1}(-3y)^{3-1} = \frac{12}{2!}x^2(-3y)^2 = \frac{12}{2 \cdot 1}x^2 \cdot 9y^2 \\
    &= 6x^2 \cdot 9y^2 = 54x^2y^2
    \end{aligned}$$

4. Find the third term of $(x - 3y)^4$ using the binomial coefficient $\binom{n}{k}$.

The formula for finding any rth term in the expansion of $(a + b)^n$ using $\binom{n}{k}$ is

$$\binom{n}{r-1} a^{n-r+1} b^{r-1}$$

In our example, $a = x$, $b = -3y$, $n = 4$, and $r = 3$:

$$= \binom{4}{3-1} x^{4-3+1}(-3y)^{3-1}$$

$$= \binom{4}{2} x^2(-3y)^2 \quad \text{since} \quad \binom{4}{2} = \frac{4!}{2!(4-2)!} = 6$$

$$= 6x^2 \cdot 9y^2 = 54x^2y^2$$

➤ **Bisect**   As a geometric term, to **bisect** means to separate or cut into two equal parts.

➤ **Bisector of an Angle**   The **bisector of an angle** is a ray that cuts the angle into two equal parts. By definition, $\overrightarrow{OB}$ is the bisector of $\angle AOC$ if $\overrightarrow{OB}$ is between $\overrightarrow{OA}$ and $\overrightarrow{OC}$ and $m\angle AOB = m\angle BOC$.

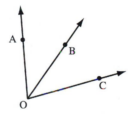

➤ **Bisector of a Segment**   The **bisector of a segment** is a segment, ray, or line that cuts the segment into two equal parts. By definition, $\overleftrightarrow{XY}$ is the bisector of $\overline{AB}$ if the intersection point M is between A and B and $AM = MB$.

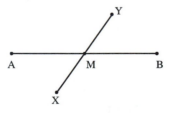

➤ **Bisector, perpendicular**   See **Perpendicular Bisector**.

➤ **Braces**   See **Parentheses**.

➤ **Brackets**   See **Parentheses**.

➤ **Branch of a Hyperbola**   See **Hyperbola**.

➤ **Calculus, articles on**

➤ **Cardinal Number**   A **cardinal number** tells how many things there are in a group. They are used to count the number in a group. The numbers 1, 2, 3, and so on are cardinal numbers. In contrast, cardinal numbers are distinguished from **ordinal numbers**. See **Ordinal Number**.

➤ **Cartesian Coordinates**   See **Coordinate Plane**.

➤ **Center of a Circle**   See **Circle**.

➤ **Center of a Hyperbola**   See **Hyperbola**.

➤ **Center of an Ellipse**   See **Ellipse**.

➤ **Central Angle of a Circle**   The **central angle** of a circle is an angle that has its vertex at the center of the circle. In the figure, $\angle AOB$ is a central angle.

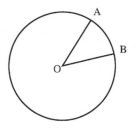

➤ **Central Angle of a Regular Polygon**   The **central angle** of a regular polygon is the angle formed by two consecutive radii. In a regular polygon of n sides, the measure of each **central angle** is $\frac{360}{n}$. See **Regular Polygon**.

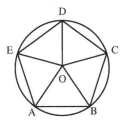

In the figure, $\overline{OD}$ and $\overline{OC}$ are consecutive radii. The angle they form, $\angle DOC$, is a central angle. Its measure is $\frac{360}{n} = \frac{360}{5} = 72$.

Example

1. Find the measure of the central angle of a regular polygon with eight sides (an octagon).

   The measure of the central angle $= \frac{360}{n} = \frac{360}{8} = 45$.

➤ **Chain Rule**   See **Derivative**, example 7.

➤ **Change a Repeating Decimal to a Common Fraction**   See **Decimal Numbers, changing a repeating decimal to a common fraction**.

➤ **Change Degrees to Radians**   See **Converting Degrees to Radians**.

➤ **Change Fractions to Decimals to Percents**   See **Fractions, change from fraction to decimal to percent**.

➤ **Change from Logarithmic Form to Exponential Form**   See **Logarithmic Form to Exponential Form, converting from**.

➤ **Change from Radical Form to Fractional Exponents**   See **Radical Form**.

➤ **Change of Base Formula**   To change a logarithm from one base to another, we apply the formula

$$\log_a N = \frac{\log_b N}{\log_b a}$$

Examples

1. Using a table of common logarithms or a calculator, find $\log_2 33$.

$$\log_2 33 = \frac{\log_{10} 33}{\log_{10} 2} = \frac{1.5185}{.3010} = 5.0449$$

See **Common Logarithms**.

2. Using a table of natural logarithms or a calculator, find $\log_2 33$.

Since $\log_e 33 = \ln 33$ and $\log_e 2 = \ln 2$, we have

$$\log_2 33 = \frac{\ln 33}{\ln 2} = \frac{3.4965}{.6931} = 5.0447$$

See **Natural Logarithms**.

The difference in the answers of examples 1 and 2 is due to round-off error. For more accuracy, more decimal places should be used.

➤ **Change Radians to Degrees**   See **Converting Radians to Degrees**.

➤ **Characteristic**   In common logarithms, the **characteristic** tells how many places the decimal point has been moved. The characteristic of a number is the same as the exponent of 10 in the scientific notation of the number. In scientific notation, the characteristic is the exponent of the 10, i.e., the 2 of 10^2. In logarithmic form, the characteristic is placed in front of the **mantissa**, i.e., the 2 of 2.7348. See **Logarithm Table (chart), how to use**.

Since common logarithms are numbers between 1.00 and 9.99, all other numbers must be written within this realm. This is accomplished by writing the numbers in scientific notation. As a result, the movement of the decimal becomes important. Thus, the purpose of the characteristic. See **Scientific Notation**.

Examples

1. Write the number 543 as a number between 1 and 10 and identify the characteristic.

Using scientific notation, we have

$$543 = 5.43 \bullet 10^2 \qquad \text{where the 2 of } 10^2 \text{ is the characteristic}$$

The 2 tells how many places the decimal point has been moved. The number 5.43 is between 1 and 10.

2. In this example, watch where the 2 of 10^2 goes.

Let $N = 543$ and calculate log 543. Then calculate the antilog.

$$N = 543$$
$$\log N = \log 543 = \log 5.43 \bullet 10^2 = \log 5.43 + \log 10^2$$
$$= \log 5.43 + 2\log 10 = .7348 + 2 = 2.7348$$

The characteristic 2 has been placed in front of the mantissa .7348. See **Logarithm Table (chart), how to use**. It tells how many places the decimal point has been moved in the number which corresponds to .7348. The mantissa .7348 corresponds with the number 5.43. See **Antilogarithm** and **Logarithm Table (chart), how to use**.

In the reverse process, the 2 tells us that the decimal point of the 5.43 must be moved 2 places. Thus, we have our original number 543 again:

$$\log N = 2.7348$$
$$\text{antilog}(\log N) = \text{antilog}(2.7348)$$
$$N = 543$$

The .7348 corresponds with 5.43, then move the decimal 2 places.

3. Negative Characteristics

A **negative characteristic** cannot be dealt with in the same manner as a positive characteristic. Putting a negative characteristic in front of the mantissa can lead to calculation mistakes. To avoid this, the following convention has been adopted: Instead of writing -4.7348 write $6.7348 - 10$ where the $6. \cdots - 10$ represents the -4.

4. Consider the number .000543. Let $N = .000543$. Find log N and then calculate the antilog.

$$N = .000543$$
$$\log N = \log .000543 = \log 5.43 \bullet 10^{-4}$$
$$= \log 5.43 + \log 10^{-4}$$
$$= \log 5.43 + (-4)\log 10 \qquad \log 10 \text{ is } 1.$$
$$= .7348 - 4 = 6.7348 - 10 \qquad \text{Write the } -4 \text{ as } 6. \cdots - 10.$$

Next, if $\log N = 6.7348 - 10$, find its antilog.

$$\log N = 6.7348 - 10$$
$$\text{antilog}(\log N) = \text{antilog } 6.7348 - 10$$
$$N = .000543$$

The .7348 corresponds with 5.43, then move the decimal -4 places left.

➤ **Charting Method of Graphing a Linear Equation**   See **Linear Equations, graphs of**, charting method.

➤ **Chord of a Circle**   The **chord** of a circle is a segment with endpoints on the circle. In the figure, each of the segments $\overline{AB}$, $\overline{CD}$, and $\overline{EF}$ are chords. Since the chord $\overline{CD}$ passes through the center of the circle, it is the diameter of the circle.

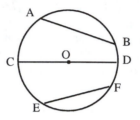

➤ **Circle**   A **circle** is the set of all points in a plane that are at a given distance from a fixed point of the plane. The fixed point is called the **center** of the circle and the given distance is called the **radius**. If the center of the circle is $C(h, k)$ and $P(x, y)$ is a point on the circle, then the standard form of the equation of the circle is

$$(x - h)^2 + (y - k)^2 = r^2$$

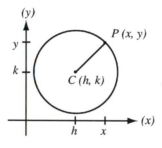

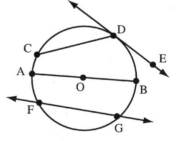

O is the **center** of the circle. $\overline{CD}$ is a **chord** of the circle.
$\overline{AB}$ is the **diameter** of the circle. $\overline{OB}$ is a **radius** of the circle.
$\overleftrightarrow{FG}$ is a **secant** to the circle. $\overleftrightarrow{DE}$ is a **tangent** to the circle.

Examples

1. Graph the circle $(x + 5)^2 + (y - 3)^2 = 16$.

   Comparing this equation to the standard form $(x - h)^2 + (y - k)^2 = r^2$, we have $h = -5, k = 3$, and $r = 4$, giving us the center at $(-5, 3)$ with a radius of $r = 4$.

First plot the points $(-5, 3)$. Then, move to the left 4 units to the point A. Also move right 4 units, to the point C. Finally, up and down to the points B and D. Then draw in the circle.

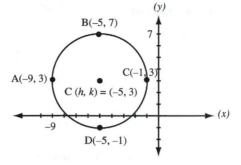

2. Graph the circle $x^2 + 10x + y^2 - 6y + 18 = 0$.

To graph the circle, we must first put the equation in standard form. This requires completing the square on the x-terms and the y-terms. See **Completing the Square**.

$$x^2 + 10x + y^2 - 6y + 18 = 0$$
$$x^2 + 10x + y^2 - 6y = -18$$
$$x^2 + 10x + 25 + y^2 - 6y + 9 = -18 + 25 + 9$$
$$(x + 5)(x + 5) + (y - 3)(y - 3) = 16$$
$$(x + 5)^2 + (y - 3)^2 = 4^2$$

Half of 10, squared is 25, then add to both sides. Also, half of -6 is -3, squared is 9, then add to both sides.

Now graph by the method described in example 1.

3. Graph $x^2 + y^2 - 16 = 0$.

First we put the equation in standard form.

$$(x - 0)^2 + (y - 0)^2 = 4^2 \qquad \text{where } (h, k) = (0, 0) \text{ and } r = 4.$$

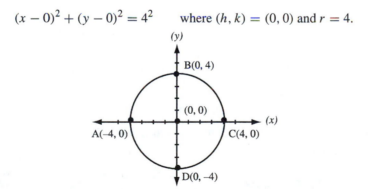

Plot the center $(0, 0)$, plot the four points left-right and up-down, then draw in the circle.

4. Find the equation of the circle given the center $C(-7, -4)$ and a point on the circle $P(1, 2)$.

To find the equation, we need to know the h, k, and r. If C is the center, then $h = -7$ and $k = -4$. To find r, we use the fact that $\overline{CP}$ is a radius. Its measure is the distance

$$r = CP = \sqrt{(x_2 - x_1)^2 + (y_2 - y_1)^2} = \sqrt{(-7 - 1)^2 + (-4 - 2)^2}$$
$$= \sqrt{64 + 36} = \sqrt{100} = 10$$

See **Distance Formula**.

Hence, we have the equation

$$(x - h)^2 + (y - k)^2 = r^2$$
$$\left(x - (-7)\right)^2 + \left(y - (-4)\right)^2 = 10^2$$
$$(x + 7)^2 + (y + 4)^2 = 100$$

➤ **Circle, area of**   The **area of a circle** with radius r may be found by applying the formula

$$A = \pi r^2$$

where π is the approximate number 3.14.

Example

---

1. Find the area of a circle with a radius of 4 inches. Letting $r = 4$ and substituting into the formula for area, we have

$$A = \pi r^2 = \pi (4)^2 = 16\pi = 16(3.14) = 50.24 \text{ in}^2$$

➤ **Circumference (of a circle)**   The distance around a circle is its **circumference**. The equation of the circumference is

$$C = 2\pi r$$

where C is the measure of the circumference, π is the number 3.14 approximately, and r is the measure of the radius.

Examples

---

1. Find the circumference of the circle in the diagram.

$$
\begin{aligned}
C &= 2\pi r \\
&= 2(3.14)(5) \\
&= 31.4 \text{ in}
\end{aligned}
$$

2. Find the circumference of the circle in the diagram.

Since the diameter is twice the radius, $r = 15$.
Hence,

$$C = 2\pi r$$
$$= 2(3.14)(15)$$
$$= 94.2 \text{ ft}$$

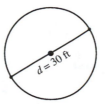

From this example, we see that the equation for the circumference of a circle can also be

$$C = \pi d$$

➤ **Circumscribed (polygon)**   A circle is **circumscribed** about a polygon if it contains each of the vertices of the polygon. In the figure, the circle O is circumscribed about the pentagon $ABCDE$.

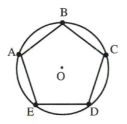

➤ **Closure Property**   See **Properties of the Real Number System**.

➤ **Coefficient**   The term **coefficient** is most often applied to the definition of a **numerical coefficient**. The numerical coefficient (or just coefficient) of the monomial $7x^2y$ is 7. The numerical coefficient of $-xy$ is -1. In the polynomial $6x^3 - 4x^2 - xy^2$, the coefficients are 6, -4, and -1, respectively.
   The term coefficient is also used in the definition of binomial coefficients and leading coefficients. See the articles under those names. Also see **Polynomial**.

➤ **Cofunction Identities**   The sine and cosine, cosecant and secant, and tangent and cotangent are each **cofunctions** of one another. The following relationships hold:

$$\sin\theta = \cos(90° - \theta) \quad \text{and} \quad \cos\theta = \sin(90° - \theta)$$
$$\csc\theta = \sec(90° - \theta) \quad \text{and} \quad \sec\theta = \csc(90° - \theta)$$
$$\tan\theta = \cot(90° - \theta) \quad \text{and} \quad \cot\theta = \tan(90° - \theta)$$

These concepts follow from the definition of complementary angles. See **Complementary Angles**. In the figure, $\triangle ABC$ is given as a right triangle with the right angle at C. Since the sum of the angles of a triangle is $180°$, we have (see **Sum of the Measures of the Angles of a Triangle**)

$$m\angle A + m\angle B + m\angle C = 180°$$
$$m\angle A + m\angle B + 90° = 180°$$
$$m\angle A + m\angle B = 90°$$
$$m\angle B = 90° - m\angle A$$

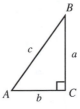

Now consider the $\sin A$ and the $\cos B$ (see related articles)

$$\sin A = \frac{a}{c} \quad \text{and} \quad \cos B = \frac{a}{c}$$

Since they both equal $\frac{a}{c}$, we have

$$\sin A = \cos B$$
$$\sin A = \cos(90° - m\angle A) \quad \text{by substituting } m\angle B = 90° - m\angle A.$$

Letting $m\angle A = \theta$, then $m\angle B = 90° - \theta$ and we have

$$\sin \theta = \cos(90° - \theta)$$

A similar proof can be done for each of the other cofunctions. The cofunction identities can also be derived using the sum and difference formulas. See the article under that heading.

Examples

1. Express $\cos 67°$, $\sec 15°42'$, and $\cot\frac{\pi}{6}$ in terms of their cofunctions.

$$\cos 67° = \sin(90° - 67°) = \sin 23°$$

$$\begin{aligned}
\sec 15°42' &= \csc\left(90° - 15°42'\right) \\
&= \csc\left(89°60' - 15°42'\right) \\
&= \csc 74°18' \qquad\qquad \textbf{See Degree-Minutes-Seconds.}
\end{aligned}$$

$$\begin{aligned}
\cot\frac{\pi}{6} &= \tan\left(\frac{\pi}{2} - \frac{\pi}{6}\right) = \tan\left(\frac{3\pi}{6} - \frac{\pi}{6}\right) \\
&= \tan\frac{2\pi}{6} = \tan\frac{\pi}{3} \qquad\qquad \textbf{See Radians.}
\end{aligned}$$

In each case, it can be seen that the two angles are complementary:

$$67° + 23° = 90° \qquad 15°42' + 74°18' = 90° \qquad \frac{\pi}{6} + \frac{\pi}{3} = \frac{\pi}{6} + \frac{2\pi}{6} = \frac{3\pi}{6} = \frac{\pi}{2}$$

2. In $\cot\frac{\pi}{6}$, convert the $\frac{\pi}{6}$ to degrees and find its cofunction. See **Converting Radians to Degrees.**

Converting $\frac{\pi}{6}$ to degrees, we have $\frac{\pi}{6} = \frac{180°}{6} = 30°$. Substituting, we have

$$\cot\frac{\pi}{6} = \cot 30° = \tan(90° - 30°) = \tan 60°$$

3. Using reference angles and cofunctions, find $\sin 115°$ and $\tan 300°$. See **Trigonometry, reference angle**. Also see **Trigonometry, signs of the six functions**.

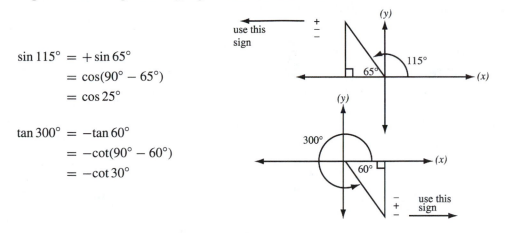

$$\sin 115° = +\sin 65°$$
$$= \cos(90° - 65°)$$
$$= \cos 25°$$

$$\tan 300° = -\tan 60°$$
$$= -\cot(90° - 60°)$$
$$= -\cot 30°$$

▶ **Collecting Like Terms** See **Combining Like Terms**.

▶ **Collinear Points** Points that lie on the same line are **collinear points**. Points A, B, and C are collinear while point D is **noncollinear**.

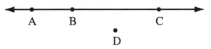

▶ **Combination** A **combination** is an arrangement of a set of elements without a specific order. If n objects (elements) are taken r at a time, then the number of combinations is

$$_nC_r = \frac{n!}{(n-r)!r!}$$ where n and r whole numbers.

The term $n!$ is read "n factorial" where

$$n! = 1 \bullet 2 \bullet 3 \bullet \cdots \bullet n \quad \text{or} \quad n! = n(n-1) \bullet (n-2) \bullet (n-3) \bullet \cdots \bullet 3 \bullet 2 \bullet 1$$

See **Factorial**. Other symbols used to represent a combination are $C(n, r)$, C_r^n and $\binom{n}{r}$.

The main difference between a permutation (see **Permutation**) and a combination is the order of the elements. In a permutation, the specific order of the elements is considered, whereas in a combination, it is not.

The number of permutations that can be determined for the numbers 1, 2, and 3 is 6:

$$1\,2\,3 \quad 1\,3\,2 \quad 2\,1\,3 \quad 2\,3\,1 \quad 3\,1\,2 \quad 3\,2\,1$$

This follows from the permutation equation for 3 objects taken 3 at a time:

$$_3P_3 = \frac{3!}{(3-3)!} = \frac{1 \bullet 2 \bullet 3}{0!} = \frac{6}{1} = 6 \quad \text{where 0! is defined to be 1.}$$

However, the number of combinations that can be determined for the numbers 1, 2, and 3 is 1. The order of 123, 132, and so on represents the same combination. This follows from the combination equation for 3 objects taken 3 at a time:

$$_3C_3 = \frac{3!}{(3-3)! \bullet 3!} = \frac{3!}{0! \bullet 3!} = \frac{1}{1} = 1$$

The following are arrangements that are represented by a permutation:

1. People sitting in a row.

2. People sitting around a table.

3. Three numbers chosen in a specific order from the sequence 1, 2, 3, 4, 5, 6.

4. The answers from a true-false test.

The following are arrangements that are represented by a combination:

1. The starting five players of a basketball team of 12 (the specific order of the five players is not necessary, a permutation if it is).

2. A card hand composed of seven cards (the specific order of the cards is not considered).

3. A group of six books chosen from a list of 11 (the specific order of each six is not considered).

Examples

1. Determine the number of ways 3 class officers can be chosen from a class of 30 students.

   Since the specific order of the choices of the class officers is not a consideration, this is a combination (not permutation) problem. Letting $n = 30$ and $r = 3$, we have

   $$_{30}C_3 = \frac{3!}{(3-3)! \bullet 3!} = \frac{30!}{27! \bullet 3!} = \frac{30 \bullet 29 \bullet 28 \bullet 27 \bullet \cdots \bullet 2 \bullet 1}{27 \bullet 26 \bullet 25 \bullet \cdots \bullet 2 \bullet 1 \bullet 1 \bullet 2 \bullet 3}$$
   $$= \frac{30 \bullet 29 \bullet 28}{1 \bullet 2 \bullet 3} = \frac{24,360}{6} = 4,060$$

2. Find the number of lines that are determined by 9 points if no 3 of the points are collinear (on the same line).

   Since two points determine a line, we want to find out how many lines can be drawn between 9 points taken 2 at a time. With $n = 9$ and $r = 2$, we have

   $$_9C_2 = \frac{9!}{(9-3)! \bullet 2!} = \frac{9!}{7! \bullet 2!}$$
   $$= \frac{9 \bullet 8 \bullet 7 \bullet \cdots \bullet 2 \bullet 1}{7 \bullet 6 \bullet \cdots \bullet 2 \bullet 1} = \frac{72}{2} = 36$$

3. How many committees of 5 women and 7 men can be formed from a group composed of 10 women and 11 men?

There are three stages to answering this problem: find the number of committees composed of women, find the number of committees composed of men, and then, using the counting principle (see **Counting Principle**), find the product of these three values:

Women

$$_{10}C_5 = \frac{10!}{(10-5)! \bullet 5!} = \frac{10!}{5! \bullet 5!}$$

$$= \frac{10 \bullet 9 \bullet 8 \bullet 7 \bullet 6}{1 \bullet 2 \bullet 3 \bullet 4 \bullet 5} = \frac{30,240}{120} = 252$$

Men

$$_{11}C_7 = \frac{11!}{(10-7)! \bullet 7!} = \frac{11!}{4! \bullet 7!}$$

$$= \frac{11 \bullet 10 \bullet 9 \bullet 8}{1 \bullet 2 \bullet 3 \bullet 4} = \frac{7,920}{24} = 330$$

Number of committees composed of 5 women and 7 men: $(252)(330) = 83,160$.

4. In how many ways can 7 cards be chosen from a standard deck of 52 cards so that 2 of the cards are from one suit and 5 of the cards are from another?

There are four stages to answering this problem: determining which two suits (of the four suits) are to be considered, finding out how many ways 2 cards can be chosen from 13 cards, finding out how many ways 5 cards can be chosen from 13 cards, and then, using the counting principle (see **Counting Principle**), finding the product of these values.

Determining which two suits of the four must be considered is a permutation problem, since we must consider specific order. Letting $n = 4$ and $r = 2$, we have (see **Permutation**)

$$_4P_2 = \frac{4!}{(4-2)!} = \frac{4!}{2!} = 12$$

Letting $n = 13$ and $r = 2$, we can find the number of ways, or combinations, of choosing 2 cards of the same suit from 13:

$$_{13}C_2 = \frac{13!}{(13-2)!2!} = \frac{13!}{11!2!} = \frac{13 \bullet 12}{1 \bullet 2} = 78$$

Letting $n = 13$ and $r = 5$, we can find the number of ways of choosing 5 cards of the same suit from 13:

$$_{13}C_5 = \frac{13!}{(13-5)!5!} = \frac{13!}{8!5!} = \frac{13 \bullet 12 \bullet 11 \bullet 10 \bullet 9}{1 \bullet 2 \bullet 3 \bullet 4 \bullet 5} = \frac{154,440}{120} = 1,287$$

The number of ways of choosing 7 cards so that 2 are from one suit and 5 are from another is

$$(12)(78)(1,287) = 1,204,632$$

5. Expand $(x + y)^5$ using combinations as coefficients.

From the Pascal triangle, we know that the coefficients for the expansion of $(x + y)^5$ are 1, 5, 10, 10, 5, and 1. See **Binomial Coefficient**. If we compare this sequence to the one generated by calculating

$_nC_r$ for $n = 5$ and r varying from 0 to 5, we notice an equivalence:

$$_5C_0 = \frac{5!}{(5-0)!0!} = \frac{5!}{5!0!} = 1 \qquad _5C_1 = \frac{5!}{(5-1)!1!} = \frac{5!}{4!1!} = 5$$

$$_5C_2 = \frac{5!}{(5-2)!2!} = \frac{5!}{3!2!} = 10 \qquad _5C_3 = \frac{5!}{(5-3)!3!} = \frac{5!}{2!3!} = 10$$

$$_5C_4 = \frac{5!}{(5-4)!4!} = \frac{5!}{1!4!} = 5 \qquad _5C_5 = \frac{5!}{(5-5)!5!} = \frac{5!}{0!5!} = 1$$

We get the same sequence. Hence, we can write the expansion of $(x + y)^5$ using combinations instead of coefficients:

$$(x + y)^5 = 1x^5 + 5x^4y + 10x^3y^2 + 10x^2y^3 + 5xy^4 + 1y^5$$
$$= {}_5C_0x^5 + {}_5C_1x^4y + {}_5C_2x^3y^2 + {}_5C_3x^2y^3 + {}_5C_4xy^4 + {}_5C_5y^5$$

Or, using the notation $\binom{n}{r}$ instead of $_nC_r$ to represent a combination, we have

$$(x + y)^5 = \binom{5}{0}x^5 + \binom{5}{1}x^4y + \binom{5}{2}x^3y^2 + \binom{5}{3}x^2y^3 + \binom{5}{4}xy^4 + \binom{5}{5}y^5$$

In general, the expansion of $(x + y)^n$ can be written with coefficients that are combinations:

$$(x + y)^n = \binom{n}{0}x^n + \binom{n}{1}x^{n-1}y + \binom{n}{2}x^{n-2}y^2 + \cdots + \binom{n}{n}y^n$$

Any rth term in the expansion of $(x + y)^n$ is represented by

$$\binom{n}{r-1}a^{n-r+1}b^{r-1}$$

where

$$\binom{n}{r-1} = \frac{n(n-1)(n-2)\cdots(n-r+2)}{(r-1)!}$$

is the coefficient of the rth term. See **Binomial Theorem**.

► **Combined Variation**   **Combined variation** is a mathematical situation in which both direct variation and inverse variation occur at the same time. Combined variation is a linear function defined by the equation $y = \frac{kxz}{w}$, or $k = \frac{yw}{xz}$ where k is a nonzero constant called the **constant of variation**. In a combined variation, y is said to vary jointly with x and z and inversely with w.

Combined variation can be written as a **proportion**. Suppose x_1, y_1, w_1, and z_1 as well as x_2, y_2, w_2, and z_2 satisfy the equation $k = \frac{yw}{xz}$. Then $k = \frac{y_1w_1}{x_1z_1}$ and $k = \frac{y_2w_2}{x_2z_2}$. Substituting, we have

$$k = k$$
$$\frac{y_1w_1}{x_1z_1} = \frac{y_2w_2}{x_2z_2}$$

Example

1. Let y vary jointly as x and z and inversely with w. If $y = 16$ when $x = 2$, $z = 8$, and $w = 9$, find y when $x = 10$, $z = 5$, and $w = 25$.

Substituting into the proportion for combined variation, we have

$$\frac{16 \bullet 9}{12 \bullet 8} = \frac{y \bullet 25}{10 \bullet 5}$$
$$9 = y$$
$$y = 9$$

For more, see **Direct Variation, Inverse Variation, Joint Variation**, or **Proportion**.

➤ **Combining Like Terms**   **Combining like terms** is a process of adding or subtracting terms that are alike. The terms differ only in their numerical coefficients. The following terms are like terms: $-6x$ and $25x$, $-6xy$ and $25xy$, $-6(x + 3)$ and $25(x + 3)$, $-6x^2yz^3$, and $25x^2yz^3$, $-6\sqrt{5}$ and $25\sqrt{5}$, and $-6\sin x$ and $25\sin x$. These terms differ only in their numerical coefficients.

To combine like terms, the concepts of positive and negative numbers must be completely clear. See **Positive and Negative Numbers**.

Examples

1. Simplify $-6x + 25x$.

$$-6x + 25x = 19x$$

To do this, a student must know that $-6 + 25 = 19$. See **Positive and Negative Numbers**.

2. Simplify $6 + 9x + 8y + 5x - 4 - 3y$.

$$6 + 9x + 8y + 5x - 4 - 3y = 9x + 5x + 8y - 3y + 6 - 4 \qquad \text{Group the like terms.}$$
$$= 14x + 5y + 2$$

To group like terms, the concepts of the **commutative** and **associative** properties are applied.

3. Simplify

$$-4 - (6y - 7) + 2y = -4 - 6y + 7 + 2y = -6y + 2y - 4 + 7 = -4y + 3$$

Multiply into the parentheses using the distributive property. See **Distributive Property**. For more, see the various articles under **Polynomial**.

➤ **Combining Radicals**   See **Radicals, combining**.

➤ **Common Denominator**   In two or more fractions, a **common denominator** is a number that can be divided by all of the denominators. If the number is the smallest possible, then it is the **lowest common denominator**.

Examples

1. Find the lowest common denominator for $\frac{1}{2}$ and $\frac{1}{3}$.

   Method 1:

   Think of numbers that can be divided by both 2 and 3. Some numbers are 12, 6, 24, and 18. Each of these is a common denominator. The smallest, however, is 6. The number 6 is the lowest common denominator.

   Method 2:

   Another way to find the lowest common denominator is to look at a list of the multiples of the denominators. The least common multiple of the list is the lowest common denominator. See **Least Common Multiple**.

   $$2, 4, \textbf{6}, 8, 10, 12, 14, \ldots$$
   $$3, \textbf{6}, 9, 12, 15, 18, \ldots$$

   In each list, the 6 is the least common multiple. There are other common multiples, i.e., 12, but 6 is the lowest. Therefore, 6 is the lowest common denominator.

2. Find the lowest common denomenator of $\frac{1}{2}$ and $\frac{1}{5}$.

   Method 1:

   Thinking of numbers that can be divided by 2 and 5, 10 is the lowest.

   Method 2:

   $$\left. \begin{array}{l} 2, 4, 6, 8, \textbf{10}, 12, \ldots \\ 5, \textbf{10}, 15, \ldots \end{array} \right\} \quad \text{10 is the lowest.}$$

3. Find the lowest common denominator in the following examples.

   a. The lowest common denominator for $\frac{1}{2}$ and $\frac{1}{4}$ is 4.

   b. The lowest common denominator for $\frac{3}{4}$ and $\frac{5}{6}$ is 24.

   c. The lowest common denominator for $\frac{1}{2}$, $\frac{1}{3}$ and $\frac{1}{4}$ is 12.

   d. The lowest common denominator for $\frac{4}{5}$, $\frac{1}{4}$, and $\frac{2}{3}$ is 60.

4. Change $\frac{1}{2}$ and $\frac{1}{3}$ to equivalent fractions with common denominators.

   The $\frac{1}{2}$ and $\frac{1}{3}$ have a common denominator of 6 (see example 1). We need to change $\frac{1}{2}$ and $\frac{1}{3}$ each to equivalent fractions with denominators of 6. See **Fractions, equivalent**.

   $$\frac{1}{2} = \frac{1}{2} \bullet \frac{3}{3} = \frac{3}{6}$$
   $$\frac{1}{3} = \frac{1}{3} \bullet \frac{2}{2} = \frac{2}{6}$$

   Hence, $\frac{3}{6}$ and $\frac{2}{6}$ are the equivalent fractions of $\frac{1}{2}$ and $\frac{1}{3}$ with common denominators of 6.

5. Change $\frac{3}{4}$ and $\frac{2}{6}$ to equivalent fractions with a common denominator.

The common denominator for 4 and 5 is 20. Hence,

$$\frac{3}{4} = \frac{3}{4} \bullet \frac{5}{5} = \frac{15}{20}$$
$$\frac{2}{5} = \frac{2}{5} \bullet \frac{4}{4} = \frac{8}{20}$$

➤ **Common Divisor**   A **common divisor** is a common factor. See **Common Factor**.

➤ **Common Factors**   A term (or number) that is a factor of each of two or more terms (or numbers) is a **common factor** of the terms (or numbers).

Examples

1. The number 4 is a common factor of the numbers 8, 12, and 16 because $8 = 4 \bullet 2$, $12 = 4 \bullet 3$, and $16 = 4 \bullet 4$.

2. The term a^2 is a common factor of the terms a^2b, a^3b^2, and a^4 because $a^2b = a^2 \bullet b$, $a^3b^2 = a^2 \bullet ab^2$, and $a^4 = a^2 \bullet a^2$.

➤ **Common Logarithms**   **Common logarithms** are logarithms with a base 10. In calculations, we usually write $\log x$ instead of $\log_{10} x$, read "log to the base 10 of x." Remember that a logarithm is an exponent. In common logarithms, it is the exponent to which 10 is raised to yield a given number. If $\log_{10} 100 = 2$, then $10^2 = 100$. See **Logarithmic Form to Exponential Form, converting from**.

To the beginning student, logarithms sometimes seem difficult and confusing. With our daily usage of calculators, it's hard to imagine logarithms were once used to simplify calculations. Because logarithms have also been found useful in calculus, in problems related to nature and finance, and in solving exponential equations, they remain with us still.

Examples

1. Find $\log 842,000$ to the nearest ten-thousandth.

$842,000 = 8.42 \times 10^5$	See **Scientific Notation**.
$\log 842,000 = \log 8.42 \times 10^5$	Take logarithms of both sides **Logarithmic**
$= \log 8.42 + \log 10^5$	**Functions**.
$= \log 8.42 + 5 \log 10$	See **Logarithms, properties of (rules) of**.
$= \log 8.42 + 5(1)$	
$= .9253 + 5$	See **Characteristic** and **Logarithm Table**
$= 5.9253$	**(chart), now to use**.

Hence, $\log 842,000 = 5.9253$.

Since a logarithm is an exponent, 5.9253 is an exponent. It is the exponent to which 10 must be raised to yield 842,000. This is to say that $10^{5.9253} = 842,000$.

To see that this is correct, try $10^{5.9253}$ in a calculator by using the y^x button or the 10^x button. Since 5.9253 is an approximation to the nearest ten-thousandth, the calculator answer (841,976.56) is an approximation. It must be rounded off to 842,000. For more accuracy, use more decimals on 5.9253.... If 5.9253121 is used, $10^{5.9253121} = 842{,}000$.

2. If $\log N = 5.9253$, find N to the nearest thousand.

$$\log N = 5.9253$$
$$\text{antilog}(\log N) = \text{antilog } 5.9253 \qquad \text{See \textbf{Antilogarithm}.}$$
$$N = 841{,}976.56 \qquad \text{Antilog from a calculator.}$$
$$\text{or} \quad N = 842{,}000 \qquad \text{Antilog from the chart rounded}$$
$$\text{to the nearest thousand.}$$

It is understood in logarithms that $=$ means $\approx$ (approximately) equal.

3. Find $\log .000642$.

$$.000642 = 6.42 \times 10^{-4}$$
$$\log .000642 = \log 6.42 \times 10^{-4}$$
$$= \log 6.42 + \log 10^{-4} = \log 6.42 + (-4) \log 10$$
$$= .8075 - 4 = 6.8075 - 10$$

The -4 tells how many places the decimal point has been moved in the number 6.42. We can not just put it in front of the .8075. To avoid problems, the convention of letting the 6. ... -10 represent the -4 has been adopted. It could also have been 16. ... -20 or 26. ... -30. Hence, $\log .000642 = 6.8075 - 10$.

The reason we cannot put the -4 in front of the .8075 follows from $\log 6.42 \times 10^{-4}$.

In exponential form, we have

$$\log 6.42 \times 10^{-4} = 10^{.8075} \times 10^{-4} = 10^{.8075 - 4}$$
$$= 10^{-3.1925} = .0006419 \qquad \text{Calculator.}$$
$$= .000642 \qquad \text{Rounded off or chart.}$$

For calculations, we need to keep the -4 separate from the .8075. Calculations can get confusing if we use the -3.1925.

➤ **Common Monomials**   See **Factoring Common Monomials**.

➤ **Common Multiple**   See **Least Common Multiple**.

➤ **Commutative Property of Addition**   The **commutative property of addition** is one of the properties of the real number system. It states that, for any real numbers a and b,

$$a + b = b + a$$

In general, this property allows for turning addition around.

Examples

1. Use the **commutative property of addition** to simplify $3 + 7x + 5$.

$$3 + 7x + 5 = 7x + 3 + 5 \qquad \text{The } 3 + 7x \text{ was turned around.}$$
$$= 7x + 8$$

2. We usually write linear equations in the form $y = mx + b$. If the equation is backwards, i.e., $y = 4 + 2x$, it can be turned around using the **commutative property of addition**, i.e., $y = 2x + 4$.

See **Properties of the Real Number System**.

➤ **Commutative Property of Multiplication**   The **commutative property of multiplication** is one of the properties of the real number system. It states that, for any real numbers a and b,

$$a \bullet b = b \bullet a$$

In general, this property allows for turning multiplication around.

Examples

1. Use the commutative property of multiplication to simplify $2 \bullet x \bullet 5$.

$$2 \bullet x \bullet 5 = 2 \bullet 5 \bullet x \qquad \text{The } x \bullet 5 \text{ was turned around.}$$
$$= 10x$$

2. We usually write terms with several variables in alphabetical order. Instead of writing yxz, we write xyz. The yx is turned around by the commutative property of multiplication.

See **Properties of the Real Number System**.

➤ **Comparing Decimals**   See **Decimal Numbers, comparing**.

➤ **Comparing Fractions**   See **Fractions, comparing**.

➤ **Complementary Angles**   If the measures of two angles have a sum of 90°, the angles are called **complementary angles**. They do not have to be adjacent angles.

Examples

1. Show that the angles in the figure are complementary.

   Adding the angles, we have

   $$m\angle A + m\angle B = 25° + 65° = 90°$$

   Since the angles have a sum of 90°, the angles are complementary.

2. If two angles are complementary and one of them has a measure of 20°, what is the measure of the other angle?

   Since both angles must have a sum of 90°, the other angle is 70°:

   $$20° + 70° = 90°$$

3. The measure of an angle is twice the measure of its complement. Find the measure of the angle and its complement.

   If we let the complement be x, then the angle is $2x$ (because it is twice the complement). Since the angles are complementary, we know they have a sum of 90°. Therefore,

   $$x + 2x = 90°$$
   $$3x = 90°$$
   $$x = 30°$$

   But x is the value of the complement and $2x$ is the value of the angle. Hence, the complement is 30° and the angle is $2 \bullet 30° = 60°$.

➤ **Complement of a Set**   The **complement of a set** A is defined as the set of elements in the universal set that are not in A, and we write $\bar{A}$ (read "A bar" or "the complement of A"). See **Universal Set**. Other symbols for the complement of A are A' (read "A prime") and $\sim A$ (read "not A"). Also see **Venn Diagram**.

Example

1. If $U = \{1, 2, 3, 4, 5, 6\}$ and $A = \{1, 2, 4\}$, find $\bar{A}$.

   Since U is the universal set, $\bar{A}$ is the set of elements in U that are not in A. Therefore,

   $$\bar{A} = \{3, 5, 6\}$$

➤ **Completing the Square**   **Completing the square** is a process that is used to solve quadratic equations, develop the quadratic formula, graph quadratic equations, or simplify conic sections. It can be applied whenever it becomes necessary to force a quadratic polynomial to factor.

   To solve an equation of the form $ax^2 + bx + c = 0$ by completing the square, the following must be applied:

1. Add, or subtract, the c term on both sides of the equation. By doing this, the variables will be on one side of the $=$ and the constants on the other.

2. The x^2 term must be cleared of its coefficient. If the a of ax^2 is present, divide through the equation by a, i.e., $\frac{ax^2}{a} + \frac{bx}{a} = \frac{-c}{a}$ to get $x^2 + \frac{b}{a}x = -\frac{c}{a}$.

3. Divide the coefficient of the x term by 2.

4. Square the number in step 3.

5. Add the number in step 4 to both sides of the equation.

6. The polynomial on the left (or right) side will now factor. Factor the polynomial and write it as the square of a binomial.

7. The equation is now in a form that can be solved. Take the square root of both sides of the equation. Remember to write $\pm$ on the square root of the constant term, i.e., $\pm\sqrt{\text{constant}}$.

8. Solve the equation for x, one time allowing the sign of the $\sqrt{\text{constant}}$ to be $+$ and another time allowing the $\sqrt{\text{constant}}$ to be $-$. There will be two real number answers if the constant $t > 0$, one answer if the constant $= 0$, and two complex number answers if the constant < 0.

Examples

1. Solve $3x^2 - 24x - 99 = 0$ by completing the square.

$$3x^2 - 24x - 99 = 0$$

$$3x^2 - 24x = 99 \qquad \text{Add 99 to both sides of the equation.}$$

$$x^2 - 8x = 33 \qquad \text{Divide through the equation by 3.}$$

$$x^2 - 8x + 16 = 33 + 16 \qquad \text{Divide the coefficient of } x \text{ by 2, square the result, then}$$

$$(x - 4)(x - 4) = 49 \qquad \text{add that to both sides, i.e., } -8 \div 2 = -4, (-4)^2 = 16.$$

$$(x - 4)^2 = 49 \qquad \text{Factor, then write as a square.}$$

$$\sqrt{(x - 4)^2} = \pm\sqrt{49} \qquad \text{Take the square root of both sides.}$$

$$x - 4 = \pm 7 \qquad \text{Finally, solve for } x.$$

$$x = 4 \pm 7$$

$$x = 4 + 7 \qquad \text{and} \qquad x = 4 - 7 \qquad \text{Let 7 be } + \text{ one time and } - \text{ the other.}$$
$$x = 11 \qquad\qquad\qquad x = -3$$

The solutions to the equation $3x^2 - 24x - 99 = 0$ are $x = 11$ and $x = -3$.

2. Solve $4 = -12y - y^2$ by completing the square.

$$4 = -12y - y^2$$

$$y^2 + 12y = -4$$

$$y^2 + 12y + 36 = -4 + 36 \qquad \left(\frac{12}{2}\right)^2 = 36$$

$$(y + 6)(y + 6) = 32$$

$$(y+6)^2 = 32$$
$$\sqrt{(y+6)^2} = \pm\sqrt{32}$$
$$y+6 = \pm\sqrt{32}$$
$$y = -6 \pm 4\sqrt{2} \qquad \sqrt{32} = \sqrt{16 \bullet 2} = 4\sqrt{2}$$

There are two answers here, $y = -6 + 4\sqrt{2}$ and $y = -6 - 4\sqrt{2}$. However, we usually write the answer as $y = -6 \pm 4\sqrt{2}$. If decimal answers are required, then

$$y = -6 \pm 4\sqrt{2}$$
$$y = -6 \pm 4(1.41)$$
$$y = -6 \pm 5.64$$

$$\begin{array}{ccc} y = -6 + 5.64 & \text{and} & y = -6 - 5.64 \\ y = -.36 & & y = -11.64 \end{array}$$

3. Solve $\frac{2}{3} + \frac{7}{3}x = -x^2$ by completing the square.

$$\frac{2}{3} + \frac{7}{3}x = -x^2$$
$$x^2 + \frac{7}{3}x = \frac{-2}{3}$$
$$x^2 + \frac{7}{3}x + \frac{49}{36} = \frac{-2}{3} + \frac{49}{36} \qquad \left(\frac{7}{3} \div 2^2\right) = \left(\frac{7}{6}\right)^2 = \frac{49}{36}$$
$$\left(x + \frac{7}{6}\right)\left(x + \frac{7}{6}\right) = \frac{-24}{36} + \frac{49}{36} \qquad \text{See \textbf{Fractions, equivalent}.}$$
$$\left(x + \frac{7}{6}\right)^2 = \frac{25}{36}$$
$$x + \frac{7}{6} = \pm\sqrt{\frac{25}{36}}$$
$$x = \frac{-7}{6} \pm \frac{5}{6}$$
$$x = \frac{-7 \pm 5}{6}$$

$$\begin{array}{ccc} x = \dfrac{-7+5}{6} & \text{and} & x = \dfrac{-7-5}{6} \\[2mm] x = \dfrac{-2}{6} & & x = \dfrac{-12}{6} \\[2mm] x = -\dfrac{1}{3} & & x = -2 \end{array}$$

4. Solve $-14x = -8 - 3x^2$ by completing the square.

$$-14x = -8 - 3x^2$$

$$3x^2 - 14x = -8$$

$$x^2 - \frac{14}{3}x = \frac{-8}{3} \qquad \text{Divide through by 3.}$$

$$x^2 - \frac{14}{3}x + \frac{49}{9} = \frac{-8}{3} + \frac{49}{9} \qquad \left(\frac{-14}{3} \div 2\right)^2 = \left(\frac{-14}{6}\right)^2 = \left(\frac{-7}{3}\right)^2 = \frac{49}{9}$$

$$\left(x - \frac{7}{3}\right)\left(x - \frac{7}{3}\right) = \frac{-24}{9} + \frac{49}{9}$$

$$\left(x - \frac{7}{3}\right)^2 = \frac{25}{9}$$

$$\sqrt{\left(x - \frac{7}{3}\right)^2} = \pm\sqrt{\frac{25}{9}}$$

$$x - \frac{7}{3} = \pm\frac{5}{3}$$

$$x = \frac{7}{3} \pm \frac{5}{3}$$

$$x = \frac{7 \pm 5}{3}$$

$$x = \frac{7}{3} + \frac{5}{3} \qquad \text{and} \qquad x = \frac{7}{3} - \frac{5}{3}$$

$$x = \frac{12}{3} \qquad\qquad\qquad\qquad x = \frac{2}{3}$$

$$x = 4$$

5. Solve $ax^2 + bx + c = 0$ by completing the square.

$$ax^2 + bx + c = 0$$

$$ax^2 + bx = -c$$

$$x^2 + \frac{b}{a}x = \frac{-c}{a} \qquad \text{Divide through by } a.$$

$$x^2 + \frac{b}{a}x + \frac{b^2}{4a^2} = \frac{b^2}{4a^2} - \frac{c}{a} \qquad \left(\frac{b}{a} \div 2\right)^2 = \left(\frac{b}{2a}\right)^2 = \frac{b^2}{4a^2}$$

$$\left(x + \frac{b}{2a}\right)\left(x + \frac{b}{2a}\right) = \frac{b^2}{4a^2} - \frac{4ac}{4a^2} \qquad \text{Common denominator, } -\frac{c \bullet 4a}{a \bullet 4a} = \frac{4ac}{4a^2}$$

$$\left(x + \frac{b}{2a}\right)^2 = \frac{b^2 - 4ac}{4a^2}$$

$$\sqrt{\left(x + \frac{b}{2a}\right)^2} = \pm\sqrt{\frac{b^2 - 4ac}{4a^2}}$$

$$x + \frac{b}{2a} = \pm \frac{\sqrt{b^2 - 4ac}}{4a^2}$$

$$x = \frac{-b}{2a} \pm \frac{\sqrt{b^2 - 4ac}}{4a^2}$$

$$x = \frac{-b \pm \sqrt{b^2 - 4ac}}{2a}$$

The solution to $ax^2 + bx + c = 0$ is the quadratic formula. It is derived by completing the square. For application, see **Quadratic Formula**.

Sometimes, as with conic sections, it becomes necessary to *complete the square on one side of the equal sign*. For an example of this, see **Ellipse**, example 3.

For more, see **Quadratic Equations, solutions of**; **Quadratic Formula**; **Quadratic Equations, graphs of**; **Ellipse**; **Hyperbola**; or **Parabola**.

➤ **Complex Fractions**   See **Fractions, complex**.

➤ **Complex Fractions, algebraic**   See **Rational Expressions, complex fractions of**.

➤ **Complex Number, absolute value**   If $z = a + bi$ is a complex number, then its **absolute value**, called the **modulus**, is

$$|z| = |a + bi| = \sqrt{a^2 + b^2}$$

Example

1. Find the absolute value of $z = 3 + 4i$.

$$|z| = |3 + 4i|$$
$$= \sqrt{3^2 + 4^2}$$
$$= \sqrt{9 + 16}$$
$$= \sqrt{25} = 5$$

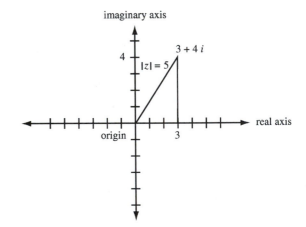

Graphing z in the complex plane, we can see that $|z| = 5$ is the hypotenuse. It is called the **modulus** of z. See **Pythagorean Theorem**. Other related articles are **Complex Number, trigonometric form**; **Complex Plane**; and **De Moivre's Theorem**.

➤ **Complex Numbers**   A **complex number** is a number that can be written in the form $a + bi$ where a and b are real numbers and i is the **imaginary number** $\sqrt{-1}$. The bi term is called a **pure imaginary number**. When $b = 0$, $a + 0i = a$. Since a is a real number, this implies that the real numbers are a subset of the complex numbers.

The impetus for such a definition arose from attempts at solving equations like $x^2 = -1$. Taking the square root of both sides gives $x = \sqrt{-1}$, which makes no sense. No real number has a square root that is negative. Even so, if we let $i = \sqrt{-1}$, we can develop a consistent algebra, the algebra of complex numbers. It turns out that there is a large application of complex numbers in engineering and physics as well as mathematics.

Examples

1. Simplify $(8 + 5i) - (2 + 3i)$.

   We combine complex numbers as though the i has no meaning:

   $$(8 + 5i) - (2 + 3i) = 8 + 5i - 2 - 3i$$

   See **Distributive Property** or **Positive and Negative Numbers**.

   $$= 6 + 2i$$

   The answer is left in the complex form $a + bi$.

2. Simplify $(12 - \sqrt{-5}) - (-3 + \sqrt{-7})$.

   $$\left(12 - \sqrt{-5}\right) - \left(-3 + \sqrt{-7}\right) = 12 - i\sqrt{5} + 3 - i\sqrt{7} = 15 - i\sqrt{5} - i\sqrt{7}$$

   To write $\sqrt{-5}$ as a *pure imaginary* number, we do the following:

   $$\sqrt{-5} = -\sqrt{(5)(-1)} = \sqrt{5}\sqrt{-1} = \sqrt{5}i = i\sqrt{5}, \qquad \text{where } \sqrt{-1} = i$$

   We write $i\sqrt{5}$ instead of $\sqrt{5}i$ since $\sqrt{5}i$ and $\sqrt{5i}$ could be confused. If it is required to write the answer in the complex form $a + bi$, factor out the i from $-i\sqrt{5} - i\sqrt{7}$.

   $$15 - i\sqrt{5} - i\sqrt{7} = 15 - i\left(\sqrt{5} + \sqrt{7}\right) = 15 - \left(\sqrt{5} + \sqrt{7}\right)i$$

3. Simplify $\sqrt{-50}$.

   $$\sqrt{-50} = \sqrt{(-1)(50)}$$
   $$= \sqrt{-1}\sqrt{50} = i\sqrt{50} \qquad \text{Take out the imaginary number.}$$
   $$= i\sqrt{25 \bullet 2} = 5i\sqrt{2} \qquad \text{Simplify the } \sqrt{50}. \text{ See } \textbf{Radicals, rules for}.$$

4. Simplify $\sqrt{\frac{-1}{3}}$.

   $$\sqrt{\frac{-1}{3}} = \frac{\sqrt{-1}}{\sqrt{3}} = \frac{i}{\sqrt{3}}$$

   This would be the answer but we cannot leave a $\sqrt{}$ in the denominator.

   $$= \frac{i}{\sqrt{3}} \bullet \frac{\sqrt{3}}{\sqrt{3}}$$
   $$= \frac{i\sqrt{3}}{\sqrt{9}} = \frac{i\sqrt{3}}{\sqrt{3}}$$

   See **Rationalizing the Denominator**.

5. Simplify $(2 - 4i)(6 + i)$.

We first multiply the binomials. See **Polynomials, multiplication of**.

$$(2 - 4i)(6 + i) = 12 + 2i - 24i - 4i^2$$
$$= 12 - 22i - 4i^2$$
$$= 12 - 22i - 4(-1) \qquad \text{Substituting } i^2 = -1.$$
$$= 12 - 22i + 4$$
$$= 16 - 22i$$

The answer should be left in the complex form $a + bi$.

Why does $i^2 = -1$? The answer is most easily seen from the following:

$$i = \sqrt{-1} = i$$
$$i^2 = \left(\sqrt{-1}\right)^2 = -1 \qquad \text{Since square and square root cancel each other,}$$
$$i^3 = i^2 i = -1 \bullet i = -i \qquad \text{i.e., } (\sqrt{4})^2 = 4.$$
$$i^4 = i^3 i = (-i)i = -i^2 = +1$$
$$i^5 = i^4 i = 1 \bullet i = i$$
$$i^6 = i^5 i = i \bullet i = i^2 = -1$$
$$i^7 = i^6 i = (-1) \bullet (i) = -i$$

The pattern continues with i, -1, $-i$, and $+1$ repeating in cycles of four. To find i^{15}, we do as follows:

$$i^{15} = i^4 i^4 i^4 i^3 \qquad\qquad\qquad i^{15} = i^{14} \bullet i = \left(i^2\right)^7 \bullet i$$
$$= 1 \bullet 1 \bullet 1 \bullet (-i) \qquad \text{or} \qquad = (-1)^7 \bullet i$$
$$= -i \qquad\qquad\qquad\qquad\qquad = -i$$

To find i^{64}, we do as follows:

$$i^{64} = \left(i^2\right)^{32} = (-1)^{32} = 1 \qquad \begin{array}{ll} (-1)^{\text{even}} = +1 & (-1)^{\text{odd}} = -1 \\ \text{i.e., } (-1)^2 = +1 & \text{i.e., } (-1)^3 = -1 \end{array}$$

6. Solve the equation $3z^2 + 24 = 0$.

$$3z^2 + 24 = 0 \qquad\qquad \text{See } \textbf{Quadratic Equations, solutions of.}$$
$$3x^2 = -24$$
$$x^2 = -8$$
$$x = \pm\sqrt{-8}$$
$$x = \sqrt{(-1)(8)}$$
$$x = \pm\sqrt{-1}\sqrt{8} \qquad \sqrt{8} = \sqrt{4 \bullet 2} = \sqrt{4} \bullet \sqrt{2} = 2\sqrt{2}$$
$$x = \pm i 2\sqrt{2}$$
$$x = \pm 2i\sqrt{2}$$

$$x = 2i\sqrt{2} \qquad \text{and} \qquad x = -2i\sqrt{2}$$

The solutions are both *pure imaginary* numbers. For help in simplifying examples 7 and 8, see **Rationalizing the Denominator**.

7. Simplify $\frac{3+i}{5i}$.

$$\frac{3+i}{5i} = \frac{3+i}{5i} \cdot \frac{i}{i} = \frac{3i+i^2}{5i^2} = \frac{3i+(-1)}{5(-1)} = \frac{-1+3i}{-5}$$

$$= \frac{-1}{-5} + \frac{3i}{-5} = \frac{1}{5} - \frac{3}{5}i$$

8. Simplify $\frac{2}{6+4i}$.

$$\frac{2}{6+4i} = \frac{2}{6+4i} \cdot \frac{6-4i}{6-4i} = \frac{12-8i}{36-16i^2}$$

$$= \frac{12-8i}{36-16(-1)} = \frac{12-8i}{36+16} = \frac{4(3-2i)}{52}$$

$$= \frac{3-2i}{13} = \frac{3}{13} - \frac{2i}{13}$$

➤ **Complex Number, trigonometric form**

The **trigonometric form** of a complex number $z = a + bi$ is $z = r(\cos\theta + i\sin\theta)$ where $a = r\cos\theta$, $b = r\sin\theta$, r is the absolute value of z with $r = \sqrt{a^2 + b^2}$, and θ is defined by $\tan\theta = \frac{b}{a}$. The hypotenuse r is called the **modulus** of z and the angle θ is called the **argument** of z.

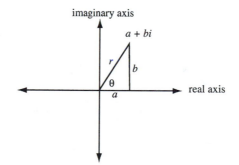

The reason $a = r\cos\theta$ and $b = r\sin\theta$ can easily be seen from a graph of $a + bi$. From the graph, we see that $\cos\theta = \frac{a}{r}$ or $a = r\cos\theta$. Also, $\sin\theta = \frac{b}{r}$ or $b = r\sin\theta$. Substituting in $a + bi$, we have

$$a + bi = r\cos\theta + r\sin\theta i$$
$$= r(\cos\theta + i\sin\theta) \qquad \text{the trigonometric form of a complex number}$$

Examples

1. Find the trigonometric form of $z = 3 - 3i$.

   What is required is to find the r and the θ of $z = r(\cos\theta + i\sin\theta)$.

To find r, we use $r = \sqrt{a^2 + b^2}$:

$$r = \sqrt{(3)^2 + (-3)^2}$$
$$= \sqrt{18}$$
$$= 3\sqrt{2}$$

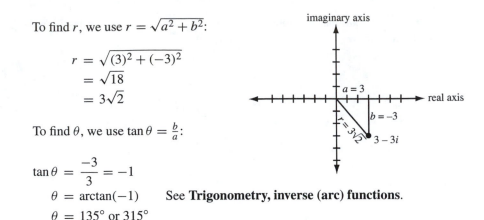

imaginary axis

$a = 3$

real axis

$b = -3$

$r = 3\sqrt{2}$

$3 - 3i$

To find θ, we use $\tan \theta = \frac{b}{a}$:

$$\tan \theta = \frac{-3}{3} = -1$$
$$\theta = \arctan(-1) \qquad \text{See \textbf{Trigonometry, inverse (arc) functions}.}$$
$$\theta = 135° \text{ or } 315°$$

We use $\theta = 315°$ since $z = 3 - 3i$ is in the fourth quadrant (see diagram). Hence,

$$z = 3 - 3i = 3\sqrt{2}(\cos 315° + i \sin 315°) \qquad \text{in degrees}$$
$$= 3\sqrt{2}\left(\cos \frac{7\pi}{4} + i \sin \frac{7\pi}{4}\right) \qquad \text{in radians}$$

2. Find the standard form, i.e., $a + bi$, of $3\sqrt{2}\left(\cos \frac{7\pi}{4} + i \sin \frac{7\pi}{4}\right)$.

   This is the reverse process of example 1. First, we see that

   $$\cos \frac{7\pi}{4} = \frac{3}{3\sqrt{2}} = \frac{1}{\sqrt{2}} = \frac{\sqrt{2}}{2} \qquad \text{and} \qquad \sin \frac{7\pi}{4} = \frac{-3}{3\sqrt{2}} = \frac{-\sqrt{2}}{2}$$

   Substituting and simplifying, we have

   $$3\sqrt{2}\left(\cos \frac{7\pi}{4} + i \sin \frac{7\pi}{4}\right) = 3\sqrt{2}\left(\frac{\sqrt{2}}{2} + i\frac{-\sqrt{2}}{2}\right) = 3\sqrt{2}\left(\frac{\sqrt{2}}{2} - \frac{i\sqrt{2}}{2}\right)$$
   $$= \frac{3\sqrt{4}}{2} - \frac{3i\sqrt{4}}{2} = \frac{6}{2} - \frac{6i}{2} = 3 - 3i$$

3. This example shows how to find the product of two complex numbers in trigonometric form.

   If $z_1 = r_1(\cos \theta_1 + i \sin \theta_1)$ and $z_2 = r_2(\cos \theta_2 + i \sin \theta_2)$, then

   $$z_1 z_2 = r_1 r_2 \left[\cos(\theta_1 + \theta_2) + i \sin(\theta_1 + \theta_2)\right]$$

   Let

   $$z_1 = 2\left(\cos \frac{\pi}{3} + i \sin \frac{\pi}{3}\right) \qquad \text{and} \qquad z_2 = 4\left(\cos \frac{2\pi}{3} + i \sin \frac{2\pi}{3}\right)$$

   Then

   $$z_1 z_2 = 2\left(\cos \frac{\pi}{3} + i \sin \frac{\pi}{3}\right) \bullet 4\left(\cos \frac{2\pi}{3} + i \sin \frac{2\pi}{3}\right)$$

$$= 8\left[\cos\left(\frac{\pi}{3} + \frac{2\pi}{3}\right) + i\sin\left(\frac{\pi}{3} + \frac{2\pi}{3}\right)\right] = 8\left[\cos\left(\frac{3\pi}{3}\right) + i\sin\left(\frac{3\pi}{3}\right)\right]$$

$$= 8(\cos\pi + i\sin\pi)$$

If it is preferable to solve the original problem with degrees rather than radians, then do so. Converting, we have $\frac{\pi}{3} = \frac{180}{3} = 60°$ and $\frac{2\pi}{3} = \frac{2 \bullet 180}{3} = \frac{360}{3} = 120°$. See **Converting from Radians to Degrees**.

4. This example shows how to find the quotient of two complex numbers in trigonometric form.

    If $z_1 = r_1(\cos\theta_1 + i\sin\theta_1)$ and $z_2 = r_2(\cos\theta_2 + i\sin\theta_2)$ then,

    $$\frac{z_1}{z_2} = \frac{r_1}{r_2}\left[\cos(\theta_1 - \theta_2) + i\sin(\theta_1 - \theta_2)\right] \qquad \text{with } z_2 \neq 0$$

    Let

    $$z_1 = 6(\cos 120° + i\sin 120°) \qquad \text{and} \qquad z_2 = 2(\cos 30° + i\sin 30°)$$

    Then

    $$\frac{z_1}{z_2} = \frac{6}{2}\left[\cos(120° - 30°) + i\sin(120° - 30°)\right] = 3(\cos 90° + i\sin 90°)$$

5. To deal with complex numbers of a trigonometric form that have been raised to an exponent, i.e., z^n, see **De Moivre's Theorem**.

➤ Complex Plane

To develop the **complex plane**, we treat the complex number $z = a + bi$ as though it were a point (a, b). We call the horizontal axis the **real axis** and the vertical axis the **imaginary axis**. The hypotenuse r is called the **modulus**. See **Complex Number, absolute value**. The angle θ is defined by $\tan\theta = \frac{a}{b}$. See **Complex Number, trigonometric form**.

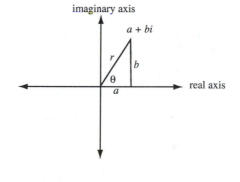

Examples

1. Graph $z = 4 - 3i$ in the complex plane.

    To graph $z = 4 - 3i$, we need to locate the point that corresponds with a movement of 4 units along the real axis and $-3i$ units along the imaginary axis.

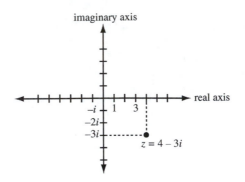

2. Graph $z = -3 - 4i$ in the complex plane.

To graph $z = -3 - 4i$, we need to locate the point that corresponds with a movement of -3 units along the real axis and $-4i$ units along the imaginary axis.

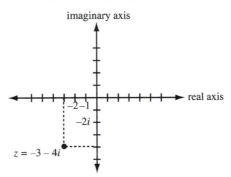

For a correlation, see **Coordinate Plane**.

➤ **Composite Functions**   If one function is composed of another function, then it is said to be a **composite function**. The composite function $f \circ g$ (read "f composition g") also described by $f(g(x))$ (read "f of g of x") where $f \circ g(x) = f(g(x))$.

Examples

1. If $f(x) = x + 1$ and $g(x) = 2x$, find $f(g(3))$.

First, calculate $g(3)$:

$$g(3) = 2(3) = 6 \qquad \text{See \textbf{Function Notation}.}$$

Next, calculate $f(6)$:

$$f(6) = 6 + 1 = 7$$

Hence, $f(g(3)) = f(6) = 7$.

It's not confusing if you first find one answer and then use that to find the other.

2. If $f(x) = 7x + 2$ and $g(x) = 3x + 1$, find $f(g(4))$.

$$g(4) = 3(4) + 1 = 13$$
$$f(13) = 7(13) + 2 = 93$$

Hence, $f(g(4)) = f(13) = 93$. This can also be written $[f \circ g](4) = 93$.

3. If $f(x) = x^2 + 1$ and $g(x) = 3x - 2$, find $g(f(-2))$.

This is reversed. First, find $f(-2)$:

$$f(-2) = (-2)^2 + 1 = 5$$

Next, find $g(5)$:

$$g(5) = 3(5) - 2 = 13$$

Hence, $g(f(-2)) = g(5) = 13$. Also, $[f \circ g](-2) = 13$.

4. Sometimes composite functions are done in general. There is no value for x.

If $f(x) = 3x$ and $g(x) = 2x + 1$, find $f(g(x))$.

$$f\big(g(x)\big) = f(2x + 1) = 3(2x + 1) \qquad \text{Substitute the } g \text{ function into the } f \text{ function.}$$
$$= 6x + 3$$

Hence, $f(g(x)) = 6x + 3$.

5. In example 4, find $g(f(x))$.

$$g\big(f(x)\big) = g(3x) = 2(3x + 1) = 6x + 2$$

Hence, $g(f(x)) = 6x + 2$.

➤ **Composite Number**   A **composite number** is a whole number (the numbers $0, 1, 2, 3, \ldots, \infty$) greater than 1 that is not a prime number. See **Prime Numbers**. A composite number can be written as a product of two or more smaller whole numbers. The first 10 composite numbers are

$$4, 6, 8, 9, 10, 12, 14, 15, 16, 18$$

The number 4 is a composite number because it can be written as the product of two smaller whole numbers, $4 = 2 \bullet 2$. The number 6 is composite because it can be written as $6 = 2 \bullet 3$.

➤ **Compound Interest**   See **Interest**.

➤ **Concentric Circles**   Circles that have the same center but radii of different lengths are **concentric circles**. The circles on an archery target are concentric circles.

➤ **Conclusion**   In a conditional statement $p \rightarrow q$ (see **Conditional Statement**) the *then-part* (q) is called the **conclusion**. For more, see **Hypothesis and Conclusion**. Also see **Logic**.

➤ **Concurrent Lines**   Three or more lines that have a point in common are called **concurrent lines**. Two lines that intersect at a point are not called concurrent.

➤ **Conditional Statement**   A statement written in *if-then* form is called a **conditional statement**. If we let p denote the *if-part* of the statement and q denote the *then-part*, then the expression $p \rightarrow q$ denotes the conditional statement (read "if p, then q" or "p implies q"). The *if-part* (p) is called the **hypothesis** and the *then-part*(q) is called the **conclusion**. See **Hypothesis and Conclusion**.

As seen in the truth table, a conditional statement is false when the hypothesis is true and the conclusion is false. Otherwise, it is true.

p	q	$p \rightarrow q$
T	T	T
T	F	F
F	T	T
F	F	T

A conditional statement is like a promise. Suppose I promise that if you mow the lawn, I will pay you 10 dollars. Then by an application of the truth table the following can occur:

1. You mowed the lawn, so I paid you 10 dollars. I kept the promise and the statement is true.

2. You mowed the lawn, but I did not pay you 10 dollars. I broke the promise and the statement is false.

3. You did not mow the lawn, but I still paid you 10 dollars. I'm just a nice person and I gave you 10 dollars. I did not break the promise and the statement remains true.

4. You did not mow the lawn, and I did not pay you 10 dollars. I kept the promise and the statement is true.

Examples

1. Write the following conditional statement in if-then form: An equilateral triangle has three congruent sides.

   In if-then form the statement is: If a triangle is an equilateral triangle, then the triangle has three congruent sides.

2. With the aid of a truth table, prove that the conditional statement $p \rightarrow (p \vee q)$ is true for all truth values of p or q.
   To prove that $p \rightarrow (p \vee q)$ is true for all truth values of p or q, we need to enter p, q, and $p \vee q$ in a truth table. See **Truth Table**, $p \vee q$. Then, applying the truth table for a conditional statement to columns 1 and 3, we get the entries in column 4. Hence, $p \rightarrow (p \vee q)$ is true for all truth values of p and q.

p	q	$p \to q$	$p \to (p \lor q)$
T	T	T	T
T	F	T	T
F	T	T	T
F	F	F	T

➤ **Cone (right circular)**
To find the **lateral area** of a right circular cone (the outside area, not including the base), we use the formula

$$L = \pi r s$$

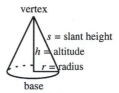

where L is the lateral area, r is the radius, and s is the slant height. For example, see **Area (lateral) of a Right Circular Cone (and total area)**.

To find the total area of a right circular cone (lateral area and area of the base), we use the formula

$$\text{T.A.} = \pi r s + \pi r^2$$

where $\pi r s$ is the lateral area and πr^2 is the area of the base (see **Circle, area of**). For an example, see **Area (lateral) of a Right Circular Cone (and total area)**.

To find the **volume** of right circular cone, we use the formula

$$V = \frac{1}{3} \pi r^2 h$$

where h is the altitude of the cone.

➤ **Congruent Angles**   Angles that have the same measure are **congruent angles**. Since $m\angle A = m\angle B = 30°$, we write $\angle A \cong \angle B$, read "angle A is congruent to angle B."

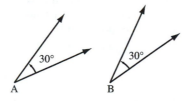

➤ **Congruent Segments**   Segments that have the same measure are **congruent segments**. Since $AB = CD = 8$ in., we write $\overline{AB} \cong \overline{CD}$, read "segment AB is congruent to segment CD." Remember that $\overline{AB}$ denotes a set of points, whereas AB denotes measure or length.

➤ **Congruent Triangles**   Loosely speaking, we can say that triangles with the same size and shape are **congruent triangles**. More specifically, we can say that two triangles are congruent if their corresponding sides and corresponding angles are congruent. This leads to a question. What is the minimum number of congruent sides and angles that will yield congruent triangles? The answer is three.

Not just any three correspondences will yield congruent triangles, however. Those that do are SSS, SAS, ASA, AAS, and SAA. Those that do not are SSA, ASS, and AAA. The last of these, AAA, is a correspondence that is used to develop similar triangles.

Sometimes AAS and SAA are not considered as part of the list since they can be proved by ASA. This follows from the fact that if both the angles are known, so is the third (by the sum of the angles of a triangle).

For an example, see **Proof in Geometry**.

➤ **Conic Sections**   The following is a more formal explanation of **conic section**. For application, see the articles under each type, **Parabola, Circle, Ellipse**, or **Hyperbola**.

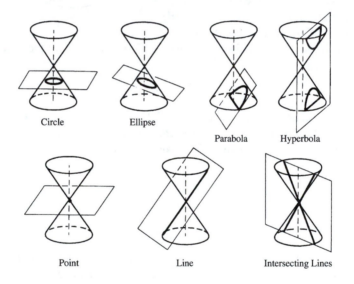

Circle   Ellipse

Parabola   Hyperbola

Point   Line   Intersecting Lines

Any plane section of a double cone is a parabola, ellipse, hyperbola, or some special case of these. In general, a **conic** is the set of all points such that its distance from a fixed point has a constant ratio to its distance from a fixed line. The fixed point is called the **focus**. The fixed line is called the **directrix**. The constant ratio is called the **eccentricity of the conic** and has the following properties:

1. If $e = 1$, the conic is a parabola.

2. If $0 < e < 1$, the conic is an ellipse.

3. If $e > 1$, the conic is a hyperbola.

The **standard form of a conic section** is

$$Ax^2 + By^2 + Cx + Dy + E = 0$$

where the coefficients are real numbers.

The following cases apply:

1. If either A or B is 0, the conic is a parabola with special cases that are lines or imaginary.

2. If A and B have the same sign, the conic is an ellipse with special cases that are a circle, a point, or imaginary.

3. If A and B have unlike signs, the conic is a hyperbola with special cases that are intersecting lines.

For applications of each, see the articles listed under **Parabola, Circle, Ellipse,** or **Hyperbola**. The lines are considered under **Linear Equation**.

➤ **Conjugate**   Binomials such as $a + b$ and $a - b$ are called **conjugates**. The product of two conjugates is the difference of two squares:

$$(a + b)(a - b) = a^2 - b^2$$

The concept of conjugates is most often applied in simplifying quotients that are composed of radicals or complex numbers.

Examples

1. The conjugate of $2x + 3$ is $2x - 3$.

2. The conjugate of $6 - 4i$ is $6 + 4i$.

For an application, see **Rationalizing the Denominator**, examples 4, 5, and 6.

➤ **Conjugate Axis**   A **conjugate axis** is one of the two axes of a parabola. It is perpendicular to the other axis, the **transverse axis**. See **Hyperbola**.

➤ **Conjunction**   In logic, the compound statement "p and q" written $p \wedge q$, where p and q represent any statement, is called the **conjunction** of p and q. If p represents the statement that "Jack is a baseball player" and q represents the statement that "Jack is a tennis player," then $p \wedge q$ represent the statement that "Jack is a baseball player and Jack is a tennis player."

A good way of determining whether a conjunction is either true or false is to make a truth table. See **Truth Table**.

The concept of conjunction is related to the concept of intersection in set theory. This is easily seen through the use of a Venn diagram. See **Venn Diagram**. If we let $B = \{$all baseball players$\}$ and $T = \{$all tennis players$\}$, then $B \cap T = \{$all people who play both baseball and tennis$\}$ is represented by the shaded area in the Venn diagram.

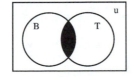

In our example above, Jack is a baseball player, so he is an element of circle B. Since he is a tennis player, he is also an element of circle T. To be an element of each at the same time, Jack must be an element of the shaded area. Therefore, Jack $\in (B \cap T)$, the shaded area.

For more, see **Truth Table** or **Logic**.

➤ **Consistent Equations**   See **Systems of Linear Equations**.

➤ **Constant**   In an expression such as $5x^2 - 3(x + 4y)$, the 5, 2, 3, and 4 are constants.

➤ **Contradiction**   A statement that is both true and false at the same time is a **contradiction**. If p is given as a fact, then the statement both p and $\sim p$ (not p) is a contradiction. For $\sim p$, see **Negation**.

Example

The following is a contradiction: The number 4 is not the number 4.

See **Logic**.

➤ **Contrapositive**   The **contrapositive** of a conditional statement is formed when the "if " and "then" of the inverse are interchanged.

$$\text{Conditional: } p \to q \qquad \text{Inverse: } \sim p \to \sim q \qquad \text{Contrapositive: } \sim q \to \sim p$$

If the conditional statement is true, the contrapositive statement is true.

Examples

Conditional: If two angles are congruent, then they have equal measure. This is a true statement.

Contrapositive: If two angles are not equal in measure, then the angles are not congruent. This also is a true statement.

See **Logic**.

➤ **Converse**   The **converse** of a conditional statement is formed by interchanging the "if " and the "then" parts of the statement (interchanging the hypothesis and the conclusion).

$$\text{Conditional: } p \to q \qquad \text{Converse: } q \to p$$

If the conditional statement is true, the converse is not necessarily true.

Examples

Conditional: If two angles are right angles, then they are supplementary. This is a true statement.

Converse: If two angles are supplementary, then they are right angles. This is not necessarily true.

See **Logic**.

➤ **Converting a Repeating Decimal to a Common Fraction**   See **Decimal Numbers, changing a repeating decimal to a common fraction**.

➤ **Converting Degrees to Radians**   To **convert from degrees to radians**, use the equation

$$1° = \frac{\pi}{180}$$

and multiply both sides of the equation by the degree measure.

    If forgotten, this equation is easy to derive. Compare two circles, one with degrees and the other with radians. It is clear that the location of 180° and the location of π are the same. From this, we have $180° = \pi$. Now, just divide both sides by 180 and we have $1° = \frac{\pi}{180}$.

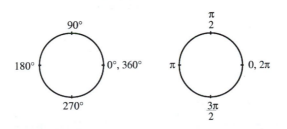

Examples

1. Convert 60° to radians.

    Start off with $1° = \frac{\pi}{180}$, then multiply both sides by 60:

$$(60) \bullet 1° = \frac{\pi}{180}(60)$$

$$60° = \frac{60\pi}{180} = \frac{6\pi}{18} = \frac{\pi}{3}$$

2. Convert 270° to radians.

$$(270) \bullet 1° = \frac{\pi}{180}(270)$$

$$270° = \frac{27\pi}{18} = \frac{3\pi}{2}$$

3. Covert 490° to radians.

$$(490) \bullet 1° = \frac{\pi}{180}(490)$$

$$490° = \frac{49\pi}{18}$$

➤ **Converting from Logarithmic Form to Exponential Form**   See **Logarithmic Form to Exponential Form, converting from.**

➤ **Converting from Radical Form to Fractional Exponents**   See **Radical Form.**

➤ **Converting Radians to Degrees**   The easiest way to **convert from radians to degrees** is to substitute 180° for π. The other is to use the conversion formula

$$1^R = \frac{180°}{\pi}$$

and multiply both sides of the equation by the radian measure.

    This equation is derived from the fact that $\pi^R = 180°$. Divide both sides by π and we have $1^R = \frac{180°}{\pi}$.

1. Convert $\frac{\pi}{3}$ to degrees by using the first method.

   Substituting $180°$ for π, we have $\frac{\pi}{3} = \frac{180°}{3} = 60°$

   Convert $\frac{\pi}{3}$ to degrees by using the second method. Start with $1^R = \frac{180°}{\pi}$ and multiply both sides by $\frac{\pi}{3}$:

   $$1^R = \frac{180°}{\pi}$$

   $$\frac{\pi}{3} \bullet 1^R = \frac{180°}{\pi} \bullet \frac{\pi}{3}$$

   $$\frac{\pi}{3} = \frac{180°}{3} = 60°$$

2. Convert $\dfrac{3\pi}{2}$ to degrees.

   $$\frac{3\pi}{2} = \frac{3(180°)}{2} = 270°$$

3. Convert $\dfrac{49\pi}{18}$ to degrees.

   $$\frac{49\pi}{18} = \frac{49(180°)}{18} = 490°$$

➤ **Coordinate Plane**   A **coordinate plane** is formed when horizontal and vertical number lines are placed perpendicular to each other at their respective zeros. The horizontal number line is called the x-**axis** and the vertical number line is called the y-**axis**. Their point of intersection is called the **origin**. The axes (plural of axis) separate the plane into four regions called **quadrants**. Each point in the **coordinate plane** is denoted by the **ordered pair** (x, y) where x and y are real numbers. The x and y are each called **coordinates**. The x-coordinate is called the **abscissa** and the y-coordinate is called the **ordinate**. The point (x, y) is called the **graph of the ordered pair**.

In applying the coordinate plane, the first thing a student needs to learn is how to **plot** or graph points. This is very similar to giving someone directions to travel.

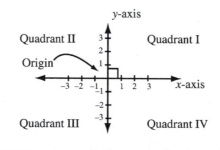

1. Plot the points $(3, 4)$.

   Since the horizontal and vertical axes are number lines, they can be numbered as in the figure. To plot the point $(3, 4)$, start at the origin, move right 3 units, then move up 4 units. This will get you to the point labeled A.

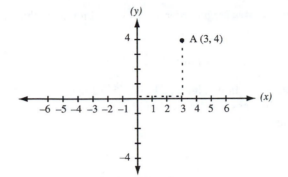

C

2. Plot the following points on the same coordinate plane: $A(4, 1)$, $B(-3, 4)$, $C(-4, -2)$, $D(1, -2)$, and $E(0, 0)$.

Get into the habit of moving in the x direction first and then the y direction. This is most useful in later topics.

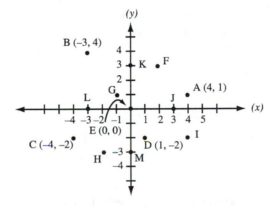

3. In the coordinate plane of example 2, the points F, G, H, I, J, K, L, and M have been graphed. What are their coordinates?

Answer: $F(2, 3)$, $G(-1, 1)$, $H(-2, -3)$, $I(4, -2)$, $J(3, 0)$, $K(0, 3)$, $L(-3, 0)$, and $M(0, -3)$.

4. The horizontal and vertical axes separate the plane into four quadrants. In the coordinate plane of example 2, list the points that are in each of the quadrants.

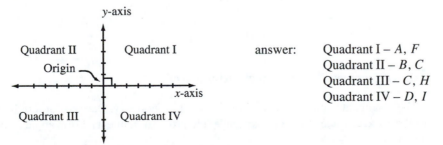

answer:

Quadrant I – A, F
Quadrant II – B, C
Quadrant III – C, H
Quadrant IV – D, I

➤ **Coordinates**   Coordinates are an ordered pair of real numbers that are used to locate a point in the coordinate plane. See **Coordinate Plane**.

➤ **Coordinate System**   See **Coordinate Plane**.

➤ **Corresponding Angles**   In the figure, ∠1 and ∠5, ∠2 and ∠6, ∠3 and ∠7, and ∠4 and ∠8 are **corresponding angles**. If the lines *l* and *m* are parallel, then ∠1 ≅ ∠5, ∠2 ≅ ∠6, ∠3 ≅ ∠7, and ∠4 ≅ ∠8. See **Congruent Angles**.

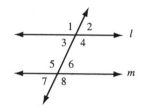

➤ **Cosecant**   The **cosecant** is one of the six functions of trigonometry. Defined for a right triangle, it is a fraction of the **hypotenuse** to the **opposite** side (reciprocal of sine). From angle A, $\csc A = \frac{c}{a}$. From angle B, $\csc B = \frac{c}{b}$.

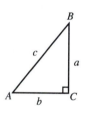

For more, see **Trigonometry, right triangle definition of**.

➤ **Cosine**   The **cosine** is one of the six functions of trigonometry. Defined for a right triangle, it is a fraction of the **adjacent** side to the **hypotenuse**. From angle A, $\cos A = \frac{b}{c}$. From angle B, $\cos B = \frac{a}{c}$.

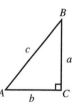

For more, see **Trigonometry, right triangle definition of**.

➤ **Cotangent**   The cotangent is one of the six functions of trigonometry. Defined for a right triangle, it is a fraction of the **adjacent** side to the **opposite** side (reciprocal of tangent). From angle A, $\cot A = \frac{b}{a}$. From angle B, $\cot B = \frac{a}{b}$.

For more, see **Trigonometry, right triangle definition of**.

➤ **Coterminal Angles** Angles that have the same terminal side are **coterminal angles**. If 360° (or 2π) is either added to or subtracted from a given angle, the result is a **coterminal angle**. Any angle has many coterminal angles. The initial side always points right and the terminal side is rotated through the plane (either counterclockwise or clockwise). See **Angle, in trigonometry**.

C

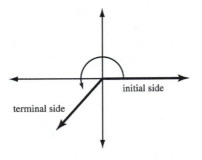

initial side

terminal side

Examples

1. Sketch an angle that is coterminal with 60°.

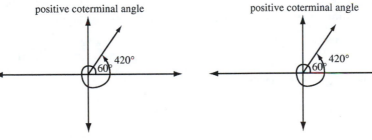

positive coterminal angle   positive coterminal angle

420°   420°

60°   60°

$$60° + 360° = 420° \qquad\qquad 60° - 360° = -300°$$

In radians, we would have

$$\frac{\pi}{3} + 2\pi = \frac{\pi}{3} + \frac{6\pi}{3} = \frac{7\pi}{3} \qquad\qquad \frac{\pi}{3} - 2\pi = \frac{\pi}{3} - \frac{6\pi}{3} = \frac{-5\pi}{3}$$

2. Find an angle that is coterminal with −210°.

$$-210° + 360° = 150° \text{ is an angle that is coterminal with } -210°$$

So is $-210° - 360° = -570°$ coterminal with −210°.

➤ **Counting Numbers**   **Counting numbers** or **natural numbers** are the numbers that are used in counting. They start with 1 and do not contain 0, any negative numbers, or any other type of number, i.e., fractions, decimals, etc.

$$N = \{1, 2, 3, \ldots\}$$

For more, see **Real Number System**.

➤ **Counting Principle**   The **counting principle** states that if two events occur in p different ways and q different ways, respectively, then their product $p \bullet q$ represents the number of ways both can occur. If there are more than two events, then the number of ways is determined by their product. If one event does not affect the outcome of the other, the events are said to be **independent**. If one event does affect the outcome of the other, the events are said to be **dependent**.

An example of independent events can be seen in the choice of three-digit numbers. Suppose we are asked to find as many three-digit numbers as is possible using the numbers 1, 2, and 3, where each number can be used more than once. The first digit of the number can be a 1, 2, or 3. Therefore, there are three choices for the first digit. Similarly, there are three choices for the second digit and three choices for the third digit. Hence, by the counting principle, there are $3 \bullet 3 \bullet 3 = 27$ possible three-digit numbers.

From a similar problem, we can see an example of dependent events. Suppose we are asked to find as many three-digit numbers as is possible using the numbers 1, 2, and 3, where each number can be used only once. There are three choices for the first digit, just as before. The second digit, however, cannot repeat the number chosen for the first digit. Therefore, there are only two choices for the second digit. This leaves only one choice for the third digit. Hence, by the counting principle, there are $3 \bullet 2 \bullet 1 = 6$ possible three-digit numbers.

➤ **Cramer's Rule**   **Cramer's rule** is a method of solving two equations in two unknowns, three equations in three unknowns, or, in general, n equations in n unknowns. In order to do the following examples, the concept of determinants must be clearly understood. See **Determinant**. For other methods of solving linear equations, see **Systems of Linear Equations, solutions of**.

Examples

1. Solve the following equations using Cramer's rule:

$$3x + 8 = -y$$
$$2(2x - y) = -14$$

To solve means to find the point of intersection of the lines. First, we need to put the equations in standard form. See **Standard Form of an Equation of a line**.

$$3x + y = -8$$
$$4x - 2y = -14$$

We can now apply these equations to the general form

$$a_1x + b_1y = c_1$$
$$a_2x + b_2y = c_2 \qquad \text{where all of the coefficients are real numbers}$$

For two equations in two unknowns, Cramer's rule is

$$x = \frac{\begin{vmatrix} c_1 & b_1 \\ c_2 & b_2 \end{vmatrix}}{\begin{vmatrix} a_1 & b_1 \\ a_2 & b_2 \end{vmatrix}} \quad \text{and} \quad y = \frac{\begin{vmatrix} a_1 & c_1 \\ a_2 & c_2 \end{vmatrix}}{\begin{vmatrix} a_1 & b_1 \\ a_2 & b_2 \end{vmatrix}} \quad \text{with} \quad \begin{vmatrix} a_1 & b_1 \\ a_2 & b_2 \end{vmatrix} \neq 0$$

The answer, (x, y), is the point of intersection of the two lines:

$$x = \frac{\begin{vmatrix} -8 & 1 \\ -14 & -2 \end{vmatrix}}{\begin{vmatrix} 3 & 1 \\ 4 & -2 \end{vmatrix}} = \frac{16 - (-14)}{-6 - 4} \qquad y = \frac{\begin{vmatrix} 3 & -8 \\ 4 & -14 \end{vmatrix}}{\begin{vmatrix} 3 & 1 \\ 4 & -2 \end{vmatrix}} = \frac{-42 - (-32)}{-6 - 4}$$

$$= \frac{30}{-10} = -3 \qquad\qquad = \frac{-10}{-10} = 1$$

The point of intersection is $(x, y) = (-3, 1)$.

An easy way to set up the numerator determinants is to replace the column of the x's and y's with the column of the answers (so to speak). The denominator determinant is the same for each. It is the columns of the x's and the y's, respectively.

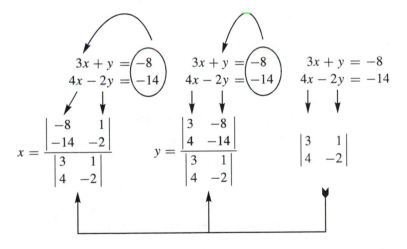

2. Using Cramer's rule, find the point of intersection of $-4y = -3x + 23$ and $2y = -3(3x + 5)$.

   In standard form, the equations are

$$3x - 4y = 23$$
$$9x + 2y = -15$$

$$x = \frac{\begin{vmatrix} 23 & -4 \\ -15 & 2 \end{vmatrix}}{\begin{vmatrix} 3 & -4 \\ 4 & 2 \end{vmatrix}} = \frac{46 - (60)}{6 - (-36)} \qquad y = \frac{\begin{vmatrix} 2 & 23 \\ 9 & -12 \end{vmatrix}}{\begin{vmatrix} 3 & -4 \\ 9 & 2 \end{vmatrix}} = \frac{-45 - (207)}{6 - (-36)}$$

$$= \frac{-14}{42} = -\frac{1}{3} \qquad\qquad\qquad\qquad = \frac{-252}{42} = -6$$

The point of intersection is $(x, y) = \left(-\frac{1}{3}, -6\right)$.

3. Using Cramer's rule, find the point of intersection of the following three lines:

$$3x = -2y + 2z + 5$$
$$-5y = -6x - 2z + 7$$
$$8z = -3x + 4y$$

To begin, we need to write the equations in standard form:

$$3x + 2y - 2z = 5$$
$$6x - 5y + 2z = 7$$
$$3x - 4y + 8z = 0$$

Then we apply Cramer's rule.

The general form for Cramer's rule is

$$a_1 x + b_1 y + c_1 z = d_1$$
$$a_2 x + b_2 y + c_2 z = d_2$$
$$a_3 x + b_3 y + c_3 z = d_3 \qquad \text{where all of the coefficients are real numbers}$$

The solutions are

$$x = \frac{\begin{vmatrix} d_1 & b_1 & c_1 \\ d_2 & b_2 & c_2 \\ d_3 & b_3 & c_3 \end{vmatrix}}{D} \qquad y = \frac{\begin{vmatrix} a_1 & d_1 & c_1 \\ a_2 & d_2 & c_2 \\ a_3 & d_3 & c_3 \end{vmatrix}}{D} \qquad z = \frac{\begin{vmatrix} a_1 & b_1 & d_1 \\ a_2 & b_2 & d_2 \\ a_3 & b_3 & d_3 \end{vmatrix}}{D}$$

where

$$D = \begin{vmatrix} a_1 & b_1 & c_1 \\ a_2 & b_2 & c_2 \\ a_3 & b_3 & c_3 \end{vmatrix} \qquad \text{and} \qquad D \neq 0$$

As with two equations, an easy way to set up the determinants is to replace the column of the x's, y's, and z's by the column of the answers (so to speak).

First, we calculate the denominator D. See **Determinant**.

$$D = \begin{vmatrix} 3 & 2 & -2 \\ 6 & -5 & 2 \\ 3 & -4 & 8 \end{vmatrix} = -120 + 12 + 48 - (30) - (-24) - (96) = -162$$

Next, we find the value of x, y, and z:

$$x = \frac{\begin{vmatrix} 5 & 2 & -2 \\ 7 & -5 & 2 \\ 0 & -4 & 8 \end{vmatrix}}{-162} = \frac{-200 + 0 + 56 - 0 - (-40) - 112}{-162} = \frac{-216}{-162} = \frac{4}{3}$$

$$y = \frac{\begin{vmatrix} 3 & 5 & -2 \\ 6 & 7 & 2 \\ 3 & 0 & 8 \end{vmatrix}}{-162} = \frac{168 + 30 + 0 - (-42) - 0 - 240}{-162} = \frac{0}{-162} = 0$$

$$z = \frac{\begin{vmatrix} 3 & 2 & 5 \\ 6 & -5 & 7 \\ 3 & -4 & 0 \end{vmatrix}}{-162} = \frac{0 + 42 + (-120) - (-75) - (-84) - 0}{-162} = \frac{81}{-162} = -\frac{1}{2}$$

The point of intersection is $(x, y, z) = \left(\frac{4}{3}, 0 - \frac{1}{2}\right)$.

➤ **Cross Multiplication**   **Cross multiplication** is a method that can be used to solve proportional equations. It is applied when a fraction is equal to a fraction. The equation $\frac{a}{b} = \frac{c}{d}$ can be solved by the cross pattern

$$\frac{a}{b} \quad \times \quad \frac{c}{d}$$

Any one of the terms a, b, c, or d can be moved in the direction of the arrows. By this process, the equation $\frac{a}{b} = \frac{c}{d}$ has four solutions:

1. $a = \frac{bc}{d}$ cross over with the b.

2. $b = \frac{ad}{c}$ switch b and c, cross over with the d, then turn the equation around.

3. $c = \frac{ad}{b}$ cross over with the d, then turn the equation around (symmetric property).

4. $d = \frac{bc}{a}$ switch d and a, cross over with the b.

Examples

1. Solve $\frac{a}{2} = \frac{9}{6}$ for a and $\frac{5}{b} = \frac{10}{4}$ for b.

$$\frac{a}{2} = \frac{9}{6} \qquad\qquad \frac{5}{b} = \frac{10}{4}$$

$$a = \frac{2(9)}{6} \qquad\qquad \frac{5(4)}{10} = b$$

$$a = \frac{18}{6} \qquad\qquad b = \frac{20}{10} \qquad \text{Symmetric property.}$$

$$a = 3 \qquad\qquad b = 2$$

2. Solve the equation $\frac{9}{6} = \frac{c}{2}$ for c and $\frac{10}{4} = \frac{5}{b}$ for b.

$$\frac{9}{6} = \frac{c}{2} \qquad\qquad \frac{10}{4} = \frac{5}{b}$$

$$\frac{2(9)}{6} = c \qquad\qquad b = \frac{4(5)}{10}$$

$$c = \frac{18}{6} \qquad\qquad b = \frac{20}{10}$$

$$c = 3 \qquad\qquad b = 2$$

Compare the problems in examples 1 and 2. They differ only by the symmetric property (if $A = B$ then $B = A$).

3. Solve the equation $\frac{5}{12} = \frac{10}{x}$ for x.

$$\frac{5}{12} = \frac{10}{x}$$

$$x = \frac{12(10)}{5} \qquad \text{Switch } x \text{ and 5, cross over with 12.}$$

$$x = 24$$

4. Solve the equation $\frac{6}{x+1} = \frac{4}{2}$ for x.

$$\frac{6}{x+1} = \frac{4}{2}$$

$$2(6) = 4(x+1) \qquad \text{Cross over with the 2 and the } x+1.$$

$$12 = 4x + 4 \qquad \text{Solve the equation. See } \textbf{Linear Equations, solutions of.}$$

$$12 - 4 = 4x$$

$$8 = 4x$$

$$\frac{8}{4} = \frac{4x}{4}$$

$$2 = x$$

$$x = 2 \qquad \text{Symmetric property.}$$

➤ **Cube Root**   The **cube root** of a number is a number that when multiplied by itself three times yields the given number. The cube root of 8 is 2 because $2 \bullet 2 \bullet 2 = 8$, and we write $\sqrt[3]{8} = 2$. Remember that the radical sign does not mean square root. The 3 denotes that we are taking the third root of 8.

Example

1. What is $\sqrt[3]{27}$, $\sqrt[3]{64}$, and $\sqrt[3]{-125}$?

$$\sqrt[3]{27} = 3 \qquad \text{because } 3 \bullet 3 \bullet 3 = 27$$
$$\sqrt[3]{64} = 4 \qquad \text{because } 4 \bullet 4 \bullet 4 = 64$$
$$\sqrt[3]{-125} = -5 \qquad \text{because } (-5) \bullet (-5) \bullet (-5) = -125$$

To simplify more complicated cube roots, see **Radicals, simplifying (reducing)** or the various articles under **Radical**.

➤ **Cubic Equations, solutions of**   Equations with a 3 as the largest exponent of the variable are called **cubic equations**. The term cubic is related to the volume of a cube with a side of length x, where $x \bullet x \bullet x = x^3$ represents the volume of the cube.

Equations with a single cubed term are solved by isolating the cubed term, taking the cube root of both sides of the equation, and then solving for the variable. Equations with more then one cubed term in them are solved by setting the equation equal to zero, factoring the equation completely, setting each factor equal to zero, and then solving each factor. See **Product Theorem of Zero**.

If a cubic equation will not factor by the usual methods of factoring, the rational roots theorem must be applied. If the equation has at least one rational root, it can be factored by applying this theorem. See **Rational Roots Theorem**.

Examples

1. Solve $x^3 + 8 = 0$.

Since there is only one cubed term, we must isolate the term, take the cube of both sides of the equation, and solve for x:

$$x^3 + 8 = 0$$
$$x^3 = -8$$
$$\sqrt[3]{x^3} = \sqrt[3]{-8} \qquad \text{The } \sqrt[3]{-8} = -2 \text{, since } (-2) \bullet (-2) \bullet (-2) = -8.$$
$$x = -2$$

Substituting $x = -2$ into the original equation, we have $(-2)^3 + 8 = 0$. Hence, $x = -2$ is a solution of the equation.

2. Solve $4x^3 - 16x = 0$.

We need to factor completely and then solve each factor:

$$4x^3 - 16x = 0 \qquad \text{See **Factoring Common Monomials**.}$$

$$4x(x^2 - 4) = 0 \qquad \text{See \textbf{Factoring the Difference of Two Squares}.}$$
$$4x(x + 2)(x - 2) = 0 \qquad \text{Also see \textbf{Product Theorem of Zero}.}$$

$$4x = 0 \quad \text{or} \quad x + 2 = 0 \quad \text{or} \quad x - 2 = 0$$
$$x = 0 \qquad\qquad x = -2 \qquad\qquad x = 2$$

There are three answers to the equation. Each substituted into the original equation yields 0. The reason for this has to do with the graph of $y = 4x^3 - 16x$. The graph will cross the x-axis in three places at $x = 0$, $x = -2$, and $x = 2$. See **Zeros of a Function or Polynomial**.

3. Solve $4x^3 = 4x^2 + 80x$.

We need to set the equation equal to 0, factor completely, and then solve each factor:

$$4x^3 = 4x^2 + 80x$$
$$4x^3 - 4x^2 - 80x = 0 \qquad\qquad \text{See \textbf{Factoring Common Monomials}.}$$
$$4x(x^2 - x - 20) = 0 \qquad\qquad \text{See \textbf{Factoring Trinomials}.}$$
$$4x(x + 4)(x - 5) = 0$$

$$4x = 0 \quad \text{or} \quad x + 4 = 0 \quad \text{or} \quad x - 5 = 0$$
$$x = 0 \qquad\qquad x = -4 \qquad\qquad x = 5$$

There are three solutions to the equation: $x = 0$, $x = -4$, and $x = 5$.

4. Solve $x^3 + x^2 = 4x + 4$ for x.

We need to set the equation equal to 0, factor completely, and then solve each factor. To factor, see **Factoring by Grouping (4-term polynomials)**.

$$x^3 + x^2 = 4x + 4$$
$$x^3 + x^2 - 4x = 0$$
$$x^2(x + 1) - 4(x + 1) = 0$$
$$(x + 1)(x^2 - 4) = 0$$
$$(x + 1)(x + 2)(x - 2) = 0$$

$$x + 1 = 0 \quad \text{or} \quad x + 2 = 0 \quad \text{or} \quad x - 2 = 0$$
$$x = -1 \qquad\qquad x = -2 \qquad\qquad x = 2$$

If the equation will not factor, apply the rational roots theorem. See **Rational Roots Theorem**.

➤ **Cylinder (right circular)**   To find the **lateral area** of a right circular cylinder (the outside area, not including the area of the top or bottom), we use the formula

$$L = 2\pi r h$$

where L is the lateral area, r is the radius, and h is the altitude or height. For an example, see **Area (lateral) of a Right Circular Cylinder (and total area)**.

To find the **total area** of a right circular cylinder (lateral area and the areas of the top and bottom), we use the formula

$$\text{T.A.} = 2\pi rh + 2\pi r^2$$

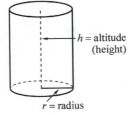

h = altitude (height)

r = radius

where $2\pi rh$ is the lateral area and $2\pi r^2$ is the combined areas of the top and bottom. The area of either the top, or bottom, is πr^2. See **Circle, area of**. For an example, see **Area (lateral) of a Right Circular Cylinder (and total area)**.

To find the **volume** of a right circular cylinder we use the formula

$$V = \pi r^2 h$$

where h is the altitude or height.

➤ **Data**   **Data** are a collection of known facts or figures, singular **datum**. If the data contain only numbers, they are called **numeric data**. Data that contain letters, numbers, and other symbols are called **alphanumeric data**.

➤ **Decimal Numbers**   For an example of **decimal numbers**, see **Decimal Numbers, definition**.

➤ **Decimal Numbers, addition of**   **Decimal numbers** are added in columns. The numbers should be written with the decimal points directly under each other. Then the numbers can be added. The decimal point of the answer is placed directly beneath the decimal points of the problem.

Examples

1. Add .4, .7, and .9.

$$
\begin{array}{r}
.4 \\
.7 \\
\underline{.9} \\
2.0
\end{array}
$$
 Line up the decimal points

Bring the decimal point straight down into the answer.

2. Add 6.3, .41, and 1.5.

$$
\begin{array}{r}
6.30 \\
.41 \\
\underline{1.50} \\
8.21
\end{array}
$$
 Fill in the empty decimal places only if you want to.

3. Add 4, .2, 81, and 7.33.

$$
\begin{array}{r}
4.00 \\
.20 \\
81.00 \\
\underline{7.33} \\
92.53
\end{array}
$$
 Numbers without a decimal point have one at the end.

4. Add 4.34, 7.2, 25, .06, and .012.

$$
\begin{array}{r}
4.340 \\
7.200 \\
25.000 \\
.060 \\
\underline{.012} \\
36.612
\end{array}
$$

➤ **Decimal Numbers, changed to a fraction**   To **change a decimal number to a fraction**, write the fraction as you say (pronounce) the decimal number. The decimal number .25 is pronounced "twenty-five hundredths." So, write it that way as a fraction, $\frac{25}{100}$. See **Decimal Numbers, definition** and **Fractions, reducing**.

Examples

In the following examples, reduce all answers.

1. Change .2 to a fraction.

$$.2 = \frac{2}{10} = \frac{1}{5}$$

    Pronounced "two tenths"

2. Change .075 to a fraction.

$$.075 = \frac{75}{1000} = \frac{3}{40}$$

    Pronounced "seventy-five thousandths"

3. Change $.33\frac{1}{3}$ to a fraction.

$$.33\frac{1}{3} = \frac{33\frac{1}{3}}{100}$$

    It is pronounced "thirty-three and one third hundredths."

$$= 33\frac{1}{3} : 100$$

    See **Fractions, division of** or **Fractions, complex**.

$$= \frac{100}{3} \bullet \frac{1}{100} = \frac{1}{3}$$

4. Change 9.875 to a fraction.

$$9.875 = 9\frac{875}{1000}$$

    Only the decimal number is written as a fraction.

$$= 9\frac{7}{8}$$

➤ **Decimal Numbers, changed to a percentage**   To **change a decimal number to a percent**, move the decimal point two places to the right.

Examples

1. Change .25 to a percent.

$$.25 = 25\%$$ We move the decimal point to the right two places. The reason it is not shown is because we don't write decimal points at the end of a number.

We don't write 25.%, simply 25%.

2. Change 1.05 to a percent.

$$1.05 = 105\%$$

3. Change $.37\frac{1}{2}$ to a percent.

$$.37\frac{1}{2} = 37\frac{1}{2}\%$$

➤ **Decimal Numbers, changing a repeating decimal to a common fraction**   To **change a repeating decimal to a common fraction**, we

1. Set the number equal to n.

2. Multiply both sides of the equation by a power of 10 that will move the repeating portion to the right side of the decimal point.

3. Subtract the two equations. The repeating portions will cancel.

4. Solve for n to get the common fraction.

Examples

1. Express $.\overline{3}$ as a common fraction.

Since $.\overline{3} = .333\ldots$, we see that the number repeats every digit. As a result, the power of 10 we multiply by is 10. If it repeated every two digits, we would multiply by 100, every three digits by 1000, and so on.

Setting the number equal to n and multiplying both sides of the equation by 10, we have

$$n = .333\ldots$$
$$10n = 3.333\ldots$$

Subtracting the two equations, we have

$$\begin{aligned}
10n &= 3.333\ldots \\
-n &= .333\ldots \\
\hline
9n &= 3 \\
\frac{9n}{9} &= \frac{3}{9} \\
n &= \frac{1}{3}
\end{aligned}$$

2. Express $.\overline{54}$ as a common fraction.

Since $.\overline{54} = .5454\ldots$, we see that the number repeats every two digits. The power of 10 we multiply by is 100.

Setting the number equal to n and multiplying both sides of the equation by 100, we have

$$n = .5454\ldots$$
$$100n = 54.5454\ldots$$

Subtracting, we have

$$
\begin{array}{r}
100n = 54.545454\ldots \\
-n = .545454\ldots \\
\hline
99n = 54 \\
\dfrac{99n}{99} = \dfrac{54}{99} \\
n = \dfrac{54}{99}
\end{array}
$$

3. Express $.58\overline{3}$ as a common fraction.

In problems where there are digits in front of the repeating portion of the decimal, we need to multiply by two, separate powers of 10. These will have repeating decimals that will line up. Then they can be subtracted.

Since $.58\overline{3} = .58333\ldots$, we see that the repeating portion is $333\ldots$. We need to move the decimal point over two places to get the repeating portion on the right side of the decimal point. We multiply by 100 to get $58.333\ldots$.

To get another number with a repeating portion of $333\ldots$, we can multiply by 1000. We get the number $583.333\ldots$.

$$n = .58333\ldots$$
$$100n = 58.333\ldots$$
$$1000n = 583.333\ldots$$

Subtracting $1000n - 100n$, we have

$$
\begin{array}{r}
1000n = 583.333\ldots \\
-100n = 58.333\ldots \\
\hline
900n = 525 \\
\dfrac{900n}{900} = \dfrac{525}{900} \\
n = \dfrac{7}{12}
\end{array}
$$

To verify that $\frac{7}{12}$ is $.58\overline{3}$, divide 7 by 12.

➤ **Decimal Numbers, comparing**   When **comparing decimals**, if the numbers have the same number of decimal places, then the larger number is the largest decimal number. If the numbers don't have the same number of decimal places, write zeros at the end until they do. Then, the larger number is the largest decimal number.

1. Which is larger: .45 or .4?

> .45 or .40   Write a 0 at the end.
> Answer: .45 is larger since 45 is larger than 40.

2. Which is larger: .599 or .6?

> .599 or .600   Write two 0's at the end.
> Answer: .6 is larger, since 600 is larger than 599.

3. Which is larger: .135 or 1.06?

   There is no need to write 0's at the end. A mixed decimal is larger than a decimal fraction. The 1.06 is larger.

4. Arrange from smallest to largest: .05, 2.3, .23, and .465.

> Writing in 0's, we need to compare .050, 2.300, .230, and .465.
> Answer: .05, .23, .465, and 2.3.

5. Arrange from smallest to largest: 5.63, .572, 4.7, 48, and .073.

> Writing in 0's, we need to compare 5.630, .572, 4.700, 48.000, and .073.
> Answer: .073, .572, 4.7, 5.63, and 48.

➤ **Decimal Numbers, definition**   A **decimal number** is a fraction. It expresses a part of a unit. However, instead of writing decimal numbers with denominators (which are powers of ten, i.e., 10, 100, 1000, etc.), we let their position determine their value. The first place to the right of the decimal is called the tenths place, the second is the hundredths, the third is the thousandths, and so on (ten-thousandths, hundred-thousandths, millionths, etc.).

In this manner the interval from 0 to 1 is divided into 10 pieces. Then each of those is divided into 10 pieces. Then each of those and so on. This process allows us a means of locating any position on the interval from 0 to 1.

To locate positions beyond 1, the concept of a **mixed decimal** is introduced. A mixed decimal is composed of a whole number and a decimal number (decimal fraction); see example 2.

1. On an interval from 0 to 1, locate the position of 0.1 or $\frac{1}{100}$, .1 or $\frac{1}{10}$, .4 or $\frac{4}{10}$, and .85 or $\frac{85}{100}$. See **Number Line**.

2. Locate the position of the mixed decimal 3.72 on a number line.

The number 3.72 is larger than 3 but less than 4. So, it is between 3 and 4.

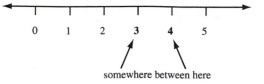

somewhere between here

To locate the position, we need to divide the interval between 3 and 4 into 10 pieces. The number 3.72 is larger than 3.7 but less than 3.8. So, it is between 3.7 and 3.8.

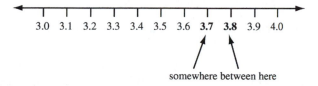

somewhere between here

To locate the position, we need to divide the interval between 3.7 and 3.8 into 10 pieces. Then we can find the position of 3.72.

3.70 3.71 **3.72** 3.73 3.74 3.75 3.76 3.77 3.78 3.79 3.80

here

➤ **Decimal Numbers, division of**   To **divide decimal numbers**, move the decimal point to the end of the divisor. Also move the decimal point the same number of places in the dividend, then place the decimal in the same location in the quotient. Now, forget about the decimal point and divide. The decimal point does not come down into the problem.

$$\text{divisor}\overline{)\text{divident}}^{\text{quotient}}$$

Examples

1. Divide .648 by .02.

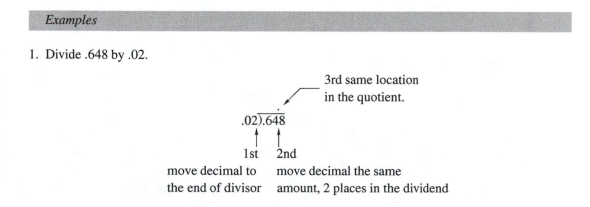

3rd same location
in the quotient.

.02)‾.648

1st   2nd

move decimal to   move decimal the same
the end of divisor   amount, 2 places in the dividend

Now the problem looks like $2\overline{)64.8}$. Dividing, we have

$$
\begin{array}{r}
32.4 \\
2\overline{)64.8} \\
\underline{6} \\
4 \\
\underline{4} \\
8 \\
\underline{8} \\
0
\end{array}
$$

2. Why does moving the decimal in example 1 work?

   Since decimals are fractions, let's divide them that way and see. See **Decimal Numbers, definition**.

$$
.648 : .02 = \frac{648}{1000} : \frac{2}{100} = \frac{\overset{324}{\cancel{648}}}{\underset{10}{\cancel{1000}}} \cdot \frac{\overset{1}{\cancel{100}}}{\underset{1}{\cancel{2}}} = \frac{324}{10} = 32\frac{4}{10} = 32.4
$$

$$
= \frac{648}{1000} \cdot \frac{\cancel{100}}{2}
$$

From here we can see that 100 is canceled in both

$$
= \frac{648}{10} \cdot \frac{1}{2}
$$

the .648 term and the .02 term. This is equivalent

$$
= 64.8 \cdot \frac{1}{2}
$$

to moving the decimal two places.

$$
= \frac{64.8}{2} \left.\right\} \longrightarrow \text{Similar to saying } 2\overline{)64.8}
$$

$$
= 32.4
$$

3. Divide 14.976 by 2.4.

   Use a caret, $\wedge$, to show the movement of the decimal.

$$
2 4_{\wedge}\overline{)14.9_{\wedge}7\,6}
$$

Now divide

$$
\begin{array}{r}
6.24 \\
24\overline{)149.76} \\
\underline{144} \\
57 \\
\underline{48} \\
96 \\
\underline{96} \\
0
\end{array}
$$

4. Divide 6.0282 by .006.

$$
\begin{array}{r}
1004.7 \\
.006_{\wedge}\overline{)6.028_{\wedge}2} \\
\underline{6} \\
00 \\
\underline{0} \\
02 \\
\underline{0} \\
28 \\
\underline{24} \\
42 \\
\underline{42} \\
0
\end{array}
$$

5. If there are no decimal places in the divisor, just bring the decimal straight up to the quotient from the dividend and divide. Divide 47.32 by 14.

$$
\begin{array}{r}
3.38 \\
14\overline{)47.32} \\
\underline{42} \\
53 \\
\underline{42} \\
112 \\
\underline{112} \\
0
\end{array}
$$

6. Divide 3 by 60.

$$
\begin{array}{r}
.05 \\
60\overline{)3.00} \\
\underline{3\ 00} \\
0
\end{array}
$$

➤ **Decimal Numbers, multiplication of**   To **multiply decimal numbers**, just multiply the numbers. The decimal point does not have to be considered until the end of the problem. Also, the decimal points do not have to be lined up. That is done in addition and subtraction only.

The decimal point is considered after the problem has been completely multiplied. At that time, the number of decimal places to the right of the decimal point should be counted up. Then, count back that many in the answer.

Examples

1. Multiply 1.2 × .34.

$$
\begin{array}{r}
.34 \\
\times\ 1.2 \\
\hline
68 \\
34 \\
\hline
.408
\end{array}
$$

First, just multiply the numbers and disregard the decimal points.

Move the decimal point three places in the answer.

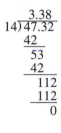

We move the decimal point three places because there are three digits to the right of the decimal points in the problem, the 3, 4, and 2.

2. Multiply 460 by 4.8.

$$
\begin{array}{r}
460 \\
\times\ 4.8 \\
\hline
3680 \\
1840 \\
\hline
2208.0
\end{array}
$$
Move the decimal point one place in the answer.

There is only one digit to the right of the decimal, the 8, so count back 1 place.

3. How does this process work? To get the answer, we need to look at fractions. Using example 1, we have

$$1.2 \times .34 = 1\frac{2}{10} \times \frac{34}{100} = \frac{12}{10} \times \frac{34}{100} = \frac{408}{1000} = .408$$

From the fraction $\frac{408}{1000}$, we see that the 3 zeros in 1000 move the decimal point three places. The 3 zeros came from the 10 and the 100. Those in turn came from the .2 and the .34, the digits to the right of the decimal. Hence, count up the digits to the right of the decimal and come back that many in the answer.

➤ **Decimal Numbers, nonterminating and nonrepeating**   A decimal number that does not terminate and does not repeat is an irrational number. See **Irrational Numbers**.

➤ **Decimal Numbers, repeating**   Decimal numbers that have a digit or group of digits that repeat are called **repeating decimals**. Repeating decimals are indicated either by a bar – or by three dots . . . . The number .3̄ or .3. . . means that the 3 repeats endlessly.
   Repeating decimals are rational numbers. They can be written as fractions. See **Decimal Numbers, changing a repeating decimal to a common fraction**.

Examples

1. Express $\frac{2}{3}$ as a repeating decimal.

$$
\begin{array}{r}
.666 = .\bar{6} \text{ or } .666\ldots \\
3\overline{)2.000} \\
\underline{18} \\
20 \\
\underline{18} \\
20 \\
\underline{18} \\
2
\end{array}
$$

2. Express $\frac{8}{11}$ as a repeating decimal.

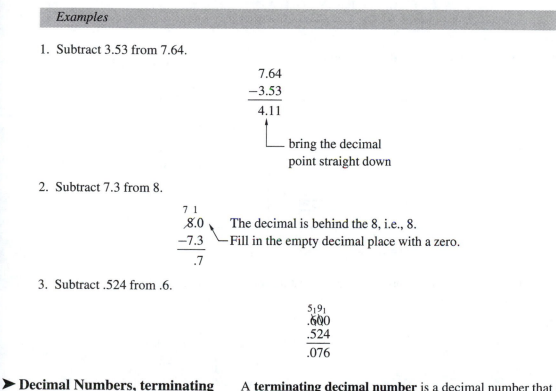

$$
\begin{array}{r}
.7272 = .\overline{72} \text{ or } .72\ldots \\
11\overline{)8.0000} \\
\underline{77} \\
30 \\
\underline{22} \\
80 \\
\underline{77} \\
30 \\
\underline{22} \\
8
\end{array}
$$

➤ **Decimal Numbers, subtraction**   Decimal numbers are subtracted in columns. The numbers should be written with the decimal points directly under each other. Then, the numbers can be subtracted. The decimal point of the answer is placed directly beneath the decimal points of the problem.

Examples

1. Subtract 3.53 from 7.64.

$$
\begin{array}{r}
7.64 \\
-3.53 \\
\hline
4.11
\end{array}
$$

bring the decimal
point straight down

2. Subtract 7.3 from 8.

$$
\begin{array}{r}
{}^{7\ 1} \\
8.0 \\
-7.3 \\
\hline
.7
\end{array}
$$
The decimal is behind the 8, i.e., 8.
Fill in the empty decimal place with a zero.

3. Subtract .524 from .6.

$$
\begin{array}{r}
{}^{5\ 9\ 1} \\
.600 \\
.524 \\
\hline
.076
\end{array}
$$

➤ **Decimal Numbers, terminating**   A **terminating decimal number** is a decimal number that does not repeat. It is a rational number. The number .75 is a terminating decimal. To convert a terminating decimal number to a fraction, see **Decimal Numbers, changed to a fraction**.

➤ **Decreasing Function**   See **Increasing and Decreasing Functions**.

➤ **Deductive Proof**   See **Deductive Reasoning** or **Proof in Geometry**.

➤ **Deductive Reasoning**   Reasoning that leads to conclusions that are based on accepted facts is called **deductive reasoning**. Such a process is used to prove theorems in geometry. The conclusion of a proof follows logically from the hypothesis. For an example, see **Proof in Geometry**.

➤ **Definite Integral**   In differential calculus, one first learns to calculate derivatives by definition and then later by rules. Similarly, in integral calculus, one first learns to calculate integrals by the **definite integral** and then later by an application of the fundamental theorem of calculus (see **Fundamental Theorem of Calculus**).

In relationship to area, if we let $f(x)$ be a continuous function on the closed interval $[a, b]$, then the definite integral defines the area of the region bounded by the graph of the function, the x-axis, and the lines $x = a$ and $x = b$. The area is found by choosing a partition of the interval $[a, b]$ (see **Partition**), and then calculating the sum of the rectangles formed by the partition and the graph of $f(x)$. As such, the definite integral is defined by

$$\int_a^b f(x)\,dx = \lim_{n \to \infty} \sum_{i=1}^{n} f(c_i)\,\Delta x, \qquad x_{i-1} \leqslant c_i \leqslant x_i$$

where the sum represents the definite integral.

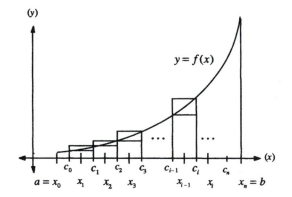

The following list can be useful when calculating the definite integral. In all, the functions are assumed to be integrable.

1. $\sum_{i-1}^{n} c = cn$

2. $\sum_{i=1}^{n} i = \dfrac{n(n+1)}{2}$

3. $\sum_{i=1}^{n} i^2 = \dfrac{n(n+1)(2n+1)}{6}$

4. $\sum_{i=1}^{n} i^3 = \dfrac{n^2(n+1)^2}{4}$

5. $\int_a^a f(x)\,dx = 0$

6. $\int_b^a f(x)\,dx = -\int_a^b f(x)\,dx$

7. $\int_a^b f(x)\,dx = \int_a^c f(x)\,dx + \int_c^b f(x)\,dx$

8. $\displaystyle\int_a^b cf(x)\,dx = c\int_a^b f(x)\,dx,$ where c is a constant

9. $\displaystyle\int_a^b [f(x) \pm g(x)]\,dx = \int_a^b f(x)\,dx \pm \int_a^b g(x)\,dx$

Example

1. Evaluate the definite integral $\int_0^3 x^2\,dx$, first by using right endpoints and then by using left endpoints.

By Right Endpoints

Choosing a regular partition, we have $\Delta x = \frac{b-a}{n} = \frac{3-0}{n} = \frac{3}{n}$.

Letting c_i represent the right endpoints of each subinterval, we have

$$c_i = a + i(\Delta x) = 0 + i\left(\frac{3}{n}\right) = \frac{3i}{n}$$

Substituting Δx and c_i into the formula for the definite integral, we have

$$\int_0^3 x^2\,dx = \lim_{n\to\infty} \sum_{i=1}^n f(c_i)\,\Delta x = \lim_{n\to\infty} \sum_{i=1}^n \left(\frac{3i}{n}\right)^2\left(\frac{3}{n}\right) = \lim_{n\to\infty} \sum_{i=1}^n \frac{27i^2}{n^3}$$

$$= \lim_{n\to\infty} \frac{27}{n^3} \sum_{i=1}^n i^2 = \lim_{n\to\infty} \frac{27}{n^3} \cdot \frac{n(n+1)(2n+1)}{6} = \lim_{n\to\infty} \frac{9(2n^2 + 3n + 1)}{2n^2}$$

$$= \frac{9}{2} \lim_{n\to\infty}\left(2 + \frac{3}{n} + \frac{1}{n^2}\right) = 9$$

By Left Endpoints

Letting c_i represent the left endpoint of each subinterval, we have

$$c_i = a + (i-1)\,\Delta x = 0 + (i-1)\left(\frac{3}{n}\right) = \frac{3i}{n} - \frac{3}{n}$$

Substituting, we have

$$\int_0^3 x^2\,dx = \lim_{n\to\infty} \sum_{i=1}^n f(c_i)\,\Delta x = \lim_{n\to\infty} \sum_{i=1}^n \left[(i-1)\left(\frac{3}{n}\right)\right]^2\left(\frac{3}{n}\right)$$

$$= \lim_{n\to\infty} \frac{3}{n} \sum_{i=1}^n (i-1)^2 \cdot \frac{9}{n^2} = \lim_{n\to\infty} \frac{27}{n^3} \sum_{i=1}^n (i^2 - 2i + 1)$$

$$= \lim_{n\to\infty} \frac{27}{n^3}\left[\frac{n(n+1)(2n+1)}{6} - 2\left(\frac{n(n+1)}{2}\right) + n\right]$$

$$= \lim_{n\to\infty} \frac{27}{n^3}\left(\frac{2n^3 - 3n^2 + n}{6}\right) = \lim_{n\to\infty}\left(\frac{18n^3}{2n^3} - \frac{27n^2}{2n^3} + \frac{9n}{2n^3}\right)$$

$$= \lim_{n\to\infty}\left(9 - \frac{27}{2n} + \frac{9}{2n^2}\right) = 9$$

➤ **Degree Conversion to Radians**   See **Converting Degrees to Radians**.

➤ **Degree Measure**   There are two methods that are used in the measurement of angles, **degree measure** and **radian measure**. See **Radian**. Degree measure is found with the use of a protractor. The measure of an angle corresponds to the number of degrees on the protractor.

➤ **Degree-Minutes-Seconds**   There are two methods for describing fractional parts of a degree. One uses minutes and seconds while the other uses decimal degrees. In the former,

$$1' = 1 \text{ minute} = \left(\frac{1}{60}\right)^\circ$$

$$1'' = 1 \text{ second} = \left(\frac{1}{60}\right)' = \left(\frac{1}{3600}\right)^\circ$$

In all, $1° = 60' = 3600''$. This means if we want an angle smaller than $1°$, we can cut $1°$ into 60 pieces. Each piece is $1'$, 1 minute. If we want even smaller divisions of a degree, we can cut each minute into 60 pieces. Each piece is $1''$, 1 second. Using seconds allows us to cut $1°$ into 3600 pieces. An angle of 72 degrees, 15 minutes, and 52 seconds is written as

$$\theta = 72°15'52''$$

The other method of describing a fractional part of a degree is to use decimal degrees. See example 1.

Examples

1. Change $72°15'52''$ to decimal degrees.

   To convert from degree-minutes-seconds to decimal degrees, divide the minutes by 60 and the seconds by 3600:

   $$72°15'52'' = 72° + \left(\frac{15}{60}\right)^\circ + \left(\frac{52}{3600}\right)^\circ \qquad \text{Multiply } 15\left(\frac{1}{60}\right) \text{ and } 52\left(\frac{1}{3600}\right).$$
   $$= 72° + .25° + .0144°$$
   $$= 72.26444°$$

2. Change $72.26444°$ to degrees-minutes-seconds.

   To convert from decimal degrees to degrees-minutes-seconds, multiply the decimal part of the number by 60 to get 15.8664. The whole number part of this number, 15, is the minutes. Then, multiply the decimal part of this number, .8664, by 60. This gives 51.984 which rounds to 52, the seconds. Hence,

   $$72.26444° = 72°15'52''$$

3. Add the angles $18°54'49''$ and $68°17'25''$.

$$\begin{array}{r} 18°54'49'' \\ + 68°17'25'' \\ \hline 86°71'74'' \end{array}$$

$= 86°72'14''$ since $60'' = 1'$

$= 87°12'14''$ since $60' = 1°$

➤ **Degree of a Polynomial**   The **degree of a polynomial** is determined by the term with the highest degree. The **degree of a term** is determined from the sum of the exponents of its variables.

Examples

1. Determine the degree of the polynomial.

$$6x + 8x^3y^2 - 5y + 12xy^2 + 4$$

Degree of terms: 1   5   1   3   0   (since $4 = 4x^0$)

Hence, the polynomial has a degree of 5.

2. Determine the degree of each polynomial.

    a. $6x - 5$ has a degree of 1

    b. $5x^3 - 9x^2 + 4x - 3$ has a degree of 3

    c. $7x^5y^2 - 5x^4y - 6x^6 + 4$ has a degree of 7

➤ **De Moivre's Theorem**   **De Moivre's theorem** is used to find powers of complex numbers. Let $z = r(\cos\theta + i\sin\theta)$. Then by De Moivre's theorem

$$z^n = \left[r(\cos\theta + i\sin\theta)\right]^n = r^n(\cos n\theta + i\sin n\theta)$$

where n is a positive integer, r is the absolute value of z with $r = \sqrt{a^2 + b^2}$, and θ is defined by $\tan\theta = \frac{b}{a}$. See **Complex Number, absolute value** or **Complex Number, trigonometric form**.

Example

1. If $z = -2 - 2i\sqrt{3}$, find z^9.

We first need to convert z to trigonometric form, $z = r(\cos\theta + i\sin\theta)$.

    a.   Find r.

$$r = \sqrt{a^2 + b^2} = \sqrt{(2)^2 + (2\sqrt{3})^2} = \sqrt{4 + 12} = 4$$

D

b.   Find θ.

$$\tan \theta = \frac{b}{a} = \frac{-2\sqrt{3}}{-2} = \sqrt{3}$$

Since $-2 - 2i\sqrt{3}$ is in the third quadrant.

$$\theta = 180 + \arctan\sqrt{3}$$
$$= \pi + \frac{\pi}{3} = \frac{4\pi}{3}$$

So

$$z = -2 - 2i\sqrt{3} = r(\cos\theta + i\sin\theta) = 4\left(\cos\frac{4\pi}{3} + i\sin\frac{4\pi}{3}\right)$$

See **Complex Number, trigonometric form**.

Now we can apply De Moivre's theorem.

$$z^9 = (-2 - 2i\sqrt{3})^9 = \left[4\left(\cos\frac{4\pi}{3} + i\sin\frac{4\pi}{3}\right)\right]^9 = 4^9\left(\cos(9)\frac{4\pi}{3} + i\sin(9)\frac{4\pi}{3}\right)$$
$$= 262,144(\cos 12\pi + i\sin 12\pi) = 262,144(1+0) = 262,144$$

➤ **Denominator**   In fractions, the **denominator** is the number below the fraction bar. The **numerator** is the number above. See **Fraction**.

➤ **Denominator, rationalizing the**   See **Rationalizing the Denominator**.

➤ **Dependent Equations**   See **Systems of Linear Equations**.

➤ **Dependent System**   See **Systems of Linear Equations**.

➤ **Dependent Variable**   In the equation $y = 3x + 4$, y is called the **dependent variable** since it is determined by x. Other examples are $y = 4x^2 - 11$ and $y = 5\sqrt{x} + 2$. The x is called the independent variable. See **Independent Variable**.

➤ **Derivative**   This article deals with the basic concepts pertaining to derivatives by definition and derivatives by rules. The examples have been broken up into two categories dealing with each.
    Simply, a **derivative** is a slope. It is the slope of all of the tangent lines to a given curve $y = f(x)$. To find the derivative, we first calculate it by applying the definition of a derivative. Then later, after patterns become apparent, we develop rules for calculating derivatives.
    The goal of finding the derivative to a given function is to apply the concepts pertaining to slope. Since the derivative is the slope of all of the tangent lines to a given curve, we can consider the rate of change of the slope to each tangent line or consider when the slope is 0. The change of slopes is dealt with in problems pertaining to **related rates** while the slope being 0 gives information about the **maximum** and **minimum** values of a function (horizontal tangent lines that have a slope of 0). This is useful for graphing and max-min problems (problems involving the maximum and minimum values of the derivative).

Derivative by Definition

Let $y = f(x)$ be a function defined on an open interval containing x. Then, the derivative of f at the point $(x, f(x))$ is

$$f'(x) = \lim_{\Delta x \to 0} \frac{f(x + \Delta x) - f(x)}{\Delta x} \qquad \text{if the limit exists}$$

The symbol $f'(x)$ is read "f prime of x." It can be replaced by the symbol $\frac{dy}{dx}$, read "the derivative of y with respect to x" or by y' read "y prime."

The formula (derivative by definition) can be developed by looking at a graph.

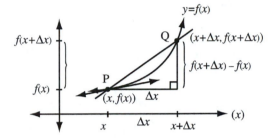

What is required is to find the slope of the tangent line that goes through the point $(x, f(x))$. However, finding the slope requires two points. We can get another point by moving a distance Δx (read "delta x," does not mean Δ times x; replace with h if you don't like using it) from x to $x + \Delta x$. The corresponding y-value to $x + \Delta x$ is $f(x + \Delta x)$. With these two points, $(x, f(x))$ and $(x + \Delta x, f(x + \Delta x))$, we can calculate the slope. However, keep in mind, this is the slope of $\overline{PQ}$, which is called the secant line (not to be confused with the secant in trigonometry; no connection).

$$m = \frac{y_2 - y_1}{x_2 - x_1} = \frac{f(x + \Delta x) - f(x)}{(x + \Delta x) - x} = \frac{f(x + \Delta x) - f(x)}{\Delta x}$$

To find the slope of the tangent line at $(x, f(x))$, we are going to "do a little trick," if you will. We are going to let Δx get smaller. When we do this, the point Q gets a little closer to P. If we continue this process, letting Δx get smaller, then Q gets closer and closer to P. As this happens, notice that the inclination (or steepness, or just slope) of $\overline{PQ}$ approaches the inclination of the tangent line at $(x, f(x))$. The slope of the secant line approaches the slope of the tangent line. We call this slope the derivative of $f(x)$ and we write it as

$$f'(x) \text{ or } \frac{dy}{dx} = \lim_{\Delta x \to 0} \frac{f(x + \Delta x) - f(x)}{\Delta x}$$

This says that the derivative is the limit as Δx approaches 0 of the slope of the secant line. See **Limits**.

Derivative by Rules

Applying the definition of a derivative can be cumbersome as well as time consuming. Fortunately, there are patterns to the process. These have been developed and proved. To go any further in calculus, a student must memorize them completely. A list of the rules is given here and various applications are given in the examples.

Algebraic Rules

1. $\dfrac{d(c)}{dx} = 0$, where c is an arbitrary constant

2. $\dfrac{d(cu)}{dx} = c\dfrac{du}{dx}$

3. $\dfrac{d(u + v + \cdots)}{dx} = \dfrac{du}{dv} + \dfrac{dv}{dx} + \cdots$

4. $\dfrac{d(x)}{dx} = 1$

5. $\dfrac{d(x^n)}{dx} = nx^{n-1}$

6. $\dfrac{d|u|}{dx} = \dfrac{u}{|u|} \bullet \dfrac{du}{dx}$, where $u \neq 0$

7. $\dfrac{d(u^n)}{dx} = nu^{n-1}\dfrac{du}{dx}$

8. $\dfrac{d(uv)}{dx} = u\dfrac{dv}{dx} + v\dfrac{du}{dx}$

9. $\dfrac{d(uvw)}{dx} = uv\dfrac{dw}{dx} + uw\dfrac{dv}{dx} + vw\dfrac{du}{dx}$

10. $\dfrac{d}{dx}\left(\dfrac{u}{v}\right) = \dfrac{v\frac{du}{dx} - u\frac{dv}{dx}}{v^2}$, where $v \neq 0$

    a. $\dfrac{d}{dx}\left(\dfrac{u}{c}\right) = \dfrac{1}{c}\dfrac{du}{dx}$, where $c \neq 0$

    b. $\dfrac{d}{dx}\left(\dfrac{c}{v}\right) = c\dfrac{d}{dx}\left(\dfrac{1}{v}\right) = \dfrac{-c}{v^2}\dfrac{dv}{dx}$, where $v \neq 0$

    c. When $c = 1$, we have $\dfrac{d}{dx}\left(\dfrac{1}{v}\right) = \dfrac{-1}{v^2}\dfrac{dv}{dx} = \dfrac{-\frac{dv}{dx}}{v^2}$, where $v \neq 0$

11. **Chain Rule**

    a. $\dfrac{dy}{dx} = \dfrac{dy}{du} \bullet \dfrac{du}{dx}$, where $y = f(u)$ and $u = g(x)$ or

    b. $\dfrac{d}{dx}f(g(x)) = f'(g(x))g'(x)$, where $y = f(g(x))$

12. **Implicit Differentiation of an Equation**

Take the derivative of each term of the equation and then solve for $\frac{dy}{dx}$.

Transcendental Rules

1. $\dfrac{d}{dx}(\log_a u) = \dfrac{1}{(\ln a)u} \bullet \dfrac{du}{dx}$, where a is a positive real number with $a \neq 1$

   or $\quad = \dfrac{1}{u}\log_a e \dfrac{du}{dx}$

2. $\dfrac{d}{dx}\ln u = \dfrac{1}{u}\dfrac{du}{dx}$

3. $\dfrac{d}{dx}(a^u) = a^u \ln a \dfrac{du}{dx}$, where a is a positive real number

4. $\dfrac{d}{dx}(e^u) = e^u \dfrac{du}{dx}$

5. $\dfrac{d}{dx}(\ln |u|) = \dfrac{1}{u}\dfrac{du}{dx}$, where $u \neq 0$

6. $\dfrac{d}{dx}(\sin u) = \cos u \dfrac{du}{dx}$

7. $\dfrac{d}{dx}(\cos u) = -\sin u \dfrac{du}{dx}$

8. $\dfrac{d}{dx}(\tan u) = \sec^2 u \dfrac{du}{dx}$

9. $\dfrac{d}{dx}(\csc u) = -\csc u \cot u \dfrac{du}{dx}$

10. $\dfrac{d}{dx}(\sec u) = \sec u \tan u \dfrac{du}{dx}$

11. $\dfrac{d}{dx}(\cot u) = -\csc^2 u \dfrac{du}{dx}$

12. $\dfrac{d}{dx}(\arcsin u) = \dfrac{\frac{du}{dx}}{\sqrt{1 - u^2}}$

13. $\dfrac{d}{dx}(\arccos u) = \dfrac{-\frac{du}{dx}}{\sqrt{1 - u^2}}$

14. $\dfrac{d}{dx}(\arctan u) = \dfrac{\frac{du}{dx}}{1 + u^2}$

15. $\dfrac{d}{dx}(\text{arccsc } u) = \dfrac{-\frac{du}{dx}}{|u|\sqrt{u^2 - 1}}$

16. $\dfrac{d}{dx}(\text{arcsec } u) = \dfrac{\frac{du}{dx}}{|u|\sqrt{u^2 - 1}}$

17. $\dfrac{d}{dx}(\text{arccot } u) = \dfrac{-\frac{du}{dx}}{1 + u^2}$

D

Examples

Derivative by Definition

1. Find the derivative by definition of $y = f(x) = x^2 + 3$.

   What is required is to apply the definition of a derivative to the function $f(x) = x^2 + 3$. With $f(x + \Delta x) = (x + \Delta x)^2 + 3$ and $f(x) = x^2 + 3$, we have

$$
\begin{aligned}
f'(x) &= \lim_{\Delta x \to 0} \frac{f(x + \Delta x) - f(x)}{\Delta x} \\
&= \lim_{\Delta x \to 0} \frac{[(x + \Delta x)^2 + 3] - (x^2 + 3)}{\Delta x} \\
&= \lim_{\Delta x \to 0} \frac{x^2 + 2x\,\Delta x + \Delta x^2 + 3 - x^2 - 3}{\Delta x} \\
&= \lim_{\Delta x \to 0} \frac{2x\,\Delta x + \Delta x^2}{\Delta x} \\
&= \lim_{\Delta x \to 0} \frac{\Delta x(2x + \Delta x)}{\Delta x} \qquad \text{Factor out } \Delta x, \text{ then cancel.} \\
& \qquad\qquad\qquad\qquad\qquad \text{Now that the denominator will not be 0, let } \Delta x \to 0. \\
&= \lim_{\Delta x \to 0} 2x + \Delta x
\end{aligned}
$$

This says that the derivative of $x^2 + 3$ is $2x$. The term $2x$ represents the slope of all of the tangent lines to the curve $y = x^2 + 3$. If you choose a particular x, you will have its corresponding slope. For example, when $x = 2$, the corresponding slope is $m = 2x = 2(2) = 4$. This means that the slope of the tangent line at the point $(2, 2^2 + 3) = (2, 7)$ is $m = 4$.

When $x = 3$, the corresponding slope is $m = 2x = 2(3) = 6$. This means that the slope of the tangent line at the point $(3, 3^2 + 3) = (3, 12)$ is $m = 6$.

Do not let the $x + \Delta x$ or $f(x + \Delta x)$ confuse you. They are just an x-value and a y-value, respectively. For clarity, consider the chart below:

x	$y = f(x) = x^2 + 3$	
2	$f(2) = 2^2 + 3 = 7$	When x is 2, y is 7.
3	$f(3) = 3^2 + 3 = 12$	
a	$f(a) = a^2 + 3$	When x is a, y is $a^2 + 3$.
x	$f(x) = x^2 + 3$	When x is x, y is $x^2 + 3$.
$x + \Delta x$	$f(x + \Delta x) = (x + \Delta x)^2 + 3$	When x is $x + \Delta x$, y is $(x + \Delta x)^2 + 3$.
	$= x^2 + 2x\,\Delta x + \Delta x^2 + 3$	

For more, see **Function Notation** or **Function**.

2. Find the derivative by definition of $y = f(x) = x^3 - x$.

   Applying the definition of a derivative with $f(x + \Delta x) = (x + \Delta x)^3 - (x + \Delta x)$ and $f(x) = x^3 - x$, we have

$$
f'(x) = \lim_{\Delta x \to 0} \frac{f(x + \Delta x) - f(x)}{\Delta x}
$$

$$= \lim_{\Delta x \to 0} \frac{[(x + \Delta x)^3 - (x + \Delta x)] - (x^3 - x)}{\Delta x}$$

$$= \lim_{\Delta x \to 0} \frac{x^3 + 3x^2 \Delta x + 3x \Delta x^2 + \Delta x^3 - x - \Delta x - x^3 + x}{\Delta x}$$

$$= \lim_{\Delta x \to 0} \frac{\Delta x(3x^2 + 3x \Delta x + \Delta x^2 - 1)}{\Delta x}$$

$$= \lim_{\Delta x \to 0} 3x^2 + 3x \Delta x + \Delta x^2 - 1 \qquad \text{After canceling, let } \Delta x \to 0.$$

$$= 3x^2 - 1$$

The derivative of $x^3 - x$ is $3x^2 - 1$. This means that the slope of all of the tangent lines to $y = x^3 - x$ is $f'(x) = 3x^2 - 1$.

Derivative by Rules

After finding several derivatives by definition, many patterns become apparent. These patterns, or rules, can be applied directly to a given function instead of going through the difficult process of finding the derivative by applying the definition. The rules must be memorized and become even more familiar than the multiplication table.

3. Find the derivative of $y = f(x) = x^3 - x$ by using the rules for derivatives.

Since there are two terms in $y = x^3 - x$, we will first apply rule 3 of the algebraic rules (take the derivative of each term). Next, we notice that both the x^3 and $-x$ terms are of the form x^n. The x^3 term has $n = 3$ and the $-x$ term has $n = 1$. Therefore, we will apply rule 5 of the algebraic rules:

$$f'(x) = \frac{d(x^3 - x)}{dx} = \frac{dx^3}{dx} - \frac{dx^1}{dx} \qquad \text{Apply rule 3.}$$

$$= 3x^{3-1} - 1x^{1-1} \qquad \text{Apply rule 5.}$$

$$= 3x^2 - x^0$$

$$= 3x^2 - 1$$

As it should be, this is the same answer we got in example 2. The derivative $3x^2 - 1$ represents the slope of all of the tangent lines to the graph of $y = x^3 - x$.

Notice that $\frac{dx^1}{dx}$ is the same as $\frac{dx}{dx}$. This means that we could have applied rule 4 instead. However, it makes no difference. As long as the rules are applied, the answer will always be correct. It's just that some approaches are shorter than others.

$$f'(x) = \frac{d(x^3 - x)}{dx} = \frac{dx^3}{dx} - \frac{dx}{dx} \qquad \text{Apply rule 3.}$$

$$= 3x^{3-1} - 1 \qquad \text{Apply rules 5 and 4.}$$

$$= 3x^2 - 1$$

4. Find the derivative of $y = 4x^3 - 2x + 3$.

The first term, $4x^3$, is of the form cu (rule 2) where the c is 4 and the u is x^3. The second term, $-2x$, is also of the form cu (rule 2) where the c is -2 and the u is x. The third term, 3, is of the form c (rule 1) where c is 3. Applying all of the rules, we have

$$f'(x) = \frac{d}{dx}(4x^3 - 2x + 3)$$

$$= \frac{d4x^3}{dx} - \frac{d2x}{dx} + \frac{d3}{dx} \qquad \text{Rule 3.}$$

$$= \frac{4dx^3}{dx} - 2\frac{dx}{dx} + 0 \qquad \text{Rules 2 and 1.}$$

$$= 4 \bullet 3x^{3-1} - 2(1) \qquad \text{Rules 5 and 4.}$$

$$= 12x^2 - 2$$

Notice that if you "just look" at $4x^3 - 2x + 3$ you can "see" $12x^2 - 2$. With practice, the derivative becomes this obvious. Also, for practice, apply the definition to $y = f(x) = 4x^3 - 2x + 3$. See if you can get $12x^2 - 2$.

5. Find the derivative of $y = x^2(x^3 - 1)^4$.

Since the x^2 term is multiplying the $(x^3 - 1)^4$ term, the equation is of the form uv (rule 8). Some people prefer to apply rule 8 by using the phrase "first times the derivatives of the last plus last times the derivative of the first." With $u = x^2$ and $v = (x^3 - 1)^4$, we have

$$f'(x) = \frac{d}{dx}x^2(x^3 - 1)^4$$

$$= x^2 \frac{d}{dx}(x^3 - 1)^4 + (x^3 - 1)^4 \frac{dx^2}{dx} \qquad \text{Rule 8.}$$

$$= x^2 \bullet 4(x^3 - 1)^3 \bullet \frac{d}{dx}(x^3 - 1) + (x^3 - 1)^4 \bullet 2x \qquad \begin{array}{l}\text{Rules 7 (see below)} \\ \text{and 5.}\end{array}$$

$$= 4x^2(x^3 - 1)^3 \bullet \left(\frac{dx^3}{dx} - \frac{d1}{dx}\right) + 2x(x^3 - 1)^4 \qquad \text{Rule 3.}$$

$$= 4x^2(x^3 - 1)^3 \bullet (3x^2 - 0) + 2x(x^3 - 1)^4 \qquad \text{Rules 5 and 1.}$$

$$= 12x^4(x^3 - 1)^3 + 2x(x^3 - 1)^4$$

Rule 7 is similar to rule 5 except that it has the derivative "tagged" on the end of it. The derivative on the end is always the derivative of the expression inside the parentheses:

$$\frac{d}{dx}(x^3 - 1)^4 = 4(x^3 - 1)^{4-1} \bullet \frac{d}{dx}(x^3 - 1)$$

6. Find the derivative of $y = \frac{6(x^2-5)^3}{x^4}$.

Referring to the numerator and denominator as the "top" and the "bottom," respectively, we see that the top is divided by the bottom (rule 10). The phrase that some people like to use when applying this rule is "the top times the derivative of the bottom minus the bottom times the derivative of the top over the bottom squared." With $u = 6(x^2 - 5)^3$ and $v = x^4$, we have

$$f'(x) = \frac{d}{dx}\left[\frac{6(x^2 - 5)^3}{x^4}\right]$$

$$= \frac{x^4 \cdot \frac{d}{dx}6(x^2 - 5)^3 - 6(x^2 - 5)^3 \cdot \frac{d}{dx}x^4}{(x^4)^2} \qquad \text{Rule 8.}$$

$$= \frac{x^4 \cdot 18(x^2 - 5)^2 \cdot \frac{d}{dx}(x^2 - 5) - 6(x^2 - 5)^3 \cdot 4x^3}{x^8} \qquad \text{Rules 7 and 5.}$$

$$= \frac{18x^4(x^2 - 5)^2 \cdot (2x - 0) - 24x^3(x^2 - 5)^3}{x^8} \qquad \text{Rules 5 and 1.}$$

$$= \frac{36x^5(x^2 - 5)^2 - 24x^3(x^2 - 5)^3}{x^8} \qquad \text{Now, just simplify.}$$

$$= \frac{12x^3(x^2 - 5)^2 \cdot [3x^2 - 2(x^2 - 5)]}{x^8} \qquad \text{Factor out } 12x^3(x^2 - 5)^2$$

$$= \frac{12(x^2 - 5)^2(3x^2 - 2x^2 + 10)}{x^5} \qquad \text{from each term.}$$

$$= \frac{12(x^2 - 5)^2(x^2 + 10)}{x^5}$$

7. Find the derivative of $y = (x^2 + 1)^4$ using the **chain rule**.

The **chain rule** is used to deal with composite functions. It can be applied when trying to find the derivative of functions that can be written in the form $f(g(x))$. See **Composite Functions**.

By definition, the **chain rule** states that if $y = f(u)$ is differentiable with respect to u and $u = g(x)$ is differentiable with respect to x, then the composite function $y = f(g(x))$ is differentiable with respect to x where

$$\frac{dy}{dx} = \frac{dy}{du} \cdot \frac{du}{dx}$$

or, in function notation,

$$\frac{df(g(x))}{dx} = f'(g(x)) \cdot g'(x)$$

To apply the chain rule, we need to decide on a substitution for u (or $g(x)$), form composite functions, calculate the derivative of each of the composite functions, multiply the derivatives of the composite functions, and then leave the answer in terms of the independent variable, i.e., substitute for u so the answer is in terms of x.

In our example, if we let $u = g(x) = x^2 + 1$, then $y = f(u) = u^4$. This is the most important step in applying the chain rule, deciding what should be u (or $g(x)$). Whatever you choose, the resulting functions must both be differentiable.

It was fine for us to choose x^2+1 as an appropriate substitution for u because $y = u^4$ and $u = x^2+1$ are both differentiable functions. Differentiating each, we have

$$\frac{dy}{du} = \frac{du^4}{du} = 4u^3 \qquad \text{and} \qquad \frac{du}{dx} = \frac{d}{dx}(x^2+1) = 2x$$

Multiplying each of the derivatives, then substituting for u, we can find $\frac{dy}{dx}$.

$$\frac{dy}{dx} = \frac{dy}{du} \bullet \frac{du}{dx} = (4u^3)(2x) = \left[4(x^2+1)^3\right](2x) = 8x(x^2+1)$$

The derivative of $y = (x^2+1)^4$ using the chain rule is $\frac{dy}{dx} = 8x(x^2+1)^3$. To verify our answer, we can find the derivative by differentiation.

$$\frac{dy}{dx} = \frac{d(x^2+1)^4}{dx} = 4(x^2+1)^3 \bullet \frac{d(x^2+1)}{dx}$$

Algebraic rule 7. Notice that rule 7 is an application of the chain rule.

$$= 4(x^2+1)^3(2x) = 8x(x^2+1)^3$$

8. Find the derivative of $x^3 - x^2y = 8 + y^2$ using **implicit differentiation**.

Implicit differentiation is usually used when an equation cannot be solved for one of the variables. If an equation can be solved for one of the variables, called an explicit solution, then the regular rules for differentiation apply.

An equation does not have to be an explicit equation in order to find its derivative. For example, suppose we wanted to find the derivative of $y - x^2 = 0$ with respect to x. Usually we would solve for y, $y = x^2$, and then find $\frac{dy}{dx}$, $\frac{dy}{dx} = \frac{dx^2}{dx} = 2x$. However, if we wanted to, we could differentiate the equation just as it was given to us. To do this, we would differentiate both sides of the equation and then solve for $\frac{dy}{dx}$:

$$y - x^2 = 0$$

$$\frac{d(y - x^2)}{dx} = \frac{d0}{dx}$$

$$\frac{dy}{dx} - \frac{dx^2}{dx} = 0 \qquad \text{Algebraic rules 3 and 1.}$$

$$\frac{dy}{dx} - 2x = 0 \qquad \text{Algebraic rule 5.}$$

$$\frac{dy}{dx} = 2x$$

Explicitly, or implicity, we get $2x$ as the derivative.

Although the variable of an equation might differ (θ and ϕ, or a and b, etc.), the steps for implicit differentiation are the same. If we consider an implicit equation in x and y, where y is a differentiable function of x, then the following steps apply:

1. Differentiate both sides of the equation with respect to x.

2. Isolate the $\frac{dy}{dx}$ terms on one side of the equal sign and all other terms on the other side of the equal sign.

3. Factor out $\frac{dy}{dx}$ from each of the $\frac{dy}{dx}$ terms (if there is more than one $\frac{dy}{dx}$ term).

4. Solve for $\frac{dy}{dx}$ by dividing both sides of the equation by the parenthetical factor in step 3 (or just solve for $\frac{dy}{dx}$ if there is only one $\frac{dy}{dx}$ term).

Simply, what is being said is this, "differentiate both sides of the equation and solve for $\frac{dy}{dx}$." If there is only one $\frac{dy}{dx}$ term, isolate it and then divide by the term on its right (the term attached to $\frac{dy}{dx}$). If there is more than one term, solve for $\frac{dy}{dx}$ as you would solve for D in the equation $-z + Dx = Dy$:

$$-z + Dx = Dy$$
$$Dx - Dy = z$$
$$D(x - y) = z$$
$$\frac{D(x - y)}{x - y} = \frac{z}{x - y}$$
$$D = \frac{z}{x - y}$$

With these steps in mind, we can now find the derivative of $x^3 - x^2y = 8 + y^2$ by using implicit differentiation:

$$x^3 - x^2y = 8 + y^2$$

$$\frac{d}{dx}(x^3 - x^2y) = \frac{d}{dx}(8 + y^2) \qquad \text{Differentiate both sides.}$$

$$\frac{dx^3}{dx} - \frac{dx^2y}{dx} = \frac{d8}{dx} + \frac{dy^2}{dx}$$

$$3x^2 - \left(x^2\frac{dy}{dx} + y\frac{dx^2}{dx}\right) = 0 + 2y\frac{dy}{dx} \qquad \text{Algebraic rules 5, 8, 1, and 7.}$$

$$3x^2 - x^2\frac{dy}{dx} - 2xy = 2y\frac{dy}{dx} \qquad \text{Algebraic rule 5.}$$

$$-x^2\frac{dy}{dx} - 2y\frac{dy}{dx} = -3x^2 + 2xy \qquad \text{Isolate the } \frac{dy}{dx} \text{ terms.}$$

$$\frac{dy}{dx}(-x^2 - 2y) = -3x^2 + 2xy \qquad \text{Factor out } \frac{dy}{dx}.$$

$$\frac{dy}{dx}\frac{(-x^2 - 2y)}{-x^2 - 2y} = \frac{-3x^2 + 2xy}{-x^2 - 2y} \qquad \text{Divide by } -x^2 - 2y.$$

$$\frac{dy}{dx} = \frac{-3x^2 + 2xy}{-x^2 - 2y}$$

$$\frac{dy}{dx} = \frac{-3x^2 + 2xy}{-(x^2 + 2y)}$$

$$\frac{dy}{dx} = \frac{3x^2 - 2xy}{x^2 + 2y}$$

➤ **Descending Order**   See **Ascending and Descending Order**.

➤ **Determinant**   A **determinant** is denoted by replacing the symbols for a square matrix (usually parentheses or brackets) with vertical lines. The following are examples of determinants:

$$\begin{vmatrix} 2 & -4 \\ -5 & 9 \end{vmatrix} \qquad \begin{vmatrix} 3 & 6 & -5 \\ -2 & 4 & 9 \\ 7 & -8 & 2 \end{vmatrix} \qquad \begin{vmatrix} 7 & 2 & -6 & 8 \\ -8 & 1 & 2 & -5 \\ 4 & 0 & 8 & 6 \\ 3 & -4 & -9 & 0 \end{vmatrix}$$

2nd-order   3rd-order   4th-order

determinant   determinant   determinant

Although similar in appearance to a matrix, a determinant differs from a matrix in that it has an actual value. The value of a determinant is a number that is assigned to the determinant. In application, the value of the determinant is used to solve systems of equations. See **Cramer's Rule**.

By definition, the value of a second-order (2 by 2 or 2×2) determinant is

$$\begin{vmatrix} a & b \\ c & d \end{vmatrix} = ab - cd$$

For example, the value of the determinant $\begin{vmatrix} 2 & -6 \\ 3 & 8 \end{vmatrix}$ is

$$\begin{vmatrix} 2 & -6 \\ 3 & 8 \end{vmatrix} = (2)(8) - (-6)(3) = 16 + 18 = 34$$

To find the value of a third-order (3×3) determinant, we apply the concept of **expansion by minors**. A **minor** is, a determinant that is formed when the row and the column of an element are deleted. See **Minor**. The expansion by minors of the determinant

$$\begin{vmatrix} a_1 & b_1 & c_1 \\ a_2 & b_2 & c_2 \\ a_3 & b_3 & c_3 \end{vmatrix}$$

can be found by combining the minors of any row and any column, usually the first row and first column. If the first row is used, then the expansion is

$$\begin{vmatrix} a_1 & b_1 & c_1 \\ a_2 & b_2 & c_2 \\ a_3 & b_3 & c_3 \end{vmatrix} = a_1 \begin{vmatrix} b_2 & c_2 \\ b_3 & c_3 \end{vmatrix} - b_1 \begin{vmatrix} a_2 & c_2 \\ a_3 & c_3 \end{vmatrix} + c_1 \begin{vmatrix} a_2 & b_2 \\ a_3 & b_3 \end{vmatrix}$$

where the sign of each term is determined by the first row in the diagram

$$\begin{vmatrix} + & - & + \\ - & + & - \\ + & - & + \end{vmatrix},$$

the $+ - +$ (see below). Also see **Minor**.

The minor of a_1 is   The minor of b_1 is   The minor of c_1 is

$$\begin{vmatrix} \cancel{a_1} & \cancel{b_1} & \cancel{c_1} \\ \cancel{a_2} & b_2 & c_2 \\ \cancel{a_3} & b_3 & c_3 \end{vmatrix} \qquad \begin{vmatrix} \cancel{a_1} & \cancel{b_1} & \cancel{c_1} \\ a_2 & \cancel{b_2} & c_2 \\ a_3 & \cancel{b_3} & c_3 \end{vmatrix} \qquad \begin{vmatrix} a_1 & b_1 & \cancel{c_1} \\ a_2 & b_2 & \cancel{c_2} \\ a_3 & b_3 & \cancel{c_3} \end{vmatrix}$$

Expanding each of the minors, we have the evaluation of the third-order determinant:

$$\begin{vmatrix} a_1 & b_1 & c_1 \\ a_2 & b_2 & c_2 \\ a_3 & b_3 & c_3 \end{vmatrix} = a_1(b_2c_3 - b_3c_2) - b_1(a_2c_3 - a_3c_2) + c_1(a_2b_3 - a_3b_2)$$

$$= a_1b_2c_3 - a_1b_3c_2 - a_2b_1c_3 + a_3b_1c_2 + a_2b_3c_1 - a_3b_2c_1$$

$$= a_1b_2c_3 - a_1b_3c_2 + a_3b_1c_2 - a_2b_1c_3 + a_2b_3c_1 - a_3b_2c_1$$

The expansions of the minors are determined by subtracting the cross products:

$$\begin{vmatrix} b_2 & c_2 \\ b_3 & c_3 \end{vmatrix} = b_2c_3 - b_3c_2 \qquad \begin{vmatrix} a_2 & c_2 \\ a_3 & c_3 \end{vmatrix} = a_2c_3 - a_3c_2 \qquad \begin{vmatrix} a_2 & b_2 \\ a_3 & b_3 \end{vmatrix} = a_2b_3 - a_3b_2$$

If the first row and first column are used to evaluate the determinant, then the positive and negative signs of the minors follow the pattern $+ - +$. These signs are determined from the following diagram:

$$\begin{vmatrix} + & - & + \\ - & + & - \\ + & - & + \end{vmatrix}$$

The diagram is developed by considering the number of the row and column of each element. If the sum of the numbers of the row and column is even, the sign is positive ($+$). If the sum of the numbers of the row and column is odd, the sign is negative ($-$). For example, the $-$ sign in the bottom row is in the third row and second column. Since the sum $3 + 2 = 5$ is an odd number, the sign is $-$. As another example, the $+$ sign in the second row is in the position of the second row and second column. Since $2 + 2 = 4$ is an even number, the sign is $+$.

Examples

1. Find the value of $\begin{vmatrix} -3 & -2 \\ 6 & 5 \end{vmatrix}$.

$$\begin{vmatrix} -3 & -2 \\ 6 & 5 \end{vmatrix} = (-3)(5) - (-2)(6) = -15 + 12 = -3$$

2. Find the value of $\begin{vmatrix} 1 & 4 & 8 \\ 0 & -3 & -2 \\ 7 & 6 & 5 \end{vmatrix}$.

Solution 1

If we use the first row with each column and apply the first row of signs in the diagram

$$\begin{vmatrix} + & - & + \\ - & + & - \\ + & - & + \end{vmatrix},$$

then we have

$$\begin{vmatrix} 1 & 4 & 8 \\ 0 & -3 & -2 \\ 7 & 6 & 5 \end{vmatrix} = +1\begin{vmatrix} -3 & -2 \\ 6 & 5 \end{vmatrix} - 4\begin{vmatrix} 0 & -2 \\ 7 & 5 \end{vmatrix} + 8\begin{vmatrix} 0 & -3 \\ 7 & 6 \end{vmatrix}$$

Minor of first row and first column $\quad \begin{vmatrix} 1 & 4 & 8 \\ 0 & -3 & -2 \\ 7 & 6 & 5 \end{vmatrix} = \begin{vmatrix} -3 & -2 \\ 6 & 5 \end{vmatrix}$

Minor of first row and second column $\quad \begin{vmatrix} 1 & 4 & 8 \\ 0 & -3 & -2 \\ 7 & 6 & 5 \end{vmatrix} = \begin{vmatrix} 0 & -2 \\ 7 & 5 \end{vmatrix}$

Minor of first row and third column $\quad \begin{vmatrix} 1 & 4 & 8 \\ 0 & -3 & -2 \\ 7 & 6 & 5 \end{vmatrix} = \begin{vmatrix} 0 & -3 \\ 7 & 6 \end{vmatrix}$

$$= 1\left[-15 - (-12)\right] - 4\left[0 - (-14)\right] + 8\left[0 - (-21)\right]$$

$$= 1(-3) - 4(14) + 8(21)$$

$$= -3 - 56 + 168 = 109$$

Solution 2

The solution can be found using any row with each of the columns. If we use the second row with each column and apply the second row in the diagram

$$\begin{vmatrix} + & - & + \\ - & + & - \\ + & - & + \end{vmatrix},$$

then we have

$$\begin{vmatrix} 1 & 4 & 8 \\ 0 & -3 & -2 \\ 7 & 6 & 5 \end{vmatrix} = -10\begin{vmatrix} 4 & 8 \\ 6 & 5 \end{vmatrix} + (-3)\begin{vmatrix} 1 & 8 \\ 7 & 5 \end{vmatrix} - (-2)\begin{vmatrix} 1 & 4 \\ 7 & 6 \end{vmatrix}$$

$$= 0 - 3(5 - 56) + 2(6 - 28) = -3(-51) + 2(-22)$$

$$= 153 - 44 = 109$$

➤ **Diameter of a Circle**   The **diameter of a circle** is a chord that passes through the center. See **Chord of a Circle**. Half of the diameter is the radius. Hence, $d = 2r$.

➤ **Difference**   **Difference** is the result obtained when two numbers are subtracted. It is the answer in a subtraction problem. When the subtrahend is subtracted from the minuend, the result is the difference or remainder.

   If 8 is subtracted from 10 we say "what is the difference of 8 and 10" or "find the difference between 8 and 10." Also, we say "what is the remainder when 8 is subtracted from 10." Hence, in a division problem, we refer to the remaining portion of the division as the **remainder**. For more, see **Subtraction of Whole Numbers**.

$$\begin{array}{r} \text{minuend} \\ - \text{ subtrahend} \\ \hline \text{difference or remainder} \end{array}$$

➤ **Different Base**   See **Change of Base Formula**.

➤ **Differentiation**   See **Derivative**.

➤ **Dihedral Angle**   A **dihedral angle** is the union of two noncoplanar half-planes with the same edge. The half-planes are called **faces**, and the **edge** is the line where the two faces meet. A dihedral angle is a spacial angle (three dimensional). The angle in the figure is the dihedral angle $A - \overleftrightarrow{CD} - B$.

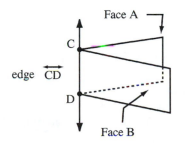

➤ **Directrix**   The **directrix** is a fixed line in the plane that is used in the definition of a parabola, an ellipse, or a hyperbola. See **Conic Sections**.

➤ **Direct Variation**   **Direct variation** is a linear function defined by the equation $y = kx$, or $k = \frac{y}{x}$ where k is a nonzero constant called the **constant of variation**. The equation $y = kx$ is read "y varies directly as x."

   Direct variation can be written as a **proportion**. Suppose both the pairs x_1 and y_1 and x_2 and y_2 satisfy the equation $y = kx$. Then, $y_1 = kx_1$ and $y_2 = kx_2$. Dividing one equation by the other, we have

$$\frac{y_1}{y_2} = \frac{kx_1}{kx_2}$$

$$\frac{y_1}{y_2} = \frac{x_1}{x_2} \qquad \text{read "} y_1 \text{ is to } y_2 \text{ as } x_1 \text{ is to } x_2 \text{."}$$

Examples

1. If y varies directly as x, and $y = 15$ when $x = 20$, find x when $y = 12$. Using the proportion equation, we have

$$\frac{y_1}{y_2} = \frac{x_1}{x_2}$$

$$\frac{15}{12} = \frac{20}{x_2}$$

$$x_2 = \frac{20(12)}{15} \qquad \text{See \textbf{Cross Multiplication}.}$$

$$x_2 = 16$$

2. Suppose y varies directly as x. If $y = 12$, then $x = -2$. Find y when $x = 4$.

$$\frac{y_1}{y_2} = \frac{x_1}{x_2}$$

$$\frac{12}{y_2} = \frac{-2}{4}$$

$$y_2 = \frac{4(12)}{-2}$$

$$y_2 = -24$$

For more, see **Combined Variation, Inverse Variation, Joint Variation,** or **Proportion**.

➤ **Discriminant**   The **discriminant,** $b^2 - 4ac$, is the expression under the radical sign in the quadratic formula. It can be used to determine the nature of the solutions of a quadratic equation.
The solution of $ax^2 + bx + c = 0$ is the quadratic formula

$$x = \frac{-b \pm \sqrt{b^2 - 4ac}}{2a}$$

From the formula, it can be seen that the $b^2 - 4ac$ can yield three different situations.

1. If $b^2 - 4ac > 0$, i.e., if it is positive, then the square root exists and the $\pm$ will render two real solutions.

2. If $b^2 - 4ac = 0$, then the square root is 0 and the $\pm$ will render one real solution.

3. If $b^2 - 4ac < 0$, i.e., if it is negative, then the square root is a complex number and the $\pm$ will render two complex solutions.

Example

1. Without solving, determine the nature of the solution of $3x^2 = 19x - 6$.

Setting the equation equal to 0, we can determine the a, b, and c.

$$3x^2 = 19x - 6$$
$$3x^2 - 19x + 6 = 0$$
$$\quad a \qquad b \quad\; c$$

Substituting these in $b^2 - 4ac$, we have

$$b^2 - 4ac = (-19)^2 - 4(3)(6)$$
$$= 361 - 72 = 289 \qquad \text{which is positive.}$$

Hence, the equation $3x^2 = 19x - 6$ has two real solutions.

If this answer had come out 0, there would have been one imaginary solution. If it had come out negative, there would have been two imaginary solutions.

D

➤ **Disjoint**   Disjoint sets are two or more sets that have no elements in common. Their intersection is empty. If $A = \{1, 2, 3\}$ and $B = \{4, 5, 6\}$, then $A \cap B = \varnothing$. In this case A and B are called disjoint sets.

➤ **Disjoint Sets**   If two sets A and B have no elements in common, they are said to be **disjointed sets**. If $A = \{1, 2, 3\}$ and $B = \{5, 6, 7\}$, then A and B are disjoint sets.

➤ **Disjunction**   In logic, the compound statement "p or q" written $p \vee q$, where p and q represent any statement is called the **disjunction** of p and q. If p represents the statement that "Jack is a baseball player" and q represents the statement that "Jack is a tennis player" then $p \vee q$ represents the statement that "Jack is a baseball player or Jack is a tennis player."

The truth or falsity of a disjunction can be determined by the use of a truth table. See **Truth Table**.

The concept of **disjunction** is related to the concept of **union** in set theory. This can be seen by the use of a **Venn diagram**. See **Venn Diagram**. If we let $B = \{$all baseball players$\}$ and $T = \{$all tennis players$\}$, then $B \cup T = \{$all people who play baseball, play tennis, or play both$\}$, represented by the entire shaded area in the Venn diagram.

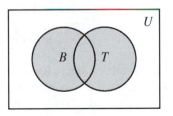

In our example above, Jack is an element of the shaded region because he is a member of both sets B and T. However, it is not required that he belong to both. The concept of disjunction (or union of sets) requires only that he should be a member of one or the other or both if the case is so determined.

➤ **Distance Formula**   The **distance formula** is used to find the distance between two points. If the two points are (x_1, y_1) and (x_2, y_2), then the distance formula is

$$d = \sqrt{(x_2 - x_1)^2 + (y_2 - y_1)^2}$$

Graphing the two points, we see that the required distance, d, is the hypotenuse of a right triangle. Therefore, by the Pythagorean theorem, we have

$$d^2 = (x_2 - x_1)^2 + (y_2 - y_1)^2$$

$$d = \sqrt{(x_2 - x_1)^2 + (y_2 - y_1)^2}$$

See **Pythagorean Theorem**.

Example

1. Find the distance between $(-7, -1)$ and $(-6, -2)$.

   Let $(x_1, y_1) = (-7, -1)$ and $(x_2, y_2) = (-6, -2)$. Then

$$d = \sqrt{(x_2 - x_1)^2 + (y_2 - y_1)^2}$$

$$= \sqrt{\left(-6 - (-7)\right)^2 + \left(-2 - (-1)\right)^2}$$ Watch for double negatives on the x_1 and y_1 terms.

$$= \sqrt{(-6 + 7)^2 + (-2 + 1)^2}$$

$$= \sqrt{(1)^2 + (-1)^2} = \sqrt{1 + 1} = \sqrt{2}$$

➤ **Distance on a Number Line**   If a and b are the coordinates of two points A and B on a number line, then the distance on the number line, between A and B, is

$$AB = |a - b| \text{ or } |b - a|$$

where a and b are real numbers. See **Number Line** or **Real Number Line**. The reason for the absolute value notation is because distance is defined as a positive value. If the difference of $a - b$ or $b - a$ comes out negative, the absolute value will make the distance positive. The three situations for the two points A and B are listed below:

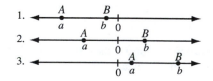

For an example in trigonometry, see **Trigonometry, graphs of the six functions**, end of example 5.

Examples

1. On the following number line, find AB, BC, and CD:

$$AB = |a - b| = \left|-5 - (-2)\right| = |-5 + 2| = |-3| = 3$$
or
$$\quad = |b - a| = \left|-2 - (-5)\right| = |-2 + 5| = |3| = 3$$
$$BC = |b - c| = \left|-2 - 2\right| = |-4| = 4$$
or
$$\quad = |c - b| = \left|-2 - (-2)\right| = |2 + 2| = |4| = 4$$
$$CD = |c - d| = |2 - 5| = |-3| = 3$$
or
$$\quad = |d - c| = |5 - 2| = |3| = 3$$

To combine the positive or negative numbers inside the absolute value signs, see **Positive and Negative Numbers**.

2. Using absolute value, rewrite the statement "the distance between x and 3 is 5."

Treating the x as a, the 3 as b, and the 5 as the distance AB, we have

$$AB = |a - b|$$
$$5 = |x - 3|$$
or $\quad |x - 3| = 5$ **See Symmetric Property**.

The equation $|x - 3| = 5$ means that the distance between x and 3 is 5.

3. Using absolute value, rewrite the statement "the distance between x and 4 is less than 6."

With $a = x$, $b = 4$, and $AB = 6$, we have

$$|a - b| < AB$$
$$|x - 4| < 6$$

The inequality $|x - 4| < 6$ means that the distance between x and 4 is less than 6.

► **Distributive Property**   The **distributive property** is one of the properties of the real number system. It states that, for any real numbers a, b, and c,

$$a(b + c) = ab + ac$$

In general, the distributive property allows multiplication into a set of parentheses.

Examples

1. Simplify $-6(4x - 2)$.

$$-6(4x - 2) = -24x + 12$$

2. Simplify $-4x^2(3x^2 - 2x - 4)$.

$$-4x^2(3x^2 - 2x - 4) = -12x^4 + 8x^3 + 16x^2$$

See **Properties of the Real Number System**.

➤ **Dividend**   In division, the **dividend** is the number to be divided. See **Quotient** or **Division of Whole Numbers**.

➤ **Dividing (division) by Zero**   **Dividing by zero** is not defined. We can see this is so by trying to divide any number, say 8, by 0. If we could divide 8 by 0, we would get a number (?) that could multiply 0 to yield 8. However, there is no such number.

$$\begin{array}{r} ? \\ 0\overline{)8} \\ \underline{8} \end{array}$$

Any number that multiplies 0 will not yield 8.

We can also see that dividing by zero is not defined by looking at the equation $\frac{8}{0} = x$. Cross multiplying, we get $8 = x \bullet 0$ which gives us $8 = 0$, which is obviously not true. Hence, dividing by zero is not defined.

➤ **Division, long**   See **Division of Whole Numbers**.

➤ **Division, long (algebraic)**   See **Long Division of Polynomials**.

➤ **Division of Complex Numbers**   See **Complex Numbers**, examples 7 and 8.

➤ **Division of Decimals**   See **Decimal Numbers, division of**.

➤ **Division of Fractions**   See **Fractions, division of**.

➤ **Division of Polynomials**   See **Long Division of Polynomials** or **Synthetic Division**.

➤ **Division of Rational Expressions**   See **Rational Expressions, multiplication and division of**.

➤ **Division of Whole Numbers**   Division of **whole numbers** is a process that determines how many times a number, called the **divisor**, goes into another number, called the **dividend**. The result is called the **quotient**. If the divisor does not go into the dividend completely, the amount left is called the **remainder**.

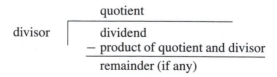

```
                        quotient
divisor    |    dividend
           − product of quotient and divisor
                remainder (if any)
```

There are four basic steps in the division of whole numbers: divide, multiply, subtract, and bring down.

1. Divide: Determine how many times the divisor goes into an equal amount of digits of the dividend. If it does not, increase the amount of digits by one. Place the number determined above the last number of the corresponding digits of the dividend.

   This is the first digit of the quotient. It will always be 9 or less.

2. Multiply: Multiply the determined digit of the quotient with the divisor. Place their product under the dividend, being careful to match up the last digit of the product with the corresponding digit of the dividend.

3. Subtract: Subtract the dividend and the product to find their difference.

4. Bring Down: Bring down the next digit of the dividend and place it at the end of the difference.

5. Repeat steps 1 through 4 using the number of step 4 to divide into.

   Remember that the divisor can never be 0. Division by 0 is not defined.

D

Examples

1. Divide 1,376 by 32. Check the answer.

Divide	Multiply	Subtract	Bring Down
$\begin{array}{r} 4 \\ 32\overline{)1376} \end{array}$	$\begin{array}{r} 4 \\ 32\overline{)1376} \\ 128 \end{array}$	$\begin{array}{r} 4 \\ 32\overline{)1376} \\ -128 \\ \hline 9 \end{array}$	$\begin{array}{r} 4 \\ 32\overline{)1376} \\ -128 \\ \hline 96 \end{array}$

Since 32 will not go into 13, see how many times it will go into 137.

Multiply 4 and 32.

Subtract 137 and 128.

Bring down the 6 from the dividend.

Divide	Multiply	Subtract	
$\begin{array}{r} 43 \\ 32\overline{)1376} \\ -128 \\ \hline 96 \end{array}$	$\begin{array}{r} 43 \\ 32\overline{)1376} \\ -128 \\ \hline 96 \\ \underline{96} \end{array}$	$\begin{array}{r} 43 \\ 32\overline{)1376} \\ -128 \\ \hline 96 \\ -96 \\ \hline 0 \end{array}$	When there is nothing left to bring down, the problem is solved.

Divide 32 into the 96.

Multiply 3 and 32.

Subtract 96 from 96.

The answer, 43, tells us how many times 32 goes into 1,376. There are forty-three 32's in 1,376.

To verify our answer, we can check it by multiplying 43 by 32 (divisor × quotient = dividend):

$$
\begin{array}{r}
43 \\
\times\,32 \\
\hline
86 \\
129 \\
\hline
1{,}376
\end{array}
$$

2. Divide 2,934 by 47. Check the answer.

$$
\begin{array}{r}
62 \\
47)\overline{2934} \\
-\,282 \\
\hline
114 \\
-\,94 \\
\hline
20
\end{array}
$$

The number 47 goes into 2,934 an amount of 62 times. However, it does not go in evenly, 20 is left over. The number 20 is called the remainder.

To verify our answer, we can check it by multiplying 62 by 47 and then adding the 20. The result should be 2,934.

$$
\begin{array}{r}
62 \\
\times\,47 \\
\hline
434 \\
248 \\
\hline
2{,}914
\end{array}
\qquad
\begin{array}{r}
2914 \\
+\,20 \\
\hline
2{,}934
\end{array}
$$

3. Divide 9,928 by 8 and check the answer.

$$
\begin{array}{r}
1241 \\
8)\overline{9928} \\
-\,8 \\
\hline
19 \\
-\,16 \\
\hline
32 \\
32 \\
\hline
08 \\
8 \\
\hline
0
\end{array}
\qquad
\begin{array}{r}
1241 \\
\times\,8 \\
\hline
9{,}928
\end{array}
$$

The number 8 goes into 9,928 an amount of 1,241 times. There are 1,241 eights in 9,928.

4. Divide 286,554 by 326 and check the answer.

$$
\begin{array}{r}
8 \\
326)\overline{286554}
\end{array}
$$

Since 326 does not divide into 286, we divide it into 2865. This can sometimes be difficult. A few cases of trial and error will always work. Think of a good number to multiply the 326. The product cannot go over 2865.

$$
\begin{array}{ccc}
326 & 326 & 326 \\
\times\,7 & \times\,8 & \times\,9 \\
\hline
2282 & 2608 & 2934
\end{array}
$$
The product of 9 and 326 is too high so we use 8.

$$
\begin{array}{r}
8 \\
326\,\overline{)\,286554} \\
2608 \\
\hline
2575
\end{array}
$$
We now need to think of a number to multiply 326 that does not go over 2575. The number 8 goes over, the number 7 does not.

$$
\begin{array}{r}
87 \\
326\,\overline{)\,286554} \\
-2608 \\
\hline
2575 \\
2282 \\
\hline
2934
\end{array}
$$

Since 326 goes into 2934 evenly, we can finish the problem.

$$
\begin{array}{r}
879 \\
326\,\overline{)\,286554} \\
-2608 \\
\hline
2575 \\
2282 \\
\hline
2934 \\
2934 \\
\hline
0
\end{array}
$$

The number 326 goes into 286,554 an amount of 879 times.

➤ **Division, synthetic**   See **Synthetic Division**.

➤ **Divisor**   In division, the **divisor** is the number that divides into the dividend. See **Division of Whole Numbers**.

➤ **Domain and Range**   In reference to a set of points (x, y) where x and y are real numbers, the **domain** is the set of all x-coordinates and the **range** is the set of all y-coordinates.

In graphing the equation $y = 2x + 1$ by the charting method (see **Linear Equations, graphs of Charting Method**), the set of x's is the domain and the set of y's is the range.

In set notation, we write the domain $\mathcal{D} = \{x \mid x \in R\}$, read "the set x such that x is an element of the real numbers." See **Set Notation**. The range $\mathcal{R} = \{y \mid y \in R\}$.

x	$y = 2x + 1$
$\vdots$	$\vdots$
-2	-3
-1	-1
0	1
1	3
2	5
$\vdots$	$\vdots$

To determine the domain of an equation, the equation must be solved for the dependent variable (usually y). Then, all possible substitutions of the independent variable (usually x) must be considered. If there are any violations, i.e., $y = \sqrt{x}$ is not defined when x is negative, then those values are not part of the domain.

Similarly, to determine the range of an equation, the equation must be solved for the independent variable (usually x). Then, all possible substitutions of the dependent variable (usually y) must be considered. If there are any violations, i.e., $x = \frac{1}{y}$ is not defined when y is 0, then those values are not part of the range.

An alternate method is to consider all of the possible substitutions for either the dependent or independent variables simply by observation. Then, state the domain and range.

Examples

1. State the domain and range of the following points: $(1, 2)$, $(-2, 3)$, $(-4, -1)$, and $(3, -5)$.

   Since the domain is the set of x's and the range is the set of y's, we have

   $$\text{The domain } \mathcal{D} = \{1, -2, -4, 3\} = \{-4, -2, 1, 3\}$$
   $$\text{The range } \mathcal{R} = \{2, 3-1, -5\} = \{-5, -1, 2, 3\}$$

2. State the domain and range for $y = \sqrt{x - 3}$.

   Since $\sqrt{x - 3}$ is defined for $x - 3 \geqslant 0$, or $x \geqslant 3$, the domain is the set of x's with $x \geqslant 3$. This can more easily be seen by looking at a chart.

x	$y = \sqrt{x - 3}$	
2	$\sqrt{2 - 3} = \sqrt{-1}$	not defined
3	$\sqrt{3 - 3} = \sqrt{0} = 0$	
4	$\sqrt{4 - 3} = \sqrt{1} = 1$	
5	$\sqrt{5 - 3} = \sqrt{2} \approx 1.41$	
$\vdots$	$\vdots$	

The x's cannot be less than 3 since the square root of a negative number is not defined. On the other hand, by observation, the y's start at 0 and continue in both directions toward positive and negative infinity. Hence, the domain is the set of x's greater than or equal to 3, while the range is the set of all real numbers.

The other way to find the range is to solve $y = \sqrt{x - 3}$ for x:

$$y = \sqrt{x - 3}$$
$$(y)^2 = \left(\sqrt{x - 3}\right)^2$$
$$y^2 = x - 3$$
$$y^2 + 3 = x$$
$$x = y^2 + 3$$

Since y^2 is defined for all values of y, the range is the set of all real numbers. Therefore, the domain $\mathcal{D} = \{x \mid x \geqslant 3\}$ and the range $\mathcal{R} = \{y \mid y \in R\}$.

3. Find the domain and range of $y = \frac{x+1}{x-2}$.

The fraction $\frac{x+1}{x-2}$ is not defined when $x - 2 = 0$, or $x = 2$. When $x = 2$ the denominator is 0. Since we can't divide by 0, the equation $y = \frac{x+1}{x-2}$ is defined for all x-values except $x = 2$. Hence, the domain is

$$\mathcal{D} = \{x \mid x \in R, \ x \neq 2\}$$

To find the range, we need to solve the equation for x so we can evaluate the y's:

$$y = \frac{x + 1}{x - 2}$$
$$y(x - 2) = x + 1$$
$$xy - 2y = x + 1$$
$$xy - x = 2y + 1$$
$$x(y - 1) = 2y + 1$$
$$x = \frac{2y + 1}{y - 1}$$

The fraction $\frac{2y+1}{y-1}$ is not defined when $y - 1 = 0$, or $y = 1$. When $y = 1$ the denominator is 0. Hence, $x = \frac{2y+1}{y-1}$ is defined for all y-values except $y = 1$. Therefore, the range is

$$\mathcal{R} = \{y \mid y \in R, \ y \neq 1\}$$

➤ **Double-Angle Formulas**   The **double-angle formulas** are one of several sets of identities used in trigonometry. See **Trigonometry, identities**. The double-angle formulas for sine, cosine, and tangent are

$$\sin 2\alpha = 2 \sin \alpha \cos \alpha$$
$$\cos 2\alpha = \cos^2 \alpha - \sin^2 \alpha = 2 \cos^2 \alpha - 1 = 1 - 2 \sin^2 \alpha$$
$$\tan 2\alpha = \frac{2 \tan \alpha}{1 - \tan^2 \alpha}$$

Remember, $\cos^2 \alpha$ means $(\cos \alpha)^2$ and $\sin 2\alpha$ is not the same as $2 \sin \alpha$. The same follows for the other functions.

The double-angle formulas are derived from the sum and difference formulas. In the following, let $\beta = \alpha$:

$$\sin(\alpha + \beta) = \sin \alpha \cos \beta + \cos \alpha \sin \beta$$
$$\sin(\alpha + \alpha) = \sin \alpha \cos \alpha + \cos \alpha \sin \alpha$$
$$\sin 2\alpha = \sin \alpha \cos \alpha + \sin \alpha \cos \alpha$$
$$\mathbf{\sin 2\alpha = 2 \sin \alpha \cos \alpha}$$

$$\cos(\alpha + \beta) = \cos \alpha \cos \beta - \sin \alpha \sin \beta$$
$$\cos(\alpha + \alpha) = \cos \alpha \cos \alpha - \sin \alpha \sin \alpha$$
$$\mathbf{\cos 2\alpha = \cos^2 \alpha - \sin^2 \alpha}$$

D

The double-angle formula for cosine is expressed in two other forms.

$$\cos 2\alpha = \cos^2 \alpha - \left(1 - \cos^2 \alpha\right) \qquad \text{From the Pythagorean identity } \cos^2 \alpha + \sin^2 \alpha = 1,$$
$$= \cos^2 \alpha - 1 + \cos^2 \alpha \qquad \text{we have } \sin^2 \alpha = 1 - \cos^2 \alpha.$$
$$\mathbf{= 2 \cos^2 \alpha - 1}$$

$$\cos 2\alpha = 1 - \sin^2 \alpha - \sin^2 \alpha \qquad \text{From the Pythagorean identity } \cos^2 \alpha + \sin^2 \alpha = 1,$$
$$\mathbf{= 1 - 2 \sin^2 \alpha} \qquad \text{we have } \cos^2 \alpha = 1 - \sin^2 \alpha.$$

$$\tan(\alpha + \beta) = \frac{\tan \alpha + \tan \beta}{1 - \tan \alpha \tan \beta}$$
$$\tan(\alpha + \alpha) = \frac{\tan \alpha + \tan \alpha}{1 - \tan \alpha \tan \alpha}$$
$$\mathbf{\tan 2\alpha = \frac{2 \tan \alpha}{1 - \tan^2 \alpha}}$$

Examples

1. Use $\sin 2\alpha$ to find $\sin 60°$.

$$\sin 2\alpha = 2 \sin \alpha \cos \alpha$$
$$\sin 2(30°) = 2 \sin 30° \cos 30° \qquad \text{Let } \alpha = 30°.$$
$$\sin 60° = 2\left(\frac{1}{2}\right)\left(\frac{\sqrt{3}}{2}\right) \qquad \text{Use the special triangle.}$$
$$\sin 60° = \frac{\sqrt{3}}{2}$$

2. Prove the identity $\cos 2t = \frac{\csc^2 t - 2}{\csc^2 t}$.

$$\cos 2t = \frac{\csc^2 t - 2}{\csc^2 t}$$

$$\cos 2t = \frac{\frac{1}{\sin^2 t} - \frac{2}{1}}{\frac{1}{\sin^2 t}} \qquad \text{See \textbf{Reciprocal Identities}.}$$

$$\cos 2t = \frac{\frac{1 - 2\sin^2 t}{\sin^2 t}}{\frac{1}{\sin^2 t}} \qquad \text{See \textbf{Rational Expressions, complex fractions of}.}$$

$$\cos 2t = \frac{1 - 2\sin^2 t}{\sin^2 t} \cdot \frac{\sin^2 t}{1}$$

$$\cos 2t = 1 - 2\sin^2 t$$

$$\cos 2t = \cos 2t \qquad \text{Use the third identity for } \cos 2\alpha.$$

For more on proof, see **Trigonometry, identities**.

3. Prove the identity $\frac{\sin 2\alpha}{2\sin^2 \alpha} = \cot \alpha$.

$$\frac{\sin 2\alpha}{2\sin^2 \alpha} = \cot \alpha$$

$$\frac{2\sin \alpha \cos \alpha}{2\sin^2 \alpha} = \cot \alpha \qquad \text{Use the identity for } \sin 2\alpha.$$

$$\frac{\cos \alpha}{\sin \alpha} = \cot \alpha$$

$$\cot \alpha = \cot \alpha \qquad \text{See \textbf{Reciprocal Identities}.}$$

4. Solve $\sin 2x = \cos x$ for $0 \leqslant x < 2\pi$. The $0 \leqslant x < 2\pi$ means, find the answer for one time around the circle only.

$$\sin 2x = \cos x$$

$$\sin 2x - \cos x = 0 \qquad \text{Set the equation equal to 0.}$$

$$2\sin x \cos x - \cos x = 0 \qquad \text{Use the identity for } \sin 2x.$$

$$\cos x(2\sin x - 1) = 0 \qquad \text{Factor, i.e., } 2AB - B = B(2A - 1); \text{ see \textbf{Factoring}.}$$

$$\cos x = 0 \qquad \text{or} \qquad 2\sin x - 1 = 0$$

$$x = \frac{\pi}{2}, \frac{3\pi}{2} \qquad\qquad\qquad 2\sin x = 1$$

$$\sin x = \frac{1}{2}$$

$$x = \frac{\pi}{6}, \frac{5\pi}{6} \qquad \text{See \textbf{Trigonometry, equations}.}$$

Hence, the solution is $x = \frac{\pi}{2}, \frac{3\pi}{2}, \frac{\pi}{6}, \frac{5\pi}{2}$.

▶ **Double Root**   If an equation has a root (solution) that is repeated two times, then we say that the equation has a **double root** or a **root of multiplicity** 2. See **Roots of Multiplicity**.

➤ *e*   The number *e* was first used as a base in natural logarithms and later became useful in differential and integral calculus, the sciences, and economics. It is defined as

$$e = \lim_{n \to \infty} \left(1 + \frac{1}{n}\right)^n \qquad \text{See **Limits**.}$$

$$e \approx 2.7182818284\ldots$$

For applications, see **Natural Logarithms**.

➤ **Eccentricity**   In conic sections, **eccentricity** is the ratio of the distance from a fixed point to the distance from a fixed line. It has the following properties:

1. If $e = 1$, the conic is a parabola.

2. If $0 < e < 1$, the conic is an ellipse.

3. If $e > 1$, the conic is a hyperbola.

The number *e* for eccentricity should not be confused with the *e* for the base of natural logarithms. See **Conic Sections**.

➤ **Ellipse**   An **ellipse** is the set of all points in the plane such that the sum of the distance from two fixed points is constant. Each of the fixed points is called a **focus**. The line through the foci (plural of focus) is the **principal axis**.

The equation of an ellipse depends on the orientation of the **major axis** (the longer axis of symmetry) and the **minor axis** (the shorter axis of symmetry). The points where the major axis intersect the ellipse are called the **vertices**, plural of **vertex**.

If the major axis is horizontal, the equation of the ellipse is either

$$\frac{x^2}{a^2} + \frac{y^2}{b^2} = 1 \qquad \text{or} \qquad \frac{(x-h)^2}{a^2} + \frac{(y-k)^2}{b^2} = 1$$

Here, and in the following definition, the first equation has a center at the origin $(0, 0)$ and the second has a center at (h, k). Either of the equations is called the standard form of the ellipse.

If the major axis is vertical, the equation of the ellipse is either

$$\frac{x^2}{a^2} + \frac{y^2}{b^2} = 1 \qquad \text{or} \qquad \frac{(x-h)^2}{b^2} + \frac{(y-k)^2}{a^2} = 1$$

The major axis is horizontal if the denominator of the x^2 term is larger than that of the y^2 term. It is vertical if the case is reversed.

In all of the equations $c^2 = a^2 - b^2$, $a \neq 0$ and $b \neq 0$. The length of the major axis is $2a$ and the length of the minor axis is $2b$. The equation of the eccentricity is

$$e = \frac{c}{a} = \frac{\sqrt{a^2 - b^2}}{a}$$

Since the sum of the distances from two fixed points is constant, $F_1 V_2 + F_2 V_2 = F_1 P + F_2 P$. The importance of this rests on the fact that $O V_2 + O V_1 = 2a = F_1 V_2 + F_2 V_2$. Hence, $F_1 P + F_2 P = 2a$ as well. It takes some thinking to realize this.

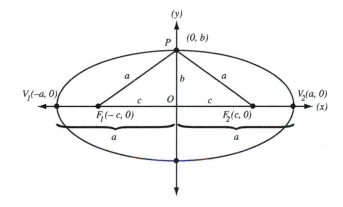

One way to see it is to recognize that the distance $F_2 V_2$ was covered twice, once by $F_1 V_2$ and then by $F_2 V_2$. If we separate the equation, we have

$$
\begin{aligned}
F_1 V_2 + F_2 V_2 &= (F_1 F_2 + F_2 V_2) + F_2 V_2 \\
&= F_1 F_2 + F_2 V_2 + F_1 V_1 && \text{Since } F_2 V_2 = F_1 V_1. \\
&= (F_1 0 + F_2 0) + F_2 V_2 + F_1 V_1 && \text{Since } F_1 F_2 = F_1 0 + F_2 0. \\
&= (F_1 V_1 + F_1 0) + (F_2 0 + F_2 V_2) && \text{Rearrange terms.} \\
&= a + a = 2a
\end{aligned}
$$

Hence, $P F_2 = a$ and $O V_2 = a$. Also $P F_1 = a$ and $O V_1 = a$. This allows the equation $c^2 = a^2 - b^2$ to be used in finding the coordinators of the foci.

Examples

1. Find the center, foci, lengths of the major and minor axes, eccentricity, and graph of $\frac{x^2}{25} + \frac{y^2}{4} = 1$.

First, we notice that the denominator of the x^2 term is larger than the denominator of the y^2 term. So, we know the major axis is horizontal. We will use the equation $\frac{x^2}{a^2} + \frac{y^2}{b^2} = 1$.

Center: Since the equation $\frac{x^2}{25} + \frac{y^2}{4} = 1$ does not have an h or k, the center is at the origin $(0, 0)$.

Foci: Comparing $\frac{x^2}{a^2} + \frac{y^2}{b^2} = 1$ and $\frac{x^2}{25} + \frac{y^2}{4} = 1$, we see that $a^2 = 25$ and $b^2 = 4$. Substituting in $c^2 = a^2 - b^2$, we have

$$c^2 = a^2 - b^2$$
$$c^2 = 25 - 4$$
$$c^2 = 21$$
$$c = \pm\sqrt{21} = 4.6 \qquad \text{Use a calculator.}$$

Hence, the foci are $(\sqrt{21}, 0)$ and $(-\sqrt{21}, 0)$.

Length of the Major Axis: Since $a^2 = 25$, $a = 5$. Therefore, the major axis is $2a = 2(5) = 10$. See **Quadratic equations, solutions of**.

Length of Minor Axis: Since $b^2 = 4$, $b = 2$. Therefore, the minor axis is $2b = 2(2) = 4$.

Eccentricity:

$$e = \frac{c}{a} = \frac{\sqrt{21}}{5} \qquad \text{Use positive } c \text{ only.}$$

Graph: Since the x^2 term has a denominator larger than the y^2 term, the major axis is horizontal.

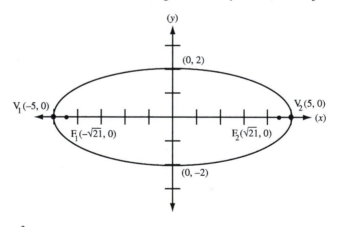

2. Graph $\frac{(x-1)^2}{4} + \frac{(y+4)^2}{9} = 1$.

Since the denominator of the $(x - 1)^2$ term is less than the denominator of the $(y + 4)^2$ term, the major axis is vertical. The center of the ellipse is $(h, k) = (1, -4)$. When you put the h and k into the point, they are always opposite in sign. Since the major axis is vertical, we are comparing with the equation

$$\frac{(x - h)^2}{b^2} + \frac{(y - k)^2}{a^2} = 1$$

From $b^2 = 4$ and $a^2 = 9$, we have $b = 2$ and $a = 3$. With all this information at hand, we can now graph the equation.

To graph the equation, we first plot the center $(1, -4)$. From the center, we plot points two units to the right and left (since $b = 2$). Also from the center, we plot points three units up and down (since $a = 3$). Draw the graph and label the points.

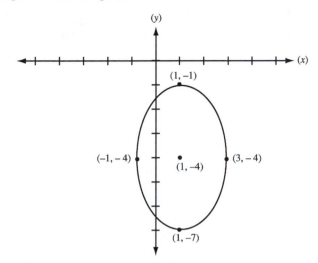

5. Write the equation $9x^2 + 4y^2 - 18x + 16y = 11$ in standard form.

This means that we want it in the form $\frac{(x-h)^2}{a^2} + \frac{(y-k)^2}{b^2} = 1$. To do this, we need to group the x terms and y terms, respectively, complete the square on them, and then divide both sides of the equation by the number on the right side of the equal sign:

$$9x^2 - 18x + 4y^2 + 16y = 11$$
$$9(x^2 - 2x) + 4(y^2 + 4y) = 11$$
$$9(x^2 - 2x + 1) + 4(y^2 + 4y + 4) = 11 + 9 + 16 \qquad \text{See below.}$$
$$9(x - 1)^2 + 4(y + 2)^2 = 36$$
$$\frac{9(x-1)^2}{36} + \frac{4(y+2)^2}{36} = \frac{36}{36}$$
$$\frac{(x-1)^2}{4} + \frac{(y+2)^2}{9} = 1$$

How the completion of the square works. Between the second and the third line, this is what happened:

$$9(x^2 - 2x) + 4(y^2 + 4y) = 11 \qquad \text{2nd line.}$$
$$9(x^2 - 2x + 1 - 1) + 4(y^2 + 4y + 4 - 4) = 11$$
$$9(x^2 - 2x + 1) - 9 + 4(y^2 + 4y + 4) - 16 = 11 \qquad 9(-1) \text{ is } -9; \ 4(-4) \text{ is } -16.$$
$$9(x^2 - 2x + 1) + 4(y^2 + 4y + 4) = 11 + 9 + 16 \qquad \text{3rd line.}$$

We complete the square here a little differently than is done when solving equations. We still take half of the coefficient of the x (or y) and square it. But then we add and subtract it, rather than add it to both sides as we would have done in solving equations.

The first three terms will now factor, but the fourth term is in the way. We remove it by multiplying it by the leading coefficient and writing it outside of the parentheses. Now we can continue to put the equation into standard form.

4. Find the equation of the ellipse with vertices $(\pm 5, 0)$ and eccentricity $\frac{1}{5}$.

Since we are given the vertices, we have $(a, 0) = (5, 0)$, or $a = 5$. From the equation for eccentricity, we have

$$e = \frac{c}{a}$$
$$\frac{1}{5} = \frac{c}{5}$$
$$c = 1$$

Substituting in $c^2 = a^2 - b^2$, we have

$$c^2 = a^2 - b^2$$
$$(1)^2 = (5)^2 - b^2$$
$$b^2 = 25 - 1$$
$$b^2 = 24$$

Since a^2 is larger than b^2, we know the major axis is horizontal. Hence, the equation of the ellipse is

$$\frac{x^2}{25} + \frac{y^2}{24} = 1$$

E

➤ **Empty Set**   The **empty set** is a set that has no elements. It is denoted by $\varnothing$ or by $\{\ \ \}$. See **Set**.

➤ **Equation, exponential**   See **Exponential Equation**.

➤ **Equation, fractional**   See **Rational Equations**.

➤ **Equation, linear**   See **Linear Equation**.

➤ **Equation, logarithmic**   See **Logarithmic Equations**.

➤ **Equation of a Line**   See **Linear Equation, finding the equation of a line**.

➤ **Equation of a Slope**   See **Slope**.

➤ **Equation, polynomial (solutions of)**   See **Rational Roots Theorem**.

➤ **Equation, quadratic**   See **Quadratic Equation, Quadratic Formula**, or **Completing the Square**.

➤ **Equations, radical**   See **Radicals, equations with**.

➤ **Equations, rational**   See **Rational Equations**.

➤ **Equations, systems of**   See **Systems of Linear Equations**.

➤ **Equations, trigonometric**   See **Trigonometry, equations**.

➤ **Equation with Fractions**   See **Rational Equations**.

➤ **Equation with Radicals**   See **Radicals, equations with**.

➤ **Equation with Square Roots**   See **Radicals, equations with**.

➤ **Equiangular Triangle**   An **equiangular triangle** is a triangle with three equal angles. Since the sum of the angles of a triangle is 180°, the measure of each of the angles of an equiangular triangle is 60°.

➤ **Equilateral Triangle**   An **equilateral triangle** is a triangle with three equal sides. Each of the angles of an equilateral triangle is 60°. This can be shown in a proof. See **Proof in Geometry**.

➤ **Equivalence**   See **Truth Table** or **Biconditional (equivalent) Statement**.

➤ **Equivalent Fractions**   See **Fractions, equivalent**.

➤ **Equivalent Statement**   See **Biconditional (equivalent) Statement**.

➤ **Euclidean Geometry**   **Euclidean geometry** is synonymous with **plane geometry**. It is so named after the Greek geometer Euclid. In the main, Euclidean geometry is a branch of mathematics that deals with sets of points in space. For more, see **Geometry**.

➤ **Even and Odd Functions**   An **even function** is a function whose graph is symmetric with respect to the y-axis. An odd function is a function whose graph is symmetric with respect to the origin. See **Symmetry**.

$$\text{A function is } \textbf{even} \text{ if } f(-x) = f(x).$$
$$\text{A function is } \textbf{odd} \text{ if } f(-x) = -f(x).$$

Examples

1. Is the function $f(x) = x^2 + 2$ even, odd, or neither?

   Since $f(-x) = (-x)^2 + 2 = x^2 + 2$ is the same as $f(x)$, the function is even.

2. Is the function $f(x) = 2x^2 - 4x$ even, odd, or neither?

   Since $f(-x) = 2(-x)^3 - 4(-x) = -2x^3 + 4x = -(2x^3 - 4x)$ is the negative of $f(x)$, the function is odd.

➤ **Even and Odd Functions, in trigonometry**   **Even and odd functions** in trigonometry are used to develop the identities for negative angles.

$$\sin(-\alpha) = -\sin\alpha \qquad \csc(-\alpha) = -\csc\alpha$$
$$\cos(-\alpha) = \cos\alpha \qquad \sec(-\alpha) = \sec\alpha$$
$$\tan(-\alpha) = -\tan\alpha \qquad \cot(-\alpha) = -\cot\alpha$$

Of the above identities, cosine and secant are even functions and the rest are odd functions. See **Even and Odd Functions**. For more, see **Trigonometry, even and odd functions (Functions or identities of negative angles)**.

➤ **Exponent**   An **exponent** or power is a number (or letter) that is written as a superscript to a **base**. It is written at the upper right of the base and, as a numeral, tells how many times the base is to multiply itself. If the base is 2 and the exponent is 3, we have

$$2^3 = 2 \bullet 2 \bullet 2$$

In algebraic form, if the base is x and the exponent is n, then the expression is x^n. It would have to await further definition for more meaning.

Examples

1. Simplify 3^4.

$$3^4 = 3 \bullet 3 \bullet 3 \bullet 3 = 81$$

2. Simplify $2^4 \bullet 3^3$.

$$2^4 \bullet 3^3 = 2 \bullet 2 \bullet 2 \bullet 2 \bullet 3 \bullet 3 \bullet 3 = 16 \bullet 27 = 432$$

The most common mistake in a problem like this is trying to multiply the 2 and 3 first. The exponents must be dealt with first. See **Order of Operations** or **Exponents, rules of**.

➤ **Exponential Equation**   An equation with one or more variables in the exponent is an **exponential equation**. The equation $9^x = 27$ is an exponential equation. For a solution of exponential equations, see **Exponential Equation, solution of**.

➤ **Exponential Equation, solution of**   There are two methods for solving exponential equations. One uses similar bases and the other employs logarithms.

In the similar base method, we need to isolate the similar bases so we can set the exponents equal to each other. If $b^x = b^n$, then $x = n$.

In the logarithmic method, we take the logarithm to the base 10 of both sides of the equation and solve for x. The logarithmic method is especially useful when similar bases cannot be isolated.

Examples

1. Solve $9^x = 81$ using similar bases.

$$9^x = 81$$
$$(3^2)^x = 3^4 \qquad \text{Write 9 and 81 in terms of the similar base 3.}$$
$$3^{2x} = 3^4 \qquad \text{Use the power-to-power rule. See \textbf{Exponents, rules of}.}$$
$$2x = 4 \qquad \text{Set exponents equal to each other.}$$
$$x = 2$$

2. Solve $9^x = 81$ using logarithms.

$$9^x = 81$$
$$\log 9^x = \log 81 \qquad \text{Take the log to the base 10 of both sides.}$$
$$x \log 9 = \log 81 \qquad \text{See \textbf{Logarithms, properties of (rules of)}.}$$
$$x = \frac{\log 81}{\log 9} \qquad \text{Solve for } x.$$
$$x = \frac{1.9084}{.9542} \qquad \text{Evaluate logs, see \textbf{Logarithm}.}$$
$$x = 2$$

3. Solve $(\frac{1}{2})^{1-x} = 16$ using similar bases.

$$\left(\frac{1}{2}\right)^{1-x} = 16$$
$$(2^{-1})^{1-x} = 2^4 \qquad \text{See \textbf{Exponents, rules of}.}$$
$$2^{-1+x} = 2^4 \qquad \text{If } b^x = b^n, \text{ then } x = n.$$
$$x - 1 = 4$$
$$x = 5$$

4. Solve $2^x = 5$.

$$2^x = 5$$
$$\log 2^x = \log 5 \qquad \text{Take the log of both sides.}$$
$$x \log 2 = \log 5$$
$$x = \frac{\log 5}{\log 2}$$
$$x = \frac{.6990}{.3010}$$
$$x = 2.3223$$

At first sight, it might seem that this equation cannot be solved using similar bases. However, $2 = 10^{.3010}$ and $5 = 10^{.6990}$, so we can use 10 as the similar base.

$$2^x = 5$$
$$(10^{.3010})^x = 10^{.6990}$$

$$10^{.3010x} = 10^{.6990}$$
$$.3010x = .6990$$
$$x = \frac{.6990}{.3010}$$
$$x = 2.3223$$

5. Solve $x^4 = 81$.

Raising each side of the equation to the 1/4 power, we have

$$x^4 = 81$$
$$\left(x^4\right)^{1/4} = (81)^{1/4}$$
$$x^{4/4} = \left(3^4\right)^{1/4}$$
$$x = 3^{4/4}$$
$$x = 3$$

E

➤ **Exponential Form**   The equation $y = b^x$ is in **exponential form**, while its equivalent equation $x = \log_b y$ is in logarithmic form. For more, see **Logarithmic Form to Exponential Form, converting from**.

➤ **Exponential Form to Logarithmic Form, converting from**   See **Logarithmic Form to Exponential Form, converting from**.

➤ **Exponential Function**   An **exponential function** is an equation defined by $y = b^x$ where b is a positive constant not equal to 1. The equations $y = 2^x$, $y = 3^{-x}$, $y = \left(\frac{1}{2}\right)^x$, $y = 3^{-x} - 4$, and $y = -3^x$ (same as $y = -(3)^x$) are exponential functions.

Examples

1. Graph $y = 2^x$ and state the domain and range.

Put the equation in a chart and choose x-values that are reasonable. See the various articles under **Graph(s)**.

x	$y = 2^x$
$\vdots$	$\vdots$
-2	$2^{-2} = \frac{1}{2^2} = \frac{1}{4}$
-1	$2^{-1} = \frac{1}{2}$
0	$2^0 = 1$
1	$2^1 = 2$
2	$2^2 = 4$
$\vdots$	$\vdots$

See **Exponents, rules of**.

The dots are a reminder that each list continues in both directions. The x's continue negatively to negative infinity and positively to positive infinity. Therefore, the domain of the function is the interval $(-\infty, \infty)$ or all x. See **Domain and Range** or **Interval Notation**.

As the x's get smaller, so do the y's. But the y's never reach 0. As the x's get larger, so do the y's. They continue to positive infinity. Hence, the range of the function is the interval $(0, \infty)$ or, in inequalities, $y > 0$.

On the left, the graph approaches the x-axis asymptotically. See **Asymptote**. On the right, it continues upward without bound.

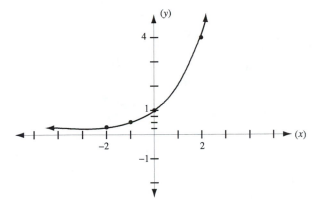

2. Graph $y = \left(\frac{1}{2}\right)^x$ and state the domain and range.

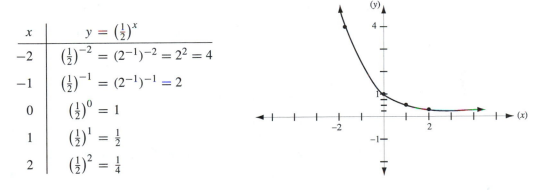

x	$y = \left(\frac{1}{2}\right)^x$
-2	$\left(\frac{1}{2}\right)^{-2} = (2^{-1})^{-2} = 2^2 = 4$
-1	$\left(\frac{1}{2}\right)^{-1} = (2^{-1})^{-1} = 2$
0	$\left(\frac{1}{2}\right)^0 = 1$
1	$\left(\frac{1}{2}\right)^1 = \frac{1}{2}$
2	$\left(\frac{1}{2}\right)^2 = \frac{1}{4}$

The domain is the interval $(-\infty, \infty)$ or $-\infty < x < \infty$.

The range is the interval $(0, \infty)$ or $0 < y < \infty$.

➤ **Exponential Graph**   See **Exponential Function**.

➤ **Exponent, rules**   See **Exponents, rules of**.

➤ **Exponents, fractional or rational**   Fractional exponents or rational exponents are used in place of radicals. Instead of writing $\sqrt{2}$ or $\sqrt[3]{5}$, we write $2^{1/2}$ or $5^{1/3}$. Instead of writing $\sqrt[3]{x^2}$, we write $x^{2/3}$.

In general,

$$b^{1/n} = \sqrt[n]{b}$$

where b is a real number and n is a natural number greater than 1. A restriction on this rule occurs when n is even. If n is even, then b must be positive or we would have an even root of a negative, i.e., $\sqrt[4]{-3}$. See **Complex Numbers**.

Also, in general,

$$b^{m/n} = \sqrt[n]{b^m}$$
$$\text{or} \quad b^{m/n} = \left(\sqrt[n]{b}\right)^m$$

where $\frac{m}{n}$ is a rational number in lowest terms, n is a natural number with $n > 1$, and b is a real number with $b \neq 0$ (prevents division by 0).

The rules for exponents apply to rational and irrational exponents as well. They are usually accepted rather than proved.

Examples

1. Evaluate $8^{2/3}$.

$$8^{2/3} = \sqrt[3]{8^2} = \sqrt[3]{64} = 4$$
$$\text{or} \quad 8^{2/3} = \left(\sqrt[3]{8}\right)^2 = (2)^2 = 4$$

With fractional exponents only, we have

$$8^{2/3} = \left(8^2\right)^{1/3} = (64)^{1/3} = 4 \qquad \text{Using the rule } (a^m)^n = a^{mn}.$$
$$\text{or} \quad 8^{2/3} = \left(8^{1/3}\right)^2 = (2)^2 = 4 \qquad \text{See } \textbf{Exponents, rules of}.$$

2. Simplify $27^{2/3}$.

$$27^{2/3} = \left(\sqrt[3]{27}\right)^2 = (3)^2 = 9$$
$$\text{or} \quad 27^{2/3} = \left(27^{1/3}\right)^2 = (3)^2 = 9$$

We took this approach because it is easier to square the cube root of 27 than it is to find the cube root of 27^2.

3. Simplify $32^{-2/5}$.

$$32^{-2/5} = \frac{1}{32^{2/5}} = \frac{1}{(32^{1/5})^2} = \frac{1}{(2)^2} = \frac{1}{4}$$

Use the rule for negative exponents and the power-to-power rule. See **Exponents, rules of**.

4. Use rational exponents to simplify $\sqrt[6]{25}$ and leave the answer in radical form. See **Radical Form**.

$$\sqrt[6]{25} = (25)^{1/6} = \left(5^2\right)^{1/6} = 5^{2/6} = 5^{1/3} = \sqrt[3]{5}$$

To do a problem like this, the 25 has to be thought of as 5^2. Then, use the power-to-power rule. See **Exponents, rules of**.

5. Simplify $x^{1/4} \bullet x^{3/4}$.

$$
\begin{aligned}
x^{1/4} \bullet x^{3/4} &= x^{1/4+3/4} && \text{Multiply similar bases; add exponents.} \\
&= x^{4/4} && \text{See \textbf{Exponents, rules of}.} \\
&= x^1 \\
&= x
\end{aligned}
$$

6. Simplify $(x^3 - 4)^{1/4}(x^3 - 4)^{3/4}$. Compare this example with example 4.

$$
(x^3 - 4)^{1/4}(x^3 - 4)^{3/4} = (x^3 - 4)^{1/4+3/4} = (x^3 - 4)^{4/4} = (x^3 - 4)^1 = x^3 - 4
$$

7. Simplify $\sqrt[5]{\dfrac{5x^{1/4}}{x^{1/3}}}$ and leave the answer with positive exponents.

$$
\begin{aligned}
\sqrt[5]{\frac{5x^{1/4}}{x^{1/3}}} &= \left(\frac{5x^{1/4}}{x^{1/3}}\right)^{1/5} \\
&= \left(5x^{1/4-1/3}\right)^{1/5} && \text{Divide similar bases; subtract exponents.} \\
&= \left(5x^{3/12-4/12}\right)^{1/5} && \text{See \textbf{Exponents, rules of}.} \\
&= \left(5x^{-1/12}\right)^{1/5} \\
&= 5^{1/5}x^{-1/60} && \text{Power-to-power rule; multiply exponents.} \\
&= \frac{5^{1/5}}{x^{1/60}} && \text{Take exponents up or down; change sign of exponent.}
\end{aligned}
$$

8. Use fractional exponents to multiply $\sqrt[3]{x^2}\,\sqrt{y}$. Leave the answer in radical form.

$$
\begin{aligned}
\sqrt[3]{x^2}\,\sqrt{y} &= x^{2/3}y^{1/2} = x^{4/6}y^{3/6} && \text{Change to a common denominator.} \\
&= \sqrt[6]{x^4} \bullet \sqrt[6]{y^3} && \text{Remember, the numerator is the power} \\
&= \sqrt[6]{x^4y^3} && \text{and the denominator is the root.}
\end{aligned}
$$

► **Exponents, negative**   If a is a real number, with $a \neq 0$, and n is a natural number, then

$$
a^{-n} = \frac{1}{a^n} \qquad \text{or} \qquad a^n = \frac{1}{a^{-n}}
$$

Both of these follow from the equation $a^n a^{-n} = a^{n-n} = a^0 = 1$. Since $a^n a^{-n} = 1$, dividing both sides of the equation by either a^n or a^{-n} yields the desired equations.

From the above equations, the following rule can be made:

Rule: If you move a number above or below the division bar, change the sign of its exponent.

Examples

1. Simplify $(6x^{-4}y^5)^{-2}$.

$$(6x^{-4}y^5)^{-2} = 6^{-2}x^8y^{-10} \qquad \text{Power-to-power rule.}$$

$$= \frac{x^8}{6^2y^{10}} \qquad \text{Move down; change sign.}$$

$$= \frac{x^8}{36y^{10}}$$

2. Simplify $\frac{-5^{-2}x^{-3}}{y^3z^{-4}}$.

$$\frac{-5^{-2}x^{-3}}{y^3z^{-4}} = \frac{-z^4}{5^2x^3y^3} \qquad \text{The "–" is not attached to the } -5^{-2}. \text{ Only the } 5^{-2} \text{ goes}$$
$$\text{down. The "–" is really } -1.$$

$$= \frac{-z^4}{25x^3y^3}$$

3. Simplify $\left(\frac{-5^{-2}x^{-3}}{y^3z^{-4}}\right)^2$.

$$\left(\frac{-5^{-2}x^{-3}}{y^3z^{-4}}\right)^2 = \frac{+5^{-4}x^{-6}}{y^6z^{-8}} \qquad \text{The "–" is separate from the } -5^{-2}, (-5^{-2}x^{-3})^2 =$$
$$(-)^2(5^{-2})^2(x^{-3})^2. \text{ The } (-) \text{ is really } (-1).$$

$$= \frac{z^8}{5^4x^6y^6}$$

$$= \frac{z^8}{625x^6y^6}$$

▶ **Exponents, rules of**   If a is a real number and m and n are natural numbers, then

1. $a^m \bullet a^n = a^{m+n}$ When multiplying similar bases, add exponents.

2. $\dfrac{a^m}{a^n} = a^{m-n}$ if $m > n$ When dividing similar bases, subtract the smaller exponent from the larger exponent.

     $\dfrac{a^m}{a^n} = \dfrac{1}{a^{n-m}}$ if $m < n$

3. $(a^m)^n = a^{mn}$ When an exponent is raised to an exponent (power-to-power), multiply the exponents.

A special case of the second rule is that (any real number)$^0 = 1$. This can be seen by letting $m = n$. When $m = n$, we have

$$\frac{a^m}{a^n} = \frac{a^m}{a^m} = 1 = a^{m-m} = a^0 \qquad \text{Hence, } a^0 = 1.$$

$$\text{or} \quad \frac{a^m}{a^n} = \frac{a^m}{a^m} = 1 = \frac{1}{a^{m-m}} = \frac{1}{a^0}$$

Since $1 = \frac{1}{a^0}$, $a^0 = 1$ by cross multiplication.
Because a is a real number, this implies, as a rule, that

$$(\text{any real number})^0 = 1$$

Examples

1. Simplify $-4x^5y \bullet 3x^2y$. Use rule 1.

$$-4x^5y \bullet 3x^2y = -12x^7y^2 \qquad \text{The } y\text{'s have exponents of 1, } y^1y^1 = y^2.$$

2. Simplify $\frac{6x^5y^2}{2xy^6}$. Use rule 2.

$$\frac{6x^5y^2}{2xy^6} = \frac{3x^4}{y^4} \qquad \text{The } x \text{ has an exponent of 1, } \frac{x^5}{x^1} = x^{5-1} = x^4.$$

3. Simplify $(-2x^2y)^4$. Use rule 3.

$$(-2x^2y)^4 = +2^4x^8y^4 \qquad \text{Treat ``}-\text{'' separately, } (-)^4 = +. \text{ This is like}$$
$$= 16x^8y^4 \qquad \begin{array}{l}(-1)(-1)(-1)(-1) = +1. \text{ The 2 has an exponent of 1,}\\ (2^1)^4 = 2^4. \text{ Same for } y, (y^1)^4 = y^4.\end{array}$$

4. Simplify $\left(\frac{-2xy^3}{3x^4y^2}\right)^3$.

$$\left(\frac{-2xy^3}{3x^4y^2}\right)^3 = \left(\frac{-2y}{3x^3}\right)^3$$
$$= \frac{-2^3y^3}{3^3x^9} = -\frac{8y^3}{27x^9} \qquad \text{The } (-)^3 = -. \text{ This is like } (-1)(-1)(-1) = -1.$$

5. Simplify $\left(\frac{x^5y^{-7}z^3}{xy^3z^{-5}}\right)^{-2}$.

$$\left(\frac{x^5y^{-7}z^3}{xy^3z^{-5}}\right)^{-2} = \frac{x^{-10}y^{14}z^{-6}}{x^{-2}y^{-6}z^{10}} = \frac{x^2y^{14}y^6}{x^{10}z^{10}z^6} \qquad \text{See } \textbf{Exponents, negative.}$$
$$= \frac{x^2y^{20}}{x^{10}z^{16}} = \frac{y^{20}}{x^8z^{16}}$$

6. Simplify $(4x^3y - 3z)^4(4x^3y - 3z)^{-4}$.

$$\left(4x^3y - 3z\right)^4\left(4x^3y - 3z\right)^{-4} = \left(4x^3y - 3z\right)^{4-4} = \left(4x^3y - 3z\right)^0 = 1$$

7. A common mistake with rule 3 is

$$\left(x^2 + y^3\right)^2 = x^4 + y^6 \qquad \text{The } (x^2 + y^3)^2 \text{ term means}$$
$$\text{incorrect} \qquad (x^2 + y^3)(x^2 + y^3), \text{ not } x^4 + y^6.$$

E

➤ **Exponents, zero**   See **Exponents, rules of**, special case.

➤ **Expression**   An **expression** is a group of terms. The terms are not set equal to anything as with an equation. An expression is not an equation. It cannot be solved.

To simplify the expression $6x + 4 - 2x + 3$ means to gather like terms. The answer is $4x + 7$. The most common mistake is trying to find an answer for x. There isn't one.

Expressions are usually used on an instructive basis. It is not until they are put into an equation that they take on meaning. If the expression above was set equal to 15, then

$$6x + 4 - 2x + 3 = 15$$
$$4x + 7 = 15$$
$$4x = 8$$
$$x = 2$$

Having first learned how to simplify the expression was paramount to solving the equation. See **Combining Like Terms**.

➤ **Exterior and Interior Angles (for lines)**   See **Interior and Exterior Angles (for lines)**.

➤ **Exterior Angle (for a triangle)**   An **exterior angle** is the angle formed when one of the sides of a triangle is extended. Since each side can be extended in two directions, there are $2(3) = 6$ exterior angles to any triangle.

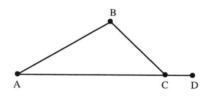

In $\triangle ABC$, if we extend side $\overline{AC}$ to D, the exterior angle, $\angle BCD$ or $\angle DCB$, is formed. Once a side is extended, two things of immediate importance are developed. The exterior angle of a triangle is supplementary to its adjacent, interior angle. In this case, $\angle BCD$ is supplementry to $\angle BCA$. Also, the measure of the exterior angle of a triangle is equal to the sum of the measures of the two remote interior angles ($\angle A$ and $\angle B$). In this case, we have $m\angle BCD = m\angle A + m\angle B$. In addition, sometimes the following is used: the exterior angle of a triangle has a measure that is greater than either of the measures of the remote interior angles.

➤ **Extraneous Solutions (roots)**   Solutions that are substituted in the original equation that do not check are **extraneous solutions**. Operations such as squaring both sides of an equation, taking the square root of both sides of an equation, or multiplying or dividing both sides of an equation by an expression that has a variable may introduce extraneous solutions.

Example

1. Solve $x - \sqrt{3 - x} = -3$ and check the solutions. See **Radicals, equations with**.

$$x - \sqrt{3 - x} = -3$$
$$x + 3 = \sqrt{3 - x}$$
$$(x + 3)^2 = \left(\sqrt{3 - x}\right)^2$$
$$x^2 + 6x + 9 = 3 - x$$
$$x^2 + 7x + 6 = 0$$
$$(x + 1)(x + 6) = 0$$

$$x + 1 = 0 \qquad \text{or} \qquad x + 6 = 0$$
$$x = -1 \qquad\qquad\qquad x = -6$$

Check the −1:

$$-1 - \sqrt{3 - (-1)} = -3$$
$$-1 - \sqrt{4} = -3$$
$$-1 - 2 = -3$$
$$-3 = -3 \qquad \text{checks}$$

Check the −6:

$$-6 - \sqrt{3 - (-6)} = -3$$
$$-6 - \sqrt{3 + 6} = -3$$
$$-6 - 3 = -3$$
$$-9 = -3 \qquad \text{does not check}$$

The −6 is an extraneous solution. It does not satisfy the original equation.

➤ **Extremes**   In the proportion $\frac{a}{b} = \frac{c}{d}$, a and d are called the **extremes**. See **Proportion**.

E

➤ Factor   If two or more numbers are multiplied, each of the numbers is called a **factor**. The same holds true for algebraic quantities or any other expressions that form a product.

Since $4 \times 3 = 12$, 4 and 3 are factors of 12.

➤ Factorial   A **factorial** is a term that is used to describe a continued product. Instead of writing $1 \bullet 2 \bullet 3 \bullet 4 \bullet 5 \bullet 6 \bullet 7 \bullet 8 \bullet 9$, we could simply write 9!, pronounced "nine factorial."

If n is a natural number, then

$$n! = 1 \bullet 2 \bullet 3 \cdots n \qquad \text{where } 0! = 1$$

The term $n!$ is read "n factorial."

Examples

1. Simplify 5!.

$$5! = 1 \bullet 2 \bullet 3 \bullet 4 \bullet 5 = 120$$

2. Simplify $\frac{8!}{5!}$.

$$\frac{8!}{5!} = \frac{1 \bullet 2 \bullet 3 \bullet 4 \bullet 5 \bullet 6 \bullet 7 \bullet 8}{1 \bullet 2 \bullet 3 \bullet 4 \bullet 5} = 6 \bullet 7 \bullet 8 = 336$$

or $\quad \dfrac{8!}{5!} = \dfrac{5! \bullet 6 \bullet 7 \bullet 8}{5!} = 6 \bullet 7 \bullet 8 = 336$

3. Simplify $\frac{(n+1)!}{(n-1)!}$.

$$\frac{(n+1)!}{(n-1)!} = \frac{1 \bullet 2 \bullet 3 \bullet \cdots \bullet (n-1) \bullet n \bullet (n+1)}{1 \bullet 2 \bullet 3 \bullet \cdots \bullet (n-1)} = n(n+1) = n^2 + n$$

➤ Factoring   **Factoring** is the reverse process of multiplication. When two or more terms have been multiplied, factoring them is a way of getting the original terms back again.

142

The factors of 6 are 2 and 3. The factors of $x^2 + 8x + 12$ are $x + 6$ and $x + 2$.

Usually students learn the following types of factoring: factoring common monomials, factoring the difference of two squares, factoring trinomials, factoring the sum and difference of two cubes, factoring four-term polynomials, and factoring six-term polynomials.

➤ **Factoring by Grouping (4-term polynomials)**   Polynomials with four terms are factored by grouping the terms in one of the following configurations: a 2 by 2 grouping, a 3 by 1 grouping, or a 1 by 3 grouping.

Examples

1. Factor $x^2 - 7x + xy - 7y$ with a 2 by 2 grouping.

   This method uses the concept of factoring common monomials twice. See **Factoring Common Monomials**.

$\underline{x^2 - 7x + xy - 7y}$	Group the terms, first pair and last pair.
$= x(x - 7) + y(x - 7)$	Factor the monomial out of each pair.
$= (x - 7)(x + y)$	Factor out the common binomial. This step is similar to factoring $xA + yA$, where $xA + yA = A(x + y)$.

   Sometimes the problem needs to be rearranged in the beginning. If the two pairs don't yield a common binomial, rearrange.

   Suppose the original problem was $x^2 - 7y - 7x + xy$. You can see that the two pairs do not each yield the same binomial. Hence, rearrange and then factor.

2. Factor $x^2 + 6x + 9 - y^2$ with a 3 by 1 grouping.

   This method uses the concept of the difference of two squares, $A^2 - B^2 = (A + B)(A - B)$. See **Factoring the Difference of Two Squares**. The A^2 part will come from the trinomial of the grouping. The trinomial must be a perfect square trinomial. See **Perfect Square Trinomial**.

$\underline{x^2 + 6x + 9} - y^2$	Group the terms, 3 by 1.
$= (x + 3)(x + 3) - y^2$	Factor the trinomial.
$= (x + 3)^2 - y^2$	Write the binomials as a square. The form now is $A^2 - B^2$
$= [(x + 3) + y][(x + 3) - y]$	which will factor. Let $A = x + 3$ and $B = y$,
$= (x + 3 + y)(x + 3 - y)$	then substitute in $(A + B)(A - B)$.

   If the terms were not in the proper order in the beginning, rearrange and try again.

3. Factor $x^2 - y^2 - 6y - 9$ with a 1 by 3 grouping.

F

Factoring with a 1 by 3 grouping is similar to that of factoring with a 3 by 1 grouping. This time, however, the trinomial will become the B^2:

$$\underline{x^2 - y^2 - 6y - 9}$$ Group the terms, 1 by 3.
$$= x^2 - (y^2 + 6y + 9)$$ Factor out the "−."
$$= x^2 - (y + 3)(y + 3)$$ Factor the trinomial.
$$= x^2 - (y + 3)^2$$ Write the binomial as a square. The form now is $A^2 - B^2$.
$$= [x + (y + 3)][x - (y + 3)]$$ Let $A = x$ and $B = y + 3$, then substitute in $(A + B)(A - B)$.
$$= (x + y + 3)(x - y - 3)$$ Multiply the "−" into the parentheses.

▶ **Factoring by Grouping (6-term polynomials)**   Polynomials with six terms are factored by grouping terms in one of the following configurations: a 3 by 3 grouping, a 3 by 2 by 1 grouping, or a 1 by 2 by 3 grouping. If the terms are not in the proper order for grouping, rearrange them.

Examples

1. Factor $a^2 - 2ab + b^2 - x^2 - 2xy - y^2$ with a 3 by 3 grouping.

   This method uses the concept of the difference of two squares, $A^2 - B^2 = (A + B)(A - B)$. See **Factoring the Difference of Two Squares**. The A^2 and B^2 terms will come from the trinomials. The trinomials must be perfect square trinomials. See **Perfect Square Trinomial**.

$$\underline{a^2 - 2ab + b^2 - x^2 - 2xy - y^2}$$ Group the terms, 3 by 3.
$$= a^2 - 2ab + b^2 - (x^2 + 2xy + y^2)$$ Factor out the "−."
$$= (a - b)(a - b) - (x + y)(x + y)$$ Factor the trinomials.
$$= (a - b)^2 - (x + y)^2$$ Write the binomials as squarer. The form is now $A^2 - B^2$.
$$= [(a - b) + (x + y)][(a - b) - (x + y)]$$ Let $A = a - b$ and $B = x + y$, then substitute in $(A + B)(A - B)$.
$$= (a - b + x + y)(a - b - x - y)$$

2. Factor $x^2 + 2xy + y^2 + 8x + 8y + 16$ with a 3 by 2 by 1 grouping.

   This method uses the concept of factoring trinomials. See **Factoring Trinomials**. The leading group of three must be a perfect square trinomial. See **Perfect Square Trinomial**.

$$\underline{x^2 + 2xy + y^2 + 8x + 8y + 16}$$ Group the terms, 3 by 2 by 1.
$$= (x + y)(x + y) + 8(x + y) + 16$$ Factor the trinomial and factor the binomial.
$$= (x + y)^2 + 8(x + y) + 16$$ Write the binomial as a square. The form is now $A^2 + 8A + 16$, which factors into $(A + 4)(A + 4)$.
$$= [(x + y) + 4][(x + y) + 4]$$ Let $A = x + y$ and substitute in $(A + 4)(A + 4)$.
$$= (x + y + 4)^2$$

In step 3, the expressions in the parentheses must be the same. If not, rearrange and try again.

3. Factor $x^2 + 2ax + 2bx + a^2 + 2ab + b^2$ with a 1 by 2 by 3 grouping.

$\underline{x^2} + \underline{2ax + 2bx} + \underline{a^2 + 2ab + b^2}$ Group the terms, 3 by 2 by 1.

$= x^2 + 2x(a + b) + (a + b)(a + b)$ Factor the binomial and trinomial.

$= x^2 + 2x(a + b) + (a + b)^2$ The form is $x^2 + 2xA + A^2 = (x + A)(x + A)$.

$= [x + (a + b)][x + (a + b)]$ Let $A = a + b$ and substitute in $(x + A)(x + A)$.

$= (x + a + b)^2$

➤ **Factoring Common Monomials**   When **factoring common monomials**, the concept of the greatest common factor is employed on the constants and the variables of each term. See **Greatest Common Factor**. As a rule, look at each term, see what is common, and factor it out.

Remember, always factor common monomials first.

Examples

1. Factor $4x + 6y$.

$$4x + 6y = 2(2x + 3y)$$

Looking at the $4x$, which is $2 \bullet 2x$, and the $6y$, which is $2 \bullet 3y$, we see that each term has a 2 in common. Factor it out. You can see that this is correct by multiplying back into the parentheses again.

Check:

$$2(2x + 3y) = 4x + 6y \qquad \text{See } \textbf{Distributive Property}.$$

2. Factor $3x^2 + 4x^5$.

$3x^2 + 4x^5 = x^2(3 + 4x^3)$ The term x^5 means $x \bullet x \bullet x \bullet x \bullet x$. When you

$= x^2(4x^3 + 3)$ take two of the x's away, three of them are left.

Check:

$$x^2(4x^3 + 3) = 4x^5 + 3x^2 = 3x^2 + 4x^5$$

See **Distributive Property**. Also see **Commutative Property of Addition**.

3. Factor $18x^3y^2 - 24xy^3$.

$$18x^3y^2 - 24xy^3 = 6xy^2(3x^2 - 4y)$$

The 18 and 24 have a 6 in common. The x^3 and x have an x in common. The y^2 and y^3 have a y^2 in common.

Check:

$$6xy^2(3x^2 - 4y) = 18x^3y^2 - 24xy^3$$

F

4. Factor $25x^3y^2z^4 - 50x^2y^3z^5 - 100x^3y^3z^3$.

$$25x^3y^2z^4 - 50x^2y^3z^5 - 100x^3y^3z^3 = 25x^2y^2z^3(xz - 2yz^2 - 4xy)$$

The following can be used as an aid in determining the common monomial:

$$\left.\begin{array}{l} 25, -50, -100 \rightarrow 25 \\ x^3, x^2, x^3 \quad\;\; \rightarrow x^2 \\ y^2, y^3, y^3 \quad\;\; \rightarrow y^2 \\ z^4, z^5, z^3 \quad\;\; \rightarrow z^3 \end{array}\right\} \quad 25x^2y^2z^3 \text{ is the monomial that is common in each term}$$

Check:

$$25x^2y^2z^3(xz - 2yz^2 - 4xy) = 25x^3y^2z^4 - 50x^2y^3z^5 - 100x^3y^3z^3$$

Remember, always factor out common monomials before you try other types of factoring.

5. Factor $3x^2 + 6x - 24$.

First, factor out the common monomial, then factor the trinomial:

$$3x^2 + 6x - 24 = 3(x^2 + 2x - 8) = 3(x + 4)(x - 2) \qquad \text{See \textbf{Factoring Trinomials}}.$$

F

➤ **Factoring Polynomials**   To **factor polynomials**, see the following articles: **Factor Theorem, Factoring Common Monomials, Factoring the Sum and Difference of Two Cubes, Factoring the Difference of Two Squares, Factoring Trinomials, Factoring by Grouping (4-term polynomials), Factoring by Grouping (6-term polynomials)**, or **FOIL**.
For solutions of polynomials by factoring, see **Rational Roots Theorem**.

➤ **Factoring Primes**   See **Prime Factorization**.

➤ **Factoring the Difference of Two Squares**   When multiplying two binomials by the FOIL method, sometimes inside and outside terms cancel. When this happens, the first and last terms are both squared terms separated by a negative. Hence, the phrase "**difference of two squares**."

$$a^2 - b^2 = (a + b)(a - b)$$

The a^2 term is split up, a and a. The b^2 term is split up, b and b.
When the $a + b$ and $a - b$ terms are multiplied, we have

$$(a + b)(a - b) = a^2 - ab + ab - b^2 = a^2 - b^2$$

See **FOIL**.

Examples

1. Factor $x^2 - 16$.

   First, write an empty pair of parentheses, $(\ )(\ )$. Next, split up the x^2 term into x and x and place these in the a's position, $(x\ )(x\ )$.

   Then, split up the 16 into 4 and 4 and place these in the b's position, $(x\ 4)(x\ 4)$. Finally, put a $+$ and $-$ in the middle of each factor, $(x + 4)(x - 4)$.

   Check:

   $$(x + 4)(x - 4) = x^2 - 4x + 4x - 16 = x^2 - 16. \qquad \text{See \textbf{FOIL}.}$$

2. Factor $25x^2 - 4$.

   Split the $25x^2$ into $5x$ and $5x$. Split the 4 into 2 and 2. Write each in the appropriate position and insert the $+$ and $-$:

   $$25x^2 - 4 = (5x + 2)(5x - 2)$$

   Check:

   $$(5x + 2)(5x - 2) = 25x^2 - 10x + 10x - 4 = 25x^2 - 4$$

3. Factor $1 - 64x^2$.

   $$1 - 64x^2 = (1 + 8x)(1 - 8x)$$

   Check:

   $$(1 + 8x)(1 - 8x) = 1 - 8x + 8x - 64x^2 = 1 - 64x^2$$

4. Factor $\frac{9}{49}x^2 - \frac{4}{25}$.

   $$\frac{9}{49}x^2 - \frac{4}{25} = \left(\frac{3}{7}x + \frac{2}{5}\right)\left(\frac{3}{7}x - \frac{2}{5}\right)$$

5. Remember, always factor out common monomials before you try any other type of factoring. See **Factoring Common Monomials**.

   Factor $50x^2 - 8$.

   $$50x^2 - 8 = 2(25x^2 - 4) \qquad \text{First take out the common monomial 2.}$$
   $$= 2(5x + 2)(5x - 2)$$

▶ Factoring the Sum and Difference of Two Cubes

To **factor the sum and difference of two cubes,** we apply the following:

$$a^3 + b^3 = (a + b)(a^2 - ab + b^2)$$
$$a^3 - b^3 = (a - b)(a^2 + ab + b^2) \qquad \text{where } a \text{ and } b \text{ are real numbers}$$

Recognizing the similarities in both equations, some people prefer to memorize one equation

$$a^3 \pm b^3 = (a \pm b)(a^2 \mp ab + b^2)$$

Examples

1. Factor $x^3 + 8y^3$.

    We first need to write $x^3 + 8y^3$ in the exact form $a^3 + b^3$. Rewriting, we have $x^3 + 8y^3 = x^3 + (2y)^3$, where a is represented by x and b is represented by $2y$. See **Exponents, rules of**, power-to-power rule. Substituting in the formula, we have

    $$a^3 + b^3 = (a + b)(a^2 - ab + b^2)$$
    $$x^3 + (2y)^3 = (x + 2y)[x^2 - x(2y) + (2y)^2]$$
    $$x + 8y^3 = (x + 2y)(x^2 - 2xy + 4y^2)$$

    Check: To check, multiply the polynomials. See **Polynomials, multiplication of**.

2. Factor $8x^3 - 27y^3$.

    Writing $8x^3 - 27y^3$ in the exact form $a^3 - b^3$, we have

    $$8x^3 - 27y^3 = (2x)^3 - (3y)^3 \qquad \text{where } a = 2x \text{ and } b = 3y$$

    Substituting in the formula, we have

    $$a^3 - b^3 = (a - b)(a^2 + ab + b^2)$$
    $$(2x)^3 - (3y)^3 = (2x - 3y)[(2x)^2 + (2x)(3y) + (3y)^2]$$
    $$8x^3 - 27y^3 = (2y - 3y)(4x^2 + 6xy + 9y^2)$$

    Check by multiplying.

3. Factor $125x^3y^6 - 1$.

    In the form $a^3 - b^3$, we have

    $$125x^3y^6 - 1 = (5xy^2)^3 - (1)^3 \qquad \text{where } a = 5xy^2 \text{ and } b = 1$$

    Substituting, we have

    $$a^3 - b^3 = (a - b)(a^2 + ab + b^2)$$
    $$(5xy^2)^3 - (1)^3 = (5xy^2 - 1)[(5xy^2)^2 + (5xy^2)(1) + (1)^2]$$
    $$125x^3y^6 - 1 = (5xy^2 - 1)(25x^2y^4 + 5xy^2 + 1)$$

4. Remember, always factor out common monomials before you try any other type of factoring. See **Factoring Common Monomials**.

Factor $24x^3 - 81y^3$.

$$24x^3 - 81y^3 = 3(8x^3 - 27y^3) \qquad \text{Factor the common monomial 3.}$$
$$= 3(2y - 3y)(4x^2 + 6xy + 9y^2) \qquad \text{See example 2.}$$

➤ **Factoring Trinomials**   A trinomial is a polynomial with three terms. To **factor a trinomial** means to figure out what its original factors are. Prior to being multiplied, the original factors were binomials. Our goal is to find out what they are.

Factoring trinomials is a process of trial and error. It does not follow a general formula. As a result, it takes practice to learn trinomial factoring. The following are general steps used in factoring trinomials:

1. Write a pair of parentheses, ( )( ).

2. Factor the first term of the trinomial (there can be many) and place these in the front of the parentheses.

3. Factor the last term of the trinomial (there can be many) and place these at the end of the parentheses.

4. Multiply the inside terms, then the outside terms, and think whether adding or subtracting them will give you the center term of the trinomial. See **FOIL**.

5. If they do, then you need to figure out the $+$ or $-$ signs of each binomial. To do this, apply one of the following:

   a. If you added to get the center term of the trinomial, then the sign of the center term goes to both binomials.

   b. If you subtracted to get the center term of the trinomial, then the sign of the center term goes to the largest product of the binomials and the other sign is opposite. Confusing enough? Try some examples.

6. Always check your answer.

Examples

1. Factor $x^2 + 6x + 8$.

   Write a pair of parentheses, ( )( ).

   The factors of x^2 are x and x.

   The factors of 8 are 2 and 4, also 1 and 8. Let's try the 2 and 4 first. If they don't work, then we'll try the 1 and 8.

   So far we have the following:

   $$(x \quad 2)(x \quad 4)$$

   Multiplying the inside and outside terms, we get $2x$ and $4x$:

The product of the inside terms is $2x$ and the product of the outside terms is $4x$.

Recognize that if we add them (the $2x$ and $4x$) we get the center term of the polynomial ($6x$). This is what we want.

Finally, since we added to get the center of the trinomial, the sign of the center (+) goes to both binomials.

$$(x + 2)(x + 4)$$

Check: We can check this by multiplying. See **FOIL**.

$$(x + 2)(x + 4) = x^2 + 4x + 2x + 8 = x^2 + 6x + 8$$

2. Factor $x^2 + 7x + 10$.

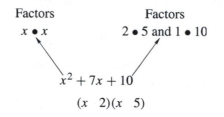

Multiplying the inside and outside gives us $2x$ and $5x$. If we add $2x$ and $5x$ we get $7x$, the center term. Therefore, the sign of the center term (+) goes to both binomials:

$$(x + 2)(x + 5)$$

Check:

$$(x + 2)(x + 5) = x^2 + 5x + 2x + 10 = x^2 + 7x + 10$$

3. Factor $x^2 - 2x - 15$.

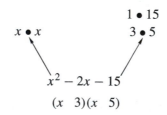

When the inside and outside terms are subtracted, we get the center term. Therefore, the sign of the center (−) goes to the largest term (5). The other binomial is opposite (+).

$$(x + 3)(x - 5)$$

Check:

$$(x + 3)(x - 5) = x^2 - 5x + 3x - 15 = x^2 - 2x - 15$$

The majority of factoring trinomials has to be done in your mind. You need to think about the product of the inside and outside terms, you always need to be aware of the center term of the trinomial, and you must think of what signs will work. With practice, the process becomes "second nature."

4. Factor $x^2 - 7x - 8$.

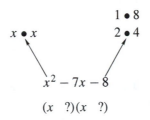

Trying the 2, 4 or the 4, 2 and adding or subtracting will not yield a 7. So, try the 1 and 8.

If the 1 and 8 are subtracted, we get the center term, so the sign of the center goes to the largest number:

$$(x + 1)(x - 8)$$

5. Factor $x^2 + x - 12$.

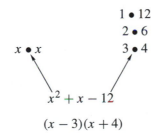

The 3 and 4 subtracted yield 1.

6. Factor $x^2 + 13x + 12$.

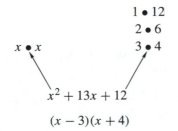

The 3 and 4 can't work because they yield 7 or 1 when added or subtracted. The 2 and 6 yield 8 or 4. We need a 13. The 1 and 12 added will give that.

Multiplying binomials can help in seeing how trinomials factor. Multiply the binomials in examples 5 and 6. Watch what happens to the terms.

7. Factor $6x^2 - 11x - 10$.

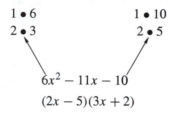

$$6x^2 - 11x - 10$$
$$(2x - 5)(3x + 2)$$

First try the $2 \bullet 3$ with the $1 \bullet 10$ and $2 \bullet 5$. If that doesn't work, try $1 \bullet 6$.

8. Factor $15x^2 + 17x - 18$.

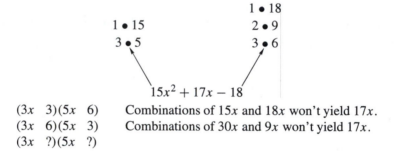

$$15x^2 + 17x - 18$$

$(3x \quad 3)(5x \quad 6)$ Combinations of $15x$ and $18x$ won't yield $17x$.
$(3x \quad 6)(5x \quad 3)$ Combinations of $30x$ and $9x$ won't yield $17x$.
$(3x \quad ?)(5x \quad ?)$

Try 2 and 9, then 9 and 2, and so on through the list. The 2 and 9 works if you subtract. The center sign $(+)$ goes to the largest product, the 9 of $27x$:

$$(3x - 2)(5x + 9)$$

Always check your answer. Sometimes you can get the center term but the last term is wrong.

9. Factor $x^2 + 5x + 6$.

Both $(x + 2)(x + 3)$ and $(x - 1)(x + 6)$ will combine to give a center term of $+5x$. Only the $(x + 2)(x + 3)$ will yield $+6$ as well.

$$x^2 + 5x + 6 = (x + 2)(x + 3)$$

10. Factor $3x^2 + 15x + 18$.

Remember, always factor out common monomials before you try any other type of factoring. See **Factoring Common Monomials**.

$$3x^2 + 15x + 18 = 3(x^2 + 5x + 6) \qquad \text{First take out the common monomial 3.}$$
$$= 3(x + 2)(x + 3)$$

F

$$x^2 - 9x + 18 = (x - 6)(x - 3)$$
$$x^2 - 6x - 27 = (x - 9)(x + 3)$$
$$x^2 + 2x - 8 = (x + 4)(x - 2)$$
$$3x^2 - 31x + 56 = (3x - 7)(x - 8)$$
$$14x^2 + 31x - 10 = (2x + 5)(7x - 2)$$

➤ **Factor Theorem**   The **factor theorem** can be used to determine the factors of a polynomial or the solutions of a polynomial if one of its factors is known. Let $f(x)$ be a polynomial. Then the binomial $x - a$ is a factor of $f(x)$ if $f(a) = 0$ and, conversely, $f(a) = 0$ if the binomial $x - a$ is a factor of $f(x)$. For more solutions of polynomials, see **Rational Roots Theorem**.

Examples

1. Show that $x + 4$ is a factor of $f(x) = x^2 + 6x + 8$.

    According to the factor theorem, $x + 4 = x - (-4)$ is a factor of $x^2 + 6x + 8$ if $f(-4) = 0$. See **Function Notation**. Remember that the number to be substituted is opposite in sign from that of the factor. Substituting, we have

    $$f(-4) = (-4)^2 + 6(-4) + 8 = 16 - 24 + 8 = 0$$

    Since $f(-4) = 0$, $x + 4$ is a factor of $x^2 + 6x + 8$. If we solve the equation $x^2 + 6x + 8 = 0$, we can see why the factor theorem works. Solving by factoring, we have

$x^2 + 6x + 8 = 0$	See **Quadratic Equations, solutions of**. Also see **Product**
$(x + 4)(x + 2) = 0$	**Theorem of Zero**.

    $$x + 4 = 0 \quad \text{or} \quad x + 2 = 0$$
    $$x = -4 \qquad\qquad x = -2$$

    Since one of the solutions is $x = -4$, we see that the substitution we did with $f(-4)$ is really a check, or verification, of -4 in the original equation. By looking at $f(-4)$, we were checking -4 as a solution. The factor theorem checks to see if the factor is a solution of the polynomial.

2. Is $x - 4$ a factor of $x^3 - 5x^2 - 2x + 24$?

    We need to substitute $+4$ in the polynomial to see if we get 0:

    $$x^3 - 5x^2 - 2x + 24 = (4)^3 - 5(4)^2 - 2(4) + 24 = 64 - 80 - 8 + 24 = 0$$

    Hence, $x - 4$ is a factor of $x^3 - 5x^2 - 2x + 24$. If we didn't get zero, then $x - 4$ would not be a factor. This would also mean that it was not a solution.

F

3. In example 2, $x - 4$ is a factor of $x^3 - 5x^2 - 2x + 24$. What are the remaining factors?

To find the remaining factors, we first observe what happens when dividing natural numbers. Divide 10 by 2:

$$\begin{array}{r} 5 \\ 2\overline{)10} \\ \underline{10} \\ 0 \end{array}$$

Since 2 divides into 10 evenly, we know that 2 and 5 are factors of 10. Similarly, if $x - 4$ could divide $x^3 - 5x^2 - 2x + 24$ evenly, we could get its factors. To divide, we can use either long division or synthetic division. See **Long Division of Polynomials** or **Synthetic Division**. We will use synthetic division.

Dividing $x^3 - 5x^2 - 2x + 24$ by $x - 4$, we have

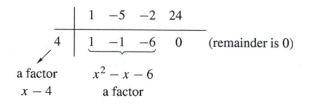

$$\begin{array}{c|cccc} & 1 & -5 & -2 & 24 \\ \hline 4 & 1 & -1 & -6 & 0 \end{array} \quad \text{(remainder is 0)}$$

a factor $x^2 - x - 6$

$x - 4$ a factor

In long division form, it looks like this

$$\begin{array}{r} x^2 - x - 6 \\ x - 4\overline{)x^3 - 5x^2 - 2x + 24} \end{array}$$

In any event, it means that $(x - 4)(x^2 - x - 6) = x^3 - 5x^2 - 2x + 24$. Factoring the trinomial, we have

$$x^2 - x - 6 = (x - 3)(x + 2)$$

Hence, the remaining factors are $x - 3$ and $x + 2$.

4. Verify that $x + 3$ is a factor of $f(x) = 2x^3 + 3x^2 - 8x + 3$ and then find the zeros of $f(x)$.

If $x + 3$ is a factor of $f(x)$, then by the factor theorem we should have $f(-3) = 0$ for the binomial $x - (-3)$:

$$f(-3) = 2(-3)^3 + 3(-3)^2 - 8(-3) + 3 = 2(-27) + 3(9) + 24 + 3 = 0$$

Since $f(-3) = 0$, $x + 3$ is a factor of $f(x)$. To find the zeros (solutions) of $f(x)$, we need to set $f(x) = 0$, factor $f(x)$, and then solve each factor. See **Zeros of a Function or Polynomial**. Since $x + 3$ is a factor of $f(x)$, we can find the other factors by dividing $f(x)$ by $x + 3$. See **Synthetic Division**.

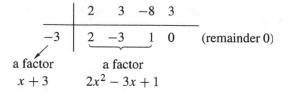

Hence, $x + 3$ and $2x^2 - 3x + 1$ are factors of $2x^3 + 3x^2 - 8x + 3$. Setting the equation equal to zero and solving, we have

$$2x^3 + 3x^2 - 8x + 3 = 0$$
$$(x + 3)(2x^2 - 3x + 1) = 0$$
$$(x + 3)(2x - 1)(x - 1) = 0$$

$x + 3 = 0$	or $\quad 2x - 1 = 0$	or $\quad x - 1 = 0$	Set each factor to zero.
$x = -3$	$x = \dfrac{1}{2}$	$x = 1$	See **Product Theorem of Zero**.

For more on finding the zeros of a polynomial, see **Rational Roots Theorem**.

➤ **Fibonacci Sequence** Named after Leonardo Fibonacci, the **Fibonacci sequence** follows the pattern

$$1, 1, 2, 3, 5, 8, 13, 21, 34, 55, 89, \ldots$$

where $F_{n+2} = F_n + F_{n+1}$ and $n \geqslant 1$.

Starting with $F_1 = 1$ and $F_2 = 1$, each term in the sequence is the sum of the previous two, i.e., $21 = 13 + 8$.

➤ **Finding the Equation of a Line** See **Linear Equation, finding the equation of a line**.

➤ **Focal Radii** If $P(x, y)$ is a point on an ellipse and F_1 and F_2 are the foci of the ellipse, then $F_1 P$ and $F_2 P$ are called the **focal radii** of the ellipse. The focal radii are the line segments joining a point on the ellipse to each of the foci.

For the ellipse $\frac{x^2}{a^2} = \frac{y^2}{b^2} = 1$, the lengths of the focal radii are $F_1 P = a + ex$ and $F_2 P = a - ex$. See **Ellipse**.

➤ **Focus** A **focus** is one or more of the fixed points of a conic section. A conic section is the set of all points such that its distance from a fixed point has a constant ratio to its distance from a fixed line. The fixed point is called the focus. The fixed line is called the **directrix**.

For more, see **Conic Sections, Parabola, Ellipse**, or **Hyperbola**.

➤ **FOIL** FOIL is an acronym for First, Outer, Inner, and Last. It is a method for multiplying binomials. If we wanted to multiply $(a + b)(c + d)$, the a would multiply the c and the d. Then, the b would multiply the c and the d. If we notice the order of multiplication, we see that ac is a product of the First letters in the parentheses, ad is the product of the Outer letters, bc is a product of the Inner letters, and bd is a product of the Last letters. Hence, FOIL.

F

Example

1. Multiply $(2x + 1)(3x + 4)$ by the FOIL method.

$$\begin{array}{cccc} & \text{First} & \text{Outer} & \text{Inner} & \text{Last} \end{array}$$
$$(2x + 1)(3x + 4) = (2x)(3x) + (2x)(4) + (1)(3x) + (1)(4)$$
$$= 6x^2 + 8x + 3x + 4 = 6x^2 + 11x + 4$$

Notice that the outer and inner products become the center term of the trinomial. This is very important when it comes to factoring, the reverse process of multiplying. For more, see **Polynomials, multiplication of** or **Perfect Square Trinomial**.

➤ Forty-Five Degree Triangle

The **forty-five degree triangle** is one of two special triangles used in trigonometry. The other is the thirty-sixty-ninety degree triangle. See **Thirty-Sixty-Ninety Degree Triangle**.

A 45 degree triangle is actually a 45-45-90 degree triangle. In a 45 degree triangle, the length of each leg is 1 and the length of the hypotenuse is $\sqrt{2}$.

To apply the 45 degree triangle, the concept of cross multiplication and similar triangles must be clear. See the articles under those names. Also see **Trigonometry, reference angle**, **Direct Variation**, or **Proportion**.

Examples

1. Find the length of the hypotenuse of a 45 degree triangle if the length of a leg is 3.

Since a 45 degree triangle is an isosceles right triangle, each of the legs are equal in length. As a result, we have three methods of finding the hypotenuse. We can compare the given triangle with the 45 degree triangle, find the sides using similar triangles, or apply the Pythagorean theorem.

To find the hypotenuse by comparing the given triangle to the 45 degree triangle, we notice that the length of each leg of the given triangle is 3 times that of the legs of the 45 degree triangle. Multiplying the sides of the 45 degree triangle by 3, we have the lengths of the sides of the given triangle, $3(1) = 3$, and the length of the hypotenuse $3(\sqrt{2}) = 3\sqrt{2}$.

To find the hypotenuse by using similar triangles, we need to set up a proportion between the given triangle and the 45 degree triangle. Letting z represent the length of the hypotenuse, we have the proportion $\frac{z}{\sqrt{2}} = \frac{3}{1}$. Solving the proportion, we have

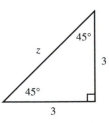

$$\frac{z}{\sqrt{2}} = \frac{3}{1}$$
$$z = 3\sqrt{2} \qquad \text{See } \textbf{Cross Multiplication.}$$

To find the hypotenuse by applying the Pythagorean theorem where z is the hypotenuse, $x = 3$, and $y = 3$, we have

$$x^2 + y^2 = z^2$$
$$3^2 + 3^2 = z^2$$
$$18 = z^2$$
$$\sqrt{z^2} = \sqrt{18}$$
$$z = \sqrt{9 \cdot 2} = \sqrt{9} \cdot \sqrt{2}$$
$$z = 3\sqrt{2}$$

See **Pythagorean Theorem**.

Distance is positive.
See **Radicals, simplifying (reducing)**.

2. Find the length of the legs of a 45 degree triangle if the length of the hypotenuse is 4.

This type of problem is more difficult to solve by comparing triangles because the 4 is compared to $\sqrt{2}$. We need to think of a number that can multiply $\sqrt{2}$ to yield 4. The number is $\frac{4}{\sqrt{2}}$ since $\frac{4}{\sqrt{2}}(\sqrt{2}) = 4$. Hence, the length of each leg is $\frac{4}{\sqrt{2}}(1) = \frac{4}{\sqrt{2}} = \frac{4\sqrt{2}}{2} = 2\sqrt{2}$. See **Rationalizing the Denominator**.

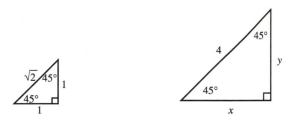

To find the legs by using similar triangles, we first set up the proportion $\frac{x}{1} = \frac{4}{\sqrt{2}}$. Solving, we have

$$\frac{x}{1} = \frac{4}{\sqrt{2}}$$
$$x = \frac{4}{\sqrt{2}} \cdot \frac{\sqrt{2}}{\sqrt{2}}$$
$$x = \frac{4\sqrt{2}}{2}$$
$$x = 2\sqrt{2}$$

See **Rationalizing the Denominator**.

Since x and y are equal, $y = 2\sqrt{2}$ as well.

To find the legs by applying the Pythagorean theorem where the hypotenuse $z = 4$ and the legs $x = y$, we have

$$x^2 + y^2 = z^2$$
$$x^2 + x^2 = 4^2$$
$$2x^2 = 16$$
$$x^2 = 8$$

See **Pythagorean Theorem**.

F

$$\sqrt{x^2} = \sqrt{8}$$
$$x = \sqrt{4 \bullet 2} = \sqrt{4} \bullet \sqrt{2}$$
$$x = 2\sqrt{2}$$

Distance is positive.

See **Radicals, simplifying (reducing)**.

➤ **Fraction**   For fraction as a rational number, see **Rational Number, definition**. Also see **Real Number System**. A **fraction** is a number expressing a part of a whole. It is written with a number above and below a bar. The number on top, **numerator**, tells how many parts we have. The number on the bottom, **denominator**, tells how many parts into which an object has been divided. In this sense, the number $\frac{5}{8}$ tells us that we have 5 parts of an object that has been divided into 8 parts.

The denominator of a fraction cannot be zero. Consider the fraction $\frac{5}{0}$. It makes no sense to say that we have 5 parts of an object that has been divided into 0 parts. For more, see **Mixed Number**, **Proper Fraction**, or **Fractions, improper**.

➤ **Fractional Exponent**   See **Exponents, fractional or rational**.

➤ **Fraction, in algebra**   For fractions related to algebra, see the various topics under **Rational Equations**.

➤ **Fractions, addition of**   To **add fractions**, the denominator must be the same. See **Common Denominator**. If the denominators are common, then add the numerators and reduce the answer, if possible. See **Fractions, reducing**.

F

Examples

1.   Add $\frac{1}{2} + \frac{1}{4}$.

$$\frac{1}{2} + \frac{1}{4} = \frac{1}{2} \bullet \frac{2}{2} + \frac{1}{4} = \frac{2}{4} + \frac{1}{4} = \frac{3}{4} \qquad \text{or} \qquad \begin{array}{r} \frac{1}{2} \bullet \frac{2}{2} = \frac{2}{4} \\ +\frac{1}{4} \\ \hline \frac{3}{4} \end{array}$$

To add fractions, the denominators must be the same. We made them the same, 4ths, by multiplying the numerator and denominator of $\frac{1}{2}$ by 2. See **Fractions, equivalent**. The $\frac{2}{4}$ then had the same denominator as the $\frac{1}{4}$. Then it was clear that if we had $\frac{2}{4}$ and $\frac{1}{4}$ more, we had $\frac{3}{4}$ altogether.

We must have a common denominator to be able to count up how much of each we have.

2. Add $\frac{1}{3} + \frac{1}{2}$.

$$\frac{1}{3} + \frac{1}{2} = \frac{1}{3} \bullet \frac{2}{2} + \frac{1}{2} \bullet \frac{3}{3} = \frac{2}{6} + \frac{3}{6} = \frac{5}{6} \quad \text{or} \quad \begin{array}{r} \frac{1}{3} \bullet \frac{2}{2} = \frac{2}{6} \\ +\frac{1}{2} \bullet \frac{3}{3} = \frac{3}{6} \\ \hline \frac{5}{6} \end{array}$$

Adding $\frac{1}{3}$ and $\frac{1}{2}$ makes no sense. Adding $\frac{2}{6}$ and $\frac{3}{6}$ does. This is why we must have common denominators.

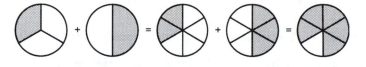

3. Add $3\frac{1}{2} + 2\frac{1}{4}$.

$$3\frac{1}{2} + 2\frac{1}{4} = 3\frac{1}{2} \bullet \frac{2}{2} + 2\frac{1}{4} = 3\frac{2}{4} + 2\frac{1}{4} = 5\frac{3}{4} \quad \text{or} \quad \begin{array}{r} 3\frac{1}{2} \bullet \frac{2}{2} = 3\frac{2}{4} \\ +2\frac{1}{4} \qquad +2\frac{1}{4} \\ \hline 5\frac{3}{4} \end{array}$$

When adding mixed fractions, i.e., fractions with a whole number and a fraction combined, concern yourself only with adding the fractions. In our example, adding $3 + 2$ isn't a problem. It's the $\frac{1}{2} + \frac{1}{4}$ that must be considered. See example 1.

4. Add $4\frac{1}{3} + 5\frac{1}{2}$.

$$4\frac{1}{3} + 5\frac{1}{2} = 4\frac{1}{3} \bullet \frac{2}{2} + 5\frac{1}{2} \bullet \frac{3}{3} = 4\frac{2}{6} + 5\frac{3}{6} = 9\frac{5}{6}$$

See example 2.

The examples from here on are variations.

5. $4 + \frac{1}{2} = 4\frac{1}{2}$

6. $\frac{1}{2} + 4 = 4\frac{1}{2}$

7. $3\frac{1}{2} + 2 = 5\frac{1}{2}$

8. $2 + 3\frac{1}{2} = 5\frac{1}{2}$

F

9. $4\frac{1}{5} + \frac{1}{2} = 4\frac{1}{5} \cdot \frac{2}{2} + \frac{1}{2} \cdot \frac{5}{5} = 4\frac{2}{10} + \frac{5}{10} = 4\frac{7}{10}$

10. $\frac{2}{3} + \frac{5}{6} = \frac{2}{3} \cdot \frac{2}{2} + \frac{5}{6} = \frac{4}{6} + \frac{5}{6} = \frac{9}{6} = 1\frac{3}{6} = 1\frac{1}{2}$ See **Fractions, reducing**.

11. $\frac{3}{4} + \frac{2}{3} = \frac{3}{4} \cdot \frac{3}{3} + \frac{2}{3} \cdot \frac{4}{4} = \frac{9}{12} + \frac{8}{12} = \frac{17}{12} = 1\frac{5}{12}$

12. $5\frac{3}{4} + 4\frac{2}{3} = 5\frac{3}{4} \cdot \frac{3}{3} + 4\frac{2}{3} \cdot \frac{4}{4} = 5\frac{9}{12} + 4\frac{8}{12} = 9\frac{17}{12} = 10\frac{5}{12}$

13. $\frac{1}{2} + \frac{1}{3} + \frac{1}{4} = \frac{1}{2} \cdot \frac{6}{6} + \frac{1}{3} \cdot \frac{4}{4} + \frac{1}{4} \cdot \frac{3}{3} = \frac{6}{12} + \frac{4}{12} + \frac{3}{12} = \frac{13}{12} = 1\frac{1}{12}$

14. $4\frac{1}{2} + 2\frac{1}{3} + 5\frac{1}{4} = 4\frac{6}{12} + 2\frac{4}{12} + 5\frac{3}{12} = 11\frac{13}{12} = 12\frac{1}{12}$

➤ **Fractions, algebraic**   See **Rational Expressions, simplifying (reducing)**.

➤ **Fractions, bigger and smaller.**   See **Fractions, comparing** or the articles under **Inequalities**.

➤ **Fractions, change from fraction to decimal to percent**   To change a fraction to a decimal, we divide the numerator by the denominator and leave the answer in decimal form. To change a decimal to a percent, we move the decimal point two places to the right.

Examples

1. Change $\frac{1}{2}$ to a decimal.

   To change $\frac{1}{2}$ to a decimal, we need to divide 1 by 2. See **Decimal Numbers, division of**.

$$\begin{array}{r} .5 \\ 2\overline{)1.0} \\ \underline{10} \\ 0 \end{array}$$

   We now have $\frac{1}{2} = .5$. The fraction $\frac{1}{2}$ has been changed to a decimal.

2. Change .5 to a percent.

   To change .5 to a percent, we need to move the decimal point two places to the right:

   $$.5 = .50 = 50\%$$ Annex (add on) as many zeros as needed.

3. Change $\frac{3}{4}$ from a fraction to a decimal and then to a percent.

   To change from a fraction to a decimal, we divide 3 by 4:

$$\begin{array}{r} .75 \\ 4\overline{)3.00} \\ \underline{28} \\ 20 \\ \underline{20} \\ 0 \end{array}$$

To change .75 to a percent, we move the decimal two places to the right:

$$.75 = .75_\smile = 75\%$$

4. Change $\frac{2}{3}$ to a percent.

First, we change $\frac{2}{3}$ to a decimal:

$$\begin{array}{r} .66 = .66\frac{2}{3} \\ 3\overline{)2.00} \\ \underline{18} \\ 20 \\ \underline{18} \\ 2 \end{array}$$ Write the remainder as a fraction after dividing through two decimal places.

Now change $.66\frac{2}{3}$ to a percent:

$$.66\frac{2}{3} = .66\frac{2}{3}_\smile = 66\frac{2}{3}\%$$

For related articles, see **Decimal Numbers, changed to a fraction**; **Decimal Numbers, changed to a percent age percentage**; and **Decimal Numbers, changing a repeating decimal to a common fraction**.

➤ **Fractions, common denominator**   See **Common Denominator**.

➤ **Fractions, comparing**   To **compare fractions**, change the given fractions to those with a common denominator. Then compare the numerators. See **Common Denominator** or **Fractions, equivalent**.

Examples

1. Which is larger, $\frac{1}{2}$ or $\frac{1}{3}$?

To compare $\frac{1}{2}$ and $\frac{1}{3}$, we need to change them to equivalent fractions with a common denominator of 6:

$$\frac{1}{2} = \frac{1}{2} \bullet \frac{3}{3} = \frac{3}{6}$$
$$\frac{1}{3} = \frac{1}{3} \bullet \frac{2}{2} = \frac{2}{6}$$

Now we need to compare $\frac{3}{6}$ and $\frac{2}{6}$. Since 3 is larger than 2, $\frac{3}{6}$ is larger than $\frac{2}{6}$.

If we use inequality signs to describe this relationship, we would have ($<$ is less than, $>$ is greates than)

$$\frac{1}{2} > \frac{1}{3} \qquad \text{since} \qquad \frac{3}{6} > \frac{2}{6}. \qquad \text{See } \textbf{Inequality}.$$

F

It must be kept in mind that $\frac{3}{6}$ means that you have 3 parts of something that has been cut into 6 pieces. Suppose a pie has been cut into 6 pieces. If you have 3 of the pieces and your friend has 2, then you have more than your friend. Your $\frac{3}{6}$ is more than the $\frac{2}{6}$.

2. Which is smaller, $\frac{2}{5}$ or $\frac{3}{4}$?

Changing to equivalent fractions, we have

$$\frac{2}{5} = \frac{2}{5} \bullet \frac{4}{4} = \frac{8}{20}$$
$$\frac{3}{4} = \frac{3}{4} \bullet \frac{5}{5} = \frac{15}{20}$$

Since $\frac{8}{20} < \frac{15}{20}, \frac{2}{5} < \frac{3}{4}$.

3. Which is larger, $\frac{2}{3}$ or $\frac{5}{6}$?

$$\frac{2}{3} \text{ or } \frac{5}{6}$$
$$\frac{4}{6} \text{ or } \frac{5}{6}$$

Since 5 is larger than 4, $\frac{2}{3}$ is smaller than $\frac{5}{6}$.

4. Arrange from smallest to largest: $\frac{1}{3}$, $\frac{2}{5}$, and $\frac{1}{2}$.

Changing to common denominators, we compare

$$\frac{10}{30}, \frac{12}{30}, \text{ and } \frac{15}{30}$$

Since 10 is smaller than 12, and 12 is smaller than 15, we have the order

$$\frac{1}{3}, \frac{2}{5}, \text{ and } \frac{1}{2}$$

► **Fractions, complex**   A **complex fraction** is a fraction with fractions in the numerator, denominator, or both. To simplify a complex fraction, we need to get it in the form $\frac{\frac{a}{b}}{\frac{c}{d}}$ and then apply the following:

$$\frac{\frac{a}{b}}{\frac{c}{d}} = \frac{a}{b} \div \frac{c}{d} = \frac{a}{b} \bullet \frac{d}{c} = \frac{ad}{bc} \qquad \text{See } \textbf{Fractions, division of.}$$

When a complex fraction is in this form, a simple rule for simplifying it is to "flip and multiply." For complex fractions in algebra, see **Rational Expressions, complex fractions of**.

F

Examples

1. Simplify $\dfrac{\frac{3}{8}}{\frac{3}{7}}$.

$$\frac{\frac{3}{8}}{\frac{3}{7}} = \frac{3}{8} \div \frac{3}{7} = \frac{3}{8} \bullet \frac{7}{3} = \frac{7}{8}$$

If you applied the rule "flip and multiply" in your mind, you would have had $\frac{21}{24} = \frac{7}{8}$. You might have even noticed that the 3's cancel and you could have gone straight to the answer $\frac{7}{8}$. Practice.

2. Simplify $\dfrac{\frac{3}{4} + \frac{7}{8}}{7 - \frac{1}{2}}$.

$$\frac{\frac{3}{4} + \frac{7}{8}}{7 - \frac{1}{2}} = \frac{\frac{6}{8} + \frac{7}{8}}{6\frac{2}{2} - \frac{1}{2}} = \frac{\frac{13}{8}}{6\frac{1}{2}} \quad \text{or} \quad = \frac{\frac{13}{8}}{\frac{13}{2}}$$

$$= \frac{13}{8} \div 6\frac{1}{2} \qquad\qquad = \frac{13}{8} \bullet \frac{2}{13}$$

$$= \frac{13}{8} \div \frac{13}{2} \qquad\qquad = \frac{2}{8}$$

$$= \frac{13}{8} \bullet \frac{2}{13} \qquad\qquad = \frac{1}{4}$$

$$= \frac{2}{8} = \frac{1}{4}$$

➤ **Fractions, division of**   To **divide fractions**, change any mixed numbers to improper fractions, then invert the divisor and multiply. This means, flip the second fraction and multiply. In the following the $\frac{1}{2}$ is the dividend, the $\frac{3}{4}$ is the divisor, and the $\frac{2}{3}$ is the quotient: $\frac{1}{2} \div \frac{3}{4} = \frac{2}{3}$. We never invert the divisor, the first fraction. When finished, reduce the answer if possible.

Examples

1. Simplify $\frac{1}{2} \div \frac{3}{4}$.

$$\frac{1}{2} \div \frac{3}{4} = \frac{1}{\underset{1}{\cancel{2}}} \bullet \frac{\overset{2}{\cancel{4}}}{3} = \frac{2}{3} \qquad \text{See \textbf{Fractions, multiplication of}.}$$

2. Simplify $\frac{2}{5} \div \frac{3}{10}$.

$$\frac{2}{3} \div \frac{3}{10} = \frac{2}{\underset{1}{\cancel{5}}} \bullet \frac{\overset{2}{\cancel{10}}}{3} = \frac{4}{3} = 1\frac{1}{3} \qquad \begin{array}{l}\text{See \textbf{Fractions, reducing}. Also}\\ \text{see \textbf{Fractions, multiplication of}.}\end{array}$$

3. Simplify $8 \div \frac{4}{5}$.

$$8 \div \frac{4}{5} = \frac{\overset{2}{\cancel{8}}}{1} \cdot \frac{5}{\underset{1}{\cancel{4}}} = \frac{10}{1} = 10$$ All numbers can be written with a 1 underneath.

4. Simplify $5\frac{3}{5} \div 7$.

$$5\frac{3}{5} \div 7 = \frac{28}{5} \div \frac{7}{1} = \frac{\overset{4}{\cancel{28}}}{5} \cdot \frac{1}{\underset{1}{\cancel{7}}} = \frac{4}{5}$$ See **Mixed Number**.

5. Simplify $4\frac{2}{5} \div 1\frac{1}{10}$.

$$= \frac{22}{5} \div \frac{11}{10} = \frac{22}{5} \cdot \frac{10}{11} = \frac{\overset{2}{\cancel{22}}}{\underset{1}{\cancel{5}}} \cdot \frac{\overset{2}{\cancel{10}}}{\underset{1}{\cancel{11}}} = \frac{4}{1} = 4$$

➤ **Fractions, equations with**   See **Rational Equations**.

➤ **Fractions, equivalent**   To change a fraction to an **equivalent (similar) fraction**, we either multiply or divide the numerator and denominator by the same number. If we multiply both the top and bottom of $\frac{3}{5}$ by 4, we get the equivalent fraction $\frac{12}{20}$. The fraction $\frac{12}{20}$ is equivalent to $\frac{3}{5}$.

Examples

1. Change $\frac{7}{25}$ to an equivalent fraction with a denominator of 100.

   To change 25 to 100, we multiply by 4. Therefore,

$$\frac{7}{25} = \frac{7}{25} \cdot \frac{4}{4} = \frac{28}{100}$$

   The fraction $\frac{28}{100}$ is equivalent to $\frac{7}{25}$.

2. Change $\frac{3}{8}$ to an equivalent fraction with a denominator of 48.

   To change 8 to 48, we multiply by 6. Therefore,

$$\frac{3}{8} = \frac{3}{8} \cdot \frac{6}{6} = \frac{18}{48}$$

   The fraction $\frac{18}{48}$ is equivalent to $\frac{3}{8}$.

3. Change $\frac{1}{5}$ to 30ths.

$$\frac{1}{5} = \frac{1}{5} \cdot \frac{6}{6} = \frac{6}{30}$$

4. Change $\frac{2}{3}$ and $\frac{3}{4}$ to 24ths.

$$\frac{2}{3} = \frac{2}{3} \cdot \frac{8}{8} = \frac{16}{24}$$
$$\frac{3}{4} = \frac{3}{4} \cdot \frac{6}{6} = \frac{18}{24}$$

➤ **Fractions, greater than or less than**   See **Fractions, comparing**.

➤ **Fractions, improper**   An **improper fraction** is a fraction whose numerator is either greater than or equal to the denominator. As a result, improper fractions either equal one or are greater than one. The following are improper fractions: $\frac{5}{5}$, $\frac{6}{5}$, $\frac{30}{2}$, and $\frac{41}{25}$.

To change an improper fraction to a mixed number, just divide and write the remainder as a fraction.

Example

1. Change $\frac{7}{2}$ to a mixed number.

Dividing 7 by 2, we have

$$\begin{array}{r} 3\frac{1}{2} \\ 2\overline{)7} \\ \underline{6} \\ 1 \end{array}$$

The improper fraction $\frac{7}{2}$ is equal to the mixed number $3\frac{1}{2}$. To change a mixed number to an improper fraction, see **Mixed Number**.

➤ **Fractions, inequalities**   See **Fractions, comparing** or the various articles under **Inequalities**.

➤ **Fractions, mixed**   See **Mixed Number**.

➤ **Fractions, multiplication of**   To **multiply fractions**, change any mixed numbers to improper fractions, then multiply and reduce. When multiplying, multiply the numerators then multiply the denominators.

Examples

1. Simplify $\frac{1}{2} \times \frac{3}{5}$.

$$\frac{1}{2} \times \frac{3}{5} = \frac{3}{10} \qquad \text{Multiply } 1 \times 3 \text{ then } 2 \times 5.$$

2. Simplify $\frac{5}{6} \times \frac{3}{8}$.

You can either multiply and then reduce:

$$\frac{5}{6} \times \frac{3}{8} = \frac{15}{48} = \frac{5}{16} \qquad \text{See **Fractions, reducing**.}$$

or you can reduce (cross cancel) and then multiply:

$$\frac{5}{\cancel{6}_2} \times \frac{\cancel{3}^1}{8} = \frac{5}{16}$$ See **Fractions, reducing**.

The 3 goes into 3 once and into 6 twice.

3. Cross cancel, then multiply $\frac{21}{10} \times \frac{15}{14}$.

$$\frac{\cancel{21}^3}{\cancel{10}_2} \times \frac{\cancel{15}^3}{\cancel{14}_2} = \frac{9}{4} = 2\frac{1}{4}$$ 7 goes into 21 and 14, 5 goes into 10 and 15.

4. Simplify $\frac{5}{8} \times 4$.

$$\frac{5}{8} \times 4 = \frac{5}{\cancel{8}_2} \times \frac{\cancel{4}^1}{1} = \frac{5}{2} = 2\frac{1}{2}$$ All numbers can be written with a 1 underneath.

5. Simplify $8 \times 5\frac{1}{4}$.

$$8 \times 5\frac{1}{4} = \frac{\cancel{8}^2}{1} \times \frac{21}{\cancel{4}_1} = \frac{42}{1} = 42$$ See **Mixed Number**.

To change $5\frac{1}{4}$ to $\frac{21}{4}$, multiply 5×4 then add 1.

6. Simplify $5\frac{1}{3} \times 4\frac{1}{2}$.

$$5\frac{1}{3} \times 4\frac{1}{2} = \frac{\cancel{16}^8}{\cancel{3}_1} \times \frac{\cancel{9}^3}{\cancel{2}_1} = \frac{24}{1} = 24$$

7. Simplify $4\frac{2}{3} \times 3\frac{3}{4}$.

$$4\frac{2}{3} \times 3\frac{3}{4} = \frac{\cancel{14}^7}{\cancel{3}_1} \times \frac{\cancel{15}^5}{\cancel{4}_2} = \frac{35}{2} = 17\frac{1}{2}$$

➤ **Fractions, proper**   A **proper fraction** is a fraction whose numerator is less than the denominator. As ax result, proper fractions are always less than one. The following are examples of proper fractions: $\frac{1}{2}, \frac{3}{4}, \frac{99}{100}$ and $\frac{1}{1000}$.

➤ **Fractions, reducing**   To **reduce a fraction** means to express it in lowest terms. Fractions that can be reduced occur in one of two forms. Either the numerator is smaller than the denominator or the numerator is larger than the denominator.

Fractions in which the numerator is smaller than the denominator can be reduced by dividing both the numerator and denominator by a common factor. If the result contains a common factor, repeat the process.

Fractions in which the numerator is larger than the denominator can be reduced by dividing the numerator by the denominator and writing the remainder, if there is one, as a fraction. If the remaining fraction has a common factor in the numerator and the denominator, reduce it further.

Examples

1. Reduce $\frac{4}{10}$.

   Since the numerator is smaller than the denominator, we need to divide each by 2, the common factor of 4 and 10. Hence,

   $$\frac{4}{10} = \frac{\overset{2}{\cancel{4}}}{\underset{5}{\cancel{10}}} = \frac{2}{5} \qquad \text{Divide 4 by 2 and divide 10 by 2.}$$

2. Reduce $\frac{12}{16}$.

   $$\frac{12}{16} = \frac{\overset{3}{\cancel{12}}}{\underset{4}{\cancel{16}}} = \frac{3}{4} \qquad \text{Divide 12 by 4 and divide 16 by 4.}$$

3. Reduce $\frac{18}{10}$.

   Since the numerator is larger than the denominator, we need to divide 18 by 10 and write the remainder as a fraction:

   $$\begin{array}{r} 1\frac{8}{10} \\ 10\overline{)18} \\ \underline{10} \\ 8 \end{array}$$

   Since the 8 and 10 have a common factor of 2, we can reduce further:

   $$1\frac{8}{10} = 1\frac{8}{10} = 1\frac{4}{5} \qquad \text{Divide 8 by 2 and divide 10 by 2.}$$

4. Reduce $\frac{35}{20}$.

   $$\begin{array}{r} 1\frac{15}{20} = 1\frac{3}{4} \\ 20\overline{)35} \\ \underline{20} \\ 15 \end{array}$$

5. Reduce $12\frac{4}{3}$.

Reduce the $\frac{4}{3}$ and add the whole number to the 12:

$$\begin{array}{r} 1\frac{1}{3} \\ 3\overline{)4} \\ \underline{3} \\ 1 \end{array}$$

Add the 1, from $1\frac{1}{3}$, to the 12 (the result is 13). Then, put the $\frac{1}{3}$ behind the 13.

$$12\frac{4}{3} = 13\frac{1}{3}$$

The reason this works is because

$$12\frac{4}{3} = 12 + \frac{4}{3} = 12 + 1\frac{1}{3} = 13\frac{1}{3}$$

6. Reduce $7\frac{5}{2}$.

Since $\frac{5}{2} = 2\frac{1}{2}$, we have $7\frac{5}{2} = 9\frac{1}{2}$. Add the 2 to the 7 and put the $\frac{1}{2}$ behind the 9.

This works because

$$7\frac{5}{2} = 7 + \frac{5}{2} = 7 + 2\frac{1}{2} = 9\frac{1}{2}.$$

➤ **Fractions, subtraction of**   To **subtract fractions**, the denominators must be the same. See **Common Denominator**. If the denominators are common, then subtract the numerators and reduce the answer, if possible. See **Fractions, reducing**. In some cases, you might need to borrow. See examples.

Examples

1. Subtract $\frac{1}{2} - \frac{1}{4}$.

$$\frac{1}{2} - \frac{1}{4} = \frac{1}{2} \bullet \frac{2}{2} - \frac{1}{4} = \frac{2}{4} - \frac{1}{4} = \frac{1}{4} \qquad \text{or} \qquad \begin{array}{r} \frac{1}{2} \bullet \frac{2}{2} = \frac{2}{4} \\ -\frac{1}{4} \qquad -\frac{1}{4} \\ \hline \frac{1}{4} \end{array}$$

To subtract the fractions, we had to have common denominators. The $\frac{1}{2}$ needed to be changed to 4ths. Multiplying the numerator and denominator by 2, we had $\frac{1}{2} \bullet \frac{2}{2} = \frac{2}{4}$. See **Fractions, equivalent**. Then it was clear what we did. We had $\frac{2}{4}$, then we took away $\frac{1}{4}$, and we were left with $\frac{1}{4}$.

2. Subtract $\frac{1}{2} - \frac{1}{3}$.

$$\frac{1}{2} - \frac{1}{3} = \frac{1}{2} \cdot \frac{3}{3} - \frac{1}{3} \cdot \frac{2}{2} = \frac{3}{6} - \frac{2}{6} = \frac{1}{6} \qquad \text{or} \qquad \begin{array}{l} \dfrac{1}{2} \cdot \dfrac{3}{3} = \dfrac{3}{6} \\[2mm] -\dfrac{1}{3} \cdot \dfrac{2}{2} = \dfrac{2}{6} \\ \hline \dfrac{1}{6} \end{array}$$

It makes no sense to subtract $\frac{1}{3}$ from $\frac{1}{2}$. However, if the fractions have the same denominator it does. Suppose a pie has been cut into six pieces. If you have 3 pieces and then give 2 pieces away, you have 1 piece left, $\frac{3}{6} - \frac{2}{6} = \frac{1}{6} - \frac{1}{3}$.

3. Subtract $9\frac{1}{2} - 4\frac{1}{3}$.

Notice the similarity to example 2.

$$9\frac{1}{2} - 4\frac{1}{3} = 9\frac{1}{2} \cdot \frac{3}{3} - 4\frac{1}{3} \cdot \frac{2}{2} = 9\frac{3}{6} - 4\frac{2}{6} = 5\frac{1}{6} \qquad \text{or} \qquad \begin{array}{l} 9\dfrac{1}{2} = 9\dfrac{1}{2} \cdot \dfrac{3}{3} = 9\dfrac{3}{6} \\[2mm] -4\dfrac{1}{3} = 4\dfrac{1}{3} \cdot \dfrac{2}{2} = 4\dfrac{2}{6} \\ \hline 5\dfrac{1}{6} \end{array}$$

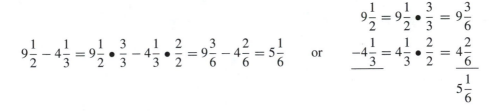

4. Subtract $8 - 2\frac{1}{3}$.

There isn't a fraction to subtract the $\frac{1}{3}$ from. So, we need to borrow from the 8. We will use the idea that 8 is the same as $7\frac{3}{3}$, since $7\frac{3}{3} = 7 + \frac{3}{3} = 7 + 1 = 8$. With this in mind,

$$8 - 2\frac{1}{3} = 7\frac{3}{3} - 2\frac{1}{3} = 5\frac{2}{3}$$

We used $\frac{3}{3}$ because the denominator of $\frac{1}{3}$ is 3.

5. Subtract $9 - 3\frac{2}{5}$.

We will use the idea that $9 = 8\frac{5}{5}$:

$$9 - 3\frac{2}{5} = 8\frac{5}{5} - 3\frac{2}{5} = 5\frac{3}{5}$$

6. Subtract $7\frac{3}{5} - 2$.

There is no fraction to subtract from the $\frac{3}{5}$. All we need to do is subtract the 2 from the 7:

$$7\frac{3}{5} - 2 = 5\frac{3}{5}$$

7. Subtract $8\frac{1}{2} - 2\frac{3}{4}$.

Changing to a common denominator, we have

$$8\frac{1}{2} - 2\frac{3}{4} = 8\frac{2}{4} - 2\frac{3}{4}$$

However, this presents a problem since we can't subtract $\frac{2}{4} - \frac{3}{4}$. To avoid this, we need to borrow from the 8. We will use the idea that $8 = 7\frac{4}{4}$. Then

$$8\frac{2}{4} - 2\frac{3}{4} = 7\frac{4}{4}8\frac{2}{4} - 2\frac{3}{4} = 7\frac{6}{4} - 2\frac{3}{4} = 5\frac{3}{4}$$

The $\frac{4}{4}$ and $\frac{2}{4}$ were added to get $\frac{6}{4}$. This allows $\frac{6}{4} - \frac{3}{4}$ to give $\frac{3}{4}$.

8. Subtract $10\frac{1}{3} - 2\frac{1}{2}$.

$$10\frac{1}{3} - 2\frac{1}{2} = 10\frac{1}{3} \cdot \frac{2}{2} - 2\frac{1}{2} \cdot \frac{3}{3} = 10\frac{2}{6} - 2\frac{3}{6}$$

$$= 9\frac{6}{6}10\frac{2}{6} - 2\frac{3}{6} \quad \text{or}$$

$$= 9\frac{8}{6} - 2\frac{3}{6}$$

$$= 7\frac{5}{6}$$

$$10\frac{1}{3} \cdot \frac{2}{2} = 9\frac{6}{6}10\frac{2}{6} = 9\frac{8}{6}$$

$$-2\frac{1}{2} \cdot \frac{3}{3} = \quad 2\frac{3}{6} = 2\frac{3}{6}$$

$$7\frac{5}{6}$$

➤ **Fractions to Decimals to Percents**   See **Fractions, change from fraction to decimal to percent**.

➤ **Frequency Table**   A **frequency table** is a chart that can be used to organize a set of data. It is used to arrange data into a form that can be easily evaluated.

Example

1. Make a frequency table for the following test scores and then determine their mean, median, and mode:

92	85	96	92	85	100	80	75	92	96
70	92	85	80	75	85	80	92	85	92
100	80	85	80	96					

Scores	Tally	Frequency	Sum
100	//	2	200
96	///	3	288
92	𝐻𝐿 /	6	552
85	𝐻𝐿 /	6	510
80	𝐻𝐿	5	400
75	//	2	150
70	/	1	70
		25	2,170

a. Mean $= \frac{\text{Sum}}{\text{Frequency}} = \frac{2,170}{25} = 86.8$.

b. Median $= 85$, since 85 is thirteenth, in the middle, of the frequency list.

c. Mode $= 92$ and 85, since these each occur most frequently, 6 times.

See **Mean (average)**, **Median**, or **Mode**.

➤ **Function**   A **function** is a relation in which each of the elements of the domain is paired with exactly one of the elements of the range. In reference to a set of points, if the domain is the set of x's and the range is the set of y's, then each x in the domain is paired with exactly one y in the range. Hence, each point has a different first coordinate. The second coordinates may or may not be the same.

In this manner, a function is a rule of correspondence that assigns to each element of one set exactly one element of another. See **Relation** or **Domain and Range**.

The equation $y = 2x + 1$ is a function. To see why, we need to look at some of its points. Consider the chart of $y = 2x + 1$ for x values of $-1, 0, 1$, and 2.

x	$y = 2x + 1$
-1	$2(-1) + 1 = -1$
0	$2(0) + 1 = 1$
1	$2(1) + 1 = 3$
2	$2(2) + 1 = 5$

The ordered pairs are $(-1, -1)$, $(0, -1)$, $(1, 3)$, and $(2, 5)$. Notice that none of the x values is the same. Each element of the domain is paired with exactly one element of the range. This is true for all values of x since each x substituted in $y = 2x + 1$ will yield a completely different y. Hence, $y = 2x + 1$ is a function.

It should be kept in mind that it is all right if two or more of the y values are the same. The following set of points is a function: $\{(3, 4), (5, 4), (7, 2), (8, 2)\}$. It is not all right if two or more of the x values are the same. The following set of points is not a function: $\{(4, 7), (4, 8), (6, 3), (2, 1)\}$.

To tell whether a graph is a function or not, we can apply the **vertical line test**. If a vertical line can be drawn intersecting the graph in more than one point, then the graph is not a function. In the following graphs, (a) and (b) are functions and (c) and (d) are not functions.

F

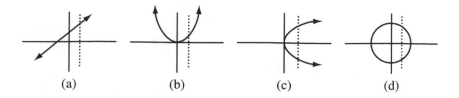

(a)   (b)   (c)   (d)

Vertical lines can cross (c) and (d) in more than one point. Hence, they are not functions. For articles related to functions, see **Relation**, **Function Notation**, or **Domain and Range**.

➤ **Function (two variables)**   A **function of two variables** is a correspondence between an ordered pair (x, y) and a unique real number $f(x, y)$. The set of ordered pairs that satisfies the correspondence is called the **domain** and the set of values corresponding to each (x, y) of the domain is called the **range**. Each x and y of the domain is called an **independent variable** and the corresponding value $z = f(x, y)$ is called the dependent variable. The equation $z = f(x, y)$ is called the function of x and y and is read "z equals f of x-y" or "z is a function of x and y." In a similar fashion, definitions of functions of three variables or more can be given.

Example

1. Evaluate $z = f(x, y) = \sqrt{4x - 2y - 6}$ for $f(x, y) = (2, -3)$ and give its domain.

To evaluate the function, we need to substitute each of the elements $x = 2$ and $y = -3$ into the equation:

$$
\begin{aligned}
z = f(x, y) &= \sqrt{4x - 2y - 6} \\
&= f(2, -3) = \sqrt{4(2) - 2(-3) - 6} \\
&= \sqrt{8 + 6 - 6} \\
&= \sqrt{8} = \sqrt{4 \bullet 2} \quad \text{See } \textbf{Radicals, simplifying (reducing)}. \\
&= 2\sqrt{2}
\end{aligned}
$$

To find the domain of the function, we use techniques similar to those described in the article **Domain and Range**. We recognize that square root is not defined for negative numbers. It is defined for $4x - 2y - 6 \geqslant 0$, positive numbers or zero. If we graph the inequality $4x - 2y - 6 \geqslant 0$ (see **Inequalities, graph in the plane**), we will find the values of x and y that when substituted into $f(x, y)$ will yield positive numbers. These are the values of the domain. Solving $4x - 2y - 6 \geqslant 0$ for y and then graphing, we have the domain of $f(x, y)$:

$$
\begin{aligned}
4x - 2y - 6 &\geqslant 0 \\
-2y &\geqslant -4x + 6 \\
\frac{-2y}{-2} &\leqslant \frac{-4}{-2} + \frac{6}{-2} \\
y &\leqslant 2x - 3
\end{aligned}
$$

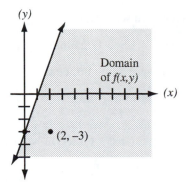

➤ **Function and Relation**   See either **Function** or **Relation**.

➤ **Function Notation**   A **function** is denoted by $y = f(x)$, read "f of x." In function notation, the equation $y = 2x + 1$ is written $f(x) = 2x + 1$. When $x = 3$, then $f(3) = 2(3) + 1 = 7$.

To get a feeling for $f(x)$, it is instrumental to see it in comparison with a chart:

x	$y = 2x + 1$	$f(x) = 2x + 1$
-1	$2(-1) + 1 = -1$	$f(-1) = 2(-1) + 1 = -1$
0	$2(0) + 1 = 1$	$f(0) = 2(0) + 1 = 1$
1	$2(1) + 1 = 3$	$f(1) = 2(1) + 1 = 3$
2	$2(2) + 1 = 5$	$f(2) = 2(2) + 1 = 5$

Each line in the chart represents a point. Each substitution in the function also represents a point. In the chart, the x column is the domain and the y column is the range. In function notation, the entries in the parentheses represent the domain and the answers represent the range.

The equation $f(2) = 5$ will now have more meaning. The 2 is an x value of the domain, the 5 is a y value of the range, and $(2, 5)$ is a point of the graph.

Example

1. If $f(x) = 3x^2 - x + 1$, find $f(-3)$.

$$f(x) = 3x^2 - x + 1$$
$$f(-3) = 3(-3)^2 - (-3) + 1 = 3 \bullet 9 + 3 + 1 = 31$$

Hence, $f(-3) = 31$, which means that $(-3, 31)$ is a point on the graph of $f(x) = 3x^2 - x + 1$.

➤ **Functions, composite**   See **Composite Functions**.

➤ **Functions, even and odd**   See **Even and Odd Functions**.

➤ **Functions, exponential**   See **Exponential Function**.

➤ **Functions, greatest integer**   See **Greatest Integer Function**.

➤ **Functions, inverse**   See **Inverse Functions**.

➤ **Functions, inverse trigonometric**   See **Trigonometry, inverse (arc) functions**.

➤ **Functions, logarithmic**   See **Logarithmic Function**.

➤ **Functions, odd**   See **Even and Odd Functions**.

➤ **Functions of Two Variables**   Just as $y = f(x)$ is a function of one variable, $z = f(x, y)$ is a function of two variables. See **Functions** or **Function Notation**.

Example

---

1. If $f(x, y) = 3x^2 y - 4x + y$, find $f(x, y)$ when $(x, y) = (2, 5)$.

$$f(x, y) = 3x^2 y - 4x + y$$
$$f(2, 5) = 3 \bullet (2)^2 \bullet 5 - 4(2) + 5 = 3 \bullet 4 \bullet 5 - 8 + 5$$
$$= 60 - 8 + 5 = 57$$

Hence, $f(2, 5) = 57$ is in the domain of the function, the 57 is in the range, and $(x, y, z) = (2, 5, 57)$ is a point on the graph of $f(x, y) = 3x^2 y - 4x + y$.

➤ **Functions or Identities of Negative Angles**   See **Trigonometry, even and odd functions (functions or identities of negative angles)**.

➤ **Functions, one-to-one**   See **One-to-One Function**.

➤ **Functions, periodic**   See **Periodic Function**.

➤ **Functions, polynomial**   See **Polynomial Functions**.

➤ **Functions, trigonometric**   See **Trigonometry, functions**.

➤ **Fundamental Theorem of Algebra**   If $P(x) = a_n x^n + a_{n-1} x^{n-1} + \cdots + a_1 x + a_0$ is a polynomial of degree greater than 0, then there is at least one complex root such that $P(r) = 0$. The root may be a real number.

The theorem does not give a means for finding the root. It merely states that it exists. Further, the theorem only applies to polynomial equations in which some or all of the coefficients are complex numbers. Some examples are $x^3 - 15x^2 + 7x - 2 = 0$, $\sqrt{5}x^{23} - \pi x^{17} + \sqrt{2} = 0$, and $x^2 - 3ix + (4+i) = 0$. For examples of polynomials with rational solutions, see **Factor Theorem**.

➤ **Fundamental Theorem of Calculus**   There are two fundamental theorems of calculus.

First Fundamental Theorem

Let $f(x)$ be a continuous function on the closed interval $[a, b]$. Then the **first fundamental theorem of calculus** states that

$$\int_a^b f(x)\, dx = F(x)\Big|_a^b = F(b) - F(a)$$

where a is the lower limit of integration, b is the upper limit of integration, and $F(x)$ is any function such that $F'(x) = f(x)$ for all values of x in the interval $[a, b]$.

The remarkable thing about the fundamental theorem is that it connects the concepts related to differentiation with those of integration. With this theorem at hand, we can evaluate a definite integral without involving the limit of a sum.

As far as the constant of integration is concerned, it is not used any further because

$$\int_a^b f(x)\, dx = F(x)\big|_a^b = \big[F(b) + C\big] - \big[F(a) + C\big] = F(b) - F(a)$$

Second Fundamental Theorem

Let $f(x)$ be a continuous function on an open interval containing a, then the **second fundamental theorem of calculus** states that for every x in the interval

$$\frac{d}{dx}\int_a^x f(t)\, dt = f(x)$$

The second fundamental theorem of calculus guarantees that if $f(x)$ is a continuous function, then it has an antiderivative.

Examples

1. Using the second fundamental theorem of calculus, find $F'(x)$ if

$$F(x) = \int_{-3}^x \left(t^3 - 4t + 7\right) dt$$

To find $F'(x)$, we need to determine $F'(x) = \frac{d}{dx}F(x)$. Using the second fundamental theorem of calculus, we have

$$F'(x) = \frac{d}{dx}F(x) = \frac{d}{dx}\int_{-3}^x \left(t^3 - 4t + 7\right) dt = x^3 - 4x + 7$$

To verify the answer, use the first fundamental theorem to find the integral of $\int_{-3}^x (t^3 - 4t + 7)\, dt$ and then differentiate the results.

The following examples use the first fundamental theorem of calculus.

2. Find the area under the curve $y = f(x) = x^2$, bounded by the x-axis, and the lines $x = 1$ and $x = 4$.

To find the area, we need to integrate $f(x)$ and then subtract the evaluations of the integral when $x = 4$ and $x = 1$:

$$A = \int_1^4 x^2\, dx = \frac{x^3}{3}\bigg|_1^4 = \frac{4^3}{3} - \frac{1^3}{3} = \frac{64}{3} - \frac{1}{3} = \frac{63}{3} = 21$$

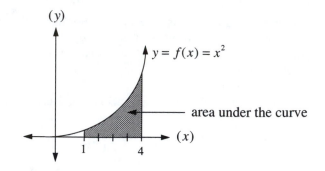

F

3. Evaluate $\int_1^4 \left(2x + \frac{1}{x^2}\right) dx$.

The method used to evaluate is the same as that used to find the area under the curve:

$$\int_1^4 \left(2x + \frac{1}{x^2}\right) dx = \int_1^4 2x\, dx + \int_1^4 \frac{1}{x^2}\, dx$$

$$= x^2\Big|_1^4 - \frac{1}{x}\Big|_1^4 \qquad \text{or} \qquad = \left(x^2 - \frac{1}{x}\right)\Big|_1^4$$

$$= (4^2 - 1^2) - \left(\frac{1}{4} - \frac{1}{1}\right) \qquad\qquad = \left(4^2 - \frac{1}{4}\right) - \left(1^2 - \frac{1}{1}\right)$$

$$= 15\frac{3}{4} \qquad\qquad\qquad\qquad = 15\frac{3}{4}$$

The evaluation of the integral(s) may be done separately or combined. The steps used in each of the integrals are shown below:

$$\int 2x\, dx = 2\int x\, dx = 2\left(\frac{x^2}{2}\right) = x^2$$

$$\int \frac{1}{x^2}\, dx = \int x^{-2}\, dx = \frac{x^{-2+1}}{-2+1} = \frac{x^{-1}}{-1} = -\frac{1}{x}$$

4. Find $\int_0^\pi \sin x\, dx$.

Using rule 8 of the rules for integration (see **Integration**), we have

$$\int_0^\pi \sin x\, dx = -\cos x\Big|_0^\pi = (-\cos \pi) - (-\cos 0)$$

$$= [-(-1)] - (-1) = 2$$

5. Calculate $\int_2^3 \frac{2x}{1+x^2}\, dx$ by using substitution.

To use substitution when integrating, we need to choose an appropriate substitution for u, differentiate u with respect to the other variable, solve the derivative for the differential of the other variable (for example, dx), evaluate u for each of the limits of integration (these become the new limits of integration), and then substitute all the new values and integrate:

$$\int_2^3 \frac{2x}{1+x^2}\,dx = \int_5^{10} \frac{2x}{u} \bullet \frac{du}{2x} \qquad \text{Let } u = 1+x^2, \text{ then } \frac{du}{dx} = 2x \text{ and } dx = \frac{du}{2x}.$$

$$= \int_5^{10} \frac{1}{u}\,du \qquad \begin{aligned} &\text{When } x = 3, \ u = 1 + (3)^2 = 10. \\ &\text{When } x = 2, \ u = 1 + (2)^2 = 5. \end{aligned}$$

$$= \ln u\,\big|_5^{10} = \ln 10 - \ln 5 = \ln \frac{10}{5} = \ln 2 \approx .693 \qquad \text{See } \textbf{Natural Logarithms}.$$

F

Gg

➤ **Geometric Mean**   The **geometric mean** or **mean proportional** of two numbers is $\sqrt{xy}$ or $-\sqrt{xy}$ where x and y are real numbers and $xy > 0$.

Example

1. Find the geometric mean of 2 and 6.

Letting $x = 2$ and $y = 6$, we have

$$\sqrt{xy} = \sqrt{2 \bullet 6} = \sqrt{12} = 2\sqrt{3} \qquad \text{See **Radicals, simplifying (reducing)**.}$$

The geometric mean of 2 and 6 is $2\sqrt{3}$ and $-2\sqrt{3}$.

This can be verified by finding the geometric means of the geometric progression 2, ?, 6. See **Geometric Means**. With $l = 6$, $a = 2$, and $n = 3$, we have

$$l = a \bullet r^{n-1}$$
$$6 = 2 \bullet r^{3-1}$$
$$6 = 2 \bullet r^{2}$$
$$3 = r^{2}$$
$$\sqrt{r^{2}} = \pm\sqrt{3}$$
$$r = \pm\sqrt{3}$$

When $r = +\sqrt{3}$, the geometric mean is $2(\sqrt{3}) = 2\sqrt{3}$.

When $r = -\sqrt{3}$, the geometric mean is $2(-\sqrt{3}) = -2\sqrt{3}$.

➤ **Geometric Means**   **Geometric means** are the terms between any two nonconsecutive terms of a geometric progression (sequence). See **Geometric Progression**. When the geometric means are found, a geometric progression is formed.

Example

1. Find the three geometric means between 5 and 80.

Since the progression formed will be a geometric progression, we can use the formula $l = a \bullet r^{n-1}$ where $l = 80$, $a = 5$, and $n = 5$. The variable n is 5 because there are five terms in the progression, 5, ?, ?, ?, 80. See **Geometric Progression**. Substituting and solving for r, we have

$$l = a \bullet r^{n-1}$$
$$80 = 5 \bullet r^{5-1}$$
$$80 = 5 \bullet r^4 \qquad \text{See } \textbf{Exponents, rules of.}$$
$$16 = r^4$$
$$\sqrt[4]{r^4} = \pm\sqrt[4]{16} \qquad \text{See } \textbf{Radicals, equations with.}$$
$$r = \pm 2$$

The equation $r^4 = 16$ can also be solved as an exponential equation

$$r^4 = 16$$
$$\left(r^4\right)^{1/4} = (16)^{1/4} \qquad \text{See } \textbf{Exponential Equation, solution of.}$$
$$r = \pm 2$$

When $r = +2$, the geometric means are

$$5(2) = 10 \qquad 5(2)^2 = 20 \qquad 5(2)^3 = 40$$

This is clearly seen in the geometric progression

$$5(2)^0, 5(2)^1, 5(2)^2, 5(2)^3, 5(2)^4 = 5, 10, 20, 40, 80$$

When $r = -2$, the geometric means are

$$5(-2) = -10 \qquad 5(-2)^2 = 20 \qquad 5(-2)^3 = -40$$

Hence, the three geometric means are 10, 20, and 40, or -10, 20, and -40.

▶ **Geometric Progression**   A **geometric progression** is a sequence in which each term is found by multiplying the preceding term by a constant. The constant is a common ratio that can be found by dividing any term by the previous term.

The nth term of a geometric progression where l is the last or nth term, a is the first term, and r is the common ratio (constant) is given by

$$l = a \bullet r^{n-1}$$

Examples

1. Which of the following sequences is not a geometric progression?

   a. $5, 10, 20, 40 \ldots$

   b. $-2, -8, -32, -128, \ldots$

   c. $2, 4, 6, 8, \ldots$

   The sequence in (a) is a geometric progression because each term divided by the previous has a common ratio of 2.

   The sequence in (b) is also a geometric progression with a constant of 4.

   The sequence in (c) is not a geometric progression because $4 \div 2 = 2$ is not the same as $6 \div 4 = 1\frac{1}{2}$.

2. Find the first four terms of the geometric progression whose first term is 12 and the common ratio is $\frac{1}{2}$.

   Each term in a geometric progression can be found by multiplying the preceding term by a constant. Since 12 is the first term of the progression, we can multiply it by $\frac{1}{2}$ to get the second term. Then we can multiply that by $\frac{1}{2}$ to get the third term, and so on.

   $$12\left(\frac{1}{2}\right) = 6 \qquad 6\left(\frac{1}{2}\right) = 3 \qquad 3\left(\frac{1}{2}\right) = \frac{3}{2}$$

   The first four terms of the progression are 12, 6, 3, and $\frac{3}{2}$.

3. Write the next three terms in the geometric progression $3, -6, 12, -24, \ldots$

   To find the next three terms in the progression, we need to know the common ratio. This can be found by dividing any two successive terms, $-6 \div 3 = -2$. Proceeding as we did in example 2, we have

   $$-24(-2) = 48 \qquad 48(-2) = -96 \qquad -96(-2) = 192$$

   The next three terms in the progression are 48, -96, and 192.

4. Find the ninth term of the geometric progression $3, -6, 12, \ldots$

   The common ratio of the progression is $-6 \div 3 = -2$, the first term is $a = 3$, and the nth term is $n = 9$. Substituting in the formula for the nth term, we have

   $$l = a \bullet r^{n-1} = 3(-2)^{9-1} = 3(-2)^8 = 3 \bullet 256 = 768$$

   The ninth term of the geometric progression $3, -6, 12, \ldots$ is 768. We can verify this by finding two more terms in example 3:

   $$192(-2) = -384 \qquad -384(-2) = 768$$

➤ **Geometric Proof**   See **Proof in Geometry**.

➤ **Geometric Sequence**   See **Geometric Progression**.

➤ **Geometric Series**   A **geometric series** is a sum of a given number of terms of a geometric progression. If a is the first term, r is the common ratio, and l is the last term, then the sum of the first n terms of a geometric series is given by

$$s = \frac{a - rl}{l - r} \quad \text{with } r \neq 1$$

Example

1. Find the sum of the first nine terms of the geometric progression $3, -6, 12 \ldots$.

   As a geometric series, we are considering the sum

   $$3 + (-6) + 12 + \cdots$$

   To find the sum of the first nine terms, we need to know the values of a, r, and l. The first term is $a = 3$, the common ratio is $r = -2$ since $-6 \div 2 = -2$, and the last term l is

   $$
   \begin{aligned}
   l &= a \bullet r^{n-1} \\
     &= 3 \bullet (-2)^{9-1} \quad \text{where the } n\text{th term } n = 9 \\
     &= 3 \bullet (-2)^8 \\
     &= 3 \bullet 256 \\
     &= 768 \quad\quad\quad\quad \text{See } \textbf{Geometric Progression}.
   \end{aligned}
   $$

   The sum is

   $$s = \frac{a - rl}{1 - r} = \frac{3 - (-2)(768)}{1 - (-2)} = \frac{3 + 1{,}536}{3} = 513$$

   We can verify this by adding the first nine terms. See **Geometric Progression**.

   $$
   \begin{aligned}
   &3 + (-6) + 12 + (-24) + 48 + (-96) + 192 + (-384) + 768 \\
   &= 3 - 6 + 12 - 24 + 48 - 96 + 192 - 384 + 768 \\
   &= 513
   \end{aligned}
   $$

➤ **Geometry**   **Geometry** is a branch of mathematics that is concerned with the properties of points, lines, and planes in space. It is based on axioms that are assumed, not proven. These in turn are used to prove certain statements or theorems. For demonstrations of proof, see **Proof in Geometry**.

As an example, suppose we wanted to prove the Pythagorean theorem, $a^2 + b^2 = c^2$. The geometric interpretation is given below:

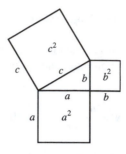

where a^2 represents the area of the square on side a, b^2 represents the area of the square on side b, and c^2 represents the area of the square on side c. The equation $a^2 + b^2 = c^2$ means that if you add the areas of the squares on the shorter sides (legs), you will get the area of the square on the longest side (hypotenuse). This occurs in a right triangle.

However, to prove this theorem, we would first have to discuss finding the area of a square and four-sided figures in general. However, before we could discuss four-sided figures, we would first have to discuss three-sided figures. However, before we could discuss three-sided figures, we would first have to discuss two-sided figures. However, before we could discuss two-sided figures, we would first have to discuss one-sided figures. And before that? Maybe points?

To prove the Pythagorean theorem, we should first discuss points, then lines, then angles, then triangles, and so on. It is this type of approach that Euclid (the founder of geometry) took in developing geometry. This method is so desirable that modern mathematics is based on it. It is called the **axiomatic method**.

➤ **Grade**   Grade, like slope, is a fraction of the rise over the run.

$$\text{grade} = \frac{\text{rise}}{\text{run}} \qquad \text{See \textbf{Slope}.}$$

The only difference is that the fraction, $\frac{\text{rise}}{\text{run}}$, is converted to a percent. As such, grade represents the percentage of steepness, so to speak, of the given situation.

Example

1. If the grade of a highway is 3%, how many horizontal feet does a car have to travel in order to have a vertical rise of 150 feet?

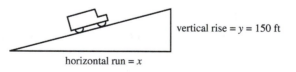

vertical rise = y = 150 ft

horizontal run = x

If we let x represent the horizontal run, then from the definition for grade we have

$$\text{grade} = \frac{\text{rise}}{\text{run}}$$
$$3\% = \frac{150}{x}$$

$$.03 = \frac{150}{x}$$

$$x = \frac{150}{.03} \qquad \text{See } \textbf{Cross Multiplication}.$$

$$x = 5,000 \text{ ft}$$

Since a mile is 5,280 feet, we see that the car has to travel a little less than a mile (5,000 ft) in order to have vertical rise of 150 ft.

$$\text{Check:} \qquad \text{grade} = \frac{\text{rise}}{\text{run}} = \frac{150}{5,000} = .03 = 3\%$$

▶ **Graph a Line**   See **Linear Equations, graphs of**. This article will include three methods of graphing: the charting method, the slope-intercept method, and the x-y intercept method.

▶ **Graphs, by charting**   See **Linear Equations, graphs of**.

▶ **Graphs, conic section**   For graphs of the various conics, see the article under each type, **Circle**, **Parabola**, **Ellipse**, or **Hyperbola**.

▶ **Graphs, exponential**   See **Exponential Function**.

▶ **Graphs, linear**   See **Linear Equations, graphs of**. This article will include three methods of graphing: the charting method, the slope-intercept method, and the x-y intercept method.

▶ **Graphs, logarithmic**   See **Logarithmic Function**.

▶ **Graphs, of inequalities in the plane**   See **Inequalities, graph in the plane**.

▶ **Graphs, of inequalities on a number line**   See **Inequalities, graph on a number line**.

▶ **Graphs, quadratic**   See **Quadratic Equations, graphs of**. This article will include two methods of graphing: the axis of symmetry method and the completion of the square method. Also see the article **Parabola**.

▶ **Graphs, translation or reflection of**   See **Translation or Reflection**. This allows for quick graphing of common equations.

▶ **Graphs, trigonometric**   See **Trigonometry, graphs of the six functions** or **Trigonometry, inverse (arc) functions**.

▶ **Greater Than and Less than**   See the articles listed under **Inequality**.

▶ **Greatest Common Factor**   The largest number that can divide evenly into two or more numbers is called the **greatest common factor of the numbers**. This should not be confused with the least common multiple. The least common multiple is the smallest number that is a multiple of two or more numbers. See **Multiple** or **Least Common Multiple**.

In algebra the concept of the greatest common factor is used to factor terms with common monomials. See **Factoring Common Monomials**.

Examples

1. Find the greatest common factor of 16 and 24.

   Think of numbers that divide evenly into both 16 and 24. Do you see that 8 is the largest that will divide into both? Since 8 is the largest number that can divide evenly into 16 and 24, it is the greatest common factor.

   In order to do a problem like this in the mind, the multiplication table must be completely memorized. See **Multiplication Table**.

   Another way to do this problem is to list the factors of 16 and 24 and then choose the greatest.

   For 16, the factors are 1, 2, 4, **8**, 16

   For 24, the factors are 1, 2, 3, 4, 6, **8**, 12, 24

   The common factors of 16 and 24 are 1, 2, 4, and 8. The greatest common factor is **8**.

   Yet another way to do the problem is to prime factor 16 and 24, then multiply the common prime factors. The product is the greatest common factor. See **Prime Factorization**.

   $$16 = 2 \bullet 8 = \mathbf{2 \bullet 2 \bullet 2} \bullet 2$$
   $$24 = 4 \bullet 6 = \mathbf{2 \bullet 2 \bullet 2} \bullet 3$$

   Since 2, 2, and 2 are common to both prime factorizations, $2 \bullet 2 \bullet 2 = 8$ is the greatest common factor.

2. Find the greatest common factor of 54 and 72.

   Thinking of all the numbers that divide evenly into both 54 and 72 can be difficult. If you came up with 18 as the largest, you're far better at it than I am.

   Listing all of the factors of 54 and 72 could become cumbersome. So, let's use prime factorization.

   $$54 = 6 \bullet 9 = \mathbf{2 \bullet 3 \bullet 3} \bullet 3$$
   $$72 = 8 \bullet 9 = 2 \bullet \mathbf{2 \bullet 2 \bullet 3 \bullet 3}$$

   The common prime factors are 2, 3, and 3. Therefore, $2 \bullet 3 \bullet 3 = 18$ is the greatest common factor.

➤ **Greatest Integer Function**   The function $f(x) = [x]$ is called the **greatest integer function**. It is sometimes referred to as a step function because its graph looks like a set of stairs.

   The term $[x]$ is called the greatest integer of x. It is sometimes referred to as "bracket x." The greatest integer of x is defined as the greatest integer not greater than x. In other words, it is the greatest integer that is less than or equal to x. Any way you say it, it's confusing. Examples usually help:

   $[5.3] = 5$, $[7.23] = 7$, $[6] = 6$, $[-1.7] = -2$ (since $-2 < -1.7$), $[-3.4] = -4$, and $[-6.35] = -7$

   There is very little utility for the greatest integer function. It is mostly used in calculus as an example of a function that is piecewise continuous.

Example

1. Graph $y = f(x) = [x]$.

x	$y = [x]$
-2	$[-2] = -2$
-1.5	$[-1.5] = -2$
-1	$[-1] = -1$
$-.5$	$[-.5] = -1$
0	$[0] = 0$
$.5$	$[.5] = 0$
1	$[1] = 1$
1.5	$[1.5] = 1$
2	$[2] = 2$
2.5	$[2.5] = 2$

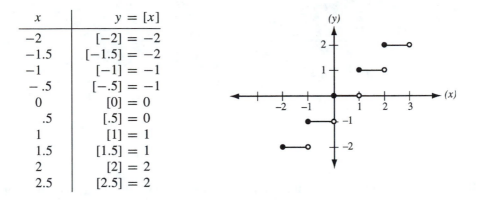

In the chart, we just chose one x value between -2 and -1, the -1.5. If we consider many x values between -2 and -1, we see that all of their bracket values are -2. Once we let $x = -1$, however, $[-1] = -1$. And all of the x values between -1 and 0 will give us y values of -1. But once we let $x = 0$, $[0] = 0$. And all of the x values between 0 and 1 will give us y values of 0. And so on.

G

➤ **Half-Angle Formulas**   The **half-angle formulas** are one of several sets of identities used in trigonometry. See **Trigonometry, identities**.

$$\sin\frac{\alpha}{2} = \pm\sqrt{\frac{1-\cos\alpha}{2}}$$

$$\cos\frac{\alpha}{2} = \pm\sqrt{\frac{1+\cos\alpha}{2}}$$

$$\tan\frac{\alpha}{2} = \pm\sqrt{\frac{1-\cos\alpha}{1+\cos\alpha}} = \frac{1-\cos\alpha}{\sin\alpha} = \frac{\sin\alpha}{1+\cos\alpha}$$

The $\pm$ signs are determined by the quadrant in which the function lies.

The half-angle formulas are derived from the double-angle formulas. See **Double-Angle Formulas**. In the following, let $A = \frac{\alpha}{2}$:

$$\cos 2A = 1 - 2\sin^2 A \qquad\qquad \cos 2A = 2\cos^2 A - 1$$

$$\cos 2\left(\frac{\alpha}{2}\right) = 1 - 2\sin^2\left(\frac{\alpha}{2}\right) \qquad \cos 2\left(\frac{\alpha}{2}\right) = 2\cos^2\left(\frac{\alpha}{2}\right) - 1$$

$$\cos\alpha = 1 - 2\sin^2\frac{\alpha}{2} \qquad\qquad \cos\alpha = 2\cos^2\frac{\alpha}{2} - 1$$

$$2\sin^2\frac{\alpha}{2} = 1 - \cos\alpha \qquad\qquad 2\cos^2\frac{\alpha}{2} = 1 + \cos\alpha$$

$$\sin^2\frac{\alpha}{2} = \frac{1-\cos\alpha}{2} \qquad\qquad \cos^2\frac{\alpha}{2} = \frac{1+\cos\alpha}{2}$$

$$\sin\frac{\alpha}{2} = \pm\sqrt{\frac{1-\cos\alpha}{2}} \qquad\qquad \cos\frac{\alpha}{2} = \pm\sqrt{\frac{1+\cos\alpha}{2}}$$

Further, from the reciprocal identity $\tan A = \frac{\sin A}{\cos A}$, we have

$$\tan\frac{\alpha}{2} = \frac{\sin\frac{\alpha}{2}}{\cos\frac{\alpha}{2}} \qquad\qquad\qquad \text{See **Reciprocal Identities**.}$$

H

$$\tan\frac{\alpha}{2} = \frac{\pm\sqrt{\frac{1-\cos\alpha}{2}}}{\pm\sqrt{\frac{1+\cos\alpha}{2}}}$$

$$\tan\frac{\alpha}{2} = \pm\sqrt{\frac{1-\cos\alpha}{1+\cos\alpha}}$$ See **Radicals, rules for** and **Fractions, complex.**

Also $\tan\dfrac{\alpha}{2} = \dfrac{1-\cos\alpha}{\sin\alpha} = \dfrac{\sin\alpha}{1+\cos\alpha}$ See example 2.

Examples

1. Find the exact value of cos 105°.

Since 105° is in the second quadrant, we will use the negative square root. See **Trigonometry, signs of the six functions**.

$$\cos 105° = \cos\left(\frac{210°}{2}\right) = -\sqrt{\frac{1+\cos 210°}{2}}$$

$$= -\sqrt{\frac{1+\frac{-\sqrt{3}}{2}}{2}}$$ See **Special Triangles.**

$$= -\sqrt{\frac{\frac{2-\sqrt{3}}{2}}{2}}$$ See **Rational Expressions, sums and differences.**

$$= -\sqrt{\frac{2-\sqrt{3}}{4}}$$ See **Rational Expressions, complex fractions of.**

$$= \frac{-\sqrt{2-\sqrt{3}}}{2}$$ See **Radicals, simplifying (reducing).**

We can verify this by using a calculator or trigonometry table. By either we know cos 105° = −.2588. Simplifying our example, we have

$$\cos 105° = \frac{-\sqrt{2-\sqrt{3}}}{2} = \frac{-\sqrt{2-1.7321}}{2} = \frac{-\sqrt{.2679}}{2} = \frac{-.5176}{2} = -.2588$$

2. Prove $\tan\frac{\alpha}{2} = \frac{1-\cos\alpha}{\sin\alpha}$.

From the identity for $\tan\frac{\alpha}{2}$, we know $\tan\frac{\alpha}{2} = \pm\sqrt{\frac{1-\cos\alpha}{1+\cos\alpha}}$.

To simplify, we will multiply the inside fraction by 1, but let $1 = \frac{1-\cos\alpha}{1-\cos\alpha}$:

$$\tan\frac{\alpha}{2} = \pm\sqrt{\frac{1-\cos\alpha}{1+\cos\alpha} \bullet 1} = \pm\sqrt{\frac{1-\cos\alpha}{1+\cos\alpha} \bullet \frac{1-\cos\alpha}{1-\cos\alpha}} = \pm\sqrt{\frac{(1-\cos\alpha)^2}{1-\cos^2\alpha}}$$

H

By the first Pythagorean identity, we know that $1 - \cos^2 \alpha = \sin^2 \alpha$. See **Pythagorean Identities**. Substituting, we have

$$\tan \frac{\alpha}{2} = \pm \sqrt{\frac{(1 - \cos \alpha)^2}{1 - \cos^2 \alpha}} = \pm \sqrt{\frac{(1 - \cos \alpha)^2}{\sin^2 \alpha}} = \pm \frac{\sqrt{(1 - \cos \alpha)^2}}{\sqrt{\sin^2 \alpha}} = \frac{1 - \cos \alpha}{\sin \alpha}$$

We dispense with the $\pm$ because $(1 - \cos \alpha)^2$ is a positive number and so is $\sin^2 \alpha$.

In retrospect, we can see why we let $1 = \frac{1 - \cos \alpha}{1 - \cos \alpha}$. To get $\sin \alpha$ in the denominator, we needed $\sqrt{\sin^2 \alpha}$. We knew this by looking at $\sqrt{\frac{1 - \cos \alpha}{1 + \cos \alpha}}$. Compare the original with the end result and you can see why.

$$\begin{array}{cc} \text{Original} & \text{End Result} \\ \pm \sqrt{\dfrac{1 - \cos \alpha}{1 + \cos \alpha}} & \dfrac{1 - \cos \alpha}{\sin \alpha} \end{array}$$

To change the $1 + \cos \alpha$ into $\sin \alpha$, we would have to multiply it by $1 - \cos \alpha$, to get $1 - \cos^2 \alpha$, which is $\sin^2 \alpha$. Of course, we can't just multiply by $1 - \cos \alpha$. We can, however, multiply by 1 and let $1 = \frac{1 - \cos \alpha}{1 - \cos \alpha}$ and see what happens. This worked out fine, since we got $(1 - \cos \alpha)^2$ in the numerator where $\sqrt{(1 - \cos \alpha)^2} = 1 - \cos \alpha$. Therefore,

$$\tan \frac{\alpha}{2} = \pm \sqrt{\frac{1 - \cos \alpha}{1 + \cos \alpha}} = \frac{1 - \cos \alpha}{\sin \alpha}$$

A similar proof can be done to show that $\tan \frac{\alpha}{2} = \frac{\sin \alpha}{1 + \cos \alpha}$. This time we let $1 = \frac{1 + \cos \alpha}{1 + \cos \alpha}$. Try it.

➤ **Half-Plane**   When a line is drawn in the plane, it divides the plane into two **half-planes**. If points A and B are on opposite sides of a line l, then A is in one half-plane and B is in the other.

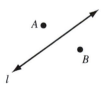

➤ **Horizontal and Vertical Lines**   **Horizontal and vertical lines** are not dealt with in the same fashion as other lines. This occurs because the slopes of each are either 0 or no slope, respectively. We will look at each separately.

Horizontal Line   From the graph we can see that a horizontal line has y-coordinates that are the same. Notice that the y-coordinate and the y-intercept are the same. We will show that the equation of a horizontal line is derived from this information.

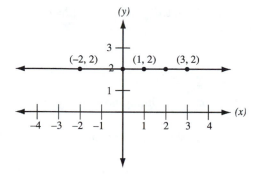

Using the points $(1, 2)$ and $(3, 2)$, let's find the equation of the horizontal line. See **Linear Equation, finding the equation of a line**, Two-Point Method. First, we find the slope:

$$m = \frac{y_2 - y_1}{x_2 - x_1} = \frac{2 - 2}{3 - 1} = \frac{0}{2} = 0$$

Now, using the point $(3, 2)$ with $m = 0$, we can find the equation. Let's use the point-slope method. See **Linear Equation, finding the equation of a line**, Point-Slope Method.

$$y_2 - y_1 = m(x_2 - x_1)$$
$$y - 2 = 0(x - 3)$$
$$y - 2 = 0$$
$$y = 2$$

Notice that the equation $y = 2$ is the same as the y-intercept. In fact, this is true for all horizontal lines. The equation of a horizontal line is the same as its y-intercept or y-coordinate. Also, since the y-coordinate of each point is the same, the slope of all horizontal lines is $m = 0$.

Vertical Lines   From the graph, we can see that a vertical line has x-coordinates that are the same. Notice that the x-coordinate and the x-intercept are the same. Similar to the approach with the horizontal line, we will show that the equation of a vertical line is derived from this information.

H

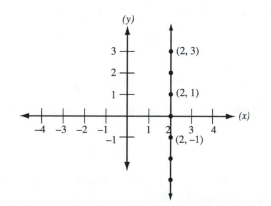

Using the points (2, 1) and (2, 3), let's find the equation of the vertical line. We first need to find the slope:

$$m = \frac{y_2 - y_1}{x_2 - x_1} = \frac{3 - 1}{2 - 2} = \frac{2}{0}$$

$$= \text{undefined or } \infty \text{ or, simply, no slope} \qquad \text{(can't divide by zero)}$$

If we try to find the equation using (2, 3) and $m = \infty$, we develop problems. The point-slope method doesn't work because $y - 3 = \infty(x - 2)$ doesn't make sense. The slope-intercept method as well can't be used since $3 = \infty(2) + b$ doesn't make sense. What does make sense is a chart of the graph.

Notice in the graph that for all values of y, the x value is always 2. Charting this, we see that all the points of the graph are accounted for. For all y, $x = 2$. Hence, $x = 2$ is the equation of the vertical line.

$x = 2$	y
⋮	⋮
2	−2
2	−1
2	0
2	1
2	2
⋮	⋮

Notice that the equation $x = 2$ is the same as the x-intercept. This is true for all vertical lines. The equation of a vertical line is the same as its x-intercept. Also, since the x-coordinate of each point is the same, the slope of all vertical lines is no slope.

Examples

1. What is the equation and slope of the horizontal line through (6, 8)?

   Since the equation of a horizontal line is the same as its y-coordinate, the equation is $y = 8$. The slope is $m = 0$.

2. What is the equation and slope of the vertical line through (6, 8)?

   Since the equation of a vertical line is the same as its x-coordinate, the equation is $x = 6$. The slope is $m = $ no slope.

3. Graph $y = -2$.

   Since the variable in the equation is y, the graph is horizontal. Its y-intercept is -2. So, we draw a horizontal line through -2 on the y-axis.

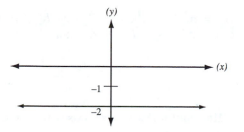

4. Graph $x = -3$.

Since the variable in the equation is x, the graph is vertical. Its x-intercept is -3. So, we draw a vertical line through -3 on the x-axis.

For slopes of horizontal and vertical lines, see **Slope**.

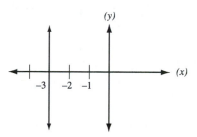

➤ **Horizontal Axis**   In the coordinate plane, the **horizontal axis** is the x-axis. The vertical axis is the y-axis. See **Coordinate Plane**.

➤ **Hyperbola**   A **hyperbola** is the set of all points in the plane such that the difference of the distances from two fixed points is a constant. Each of the two points is called a **focus**. The line passing through the foci is called the **principal axis**. On each branch of the hyperbola, the point nearest the center is the vertex, plural vertices.

The equation of a hyperbola depends on the orientation of the **transverse axis** (the segment of length $2a$ joining the vertices) and the **conjugate axis** (the segment of length $2b$ perpendicular to the transverse axis).

If the transverse axis is horizontal, the equation of the hyperbola is either

$$\frac{x^2}{a^2} - \frac{y^2}{b^2} = 1 \quad \text{or} \quad \frac{(x-h)^2}{a^2} - \frac{(y-k)^2}{b^2} = 1$$

Here, and in the following definition, the first equation has a center at the origin $(0, 0)$ and the second has a center at (h, k). Either of the equations is called the standard form of the hyperbola. The hyperbola has not been rotated.

If the transverse axis is vertical, the equation of the hyperbola is either

$$\frac{y^2}{a^2} - \frac{x^2}{b^2} = 1 \quad \text{or} \quad \frac{(y-k)^2}{a^2} - \frac{(x-h)^2}{b^2} = 1$$

The transverse axis is horizontal if the x^2 term is the leading (positive) term of the equation. It is vertical if the y^2 term is the leading (positive) term of the equation.

In all of the equations $c^2 = a^2 + b^2$ where $a^2 \neq 0$ and $b^2 \neq 0$, the length of the transverse axis is $2a$ and the length of the conjugate axis is $2b$. The equation of eccentricity is

$$e = \frac{c}{a} = \frac{\sqrt{a^2 + b^2}}{a}$$

Horizontal Transverse Axis with Center $(0, 0)$

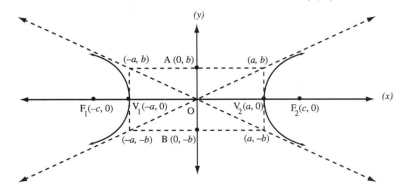

The **vertices** are V_1 and V_2. The **foci** (plural of focus) are F_1 and F_2. The **transverse axis** is $\overline{V_1 V_2}$. The **conjugate axis** is $\overline{AB}$. The **center** of the hyperbola is the point O. There are two **branches** to the hyperbola. Each branch recedes from each vertex and approaches the **asymptotes**. These are the lines the branches of the hyperbola approach as the graph recedes from the center. The asymptotes are the extended diagonals of the rectangle with vertices $(-a, -b)$, $(-a, b)$, (a, b), and $(a, -b)$. The **equations of the asymptotes** are $y = \frac{b}{a}x$ and $y = -\frac{b}{a}x$.

Vertical Transverse Axis with Center $(0, 0)$

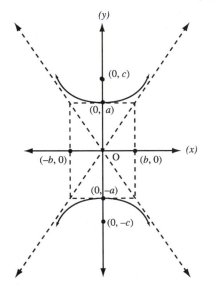

The vertices are $(0, a)$ and $(0, -a)$. The foci are $(0, c)$ and $(0, -c)$. The equations of the asymptotes are $y = \frac{a}{b}x$ and $y = -\frac{b}{a}x$.

Examples

1. Find the vertices, foci, length of the transverse axis, length of the conjugate axis, eccentricity, equation of the asymptotes, and graph of $\frac{x^2}{25} - \frac{y^2}{9} = 1$.

Since the x^2 term is the leading term of the equation, we know the transverse axis is horizontal. Also, comparing $\frac{x^2}{a^2} - \frac{y^2}{b^2} = 1$ with $\frac{x^2}{25} - \frac{y^2}{9} = 1$, we know that $a^2 = 25$ and $b^2 = 9$.

Vertices: Since $a^2 = 25$, $a = 5$. Therefore, the vertices are $(5, 0)$ and $(-5, 0)$.

Foci: From $c^2 = a^2 + b^2$ we have $c^2 = 25 + 9 = 34$ or $c = \sqrt{34}$. Therefore, the foci are $(\sqrt{34}, 0)$ and $(-\sqrt{34}, 0)$. In decimal form, they are approximately $(5.8, 0)$ and $(-5.8, 0)$.

Transverse Axis: The length of the transverse axis is $2a = 2(5) = 10$.

Conjugate Axis: Since $b^2 = 9$, $b = 3$ and the length of the conjugate axis is $2b = 2(3) = 6$.

Eccentricity: The eccentricity is $e = \frac{c}{a} = \frac{\sqrt{34}}{5}$.

Asymptotes: The equations of the asymptotes are

$$y = \frac{b}{a}x = \frac{3}{5}x \qquad \text{and} \qquad y = -\frac{b}{a}x = -\frac{3}{5}x$$

Graph: Use the $a = 5$ and the $b = 3$ to graph the rectangle. Then draw the asymptotes and sketch in the graph. Once the graph is drawn, plot the foci.

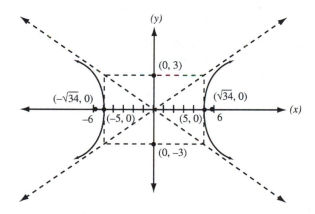

2. Draw the graph of $\frac{(y-1)^2}{16} - \frac{(x+2)^2}{9} = 1$.

Since the y^2 term is the leading term of the equation, we know the transverse axis is vertical. Comparing $\frac{(y-1)^2}{16} - \frac{(x+2)^2}{9} = 1$ with $\frac{(y-k)^2}{a^2} - \frac{(x-h)^2}{b^2} = 1$ we know that $a^2 = 16$ or $a = 4$, $b^2 = 9$ or $b = 3$, and the center $(h, k) = (-2, 1)$. This is because $y - k$ and $x - h$ have negative signs. When we take the h and k out of $y - 1$ and $x + 2$, they will be opposite in sign, hence, $(-2, -1)$.

To graph the equation, we first plot the center, then plot the a up 4 and down 4 from the center since the transverse axis is vertical, then plot the b left 3 and right 3 from the center, then draw in the rectangle, then draw in the asymptotes (extended diagonals of the rectangle), and then sketch the graph.

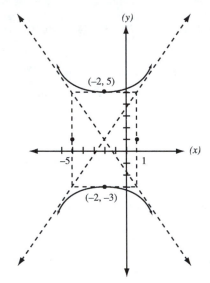

3. Write $16x^2 - 9y^2 + 64x + 18y + 199 = 0$ in standard form.

To write the equation in standard form means we want it in the form

$$\frac{(x-h)^2}{a^2} - \frac{(y-k)^2}{b^2} = 1 \qquad \text{or} \qquad \frac{(y-k)^2}{a^2} - \frac{(x-h)^2}{b^2} = 1$$

To do this, we need to group the x terms and y terms, respectively, complete the square on them, and then divide both sides of the equation by the number on the right side of the equal sign. See **Completing the Square**.

$$16x^2 - 9y^2 + 64x + 18y + 199 = 0$$
$$16x^2 + 64x - 9y^2 + 18y = -199$$
$$16(x^2 + 4x) - 9(y^2 - 2y) = -199$$
$$16(x^2 + 4x + 4) - 9(y^2 - 2y + 1) = -199 + 64 - 9 \qquad \text{See below.}$$
$$16(x+2)^2 - 9(y-1)^2 = -144$$
$$\frac{16(x+2)^2}{-144} - \frac{-9(y-1)^2}{-144} = \frac{-144}{-144}$$
$$\frac{-(x+2)^2}{9} + \frac{(y-1)^2}{16} = 1$$
$$\frac{(y-1)^2}{16} - \frac{(x+2)^2}{9} = 1 \qquad \qquad \text{This is the equation we graphed in example 2.}$$

How does completion of the square work? Between the third and fourth lines this is what happened

$$16(x^2 + 4x) - 9(y^2 - 2y) = -199 \qquad \text{3rd line}$$

$$16(x^2 + 4x + 4 - 4) - 9(y^2 - 2y + 1 - 1) = -199$$

$$16(x^2 + 4x + 4) - 64 - 9(y^2 - 2y + 1) + 9 = -199$$

$$9(x^2 + 4x + 4) - 9(y^2 - 2y + 1) = -199 + 64 - 9 \qquad \text{4th line}$$

We complete the square here a little differently than is done when solving equations. We still take half of the coefficient of the x (or y) and square it. But we add and subtract it, rather than add it to both sides as we would have done in solving equations.

The first three terms will now factor, but the fourth term is in the way. We remove it by multiplying it by the leading coefficient and writing it outside of the parenthesis. Now we can continue to put the equation into standard form.

4. Find the equation of the hyperbola with a horizontal transverse axis of length 8, eccentricity of $\frac{5}{4}$, and center $(0, 0)$.

    From the transverse axis, we have $2a = 8$ or $a = 4$. From the eccentricity, we have $e = \frac{c}{a} = \frac{5}{4}$, so $c = 5$. From the equation $c^2 = a^2 + b^2$, we can find b. Substituting, we have

    $$c^2 = a^2 + b^2$$
    $$5^2 = 4^2 + b^2$$
    $$b^2 = 25 - 16$$
    $$b^2 = 9$$
    $$b = 3$$

Since the center is $(0, 0)$, we know $h = 0$ and $k = 0$. Hence, we have the equation

$$\frac{x^2}{a^2} - \frac{y^2}{b^2} = 1$$

$$\frac{x^2}{4^2} - \frac{y^2}{3^2} = 1$$

$$\frac{x^2}{16} - \frac{y^2}{9} = 1$$

The equation $y = \frac{k}{x}$ or $xy = k$ with $k \neq 0$ is called a rectangular hyperbola. Its center is the origin and its asymptotes are the x- and y-axes.

5. Graph the rectangular hyperbola $xy = 4$.

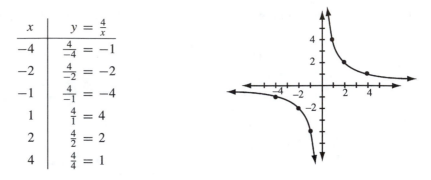

x	$y = \frac{4}{x}$
-4	$\frac{4}{-4} = -1$
-2	$\frac{4}{-2} = -2$
-1	$\frac{4}{-1} = -4$
1	$\frac{4}{1} = 4$
2	$\frac{4}{2} = 2$
4	$\frac{4}{4} = 1$

➤ **Hypothesis and Conclusion**   In a conditional statement $p \rightarrow q$ (see **Conditional Statement**), the if-part (p) is called the hypothesis and the then-part (q) is called the **conclusion**. The hypothesis describes the facts or figures related to the statement, whereas the conclusion describes what is to be proven about the facts or figures. In a deductive proof, the hypothesis is the **given** and the conclusion is the **prove**. For examples, see **Proof in Geometry**.

Examples

1. State the hypothesis and conclusion of the following statement: If it rains, then the tennis match will be postponed.

   Since the hypothesis is the if-part of the statement and the conclusion is the then-part, we have

   Hypothesis: If it rains

   Conclusion: then the tennis match will be postponed

2. State the hypothesis and conclusion of the following statement: An isosceles triangle has two congruent sides.

   We first need to write the statement in if-then form, then we can determine the hypothesis and conclusion. In if-then form the statement is: If a triangle is isosceles, then it has two congruent sides.

   Hypothesis: If a triangle is isosceles

   Conclusion: then it has two congruent sides

H

➤ **Identities in Trigonometry**   See **Trigonometry, identities**.

➤ **Identity, additive**   See **Additive Identity** or **Properties of the Real Number System**.

➤ **Identity, multiplicative**   See **Multiplicative Identity** or **Properties of the Real Number System**.

➤ **If and Only If Statement**   An **if and only if statement** is a biconditional statement in which both the conditional statement and its converse are written in one statement. In a biconditional statement, either the conditional statement and its converse are both true, or both false. See **Biconditional (equivalent) Statement**.

➤ **If-Then Statement**   See **Conditional Statement**.

➤ **Imaginary Axis**   The **imaginary axis** is the vertical axis in the complex plane. The **horizontal axis** is the real axis. See **Complex Plane**.

➤ **Imaginary Number**   Complex numbers such as $-5 - 2i$, $-6i\sqrt{2}$, or $-6\sqrt{2}i$ are **imaginary numbers** where $i = \sqrt{-1}$. See **Complex Numbers** or **Complex Plane**.

➤ **Implicit Differentiation**   See **Derivative**, example 8.

➤ **Improper Fraction**   An **improper fraction** is a fraction whose numerator is either greater than or equal to the denominator. As a result, improper fractions either equal one or are greater than one. The following are improper fractions: $\frac{5}{6}$, $\frac{6}{5}$, $\frac{30}{2}$, and $\frac{41}{25}$. For more, see **Fractions, improper**.

➤ **Inconsistent System**   If a system of linear equations has no solution, then it is an **inconsistent system**. If the system is composed of two lines, then what this definition means is that the lines don't cross each other. In other words, they are parallel. See **Systems of Linear Equations**.

➤ **Increasing and Decreasing Functions**   A function is **increasing** on an interval if, for any x_1 and x_2 in the interval, $f(x_1) < f(x_2)$ when $x_1 < x_2$. A function is **decreasing** on an interval if, for any x_1 and x_2 in the interval, $f(x_1) > f(x_2)$ when $x_1 < x_2$.

Simply, this means that as you look at a graph from left to right and it goes up, then it is increasing. If it goes down, then it is decreasing. The following graph is increasing from 0 to 4 and decreasing from 4 to 8.

I

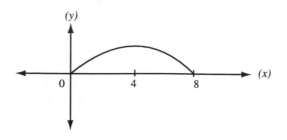

➤ **Indefinite Integral**   See **Integration**.

➤ **Independent System**   If a system of linear equations has one solution, then it is an **independent system**. If the system is composed of two lines, then what this definition means is that the lines cross each other. In other words, they intersect. See **Systems of Linear Equations**.

➤ **Independent Variable**   In the equation $y = 3x + 4$, the x is called the **independent variable**. It is independent of y. In contrast, the y variable is called the **dependent variable**. Its value depends on the value of x. When calculating values of the equation, the x is chosen independently and the value of y is then determined. It depends on the value of x.

➤ **Inequalities, absolute value**   See **Absolute Value Inequalities**.

➤ **Inequalities, graph in the plane**

To graph **inequalities** in the plane, we can use individual values of y to describe the shading of the graph, or we can just shade the graph by using the $>$ or $<$ signs. In this article, we will use the latter.

If we are asked to graph the inequality $y \geqslant 2x + 1$, we are being asked to describe what area of the plane is greater than the line $2x + 1$ and includes the line $2x + 1$. To do this, we need to first graph the line and then shade in the area that is greater than the line.

Graphing the line by the slope-intercept method, we have

To graph the equation by the slope-intercept method, see **Linear Equations, graphs of**. The graph of $y \leqslant 2x + 1$ is

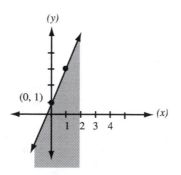

From this, we see that if we want to graph >, shade above the line. If we want to graph <, shade below the line.

In the following the > and < signs are described for all possible orientations of a line.

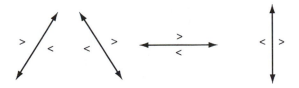

The only line that does not follow the above (>) and below (<) is the vertical. For the vertical, it is left (<) and right (>) just as the directions are on a number line.

Examples

1. Graph the inequality $y < 3x + 2$ in the plane.

Since the inequality sign is strictly less than, the graph does not include the line. To point this out, we use a dotted line instead of a solid line. This is similar to graphing on a number line where we use an open dot instead of a solid dot or a parenthesis instead of a bracket.

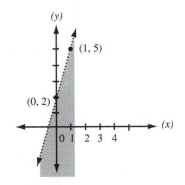

2. Graph $-2(x + 2y) + 3y \leqslant 1$ in the plane.

We first need to solve the inequality for y. See **Inequalities, linear solutions of**.

$$-2(x + 2y) + 3y \leqslant 1$$
$$-2x - 4y + 3y \leqslant 1$$
$$-y \leqslant 2x + 1$$
$$\frac{-y}{-1} \oslash \frac{2x}{-1} + \frac{1}{-1} \qquad \text{Remember to change the direction}$$
$$y \geqslant -2x - 1 \qquad \text{of the inequality sign when you divide by a negative.}$$

Now we can graph $y \geqslant -2x - 1$.

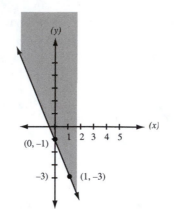

3. Graph $x < 2$ in the plane. Then graph $y < 2$.

We first graph the line $x = 2$. See **Linear Equations, graphs of**. Remember, since the inequality is strictly less than, the line is dotted. The shade is left.

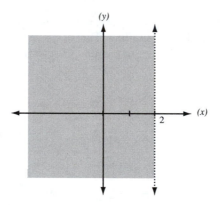

For $y < 2$, we have

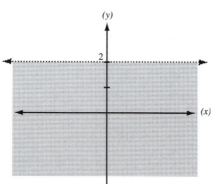

4. Graph $y < x^2 - 6x + 5$ in the plane.

If you have come to this example from the cross reference **Inequalities, quadratic graphs in the plane**, there are a couple of things we should mention first. We are using the concept that if an inequality is less than, we shade below the graph of the equation. If greater than, we shade above. Also, if an inequality sign is strictly $<$ or $>$, we draw the graph with a dotted curve. If $\leqslant$ or $\geqslant$, we draw a solid curve.

To graph the inequality $y < x^2 - 6x + 5$, we first need to graph the equation $y = x^2 - 6x + 5$. See **Quadratic Equations, graphs of**. We will use the axis of symmetry method.

x-intercept	**Vertex**	
$x^2 - 6x + 5 = 0$	$x = \dfrac{-b}{2a}$	$y = x^2 - 6x + 5$
$(x-1)(x-5) = 0$	$x = \dfrac{-(-6)}{2(1)}$	$y = 3^2 - 6(3) + 5$
$x = 1 \quad x = 5$	$x = 3$	$y = -4$

$V(3, -4)$

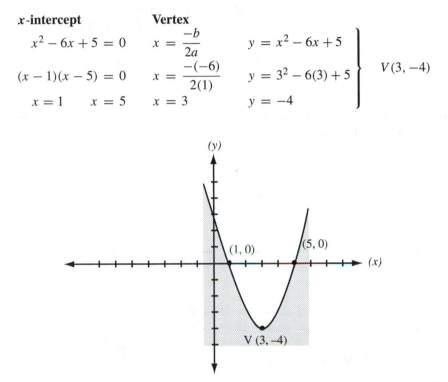

If we had been asked to graph $y > x^2 - 6x + 5$, the inside of the curve would have been shaded.

▶ **Inequalities, graph on a number line**   If we consider the number 2 and then consider the numbers bigger than 2, how could we speak about all of them in a simple way? This is one of the concepts that inequalities address.

By ourselves we would have to list the numbers 2, 3, 4, 5, on up to infinity. We would also have to list all of the fractions between 2 and 3, between 3 and 4, and so on. Further, we would have to list all of the irrational numbers between all of the fractions. Or we could simply write $x \geqslant 2$.

The inequality $x \geqslant 2$, read "x is greater than or equal to 2," means two things. It means $x = 2$ or it means $x > 2$, read "x is greater than 2." If we were to graph $x \geqslant 2$ on a number line, we would have where the solid dot at 2 means we include 2 in the graph and the ray means we include all of the numbers to the right of 2. In this manner, the x in $x \geqslant 2$ represents 2 and any number to the right of 2. If we did not want to include 2 in our solution, we would write $x > 2$. Graphed, we would have

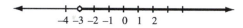

where the open dot at 2 means we do not include 2 in the graph and the ray means we include all of the numbers to the right of 2. See **Number Line**, **Inequality Signs**, or **Ray**.

In this article, we will describe the points on the number line with either an open dot ($<$ or $>$) or a solid dot ($\leqslant$ or $\geqslant$). For usage of parentheses and brackets, see **Interval Notation**.

Examples

1. Graph the inequality $x > -3$ on a number line.

   The graph will have an open dot at -3 and a ray pointing right. We use an open dot because the inequality is strictly greater than. If it were $\geqslant$, we would use a solid dot.

   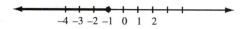

2. Graph the inequality $x \leqslant -1$ on a number line.

   The graph will have a solid dot at -1 and a ray pointing left.

   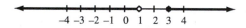

3. Graph $2 < x$ on a number line.

   This means that 2 is less than any number x. In reverse, it means that x is greater than the number 2. Writing it this way, we have $x > 2$. Hence, when we turn an inequality around we also turn the inequality sign around, i.e., $3 < x$ or $x > 3$, $-2 \leqslant x$ or $x \geqslant -2$. For more, see **Inequalities, linear solutions of** or **Inequalities, systems of**.

4. Graph on a number line: $x < 1$ or $x \geqslant 3$. Then graph $x < 1$ and $x \geqslant 3$.

   The "or" means that x can be in $x < 1$, it can be in $x \geqslant 3$, or it can be in both. To be in both, see example 5. See **Union and Intersection**.

   Simply, we just need to graph the two inequalities on the same number line.

   To graph $x < 1$ and $x \geqslant 3$ means that we want to graph the x's that are in $x < 1$ and at the same time are in $x \geqslant 3$. But nothing in $x < 1$ is in $x \geqslant 3$. Hence, the solution is empty, or $\varnothing$. To graph it, we just draw a number line that is empty.

5. Graph $x \geqslant -2$ or $x < 4$. Then graph $x \geqslant -2$ and $x < 4$. Graph on a number line.

As stated in example 4, the "or" means that x can be in $x \geqslant -2$, it can be in $x < 4$, or it can be in both. The x's are in both where the two inequalities overlap. See **Union and Intersection**.

The $x \geqslant -2$ includes the 4 in the graph. Otherwise, it would have been an open dot. The graph of $x \geqslant -2$ or $x < 4$ is the entire number line.

To graph $x \geqslant -2$ and $x < 4$ means that we want to graph the x's that are in $x \geqslant -2$ and at the same time are in $x < 4$. This occurs where the two inequalities overlap. If the inequalities didn't overlap, the solution would have been empty.

➤ **Inequalities, linear graphs of in the plane.** See **Inequalities, graph on a number line** or **Inequalities, graph in the plane**.

➤ **Inequalities, linear solutions of** We solve **linear inequalities** the same as we do equations except for one step; if we divide (or multiply) by a negative, we change the direction of the inequality sign. Otherwise, solve just as with equations. See **Linear Equations, solutions of**.

Examples

1. Solve the inequality $5(x - 4) \geqslant 15$.

$$5(x - 4) \geqslant 15$$
$$5x - 20 \geqslant 15 \qquad \text{Parentheses}$$
$$5x \geqslant 15 + 20 \qquad \text{Move}$$
$$5x \geqslant 35 \qquad \text{Combine}$$
$$\frac{5x}{5} \geqslant \frac{35}{5}$$
$$x \geqslant 7 \qquad \text{Divide}$$

Since we didn't divide (or multiply) by a negative, the solution of this example is the same as solving an equation.

2. Solve the inequality $2(3x + 4) - 2 < 3(1 + 3x)$. Give the answer in set notation, interval notation, and graph the solution set.

$$2(3x + 4) - 2 < 3(1 + 3x)$$
$$6x + 8 - 2 < 3 + 9x$$
$$6x - 9x < 3 - 8 + 2$$
$$-3x < -3$$
$$\frac{-3x}{-3} \oslash \frac{-3}{-3} \qquad \text{Since we divided by a negative,}$$
$$x > 1 \qquad\qquad \text{the } -3, \text{ we change the direction of the inequality sign.}$$

In set notation, the answer is $\{x \mid x > 1\}$. See **Set Notation**. In interval notation, the answer is $(1, \infty)$. See **Interval Notation**. The graph of the solution set is

See **Inequalities, graph on a number line**.

➤ **Inequalities, polynomial**   See **Polynomial Inequalities**. This article is an expansion beyond quadratic inequalities.

➤ **Inequalities, quadratic graphs in the plane**   See **Inequalities, graph in the plane**, example 4.

➤ **Inequalities, quadratic graphs on a number line**   See **Inequalities, quadratic solutions of**.

➤ **Inequalities, quadratic solutions of**   **Quadratic inequalities** can be solved by graphing in the plane, by applying linear inequalities algebraically, or by testing intervals. In this article, we will discuss the latter.

To solve inequalities by testing intervals, we need to solve the equation and then test the intervals determined by the solutions. Also see **Polynomial Inequalities**.

> *Examples*

1. Find the solution set of $x^2 - x - 2 \geqslant 0$. Also give the answer in interval notation.

We first need to solve the equation $x^2 - x - 2 = 0$. See **Quadratic Equations, solutions of**.

$$x^2 - x - 2 = 0$$
$$(x + 1)(x - 2) = 0$$
$$x = -1 \text{ or } x = 2$$

Next, we need to locate the solutions on a number line

The solutions divide the number line into three parts: part A which is $x \leqslant -1$, part B which is $-1 \leqslant x \leqslant 2$, and part C which is $x \geqslant 2$.

The solution of $x^2 - x - 2 \geqslant 0$ is either one or more of these parts, or it is the empty set. To find out, we need to test each of the parts. This is done by choosing any number in each of the parts, substituting it into $x^2 - x - 2 \geqslant 0$, and seeing whether the result is correct or not.

Part A: In part A, let's choose $x = -2$ for our test. Then

$$x^2 - x - 2 \geqslant 0$$
$$(-2)^2 - (-2) - 2 \geqslant 0$$
$$4 + 2 - 2 \geqslant 0$$
$$4 \geqslant 0 \qquad \text{which is true}$$

Part B: In part *B*, let's choose $x = 0$ for our test. Then

$$x^2 - x - 2 \geqslant 0$$
$$(0)^2 - (0) - 2 \geqslant 0$$
$$-2 \geqslant 0 \qquad \text{which is false}$$

Part C: In part *C*, let's choose $x = 3$ for our test. Then

$$x^2 - x - 2 \geqslant 0$$
$$(3)^2 - (3) - 2 \geqslant 0$$
$$9 - 5 \geqslant 0$$
$$4 \geqslant 0 \qquad \text{which is true}$$

Since part *A* and part *B* are true, $x \leqslant -1$ or $x \geqslant 2$ are solutions of $x^2 - x - 2 \geqslant 0$. In set notation, the answer is $\{x \mid x \leqslant -1\}$ or $\{x \mid x \geqslant 2\}$. Remember to use an "or" not an "and." An "or" is used for "union" of sets and an "and" is used for "intersection" of sets. See **Union and Intersection** or **Set Notation**. In interval notation, the answer is $(-\infty, -1] \cup [2, \infty)$. See **Interval Notation**.

If the original problem had been $x^2 - x - 2 \leqslant 0$, then part *A* would have been $4 \leqslant 0$ which is not true, part *B* would have been $-2 \leqslant 0$ which is true, and part *C* would have been $4 \leqslant 0$ which is not true. Hence, the answer for $x^2 - x - 2 \leqslant 0$ is part *B*, $-1 \leqslant x \leqslant 2$.

The graph of $x^2 - x - 2 \geqslant 0$ is

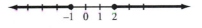

The graph of $x^2 - x - 2 \leqslant 0$ is

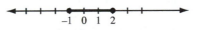

2. Solve the inequality $x^3 - x^2 - 6x < 0$. Give the answer in set notation, interval notation, and then graph the solution set.

This equation is relatively easy to solve. If the equation you are trying to solve is more complicated, see **Factor Theorem**. To solve this equation, see **Solving Polynomial Equations**.

$$x^3 - x^2 - 6x = 0$$
$$x(x^2 - x - 6) = 0$$
$$x(x + 2)(x - 3) = 0$$
$$x = 0 \qquad x + 2 = 0 \qquad x - 3 = 0$$
$$x = -2 \qquad x = 3$$

Locating the solutions on a number line, we have

where we use open dots instead of solid dots because the inequality $x^3 - x^2 - 6x$ is strictly less than 0, i.e., $<$ not $\leqslant$. See **Inequalities, graph on a number line**.

The solutions divide the number line into four parts: Part A which is $x < -2$, Part B which is $-2 < x < 0$, Part C which is $0 < x < 3$, and Part D which is $x > 3$.

Choosing x values in each part, we have

Part A: In part A, let's choose -3 for our test. Then

$$(-3)^3 - (-3)^2 - 6(-3) < 0$$
$$-27 - 9 + 18 < 0$$
$$-18 < 0 \qquad \text{which is true}$$

Part B: In part B, let's choose -1 for our test. Then

$$(-1)^3 - (-1)^2 - 6(-1) < 0$$
$$-1 - 1 + 6 < 0$$
$$4 < 0 \qquad \text{which is false}$$

Part C: In part C, let's choose 1 for our test. Then

$$(1)^3 - (1)^2 - 6(1) < 0$$
$$1 - 1 - 6 < 0$$
$$-6 < 0 \qquad \text{which is true}$$

Part D: In part D, let's choose 4 for our test. Then

$$(4)^3 - (4)^2 - 6(4) < 0$$
$$64 - 16 - 24 < 0$$
$$24 < 0 \qquad \text{which is false}$$

Since part A and part C are true, $x < -2$ and $0 < x < 3$ are solutions of $x^3 - x^2 - 6x < 0$. In set notation, the answer is $\{x \mid x < -2 \cup 0 < x < 3\}$. In interval notation, the answer is $(-\infty, -2) \cup (0, 3)$. The graph is

For more, see **Polynomial Inequalities**.

▶ **Inequalities, systems of**   **Systems of inequalities** are two or more inequalities that are solved at the same time. What is required is to find points that will satisfy each of the inequalities simultaneously. To find these points, we need to graph the inequalities in the same plane and determine there common area of intersection. Any of the points in this area will satisfy each of the inequalities.

Examples

1. Solve the following system of inequalities by graphing:

$$x + y \geqslant 2$$
$$-2x \geqslant -y - 2$$

In order to graph the inequalities, they must first be solved for y. Then, using the slope-intercept method, we will graph the lines and determine their common area of intersection. See **Linear Equations, graphs of** or **Inequalities, graph in the plane**.

$$x + y \geqslant 2 \qquad -2x \geqslant -y - 2$$
$$y \geqslant -x + 2 \qquad y \geqslant 2x - 2$$

For help in solving these equations, see **Linear Equations, solutions of**.

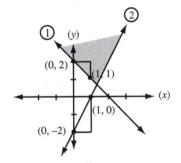

After each line is graphed, label it ① or ② after the first and second equation. The shading is then clearer. We want to shade above ($\geqslant$) line ① and above ($\geqslant$) line ②.

The following is given as an aid in determining which area to shade. In each pair of inequality signs, the top sign represents line ① and the bottom sign represents line ②.

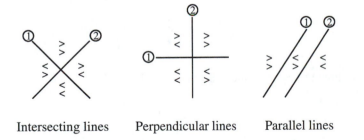

Intersecting lines   Perpendicular lines   Parallel lines

Think through each case in your mind and be certain why each is correct. Then consider cases where the ② and ① are reversed. Notice that it is only the alternate signs that change direction.

2. Solve the following system of inequalities by graphing:

$$y > 2$$
$$x \leqslant 3$$

To graph $y = 2$ and $x = 3$, see **Linear Equations, graphs of**. To graph $y > 2$ or $y \leqslant 3$, see **Inequalities, graph in the plane**.

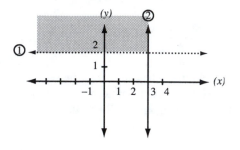

3. Solve the following system of inequalities by graphing:

$$y > -2x + 3$$
$$y \leqslant 3x + 1$$
$$y \geqslant \frac{2}{3}x - 1$$

As the equations are graphed, remember to label them ①, ②, and ③. When shading, first shade lines ① and ② as we have done previously. Then see what is common between that and the shading of line ③.

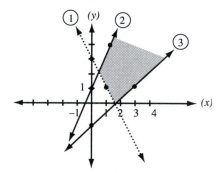

4. Solve the following system of inequalities by graphing:

$$y > x^2 - 1$$
$$x^2 + y^2 \leqslant 25$$

For information on graphing the first equation, see **Quadratic Equations, graphs of**. For information on the second equation, see **Circle**.

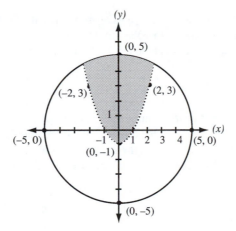

The graph of the parabola is shaded above since the inequality sign is $>$. If we solve the equation of the circle for y, we have $y \leqslant \pm\sqrt{25 - x^2}$; therefore, we shade inside the circle.

➤ **Inequality**   An **inequality** is a means of describing parts of a number line. Consider the number 3 on a number line. It divides the line into two parts, the part to the right of 3 and the part to the left of 3.

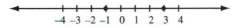

The part to the right of 3 is composed of numbers that are greater than 3. We use the symbol $>$ to mean **greater than** and we write $x > 3$. The x is any number greater than 3. The part to the left of 3 is composed of numbers that are less than 3. We use the symbol $<$ to mean **less than** and we write $x < 3$. The x in this case is any number less than 3.

If we wanted to include the 3, we would use the symbols $\geqslant$ or $\leqslant$. To the right and including 3, we would write $x \geqslant 3$. To the left and including 3, we would write $x \leqslant 3$.

Now consider two points on the number line, say -1 and 3. To describe the points to the left and including -1, we write $x \leqslant -1$. To describe the points to the right and including 3, we write $x \geqslant 3$. To describe the points between -1 and 3, and including -1 and 3, we write $-1 \leqslant x \leqslant 3$. The x is any number between, and including, the -1 and 3.

An inequality such as $-1 \leqslant x \leqslant 3$ is actually two inequalities. It says $-1 \leqslant x$ and it also says $x \leqslant 3$. The $-1 \leqslant x$ means that the smallest x can be is -1. In other words, the x is any number greater than or including -1, which we write as $x \geqslant -1$. Hence, we see that $-1 \leqslant x \leqslant 3$ means that x is any number greater than -1 or less than 3. It can also equal -1 and 3.

For more, see **Inequalities, graph on a number line**.

I

➤ **Inequality Signs**   The symbols $<$, $\leqslant$, $>$, and $\geqslant$ are inequality signs. The first two symbols mean less than, or, less than or equal to. The last two symbols mean greater than, or, greater than or equal to. One way to remember the symbols is to think of a number line.

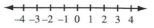

Notice that the right-hand tip of the number line ends with the symbol $>$. And as we go up to the right, the numbers get greater. In contrast, notice that the left-hand tip of the number line ends with the symbol $<$. And as we go down to the left, the numbers get less.

For an application, see **Inequalities, graph on a number line**. For a better understanding of inequalities, see **Inequality**.

➤ **Infinity**   A number line extends indefinitely in either direction. There is no upper bound and there is no lower bound. On the positive side, we say a number line continues indefinitely toward positive infinity, written $+\infty$, and, on the negative side, we say a number line continues indefinitely toward negative infinity, written $-\infty$. The symbols $+\infty$ and $-\infty$ are not numbers. They only describe the indefinite nature of no upper bound or no lower bound.

➤ **Initial Side**   The **initial side** of an angle is a ray that coincides with the positive x-axis and has its endpoint at the origin. See **Angle, in trigonometry**.

➤ **Inscribed Angle**   An **inscribed angle** is an angle whose vertex is on a circle and whose sides form chords with the circle.

In the following figures, $\angle ABC$ and $\angle DEF$ are inscribed angles.

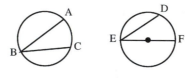

➤ **Inscribed (polygon)**   A circle is **inscribed** in a polygon if it is tangent to each of the sides of the polygon. In the figure, the circle O is inscribed in the pentagon $ABCDE$. It is tangent to the pentagon at points $FGHIJ$.

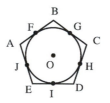

➤ **Integers**   **Integers** are the set of numbers Z such that

$$Z = \{-\infty, \ldots, -2, -1, 0, 1, 2, \ldots, +\infty\}$$

The set of integers is composed of the positive and negative counting numbers and zero. There are no fractions or irrational square roots included in this set.

The integers are the third in line of the subsets of real numbers. The first two are the counting or natural numbers and the whole numbers, then come the integers. The next subsets after the integers are the rational numbers and the irrational numbers. Above the real numbers are the complex numbers. See the articles under each of these headings. For more, see **Number Line**, **Real Number System**, or **Real Number Line**.

▶ **Integration**   This article deals with the basic concepts pertaining to indefinite integrals. To integrate problems like $\int_1^4 x^2\,dx$, see **Fundamental Theorem of Calculus**.

Simply, **integration** is the opposite operation of differentiation (antidifferentiation). By comparison, multiplication is the opposite operation of division, or square is the opposite operation of square root. Whereas the reverse operation for these and most other mathematical concepts are relatively easy, the process of integration, quite often, is not.

One of the reasons integration can be difficult, if not possible at all, can be seen by looking at the derivatives of $y = x^2$ and $y = x^2 + 3$:

$$\frac{d(x^2)}{dx} = 2x \qquad \frac{d(x^2 + 3)}{dx} = 2x + 0 = 2x$$

The derivative of either function is $2x$. How then could a reverse operation applied to $2x$ yield two different results, x^2 and $x^2 + 3$? Well, it can't.

The best we can say is that the original function might have had a constant, or maybe not. Therefore, we can say that the reverse operation applied to $2x$ will yield $x^2 + c$ where c is an arbitrary constant. We cannot know exactly what c is unless we are given further information.

If we call the reverse operation "integration" or the "integral" of the function, then the integral of $2x$ is $x^2 + c$ and we write

$$\int 2x\,dx = x^2 + c$$

This is read "the integral of $2x$ with respect to x." The $\int$ symbol denotes an integral and the dx symbol, called the **differential of integration**, denotes the variable in which the integral is being calculated.

This can also be seen by considering the differentials of $\frac{d(x^2)}{dx} = 2x$. If we cross multiply by dx and take the integral of both sides of the equation, we have

$$2x\,dx = dx^2$$
$$\int 2x\,dx = \int dx^2$$
$$\int 2x\,dx = x^2 + c \qquad \text{The opposite operations cancel each other.}$$

Again, we use c since we don't know if there was a constant to begin with, or not.

In general, let us consider the derivative $\frac{dy}{dx} = f'(x)$ where $y = f(x)$. Cross multiplying and integrating both sides, we have

$$f'(x)dx = dy$$
$$f'(x)\,dx = df(x)$$

$$\int f'(x)\, dx = \int df(x)$$

$$\int f'(x)\, dx = f(x) + c$$

where c is an arbitrary constant. This equation says that the integral of the derivative of a function yields the function plus a constant. Since c is unknown, the expression $f(x) + c$ is called the **indefinite integral** of $f'(x)$.

Unlike differentiation, integration does not have a general rule with which to "crank out" the answer. Derivatives by definition, although tedious, generally yield the desired answer. On the other hand, integration must be approached from a different point of view. We need to think of a function that, when differentiated, will yield the given expression.

As an example, if we want to know $\int 2x\, dx$, we need to think of a function (x^2) that, when differentiated, will yield the given expression $(2x)$. The integral of $2x$ is x^2 because the derivative of x^2 is $2x$, disregarding the constant. If the constant is considered, then the integral of $2x$ is $x^2 + c$ because the derivative of $x^2 + c$ is $2x$.

Considering the variation of derivatives, recognizing the correct form for integration can be difficult. To simplify the process, tables of integrals have been constructed to aid in the recognition of standard forms. If a particular problem does not fit the form, various methods have been developed that will reduce the problem to one that is recognizable. The list given in this book is a list of basic integration rules (or forms). For a more extensive list, refer to the back of any reputable calculus text.

Rules (forms) for Integration

1. $\displaystyle\int du = u + c$

2. $\displaystyle\int K f(u)\, du = K \int f(u)\, du$, where K is a constant

3. $\displaystyle\int [f(u) \pm g(u) \pm h(u)]\, du = \int f(u)\, du \pm \int g(u)\, du \pm \int h(u)\, du$

4. $\displaystyle\int u^n\, du = \frac{u^{n+1}}{n+1} + c$, where $n \neq -1$ (called the power rule)

5. $\displaystyle\int \frac{1}{u}\, du = \int \frac{du}{u} = \ln|u| + c$

6. $\displaystyle\int e^u\, du = e^u + c$

7. $\displaystyle\int a^u\, du = \frac{a^u}{\ln a} + c$

8. $\displaystyle\int \sin u\, du = -\cos u + c$

9. $\displaystyle\int \cos u\, du = \sin u + c$

I

10. $\displaystyle\int \tan u \, du = -\ln|\cos u| + c$

11. $\displaystyle\int \csc u \, du = -\ln|\csc u + \cot u| + c$

12. $\displaystyle\int \sec u \, du = \ln|\sec u + \tan u| + c$

13. $\displaystyle\int \cot u \, du = \ln|\sin u| + c$

14. $\displaystyle\int \sec^2 u \, du = \tan u + c$

15. $\displaystyle\int \csc^2 u \, du = -\cot u + c$

16. $\displaystyle\int \sec u \tan u \, du = \sec u + c$

17. $\displaystyle\int \csc u \cot u \, du = -\csc u + c$

18. $\displaystyle\int \frac{du}{u^2 + a^2} = \frac{1}{a}\arctan\frac{u}{a} + c$

19. $\displaystyle\int \frac{du}{u^2 - a^2} = \frac{1}{2a}\ln\frac{u - a}{u + a} + c, \text{ where } u^2 > a^2$

20. $\displaystyle\int \frac{du}{u^2 - a^2} = \frac{1}{2a}\ln\frac{a + u}{a - u} + c, \text{ where } u^2 < a^2$

21. $\displaystyle\int \frac{du}{\sqrt{a^2 - u^2}} = \arcsin\frac{u}{a} + c$

22. $\displaystyle\int \frac{du}{u\sqrt{u^2 - a^2}} = \frac{1}{a}\operatorname{arcsec}\frac{|u|}{a} + c$

23. $\displaystyle\int \sqrt{a^2 - u^2} \, du = \frac{u}{2}\sqrt{a^2 - u^2} + \frac{a^2}{2}\arcsin\frac{u}{a} + c$

Examples

1. Find $\int 5x \, dx$.

   What is required is to find an expression whose derivative is $5x$. Since 5 is a constant, we know by rule 2 that

   $$\int 5x \, dx = 5 \int x \, dx$$

   Next, since x can be written as x^1, we can apply the power rule, rule 4, where we let $u^n = x^1$:

   $$5 \int x \, dx = 5\left(\frac{x^{1+1}}{1+1}\right) + c = 5\left(\frac{x^2}{2}\right) + c = \frac{5}{2}x^2 + c$$

To verify that $\frac{5}{2}x^2 + c$ is the required answer, we can take its derivative:

$$\frac{d}{dx}\left(\frac{5}{2}x^2 + c\right) = \frac{d}{dx}\left(\frac{5}{2}x^2\right) + \frac{dc}{dx} = \frac{5}{2}\frac{dx^2}{dx} + 0 = \frac{10x}{2} = 5x$$

2. Find $\int \frac{1}{x^4}\, dx$.

Since $\frac{1}{x^4} = x^{-4}$, we see that we can apply the power rule (rule 4):

$$\int \frac{1}{x^4}\, dx = \int x^{-4}\, dx = \frac{x^{-4+1}}{-4+1} + c = \frac{x^{-3}}{-3} + c = \frac{-1}{3x^3} + c$$

As can readily be seen, the derivative of $\frac{-1}{3x^3} + c$ is $\frac{1}{x^4}$.

3. Find $\int \sqrt{x}\, dx$.

Since radicals can be written in exponential form, the $\sqrt{x}$ can be written as $x^{1/2}$. This allows us to use the power rule (rule 4):

$$\int \sqrt{x}\, dx = \int x^{1/2}\, dx = \frac{x^{1/2+1}}{\frac{1}{2}+1} + c = \frac{x^{3/2}}{\frac{3}{2}} + c = \frac{2}{3}x^{3/2} + c$$

4. Find $\int (x + 3)\, dx$.

Within the parentheses there is a sum. By rule 3, the integral of a sum is the sum of the integrals. Hence,

$$\int (x + 3)\, dx = \int x\, dx + \int 3\, dx$$

The method of calculating the first of these integrals was covered in example 1. The second integral uses both rules 2 and 1, respectively. Hence, we have

$$\int (x + 3)\, dx = \int x\, dx + \int 3\, dx = \frac{x^{1+1}}{1+1} + c_1 + 3\int dx$$

$$= \frac{x^2}{2} + c_1 + 3x + c_2 \qquad \text{since } \int dx = x + c$$

$$= \frac{1}{2}x^2 + 3x + c \qquad \text{where } c = c_1 + c_2$$

Since the value of each of the constants is arbitrary, it's just as arbitrary to let $c = c_1 + c_2$. We do this only to indicate that there might have been an original constant. If each of the constants had values of 5 and 3, respectively, the derivative of their combined value, 8, would still be 0.

5. Find $\int (4x^3 - 5x^2 - 3x + 2)\, dx$.

Applying rules 1–4, we have

$$\int (4x^3 - 5x^2 - 3x + 2)\, dx = \int 4x^3\, dx - \int 5x^2\, dx - \int 3x\, dx + \int 2\, dx$$

$$= 4 \int x^3 \, dx - 5 \int x^2 \, dx - 3 \int x \, dx + 2 \int dx$$

$$= 4 \left(\frac{x^4}{4} \right) - 5 \left(\frac{x^3}{3} \right) - 3 \left(\frac{x^2}{2} \right) + 2x + c$$

$$= x^4 - \frac{5}{3} x^3 - \frac{3}{2} x^2 + 2x + c$$

The c in this example represents the combined constants of each of the integrals.

6. Find $\int \frac{x-1}{\sqrt{x}} \, dx$.

At first sight, this integral might seem too difficult. However, if we change $\sqrt{x}$ to $x^{1/2}$ and use the concept of splitting fractions, $\frac{a+b}{c} = \frac{a}{c} + \frac{b}{c}$, we have

$$\int \frac{x-1}{\sqrt{x}} \, dx = \int \frac{x-1}{x^{1/2}} \, dx = \int \frac{x}{x^{1/2}} \, dx - \int \frac{1}{x^{1/2}} \, dx = \int x^{1/2} \, dx - \int x^{-1/2} \, dx$$

$$= \frac{x^{1/2+1}}{\frac{1}{2}+1} - \frac{x^{-1/2+1}}{-\frac{1}{2}+1} + c = \frac{x^{3/2}}{\frac{3}{2}} - \frac{x^{1/2}}{\frac{1}{2}} + c = \frac{2}{3} x^{3/2} - 2x^{1/2} + c$$

7. Find $\int x(a + bx^3)^2 \, dx$ where a and b are constants.

This integral becomes recognizable if we square the binomial, multiply x into the resulting trinomial, and then integrate each of the terms. The answer is

$$\frac{a^2 x^2}{2} + \frac{2abx^5}{5} + \frac{b^2 x^8}{8} + c$$

8. Find $\int \frac{x^3 \, dx}{\sqrt{a^4 + x^4}}$ where a is a constant.

This problem will reduce to a simple one if we use the concept of **substitution**. To apply this method, we need to choose an appropriate substitution, take its derivative, solve for the differential of integration, make a substitution for the differential, reduce the resulting expression, and then calculate the integral. After the integral has been found, resubstitute the original values and simplify.

If we let $u = a^4 + x^4$, then $\frac{du}{dx} = 4x^3$, $dx = \frac{du}{4x^3}$, and we have

$$\int \frac{x^3 \, dx}{\sqrt{a^4 + x^4}} = \int \frac{x^3}{\sqrt{u}} \bullet \frac{du}{4x^3} = \int \frac{du}{4\sqrt{a}} = \frac{1}{4} \int u^{-1/2} \, du = \frac{1}{4} \left(\frac{u^{1/2}}{\frac{1}{2}} \right) + c = \frac{1}{4} \bullet \frac{2}{1} u^{1/2} + c$$

$$= \frac{1}{2} u^{1/2} + c = \frac{1}{2} (a^4 + x^4)^{1/2} + c \qquad \text{by resubstituting } u = a^4 + x^4$$

9. Find $\int \frac{x^2 + 1}{\sqrt{x^3 + 3x}} \, dx$.

If we let $u = x^3 + 3x$, then $\frac{du}{dx} = 3x^2 + 3 = 3(x^2 + 1)$, $dx = \frac{du}{3(x^2+1)} dx$, and we have

$$\int \frac{x^2+1}{\sqrt{x^3+3x}} dx = \int \frac{x^2+1}{\sqrt{u}} \bullet \frac{du}{3(x^2+1)} = \int \frac{du}{3u^{1/2}} = \frac{1}{3} \int u^{-1/2} du = \frac{1}{3} \left(\frac{u^{1/2}}{\frac{1}{2}} \right) + c$$

$$= \frac{2}{3} u^{1/2} + c = \frac{2}{3}(x^3+3x)^{1/2} + c \qquad \text{by resubstituting } u = x^3 + 3x$$

10. Find $\int \sin ax \cos ax \, dx$ where a is a constant.

If we let $u = \sin ax$, then $\frac{du}{dx} = \cos ax \frac{dax}{dx} = a \cos ax$, $dx = \frac{du}{a \cos ax}$, and we have

$$\int \sin ax \cos ax \, dx = \int u \cos ax \bullet \frac{du}{a \cos ax} = \int \frac{u}{a} du = \frac{1}{a} \int u \, du$$

$$= \frac{1}{a} \left(\frac{u^2}{2} \right) + c = \frac{\sin^2 ax}{2a} + c \qquad \text{by resubstituting } u = \sin ax$$

► **Intercept**   An **intercept** is a point where a graph crosses either the x-axis or the y-axis. If the graph crosses the x-axis, the point of intersection is called the x-intercept. If the graph crosses the y-axis, the point of intersection is called the y-intercept.

► **Interest**   This article covers the two most commonly used types of interest: **simple interest** and **compound interest**. Further, compound interest is divided into compound interest compounded annually and compound interest compounded n times a year (in example 6, the concept of continuously compounded interest is discussed). The formulas for each are listed below:

Simple interest	Compound interest (annual)	Compound interest (n times a year)
$I = Prt$	$A = P(1+r)^t$	$A = P\left(1 + \dfrac{r}{n}\right)^{nt}$

where I represents the amount of interest earned after t years, P is the principal (original amount of money), r is the annual rate of interest (written as a decimal), t is the time in years, A is the total amount accumulated after t years, and n is the number of times per year that the principal is compounded.

When money is either borrowed or invested, there is usually an amount charged for the sevice. The amount charged for the service is called **interest**. It is the amount of money paid for the usage of the original amount of money. The original amount of money is called the **principal**. If interest is paid only on the principal, it is called **simple interest**. If the interest is added to the principal and then calculated again at a later date, the interest is called **compound interest**.

Interest is usually calculated as a percentage of the principal, called the **rate of interest**, or interest rate. The rate of interest can either be fixed (constant throughout the agreement) or variable (changes throughout the agreement).

Examples

1. If \$1,235 is borrowed for 3 years at an interest rate of 8%, find the interest and the amount due.

Since the interest is not added onto the principal during the time of the loan, this is a simple interest problem. If it were a compound interest problem, the words "compound interest" would have been used in the problem, or the idea of adding on the interest to the principal would have been implied.

To find the simple interest, we use the equation $I = Prt$:

$$I = Prt$$
$$= (1,235)(.08)(3) \qquad \text{Change 8\% to .08.}$$
$$= \$296.40$$

The interest charged during the 3-year period is $296.40.

To find the amount due after the 3 years have ended, we need to add the interest ($296.40) to the principal ($1,235):

$1,235.00   principal
$296.40   interest
$1,531.40   amount due at the end of 3 years

2. Suppose, in example 1, the amount due is known but the interest rate is not. How would we find the rate? Problem: If $1,531.40 is the amount due after $1,235 is borrowed for 3 years, what is the rate charged?

Again, since there is no mention of compound interest, we use the equation $I = Prt$. To find the rate, we first need to solve the equation for r. See **Linear Equations, solutions of**.

$$I = Prt$$
$$\frac{I}{Pt} = \frac{Prt}{Pt} \qquad \text{Divide both sides by } Pt.$$
$$\frac{I}{Pt} = r$$
$$r = \frac{I}{Pt}$$

In order to solve this equation, we first need to determine the value of I. Since the amount due ($1,531.40) is the sum of the principal ($1,235) and the interest (I), we can subtract the principal from the amount due to get the interest:

$1,531.40   amount due
−$1,235.00   principal
$296.40   interest paid after 3 years

Now we can find the rate:

$$r = \frac{I}{Pt}$$
$$= \frac{\$296.40}{(\$1,235)(3)}$$
$$= .08$$
$$= 8\% \qquad \text{Move the decimal point two places to change to a percent.}$$

I

3. If $1,235 is borrowed for 3 years at an interest rate of 8% compounded annually, find the amount due and the interest charged at the end of the 3-year period.

As stated in the problem, this is a compound interest problem compounded annually. So, we will use the equation $A = P(1 + r)^t$. Since A represents the amount due, we want to find A:

$$A = P(1 + r)^t$$
$$= (\$1,235)(1 + .08)^3$$
$$= (\$1,235)(1.08)^3$$
$$= (1,235)(1.259712)$$
$$= \$1,555.74$$

The amount due at the end of 3 years is $1,555.74.

Since interest is the amount of money charged for the usage of money, the difference between the amount due ($1,555.74) and the principal ($1,235) is the interest paid:

$1,555.74   amount due
−$1,235.00   principal
$320.74   interest at the end of 3 years

To verify our answer, let's calculate the problem through each of the 3 years. We will calculate the interest for the first year and add it onto the principal. Then, using the sum as the new principal, we will calculate the interest on that amount and add the interest to the new principal, and so on.

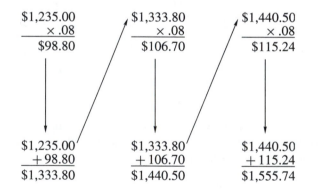

$1,235.00 $1,333.80 $1,440.50
× .08   × .08   × .08
$98.80 $106.70 $115.24

$1,235.00 $1,333.80 $1,440.50
+98.80   +106.70   +115.24
$1,333.80 $1,440.50 $1,555.74

Hence, we see that we get the same amount ($1,555.74).

Notice the difference in the amount of interest paid in example 1 and example 3. With simple interest, the amount charged for the use of the money is $296.40, whereas with compound interest compounded annually it is $320.74.

4. Suppose, in example 3, the amount due is known but the time of the loan is not. How would we find the time? Problem: If $1,555.74 is the amount due after $1,235 is borrowed at an interest rate of 8% compounded annually, what is the time of the loan?

Since the loan is compounded annually, we will use the equation $A = P(1 + r)^t$. If we had been trying to find P or r, we could have used algebraic methods to find either. But how can we find t? To solve an equation for an exponent requires the solution of an exponential equation; see **Exponential Equation, solution of**. Solving for t, we have

$$A = P(1+r)^t$$
$$1{,}555.74 = (1{,}235)(1+.08)^t$$
$$1{,}555.74 = (1{,}235)(1.08)^t$$
$$\frac{1{,}555.74}{1{,}235} = \frac{(1{,}235)(1.08)^t}{1{,}235}$$
$$1{,}25971 = (1.08)^t$$
$$\log(1{,}25971) = \log(1.08)^t \qquad \text{Take the log of both sides.}$$
$$\log 1{,}25971 = t\log 1.08$$
$$\frac{\log 1{,}25971}{\log 1.08} = \frac{t\log 1.08}{\log 1.08}$$
$$2.999979 = t \qquad \text{Calculate the logs and divide.}$$
$$t = 3 \qquad \text{Round off.}$$

The time is, as we expected, 3 years.

5. If $2,000 is placed in a savings account at 6% a year compounded quarterly, how much is in the account after 3 years?

Since the interest is compounded quarterly (four times a year), we will use the equation $A = P\left(1+\frac{r}{n}\right)^{nt}$. Substituting, we have

$$A = P\left(1+\frac{r}{n}\right)^{nt}$$
$$= (2{,}000)\left(1+\frac{.06}{4}\right)^{(4)(3)}$$
$$= (2{,}000)(1+.015)^{12}$$
$$= (2{,}000)(1.015)^{12}$$
$$= (2{,}000)(1.1956)$$
$$= \$2{,}391.20$$

The amount in the account after 3 years is $2,391.20.

If a problem you are doing requires finding the values of n or t, see example 4. Otherwise, solve for any of the other letters using algebraic methods.

6. If $1,235 is borrowed for 3 years at an interest rate of 8% compounded daily, find the amount due and the interest charged at the end of the 3-year period.

Since the interest is compounded daily (365 times a year), we will use the equation $A = P\left(1+\frac{r}{n}\right)^{nt}$:

I

$$A = P\left(1 + \frac{r}{n}\right)^{nt}$$

$$= (1{,}235)\left(1 + \frac{.08}{365}\right)^{(365)(3)}$$

$$= (1{,}235)(1 + .0002192)^{1095}$$

$$= (1{,}235)(1.0002192)^{1095}$$

$$= (1{,}235)(1.27125)$$

$$= \$1{,}569.99$$

The amount due at the end of 3 years is $1,569.99.

As in previous examples, to find the interest paid, we subtract the principal from the amount due:

$$
\begin{array}{ll}
\$1{,}569.99 & \text{amount due} \\
-\$1{,}235.00 & \text{principal} \\
\hline
\$334.99 & \text{interest at the end of 3 years}
\end{array}
$$

If you are solving a problem that requires finding n or t, see example 4. Otherwise, use algebraic solutions for any of the other variables.

If you are solving a problem that must be continuously compounded, use the formula $A = Pe^{rt}$ where e is the base from natural logarithms and $e = 2.7182818$. As an example, solving our current exercise, we have

$$A = Pe^{rt}$$

$$= (1{,}235)(2.7182818)^{(.08)(3)}$$

$$= (1{,}235)(2.7182818)^{.24}$$

$$= (1{,}235)(1.2712492)$$

$$= \$1{,}569.99$$

Other than the digits from the thousandths place on, we see that interest compounded daily or interest compounded continuously, in our example, yields the same amount. Most credit institutions use one or the other of these two methods in calculating their loans.

► Interior and Exterior Angles (for lines)

If two lines are cut (intersected) by a transversal (another line), the angles formed on the inside of the intersection are called **interior angles**. The angles formed on the outside are called **exterior angles**.

In the figure below, the angles 3, 4, 5, and 6 are interior angles and the angles 1, 2, 7, and 8 are exterior angles.

For more, see **Parallel Lines (geometric)**.

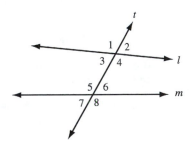

➤ **Interior and Exterior Angles of a Triangle**   If the sides of a triangle are extended, the angles formed on the outside are **exterior angles** and the angles inside the triangle are **interior angles**.

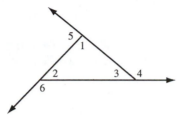

The angles 4, 5, and 6 are exterior angles. The angles 1, 2, and 3 are interior angles. The adjacent interior and exterior angles are supplementary.

➤ **Interior Angles on the Same Side of the Transversal**   If two lines are cut (intersected) by a transversal (another line), the angles on the same side of the transversal and interior to the two lines are referred to as **interior angles on the same side of the transversal**.

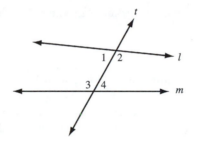

The angle pairs ∠2 and ∠4 and ∠1 and ∠3 are interior angles on the same side of the transversal. If ∠2 and ∠4 are supplementary or ∠1 and ∠3 are supplementary, then $l \parallel m$, read "l is parallel to m." See **Parallel Lines (geometric)**.

➤ **Interpolation, linear**   **Linear interpolation** is a method used to calculate an unknown value between two known values. It allows for more accuracy. The method is developed from similar triangles that are projected from three points on a line, two known and one unknown. A proportion is then set up.

There are four examples in this article, two for logarithms and two for trigonometry.

Examples

1. Using a logarithm table, approximate the value of log 4.683. See **Logarithm Table (chart), how to use**.

We are asked to approximate the value because the entries in the table are four-digit approximations. Looking up the log 4.683, we see it is between two numbers, log 4.68 = .6702 and log 4.69 = .6712. Using these values, we can set up the following table and then calculate the proportion of differences:

I

$$\text{Number} \quad \text{Mantissa}$$

$$.01 \left[.003 \left[\begin{matrix} 4.68 \\ 4.683 \\ 4.69 \end{matrix} \quad \begin{matrix} .6702 \\ ? \\ .6712 \end{matrix} \right] d \right] .0010$$

where .003 is the difference $4.683 - 4.68$, .01 is the difference $4.69 - 4.68$, d is the difference between the unknown and .6702, and .0010 is the difference $.6712 - .6702$. The proportion is

$$\frac{.003}{.01} = \frac{d}{.0010}$$

$$d = \frac{(.001)(.003)}{.01} \qquad \text{Cross multiplication and symmetric property.}$$

$$d = .0003$$

Since the values in the table are increasing from .6702 to .6712, the d is added onto .6702. If the case were reversed, we would subtract. Adding $d = .0003$ to .6702, we have $.6702 + .0003 = .6705$. Hence, $\log 4.683 = .6705$. More formally, we have

$$\log 4.683 = \log 4.68 + d = .6702 + .0003 = .6705$$

See **Common Logarithms**.

To verify that $\log 4.683 = .6705$, we can change the logarithm into exponential form and check with a calculator. See **Logarithmic Form to Exponential Form, converting from**. In exponential form, $\log_{10} 4.683 = .6705$ converts to $10^{.6705} = 4.683$. We want to see if $10^{.6705} = 4.683$. To do this, we can use the y^x button. First enter 10 in the calculator and press the y^x button, then enter .6705 and press y^x again (or the $=$). Rounded off, we get 4.683.

2. Find the antilog .6705 using a logarithm table.

   What is required is to find the number to which antilog .6705 corresponds. We are finding a number, not a logarithm.

   Looking through the table for .6705, we see that it is not there. However, it is between .6702 and .6712. The .6702 corresponds with the number 4.68 and the .6712 corresponds with 4.69. The number that corresponds with .6705 is somewhere between those. We can interpolate to calculate its value. We set up a table and then calculate the proportion of differences:

$$\text{Mantissa} \quad \text{Number}$$

$$.001 \left[.0003 \left[\begin{matrix} .6702 \\ .6705 \\ .6712 \end{matrix} \quad \begin{matrix} 4.68 \\ ? \\ 4.69 \end{matrix} \right] c \right] .10 \qquad \begin{matrix} \text{We can use } c \text{ so we will not confuse} \\ \text{this difference with } d. \end{matrix}$$

$$\frac{.0003}{.001} = \frac{c}{.01}$$

$$c = \frac{(.01)(.0003)}{.001}$$

$$c = .003$$

Since the values in the table are increasing from 4.68 to 4.69, the c is added onto 4.68. If the case were reversed, we would subtract. Adding $c = .003$ to 4.68, we have $4.68 + .003 = 4.683$. Hence, antilog $.6705 = 4.683$. More formally, we have

$$\text{antilog } .6705 = \text{antilog } .6702 + c = 4.68 + .003 = 4.683$$

See **Antilogarithm**.

As with logarithms, interpolation in trigonometry is a method to calculate an unknown value between two known values. We use the three to set up a table, calculate a proportion of differences, and then state the unknown.

3. Find $\cos 38°17'$ using trigonometric tables. See **Trigonometry Table, how to use**.

Looking through the table, we can see that there is no entry for $\cos 38°17'$. It is, however, between two other values, $\cos 38°10' = .7862$ and $\cos 38°20' = .7844$. Using these, we can set up the following table and calculate the proportion of differences:

$$10'\left[\,7'\left[\begin{matrix}\cos 38°10' & .7862 \\ \cos 38°17' & ? \\ \cos 38°20' & .7844\end{matrix}\right]d\,\right].0018$$

where $7'$ is the difference $38°17' - 38°10'$, $10'$ is the difference $38°20' - 38°10'$, d is the difference between the unknown and .7862, and .0018 is the difference $.7862 - .7844$. Notice that the unknown is less than .7862. Once we know d, we will subtract it from .7862. This would have been clear from the difference if we had subtracted $.7844 - .7862 = -.0018$. However, doing this sometimes causes confusion. The proportion is

$$\frac{7}{10} = \frac{d}{.0018}$$
$$d = \frac{7(.0018)}{10} \qquad \text{Cross multiply and symmetric property.}$$
$$d = .00126 \approx .0013$$

Since the values in the table are decreasing from .7862 to .7844, the d is subtracted from .7862. If the case were reversed, we would add. Subtracting d from .7862, we have $.7862 - .0013 = .7849$. Hence, $\cos 38°17' = .7849$.

4. If $\cos x = .7849$, find the value of x to the nearest minute. See **Trigonometry, inverse (arc) functions**.

We are required to find an angle x that has a cosine value of .7849. Looking through the table, we see that .7849 is between .7844 and .7862. The .7844 corresponds with $\cos 38°20'$ and the .7862 corresponds with $\cos 38°10'$. The number that corresponds with .7849 is somewhere between those. We can interpolate to calculate its value. Setting up a table and then calculating a proportion of difference, we have

$$10'\left[\,d\left[\begin{matrix}\cos 38°10' & .7862 \\ \cos x & .7849 \\ \cos 38°20' & .7844\end{matrix}\right].0013\,\right].0018$$

I

$$\frac{d'}{10} = \frac{.0013}{.0018}$$

$$d' = \frac{10(.0013)}{.0018}$$

$$d' = 7.\overline{22}' \approx 7'$$

Since the values in the table are increasing from $38°10'$ to $38°20'$, the d is added onto the $38°10'$. If the case were reversed, we would subtract. Adding $d = 7'$ to $48°10'$, we have $48°10' + 7' = 48°17'$. Hence, $x = 48°17'$.

▶ **Intersect**   If two lines or curves meet, they are said to **intersect**. Lines or curves that intersect do so at a common point, the point of intersection.

▶ **Intersection of Sets**   See **Union and Intersection**.

▶ **Intersection of Two Lines**   See **Systems of Linear Equations**.

▶ **Interval**   An **interval** is a part of a number line. If it includes the endpoints, it is a **closed interval**. If it does not include the endpoints, it is an **open interval.** Intervals are drawn with either a solid or an open dot (point) or with a bracket or parenthesis. Example (a) contains closed intervals and example (b) contains open intervals.

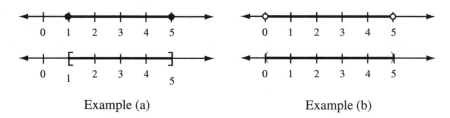

Example (a)   Example (b)

Since the closed interval contains its endpoints, it is also referred to as a segment.

▶ **Interval Notation**   **Interval notation** is a means of describing intervals using parentheses and brackets rather than inequalities. In interval notation, the **open interval** $a < x < b$ is written (a, b). This should not be confused with a point (a, b). The **closed interval** $a \leqslant x \leqslant b$ is written $[a, b]$. See **Set Notation**.

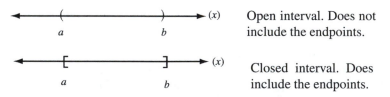

Open interval. Does not include the endpoints.

Closed interval. Does include the endpoints.

Consider the number a on a number line. Using interval notation, describe the numbers to the right of a, to the left of a, and including a.

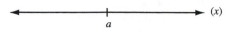

Numbers to the right of a are $x > a$. In interval notation, this is written (a, ∞). It does not include a but continues to the right.

Numbers to the right of a and including a are $x \geq a$. In interval notation, this is written $[a, \infty)$. It includes a and continues right.

Numbers to the left of a are $x < a$. In interval notation, this is written $(-\infty, a)$. It does not include a and continues to the left.

Numbers to the left of a and including a are $x \leq a$. In interval notation, this is written $(-\infty, a]$. It includes a and continues left.

Numbers to the right of a, left of a, and including a; in other words, the whole number line is written $(-\infty, \infty)$.

Examples

1. Describe the following graphs with interval notation.

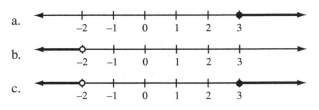

Answers: a. $[3, \infty)$. b. $(-\infty, -2)$. c. $(-\infty, -2) \cup [3, \infty)$.

To graph using interval notation, use brackets and parentheses instead of solid dots and open dots. For example,

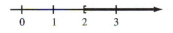

means $x \geq 2$ or $[2, \infty)$.

➤ **Inverse**   The **inverse** of a conditional statement is formed by negating both the if-part and the then-part.

$$\text{Conditional:} \quad p \rightarrow q \qquad \text{Inverse:} \quad \sim p \rightarrow \sim q$$

If the conditional statement is true, the inverse is not necessarily true.

Conditional: If two angles are complements of the same angle, then they are congruent. This is a true statement.

Inverse: If two angles are not complements of the same angle, then they are not congruent. This is not necessarily true. Two angles with measures of $80°$ are not complements of an angle of $20°$. Yet they are congruent.

For more, see **Logic**.

➤ **Inverse, additive**   See **Additive Inverse**.

➤ **Inverse (arc) Functions**   See **Trigonometry, inverse (arc) functions**.

I

▶ **Inverse Functions**   Two functions f and g are **inverse functions** if and only if $f(g(x)) = x$ for every x in the domain of g and $g(f(x)) = x$ for every x in the domain of f. Let the inverse g be denoted by f^{-1}, read "f inverse." Then $f(f^{-1}(x)) = x$ and $f^{-1}(f(x)) = x$ where the domain of f is equal to the range of f^{-1} and the domain of f^{-1} is equal to the range of f.

These concepts can easily be understood but only if composite functions are clear first. See **Composite Functions**.

Examples

1. Show that $f(x) = 2x + 6$ and $g(x) = \frac{1}{2}x - 3$ are inverse functions.

   If f and g are inverse functions, then, by definition, we must show $f(g(x)) = x$ and $g(f(x)) = x$:

   $$f\big(g(x)\big) = f\left(\frac{1}{2}x - 3\right) = 2\left(\frac{1}{2}x - 3\right) + 6 \qquad g\big(f(x)\big) = g(2x + 6) = \frac{1}{2}(2x + 6) - 3$$
   $$= x - 6 + 6 = x \qquad\qquad\qquad\qquad = x + 3 - 3 = x$$

   If solving the equations is not clear, see **Linear Equations, solutions of**.

   Since we have shown that $f(g(x)) = x$ and $g(f(x)) = x$, then f and g are inverse functions.

2. If $f(x) = 2x + 6$, find $f^{-1}(x)$.

   The notation f^{-1}, read "f inverse" or "inverse of f," represents the inverse of a function f. It should not be confused with $f^{-1} = \frac{1}{f}$. The -1 is used to indicate an inverse only, not a negative exponent.

   In example 1, we have already shown that $g(x)$ is an inverse of $f(x) = 2x + 6$. Since this is so, $f^{-1}(x) = g(x) = \frac{1}{2}x - 3$ and we are done. We have found $f^{-1}(x)$ where $f^{-1}(x) = \frac{1}{2}x - 3$. Of course, we already knew the answer. To show that this is so, we can apply the following: solve the equation $y = f(x)$ for x, then interchange the x and the y. The resulting equation will be the inverse $y = f^{-1}(x)$. Applying this to $y = f(x) = 2x + 6$, we have

   $$y = 2x + 6$$
   $$y - 6 = 2x$$
   $$\frac{y - 6}{2} = \frac{2x}{2}$$
   $$\frac{1}{2}y - 3 = x$$
   $$x = \frac{1}{2}y - 3 \qquad \text{Solve for } x.$$
   $$y = \frac{1}{2}x - 3 \qquad \text{Interchange } x \text{ and } y, \text{ nothing else, just the } x \text{ and } y.$$

   Hence, we have the inverse $y = f^{-1}(x) = \frac{1}{2}x - 3$. We know this is the inverse because we proved it in example 1.

3. Let $y = f(x) = 2x - 3$. Find $f^{-1}(x)$ and show that f and f^{-1} are inverses.

To find $f^{-1}(x)$, we solve $y = 2x - 3$ for x and then interchange the x and y:

$$y = 2x - 3$$
$$y + 3 = 2x$$
$$\frac{y + 3}{2} = \frac{2x}{2}$$
$$x = \frac{1}{2}y + \frac{3}{2}$$
$$y = \frac{1}{2}x + \frac{3}{2}$$

Therefore, $y = f^{-1}(x) = \frac{1}{2}x + \frac{3}{2}$.

To show that f and f^{-1} are inverses, we need to prove that $f(f^{-1}(x)) = x$ and $f^{-1}(f(x)) = x$:

$$f(f^{-1}(x)) = f\left(\frac{1}{2}x + \frac{3}{2}\right) \qquad f^{-1}(f(x)) = f^{-1}(2x - 3)$$
$$= 2\left(\frac{1}{2}x + \frac{3}{2}\right) - 3 \qquad = \frac{1}{2}(2x - 3) + \frac{3}{2}$$
$$= x + 3 - 3 = x \qquad = x - \frac{3}{2} + \frac{3}{2} = x$$

Hence, f and f^{-1} are inverses.

4. Show by graphing that f and f^{-1} from example 3 are inverse functions. Graphing each, we have

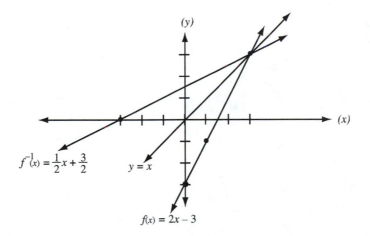

If we fold the graph along the line $y = x$, we see that the graph of $f(x)$ will coincide with that of $f^{-1}(x)$. This shows that the x and y of each point are interchangeable. Looking at a chart can verify this.

Chart $f(x) = 2x - 3$ for $x = 0, 1, 2$

x	$f(x) = 2x - 3$
0	-3
1	-1
2	1

Chart $f^{-1} = \frac{1}{2}x + \frac{3}{2}$ for $x = -3, -1, 1$

x	$f^{-1}(x) = \frac{1}{2}x + \frac{3}{2}$
-3	0
-1	1
1	2

The points $(0, -3)$, $(1, -1)$, and $(2, 1)$ coincide with $(-3, 0)$, $(-1, 1)$, and $(1, 2)$. Compare this with the definition at the beginning of this article.

From our example, we see that the domain of f is equal to the range of f^{-1} and the domain of f^{-1} is equal to the range of f.

5. Find the inverse of the following points: $(-2, -3)$, $(0, 1)$, $(2, 5)$, and $(3, 7)$. Then state whether the inverse is a function.

In light of example 4, we see that the inverse of a set of points can be found by interchanging the elements. Hence, the points $(-3, -2)$, $(1, 0)$, $(5, 2)$, and $(7, 3)$ are the inverse of the given points.

To see whether the inverse is a function or not, we need to look at the first elements of each point. Since each x value is different, the inverse is a function. See **Function**.

➤ **Inverse, multiplicative**   See **Multiplicative Inverse**.

➤ **Inverse Variation**   **Inverse variation** is a function defined by the equation $xy = k$, or $y = \frac{k}{x}$, where k is a nonzero constant called the **constant of variation**. In an inverse variation, y is said to vary inversely as x.

Inverse variation can be written as a **proportion**. Suppose both the pairs x_1 and y_1 and x_2 and y_2 satisfy the equation $xy = k$. Then $x_1 y_1 = k$ and $x_2 y_2 = k$. Substituting, we have

$$x_1 y_1 = x_2 y_2$$
$$\frac{x_1}{x_2} = \frac{y_2}{y_1}$$

Examples

1. If y varies inversely as x, and $y = 25$ when $x = 10$, find x when $y = 5$. Letting $x_1 = 10$, $y_1 = 25$, $x_2 = x$, and $y_2 = 5$, we have

$$\frac{x_1}{x_2} = \frac{y_2}{y_1}$$
$$\frac{10}{x} = \frac{5}{25}$$
$$x = \frac{25(10)}{5} \qquad \text{See \textbf{Cross Multiplication}.}$$
$$x = 50$$

I

2. Suppose y varies inversely as x. If $y = 32$ when $x = -3$, find y when $x = -8$. Letting $x_1 = -3$, $y_1 = 32$, $x_2 = -8$, and $y_2 = y$, we have

$$\frac{x_1}{x_2} = \frac{y_2}{y_1}$$
$$\frac{-3}{-8} = \frac{y}{32}$$
$$y = \frac{32(-3)}{-8}$$
$$y = 12$$

For more, see **Direct Variation**, **Combined Variation**, **Joint Variation**, or **Proportion**.

➤ **Irrational Equations**   See **Radicals, equations with**.

➤ **Irrational Numbers**   **Irrational numbers** are the set of numbers I such that

$$I = \{\text{nonterminating, nonrepeating decimals}\}$$

Some examples of irrational numbers are $\sqrt{3}$, π, e, and $-\sqrt{21}$. None of the rational numbers are irrational numbers. Since irrational numbers do not repeat, they are not among the decimals that can be changed to fractions. See **Decimal Numbers, changing a repeating decimal to a common fraction**. For more, see **Real Number System**, **Root**, **Square Root**, or the various articles under **Radical**.

➤ **Isosceles Triangle**   An **isosceles triangle** is a triangle with two congruent sides. Since the sides are congruent, the base angles are also congruent. This can be proven as a theorem. See **Proof in Geometry**.

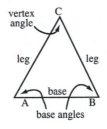

I

Jj

➤ **Joint Variation**   **Joint variation** is a linear function defined by the equation $y = kxz$, or $\frac{y}{xz} = k$, where k is a nonzero constant called the **constant of variation**. In a joint variation, y is said to vary jointly as x and z.

Joint variation can be written as a **proportion**. Suppose x_1, y_1, and z_1 and x_2, y_2, and z_2 satisfy the equation $\frac{y}{xz} = k$. Then

$$\frac{y_1}{x_1 z_1} = k \qquad \text{and} \qquad \frac{y_2}{x_2 z_2} = k$$

therefore, $\qquad \dfrac{y_1}{x_1 z_1} = \dfrac{y_2}{x_2 z_2} \qquad\qquad \dfrac{y_1}{y_2} = \dfrac{x_1 z_1}{x_2 z_2}$

Example

1. Let y vary jointly as x and z. If $y = -36$ when $x = 4$ and $z = 3$, find y when $x = -8$ and $z = 2$. Letting $y_1 = -36$, $x_1 = 4$, $z_1 = 3$, $y_2 = y$, $x_2 = -8$, and $z_2 = 2$ substituting, we have

$$\frac{-36}{4 \bullet 3} = \frac{y}{-8 \bullet 2}$$

$$\frac{-36}{12} = \frac{y}{-16}$$

$$y = \frac{(-16)(-36)}{12} \qquad \text{See \textbf{Cross Multiplication}.}$$

$$y = 48$$

For more, see **Direct Variation, Combined Variation, Inverse Variation,** or **Proportion.**

J

Kilometer   See **Appendix D, Table of Conversions**.

L₁

➤ **Latus Rectum**   For the **parabola**, the **latus rectum** is the line segment through the focus and perpendicular to the axis of symmetry. Its endpoints are on the parabola. If p is the distance from the vertex to the focus, the length of the latus rectum is $4p$.

For the **ellipse**, the **latus rectum** is the line segment through either focus and perpendicular to the major axis. Its endpoints are on the ellipse. In relation to the standard form of the equation of an ellipse, the length of the latus rectum is $\frac{2b^2}{a}$.

For the **hyperbola**, the **latus rectum** is the line segment through either focus and perpendicular to the transverse axis. Its endpoints are on the hyperbola. In relation to the standard form of the equation of the hyperbola, the length of the latus rectum is $\frac{2b^2}{a}$.

➤ **Law of Cosines**   The **law of cosines** is one of two sets of equations used to solve triangles that are not right triangles. It is used when two sides and the included angle, or three sides of the triangle are known. The other set of equations, law of sines, cannot be used in these cases. It is used when the following are known: two angles and a side, included or not included, and two sides and a not included angle. See **Law of Sines**.

If triangle ABC is any triangle, then

$$a^2 = b^2 + c^2 - 2bc \cos A$$
$$b^2 = a^2 + c^2 - 2ac \cos B$$
$$c^2 = a^2 + b^2 - 2ab \cos C$$

Rather than memorizing all three equations, it is easier to memorize the following: the square of one side of a triangle is equal to the sum of the squares of the other two sides, minus twice their product and the cosine of their included angle.

At first sight, this might also seem difficult, but look at the equations. Notice that the square of any letter is equal to the sum of the squares of the other two letters, minus twice the product of those letters and, tagged on at the end, the cosine of their included angle (b and c include $\angle A$, a and c include $\angle B$, and a and b include $\angle C$). Or, if you prefer, the angle at the end is the same as the letter in the front of the equation (A with a, B with b, and C with c).

Using these ideas, let's set up the equation for side x in the triangle below. Right off, we know $x^2 = 5^2 + 8^2$ will be the beginning. Then, substitute the product of 5 and 8 and the cosine of their included angle. Hence, the equation is $x^2 = 5^2 + 8^2 - 2(5)(8) \cos 35°$. With a little practice it becomes easy.

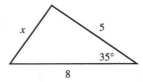

Notice that we get the same equation no matter what position the triangle is in.

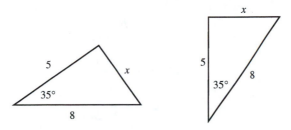

The similarity between the Pythagorean theorem and the law of cosines is no accident. The Pythagorean theorem is used to derive the law of cosines.

It is worthwhile noting that in the law of cosines, if we let the included angle be equal to 90°, the $\cos 90° = 0$. Therefore, the end term is 0 and we have the Pythagorean theorem, just as you would expect for a right triangle.

Examples

1. Solve the equation in the opening explanation for x.

$$x^2 = 5^2 + 8^2 - 2(5)(8) \cos 35°$$
$$x^2 = 25 + 64 - 80(.8192) \qquad \text{See \textbf{Appendix C} or use a calculator.}$$
$$x^2 = 89 - 65.54$$
$$x^2 = 23.46$$
$$\sqrt{x^2} = \sqrt{23.46} \qquad \text{See \textbf{Quadratic Equations, solutions of}.}$$
$$x = 4.84$$

We use the positive square root because x is the length of a side, and distance is always positive.

2. Find the length of c if $a = 13$, $b = 14$, and $m\angle C = 150°$.

First draw the triangle, then set up the equation.

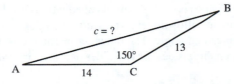

L

$$c^2 = 13^2 + 14^2 - 2(13)(14)\cos 150°$$
$$c^2 = 169 + 196 - 364(-.8660) \qquad \cos 150° = -\cos 30° = -.8660$$
$$c^2 = 365 + 315.22$$
$$c^2 = 680.22 \qquad\qquad \text{See \textbf{Trigonometry, reference angle}.}$$
$$\sqrt{c^2} = \sqrt{680.22}$$
$$c = 26.08$$

3. Find $m\angle B$ if $a = 5$, $b = 6$, and $c = 7$. When finding an angle using the law of cosines, some prefer to use equations that have been solved for the cosines. In $\triangle ABC$, we have

$$\cos A = \frac{b^2 + c^2 - a^2}{2bc}$$

$$\cos B = \frac{a^2 + c^2 - b^2}{2ac}$$

$$\cos C = \frac{a^2 + b^2 - c^2}{2ab}$$

However, what we have learned so far is sufficient. Since we have all three sides, the missing variable will be the angle. Set up the equation with the angle at the end and its corresponding side at the beginning, then solve for the angle.

$$6^2 = 5^2 + 7^2 - 2(5)(7)\cos B$$
$$36 = 25 + 49 - 70\cos B$$
$$36 - 25 - 49 = -70\cos B$$
$$-38 = -70\cos B$$
$$\frac{-38}{-70} = \frac{-70\cos B}{-70}$$
$$.5429 = \cos B$$
$$\cos B = .5429$$
$$m\angle B = 57.1°$$

The most common mistake is combining $25 + 49 - 70$. The -70 is part of the term $-70\cos B$.

See **Trigonometry, inverse (arc) functions**.

➤ **Law of Sines**   The **law of sines** is one of two sets of equations used to solve triangles that are not right triangles. It is used when the following are known: two angles and a side, included or not included, and two sides and a not included angle. The other set of equations, **law of cosines**, cannot be used in these cases. It is used when three sides, or two sides and their included angle, are known. See **Law of Cosines**.

L

If triangle ABC is any triangle, then

$$\frac{\sin A}{a} = \frac{\sin B}{b} = \frac{\sin C}{c}$$

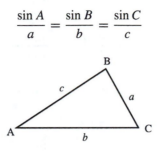

There are actually three equations here

$$\frac{\sin A}{a} = \frac{\sin B}{b} \qquad \frac{\sin B}{b} = \frac{\sin C}{c} \qquad \frac{\sin A}{a} = \frac{\sin C}{c}$$

To use the law of sines, three of the four parts of any equation must be known.

The law of sines is derived from the area formula in trigonometry. See **Trigonometry, area formula**. In non-right $\triangle ABC$ with sides a, b, and c, the area is one half the product of the lengths of any two sides and the sine of their included angle. The area K from each of the three vertices is

$$K = \frac{1}{2}bc \sin A \qquad K = \frac{1}{2}ac \sin B \qquad K = \frac{1}{2}ab \sin C$$

Since the area in each of the equations is the same, we can set the equations equal to each other and simplify by dividing through by $\frac{1}{2}abc$:

$$\frac{1}{2}bc \sin A = \frac{1}{2}ac \sin B = \frac{1}{2}ab \sin C$$

$$\frac{\frac{1}{2}bc \sin A}{\frac{1}{2}abc} = \frac{\frac{1}{2}ac \sin B}{\frac{1}{2}abc} = \frac{\frac{1}{2}ab \sin C}{\frac{1}{2}abc}$$

$$\frac{\sin A}{a} = \frac{\sin B}{b} = \frac{\sin C}{c}$$

Examples

1. Find the length of b if $a = 7$, $m\angle A = 48°$, and $m\angle B = 57°$.

First draw the triangle, then decide which equation to use. We will use $\frac{\sin A}{a} = \frac{\sin B}{b}$.

L

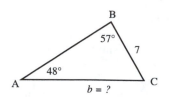

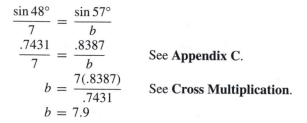

$$\frac{\sin 48°}{7} = \frac{\sin 57°}{b}$$

$$\frac{.7431}{7} = \frac{.8387}{b}$$ See **Appendix C.**

$$b = \frac{7(.8387)}{.7431}$$ See **Cross Multiplication.**

$$b = 7.9$$

2. Find the length of b if $m\angle C = 75°$, $m\angle A = 48°$, and $a = 7$.

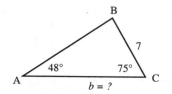

After we draw the triangle, we recognize that we have a problem. None of the three possible equations matches our information. Neither does the information match the requirements for the law of cosines. We recognize, however, that if two angles of a triangle are known, so is the third. The sum of the angles of a triangle is equal to 180°. Therefore, $m\angle B = 180° - 48° - 75° = 57°$. See **Sum of the Measures of the Angles of a Triangle**.

With this new piece of information, we can see that example 2 is exactly the same as example 1. It just seemed complicated.

By now, you have probably noticed that to use the law of sines we must match an angle with a side, and then another angle with a side. The fractions of the law of sines are set up in pairs, A with a, B with b, and C with c. It is for this reason that example 2, at first, seemed impossible. We had A (48°) with a (7), but neither of the other pairs. Once we knew angle B, however, we had another pair B (47°) with b (unknown). We then had three of the four parts of the equation.

➤ **Law of Sines, ambiguous case**   The **ambiguous case** exists only for the law of sines and, for that, only when we are considering two sides and a not included angle. In other words, for two sides and an angle opposite one of them. Otherwise, don't worry about it.

In the ambiguous case, one of three situations can occur: either no triangle exists, one triangle exists, or possibly two triangles exist. These situations occur for several reasons. If the side opposite the angle is less than the altitude, it is too short to reach the third side (no triangle can exists). If the side opposite the angle is long enough to reach the third side, then several situations can occur: it can be equal to the altitude (one triangle exists), it can be greater than the altitude but less than the other given side (two triangles exists), or it can be greater than the other given side (one triangle exists).

Whether the given angle is acute or obtuse also makes a difference. What we have discussed so far was for an acute angle (four cases). For an obtuse angle, there are two further cases. The side opposite the angle can be less than the other given side (no triangle exists). Or, the side opposite the angle can be greater than the other given side (one triangle exists). In all, there are six cases.

Before we look at sketches of the six cases, we need to represent the altitude of a triangle by using trigonometry. If $\triangle ABC$ is a right triangle, then

$$\sin A = \frac{h}{b}$$
$$b \sin A = h$$
$$h = b \sin A$$

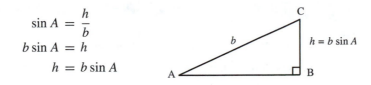

When we refer to the altitude h, we can call it "$b \sin A$" instead of h.

For an acute triangle with sides a, b, and an angle A opposite side a, there are four cases.

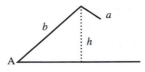

If $a < b \sin A$, then no triangle (solution) exists.

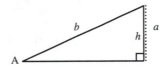

If $a = b \sin A$, then one right triangle (solution) exists.

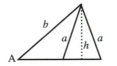

If $b \sin A < a < b$, then two triangles (solutions) exist.

If $a > b$, then one triangle (solution) exists.

For an obtuse triangle with sides a, b, and an angle A opposite side a, there are two cases. We compare a with b in these cases, not with h.

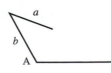

If $a \leqslant b$, then no triangle (solution) exists.

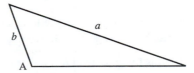

If $a > b$, then one triangle (solution) exists.

Another way to look at this is to think of the length of a getting longer and longer. We will do this for an acute angle.

L

Looking at the diagrams, we can see that if a is shorter than h, it can't reach the other side. Let a get longer. Once it gets long enough to just touch the other side, it will form a right angle, and hence a right triangle. Let a get longer. Now a will pass through the other side and, if it is shorter than b, it can swing to either side to form two triangles. Let a get longer still. Once a is longer, than b it can't swing to two sides anymore. It can only go to one side, thus forming one triangle. Try thinking this way when looking at the obtuse case.

Examples

1. Find the remaining sides and angles of $\triangle ABC$ where $m\angle A = 52°$, $b = 10$, and $a = 2$.

First make a sketch. This immediately shows us that there is no answer.

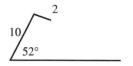

To verify this, we need to look at $b \sin A$ and compare it to a:

$$b \sin A = 10 \sin 52° = 10(.7880)$$
$$= 7.88 \qquad \text{which is bigger than } a$$

Since $a < b \sin A$, i.e., $2 < 7.88$, no triangle exists.

2. In $\triangle ABC$ if $m\angle A = 40°$, $b = 12$, and $a = 9$, find $\angle C$.

First make a sketch.

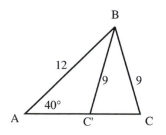

Once we draw the 40° angle with the long side of 12, it's obvious that 9 isn't long enough to make just one triangle. So we end up with two triangles.

We can verify this by looking at $b \sin A$:

$$b \sin A = 12 \sin 40° = 12(.6428) = 7.71$$

Since $b \sin A < a < b$, i.e., since $7.71 < 9 < 12$, there are two triangles.

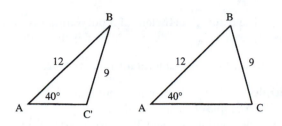

We are looking for two answers, $\angle C$ and $\angle C'$, or $\angle BC'A$. First, let's set up the law of sines for $\angle C$:

$$\frac{\sin 40°}{9} = \frac{\sin C}{12}$$ See **Law of Sines.**

$$\sin C = \frac{12 \sin 40°}{9}$$

$$\sin C = \frac{12(.6428)}{9}$$

$$\sin C = .8571$$ See **Trigonometry, inverse (arc) functions.**

$$C = 58.99°$$

To find $\angle BC'A$, we could set up the law of sines for $\triangle ABC'$, or use supplementary angles. Since $\triangle BCC'$ is isosceles, $\angle C$ and $\angle BC'C$ are the same. Hence, we have

$$m\angle BC'A + m\angle BC'C = 180°$$
$$m\angle BC'A + 58.99° = 180°$$
$$m\angle BC'A = 180° - 58.99°$$
$$m\angle BC'A = 121.01°$$

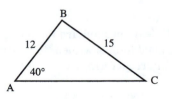

The simplest way to say it is: "Once you find one angle, subtract it from $180°$ and you have the other angle."

3. In example 2, we had two answers because a was less than b. Let's do the problem again except this time let $a = 15$. Make a sketch of the new triangle. Looking at the triangle, we see there is no way we can have two triangles.

This is easy to verify. Since $a > b$, i.e., since $15 > 12$, one solution exists.

L

➤ **Leading Coefficient**   The **leading coefficient of a polynomial** is the coefficient of the first term of the polynomial. In the expression $6x^3y - 4xy^2 - 7$, the 6 is the leading coefficient. See **Coefficient**.

➤ **Least Common Denominator**   See **Common Denominator**.

➤ **Least Common Multiple**   The least common multiple of two or more numbers is the smallest counting number that is a multiple of each.

To find the least common multiple of 2 and 3, we first need to list their multiples:

$$2, 4, \mathbf{6}, 8, 10, \underline{12}, 14, 16, 18, 20, \ldots$$
$$3, \mathbf{6}, 9, \underline{12}, 15, 18, 21, \ldots$$

Both 6 and 12 are common multiples of 2 and 3, but 6 is the least common multiple.

Least common multiples can be found for more than two numbers. To find the least common multiple of 2, 3, and 6, we need to list the multiples of 6:

$$6, \underline{12}, 18, 24, \ldots$$

Looking at the three lists of multiples, we can see that 12 is the least common multiple.

Another way of finding the least common multiple of two or more numbers is to use the prime factorization of each. See **Prime Factorization**. The factors of the least common multiple are the prime factors with the largest exponent. It is the product of these that is the least common multiple. To find the least common multiple of 12 and 18, we first need to prime factor each:

$$12 = 2 \bullet 6 = 2 \bullet 2 \bullet 3 = 2^2 \bullet 3$$
$$18 = 6 \bullet 3 = 2 \bullet 3 \bullet 3 = 2 \bullet 3^2$$

The prime factors with the largest exponents are 2^2 and 3^2. Hence, the least common multiple is $2^2 \bullet 3^2 = 4 \bullet 9 = 36$.

We can verify this by finding the least common multiple using the multiples of 12 and 18:

$$12, 24, \mathbf{36}, 48, \ldots$$
$$18, \mathbf{36}, 54, \ldots$$

The least common multiple is 36.

➤ **Leave in Slope-Intercept Form**   To **leave in slope-intercept form** means to write a linear equation in the form $y = mx + b$. The other form that an equation can be left in is **standard form**, $Ax + By = C$. Sometimes standard form is written $Ax + By + C = 0$. For examples, see **Slope-Intercept Form of an Equation of a Line**.

➤ **Leave in Standard Form**   To **leave in standard form** means to write a linear equation in the form $Ax + By = C$. Sometimes standard form is written $Ax + By + C = 0$. The other form that a linear equation can be left in is **slope-intercept form**, $y = mx + b$. For examples, see **Slope-Intercept Form of an Equation of a Line**.

➤ **Like Terms**   See **Combining Like Terms** or **Term**.

➤ **Limits**   In this article, we will cover three of the more frequent types of limits: **limits** calculated by direct substitution, **limits** in which the denominator becomes 0, and **limits** in which the variable approaches infinity. The most important of these are limits in which the denominator becomes zero. This type of limit is used to derive derivatives by definition. See **Derivative**.

The following is a list of basic limits where $f(x)$ and $g(x)$ are functions, a is a constant, and x is a variable.

1. $\lim_{x \to a} c = c$, where c is a constant

2. $\lim_{x \to a} x = a$

3. $\lim_{x \to a} x^n = a^n$

4. $\lim_{x \to a} cf(x) = c \lim_{x \to a} f(x)$

5. $\lim_{x \to a} f(x) + g(x) = \lim_{x \to a} f(x) + \lim_{x \to a} g(x)$

6. $\lim_{x \to a} f(x) \bullet g(x) = \lim_{x \to a} f(x) \bullet \lim_{x \to a} g(x)$

7. $\lim_{x \to a} \dfrac{f(x)}{g(x)} = \dfrac{\lim_{x \to a} f(x)}{\lim_{x \to a} g(x)}$, where $\lim_{x \to a} g(x) \neq 0$

8. $\lim_{x \to \infty} \dfrac{1}{x} = 0$

9. $\lim_{x \to 0} \dfrac{1}{x} = \infty$

Examples

1. Find $\lim_{x \to 2} 3x + 1$.

   By direct substitution, we have $\lim_{x \to 2} 3x + 1 = 3(2) + 1 = 7$. Simply stated, the limit means that as x gets closer and closer to 2, the expression $3x + 1$ gets closer and closer to 7. We can see this with a chart where $x \to 2$, read "x approaches 2."

x	$3x + 1$
1.5	$3(1.5) + 1 = 5.5$
1.8	$3(1.8) + 1 = 6.4$
1.9	$3(1.9) + 1 = 6.7$
1.99	$3(1.99) + 1 = 6.97$
$\vdots$	$\vdots$
2	$3(2) + 1 = 7$

   If we want to calculate the limit, however, we need not concern ourselves with x approaching 2. We need only know what happens when $x = 2$. So, just substitute x directly into the expression.

2. Find $\lim_{x \to 5} x^2 + 3$.

   By direct substitution, we have $\lim_{x \to 5} x^2 + 3 = 5^2 + 3 = 28$.

L

3. Find $\lim\limits_{x \to -3} x^3 + x$.

By direct substitution, we have

$$\lim_{x \to -3} x^3 + x = (-3)^3 + (-3) = -27 - 3 = -30$$

4. Find $\lim\limits_{x \to -2} 5x^2 - 2x + 3$.

By direct substitution, we have

$$\lim_{x \to -2} 5x^2 - 2x + 3 = 5(-2)^2 - 2(-2) + 3 = 20 + 4 + 3 = 27$$

5. Find $\lim\limits_{x \to 0} \sqrt{3x^2 + 9}$.

By direct substitution, we have

$$\lim_{x \to 0} \sqrt{3x^2 + 9} = \sqrt{3(0)^2 + 9} = \sqrt{0 + 9} = 3$$

6. Find $\lim\limits_{x \to 1} \frac{x^2 + 3x + 4}{x + 1}$.

By direct substitution, we have

$$\lim_{x \to 1} \frac{x^2 + 3x + 4}{x + 1} = \frac{1^2 + 3(1) + 4}{1 + 1} = \frac{1 + 3 + 4}{2} = 4$$

7. Find $\lim\limits_{x \to 2} \frac{x^2 - 4}{x - 2}$.

By direct substitution, we have

$$\lim_{x \to 2} \frac{x^2 - 4}{x - 2} = \frac{2^2 - 4}{2 - 2} = \frac{0}{0} = ?$$

We innocently approached the problem by using direct substitution. However, we got an answer that doesn't make any sense. For this type of problem, we need to factor, cancel, and then try direct substitution:

$$\lim_{x \to 2} \frac{x^2 - 4}{x - 2} = \lim_{x \to 2} \frac{(x + 2)(x - 2)}{(x - 2)} = \lim_{x \to 2} x + 2 = 2 + 2 = 4$$

8. Find $\lim\limits_{x \to 5} \frac{x^2 - 3x - 10}{x - 5}$.

Direct substitution will give us a denominator of 0. Therefore, we need to factor, cancel, then use direct substitution:

$$\lim_{x \to 5} \frac{x^2 - 3x - 10}{x - 5} = \lim_{x \to 5} \frac{(x + 2)(x - 5)}{(x - 5)} = \lim_{x \to 5} (x + 2) = 5 + 2 = 7$$

See **Factoring Trinomials**.

9. Find $\lim\limits_{x \to -2} \frac{x^3+8}{x+2}$.

Direct substitution will give a denominator of 0. Therefore, we need to factor, cancel, then use direct substitution:

$$\lim_{x \to -2} \frac{x^3+8}{x+2} = \lim_{x \to -2} \frac{(x+2)(x^2-2x+4)}{(x+2)} = \lim_{x \to -2} x^2 - 2x + 4$$
$$= (-2)^2 - 2(-2) + 4 = 4 + 4 + 4 = 12$$

See **Factoring the Sum and Difference of Two Cubes**.

10. Find $\lim\limits_{x \to \infty} \frac{6-5x^2}{3x+4x^2}$.

By direct substitution, we see that we will get $\frac{\infty}{\infty}$. Do not factor. Factoring works when the denominator becomes 0. When $x \to \infty$, we need to divide the numerator and denominator by the term with the highest power in the denominator. This will leave fractions that approach 0 when $x \to \infty$:

$$\lim_{x \to \infty} \frac{6-5x^2}{3x+4x^2} \frac{\frac{1}{x^2}}{\frac{1}{x^2}} = \lim_{x \to \infty} \frac{\frac{6-5x^2}{x^2}}{\frac{3x+4x^2}{x^2}} = \lim_{x \to \infty} \frac{\frac{6}{x^2} - \frac{5x^2}{x^2}}{\frac{3x}{x^2} + \frac{4x^2}{x^2}} = \lim_{x \to \infty} \frac{\frac{6}{x^2} - 5}{\frac{3x}{x} + 4} = \frac{0-5}{0+4} = \frac{-5}{4}$$

Dividing both the numerator and the denominator by x^2 is the same as multiplying by $\frac{\frac{1}{x^2}}{\frac{1}{x^2}}$. Once each $\frac{1}{x^2}$ is distributed, we end up with fractions that have x's in the denominator. Herein lies the reason this method works. When $x \to \infty$, each fraction becomes 0.

We can see why this happens by considering $\lim_{x \to \infty} \frac{1}{x}$. As x gets larger and larger, the fraction $\frac{1}{x}$ gets smaller and smaller. We can see this with a chart.

x	$\frac{1}{x}$
10	$\frac{1}{10}$
100	$\frac{1}{100}$
1000	$\frac{1}{1000}$
10,000	$\frac{1}{10,000}$
$\vdots$	$\vdots$

The same is true for $\lim_{x \to \infty} \frac{1}{x^2}$ or $\lim_{x \to \infty} \frac{-6}{x^3}$. They both become 0 as x gets larger, when $x \to \infty$.

11. Find $\lim\limits_{x \to \infty} \frac{5x^2-2}{3x^3+2x^2}$.

By direct substitution, we see that we will get $\frac{\infty}{\infty}$. The term with the highest power in the denominator is x^3. So, we will divide both the numerator and denominator by x^3, which is the same as

L

multiplying by $\frac{\frac{1}{x^3}}{\frac{1}{x^3}}$:

$$\lim_{x\to\infty} \frac{5x^2 - 2}{3x^3 + 2x^2} = \lim_{x\to\infty} \frac{5x^2 - 2}{3x^3 + 2x^2} \frac{\frac{1}{x^3}}{\frac{1}{x^3}} + \lim_{x\to\infty} \frac{\frac{5x^2-2}{x^3}}{\frac{3x^3+2x^2}{x^3}}$$

$$= \lim_{x\to\infty} \frac{\frac{5x^2}{x^3} - \frac{2}{x^3}}{\frac{3x^3}{x^3} + \frac{2x^2}{x^3}} = \lim_{x\to\infty} \frac{\frac{5}{x} - \frac{2}{x^3}}{3 + \frac{2}{x}} = \frac{0 - 0}{3 + 0} = 0$$

➤ **Linear Combination Method**   The **linear combination method** is one of several methods used to solve two equations in two unknowns. See **Systems of Linear Equations**.

➤ **Linear Equation**   A **linear equation** is an equation that pertains to a line. When solved for y, a linear equation takes on the form $y = mx + b$. When graphed, a linear equation is the graph of a line. For more, see the headings under **Linear Equation**.

➤ **Linear Equation, finding the equation of a line**   There are three methods for **finding the equation of a line**: the slope-intercept method, which has the same name as one of the methods of graphing a linear equation, the point-slope method, and the two-point method. These specific names may not be the names used in some algebra texts; they are, however, the concepts taught in all algebra texts. For information relating to equations of horizontal and vertical lines, see **Horizontal and Vertical Lines**.

Slope-Intercept Method

To find the equation of a line by the **slope-intercept method**, it is required, at least, that a point and the slope of the line are known. The method is as follows: substitute the point and the slope into $y = mx + b$ and solve for the intercept b, then substitute the slope and the intercept into $y = mx + b$ and leave the equation either in slope-intercept form ($y = mx + b$) or standard form ($Ax + By = C$, or $Ax + By + C = 0$ depending on the text).

> *Examples*

1. Find the equation of the line through the point $(3, 4)$ with a slope of $m = 5$ by the slope-intercept method.

   When we are given a point, we are given two pieces of information, an x value and a y value. To find the equation of the line, we do two things: substitute the x, y, and m into $y = mx + b$ and solve for b, then substitute m and b into $y = mx + b$ to get the equation of the line.

   Substitute x, y, and m:

   $$y = mx + b$$
   $$4 = 5(3) + b$$
   $$4 = 15 + b$$
   $$4 - 15 = b$$
   $$-11 = b$$
   $$b = -11 \qquad \text{See } \textbf{Linear Equations, solutions of.}$$

Substitute m and b:

$$y = mx + b$$
$$y = 5x - 11$$

In slope-intercept form, the answer is $y = 5x - 11$. In standard form, the answer is $5x - y = 11$. The slope-intercept method leaves the answer in slope-intercept form. If you need the answer in standard form (we will use the standard form $Ax + By = C$), move the x and y terms to the left side of the equal sign; if there is a negative on the x term divide (or multiply) the equation by -1, then clear the x and y terms of any fractions. For examples, see **Standard Form of an Equation of a Line**.

$$y = 5x - 11$$
$$-5x + y = -11$$
$$\frac{-5x}{-1} + \frac{y}{-1} = \frac{-11}{-1}$$
$$5x - y = 11$$

2. Find the equation of the line through the point $(-2, -4)$ with a slope of $m = -3$ by the slope-intercept method.

Substitute x, y, and m into $y = mx + b$ and solve for b:

$$y = mx + b$$
$$-4 = -3(-2) + b$$
$$-4 = 6 + b$$
$$-4 - 6 = b$$
$$-10 = b$$
$$b = -10$$

Then substitute m and b into $y = mx + b$:

$$y = mx + b$$
$$y = -3x - 10$$

In slope-intercept form, the answer is $y = -3x - 10$. In standard form, the answer is $3x + y = -10$.

Point-Slope Method

To find the equation of a line by the **point-slope method**, it is also required, at least, that a point and the slope of the line are known. The method is as follows: substitute the point and the slope into $y - y_1 = m(x - x_1)$ and then either solve for y to leave the equation in slope-intercept form or leave the equation in standard form. The equation $y - y_1 = m(x - x_1)$ is called the **point-slope form** of the equation of a line.

L

The point-slope form of the equation of a line is derived from the slope equation $m = \frac{y_2 - y_1}{x_2 - x_1}$. See **Slope**. Cross multiplying by $x_2 - x_1$, we have

$$m = \frac{y_2 - y_1}{x_2 - x_1}$$
$$m(x_2 - x_1) = y_2 - y_1$$
$$y_2 - y_1 = m(x_2 - x_1) \qquad \text{See \textbf{Symmetric Property}.}$$
$$y - y_1 = m(x - x_1)$$

The subscripts of 2 are dropped in the last equation because they are no longer needed. When we derived the slope equation, they had a purpose. They were used to tell the differences between one point and another. We could have used (x_1, y_1) and (x, y) to derive the slope equation but we usually reserve (x, y) to represent a general point and use the subscripts to represent a particular point. In $y - y_1 = m(x - x_1)$, the subscripts of 2 aren't needed because we are thinking of the x and y in general and the x_1 and y_1 in particular.

> ### Examples (continued)

3. Find the equation of the line through the point $(3, 4)$ with a slope of $m = 5$ by the point-slope method.

The point-slope equation must be memorized. To find the equation of the line by the point-slope method, we first write down the point-slope equation, substitute m, x_1, and y_1, and then leave the equation in one of the two forms, slope-intercept form or standard form:

$$y - y_1 = m(x - x_1)$$
$$y - 4 = 5(x - 3)$$
$$y - 4 = 5x - 15$$
$$y = 5x - 15 + 4$$
$$y = 5x - 11$$

We solve for y if we want the equation in slope-intercept form, or we also can leave the equation in standard form. For help solving the equations, see **Linear Equations, solutions of**.

$$y - 4 = 5x - 15$$
$$-5x + y = 4 - 15$$
$$-5x + y = -11$$
$$\frac{-5x}{-1} + \frac{y}{-1} = \frac{-11}{-1}$$
$$5x - y = 11$$

Whether equations are found by the slope-intercept method or by the point-slope method, the answers are still the same. Compare example 1 and example 3. The reason we have different methods for finding equations has to do with application. Different approaches to a problem will require a different method.

L

4. Find the equation of the line through the point $(-2, -4)$ with a slope of $m = -3$ by the point-slope method.

Writing down the point-slope equation and substituting m, x_1, and y_1, we have

$$y - y_1 = m(x - x_1)$$
$$y - (-4) = -3\big[x - (-2)\big] \qquad \text{Be careful with the double negatives.}$$
$$y + 4 = -3(x + 2)$$
$$y + 4 = -3x - 6$$
$$y = -3x - 6 - 4$$
$$y = -3x - 10 \qquad \text{Slope-intercept form.}$$

or

$$y + 4 = -3x - 6$$
$$3x + y = -6 - 4$$
$$3x + y = -10 \qquad \text{Standard form.}$$

Two-Point Method

The **two-point method** is not really a method completely on its own. It uses slope-intercept and point-slope methods once the slope is known.

Whereas the slope-intercept and point-slope methods need minimum information of a point and a slope, the two-point method, as the name implies, needs two points.

In order to use the slope-intercept or the point-slope methods, we have to know the value of m. If we are given two points, we can't use either method because we don't know the value of m. However, we can calculate the value of m using the slope equation where $m = \frac{y_2 - y_1}{x_2 - x_1}$. See **Slope**.

Hence, with two points we can find m. And if we have m and a point (since we have two points), we can find the equation of the line using either the slope-intercept method or the point-slope method.

The method for finding the equation of a line by the two-point method is as follows: substitute the two points into the slope equation to find m, then use one of the points and m to find the equation of the line by the slope-intercept method or by the point-slope method.

Examples (continued)

5. Find the equation of the line through the points $(3, 4)$ and $(2, -1)$ by the two-point method.

We first need to find m, then using m and either point we can find the equation.

Finding m, we have

$$m = \frac{y_2 - y_1}{x_2 - x_1} = \frac{-1 - 4}{2 - 3} = \frac{-5}{-1} = 5$$

When substituting into the slope equation, we can let $(3, 4) = (x_1, y_1)$ and $(2, -1) = (x_2, y_2)$ or $(3, 4) = (x_2, y_2)$ and $(2, -1) = (x_1, y_1)$. Either way, $m = 5$. Try it.

L

Now with $m = 5$ and a point $(3, 4)$ or $(2, -1)$, we can find the equation of the line by either the slope-intercept method or the point-slope method. In examples 1 and 3, we found the equation using $(3, 4)$ and $m = 5$. Let's find the equation using $(2, -1)$ and $m = 5$ by the point-slope method. Try finding the equation by the slope-intercept method.

$$y - y_1 = m(x - x_1)$$
$$y - (-1) = 5(x - 2)$$
$$y + 1 = 5x - 10$$
$$y = 5x - 11$$

This is exactly what we get by the other methods.

Sometimes it appears as though not enough information is given to solve a problem. That is because the information is disguised. As an example, if we are given an x-intercept of 5 and a y-intercept of 10 and then asked to find the equation, it appears as though there isn't enough information. But the x-intercept is a point. It's the point where the graph crosses the x-axis. It crosses at $(5, 0)$. Similarly, the y-intercept is $(0, 10)$. Hence, we have two points. Find m and go on from there.

If we are going to find the equation of a line using $y = mx + b$, we need to know the m and b. Usually, we use m and a point to find b. Since b is the point where the graph crosses the y-axis, if we know the y-intercept we know b. See **Linear Equations, graphs of**. Hence, if the y-intercept is 10, then $b = 10$.

In the example above, using $(5, 0)$ and $(0, 10)$, we can find m:

$$m = \frac{10 - 0}{0 - 5} = -2$$

Since $b = 10$ and $m = -2$, we have the equation $y = -2x + 10$.

As one more example, suppose we are given $b = 7$ and $m = -4$. Since these are both the pieces of information we need, we have the equation $y = -4x + 7$.

For information relating to equations of horizontal and vertical lines, see **Horizontal and Vertical Lines**.

➤ **Linear Equations, graphs of**   There are three methods of **graphing linear equations**: graphing by the charting method, graphing by the slope-intercept method, and graphing by the x-y intercept method (read "x hyphen y," not "x minus y"). Although these specific names may not be the names used in some algebra texts, they are the concepts taught in all algebra texts. For information relating to graphs of horizontal and vertical lines, see **Horizontal and Vertical Lines**.

Charting Method

The **charting method** is the most cumbersome of graphing techniques. It is generally used to get an idea of what an equation looks like, but because of its awkward nature, simpler methods are sought.

To apply the charting method, we must first solve the given equation for y, put it in a chart, choose values for x, calculate the values of y, plot the points, sketch the graph, and label the graph.

Examples

1. Graph $-2 = 4x - 2y$ by charting.

   We first need to solve the equation for y:

   $$-2 = 4x - 2y$$
   $$2y = 4x + 2$$
   $$\frac{2y}{2} = \frac{4x}{2} + \frac{2}{2}$$
   $$y = 2x + 1 \qquad \text{See } \textbf{Linear Equations, solutions of.}$$

   Next, we put the equation in a chart, choose some x values, and calculate the y values. Choose any x values you want. We will use 0, 1, and 2.

x	$y = 2x + 1$
0	$2(0) + 1 = 0 + 1 = 1$
1	$2(1) + 1 = 2 + 1 = 3$
2	$2(2) + 1 = 4 + 1 = 5$

   To calculate the y values, we substitute each x into $2x + 1$ and simplify.

   The x is the **independent variable** and the y is the **dependent variable**. We choose the x values independently. Since we then substitute each x into $2x + 1$, the y values each depend on the x values.

   The whole list (set) of x's is called the **domain**; the whole list (set) of y's is called the **range**. We choose the domain but we calculate the range.

   We can now plot the points, sketch the graph, and label the graph.

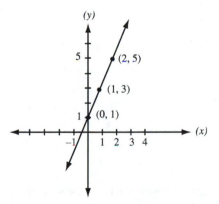

2. Graph $9x = -3y + 6$ by charting.

   Solving the equation for y, we have

   $$9x = -3y + 6$$
   $$3y = -9x + 6$$

L

$$\frac{3y}{3} = \frac{-9x}{3} + \frac{6}{3}$$
$$y = -3x + 2$$

Putting the equation in a chart and choosing x values of 0, 1, and 2, we have

x	$y = -3x + 2$
0	$-3(0) + 2 = 0 + 2 = 2$
1	$-3(1) + 2 = -3 + 2 = -1$
2	$-3(2) + 2 = -6 + 2 = -4$

Graphing, we have

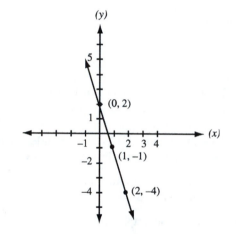

3. Graph $2x + 3y = -3$ by charting.

Solving the equation, we have

$$2x + 3y = -3$$
$$3y = -2x - 3$$
$$\frac{3y}{3} = \frac{-2x}{3} - \frac{3}{3}$$
$$y = -\frac{2}{3}x - 1$$

We need to choose x values that will not leave us with fractions. There is nothing wrong with fractions; they're just harder to plot when graphing.

x	$y = -\frac{2}{3}x - 1$
0	$-\frac{2}{3}(0) - 1 = 0 - 1 = -1$
3	$-\frac{2}{3}(3) - 1 = -2 - 1 = -3$
6	$-\frac{2}{3}(6) - 1 = -4 - 1 = -5$

Choose x values that are multiples of the denominator.

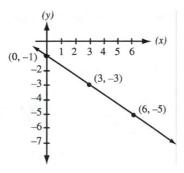

Slope-Intercept Method

The **slope-intercept method** is the most commonly used method of graphing linear equations. It can be derived by looking at a chart.

Look at the chart of example 1. Notice that as the x values each increase by 1, the y values each increase by 2, which is the number in front of the x in $y = 2x + 1$. Look at the chart of example 2. The same thing happens. As the x's increase by 1, the y's decrease by -3, which is the coefficient of x in $y = -3x + 2$. This coefficient is called the slope of the equation.

Now look at the graph of example 1. Notice that it crosses the y-axis at 1, the number at the end of $y = 2x + 1$. This also happens in example 2. The graph in example 2 crosses the y-axis at 2, which is the constant term in $y = -3x + 2$. Just by looking at an equation, we know what its y-intercept is.

When graphing an equation, the only two pieces of information we need are the y-intercept and the slope. The y-intercept tells us where the graph will cross the y-axis. The slope tells us how to get to another point. For $y = 2x + 1$, we start at 1 on the y-axis, then go over 1 and up 2 to get to the next point. Then again we can go over 1 and up 2 to get to the next point, and so on.

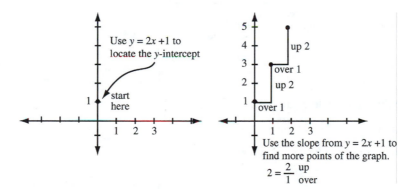

Once the points have been located, connect the points and label the graph. The result is the same as example 1.

Similarly, we could graph example 2 by this method. From the equation $y = -3x + 2$, we know that the y-intercept is 2 and other points can be found by graphing over 1 down -3, over 1 down -3, and so on. Look at the graph and convince yourself that this is the case.

L

4. Graph $y = 3x - 1$ by the slope-intercept method.

   The y-intercept is -1. From there we use the slope to go over 1 and up 3.

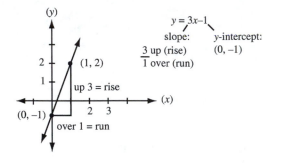

   If the rise is positive, we go up; if negative, we go down.

5. Graph $y = -\frac{2}{3}x - 1$ by the slope-intercept method.

   The negative in $-\frac{2}{3}$ can be put in either the numerator or the denominator. If it is put in the numerator, we go over 3 and down -2. If it is put in the denominator, we go left -3 and up 2. Try it both ways to see that it works. We usually use the former.

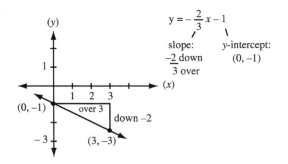

   Compare this graph with example 3.

6. Graph $-12x = -3y - 6$ by the slope-intercept method.

   We first need to solve the equation for y:

$$-12x = -3y - 6$$
$$3y = 12x - 6$$
$$\frac{3y}{3} = \frac{12x}{3} - \frac{6}{3}$$
$$y = 4x - 2$$

   From -2 we go over 1 and up 4.

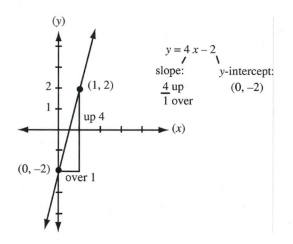

x-y Intercept Method

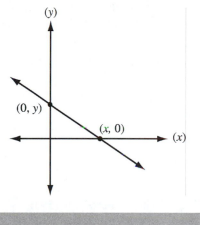

The **x-y intercept** (read "*x, y* intercept") **method** is most useful when the equation is in standard form. See **Standard Form of an Equation of a Line**. The method is derived from the observation that all lines, except the horizontal and vertical lines, cross both the *x*- and *y*-axes (plural of axis). When a line crosses the *x*-axis, the value of *y* is 0. When a line crosses the *y*-axis, the value of *x* is 0. So in the equation if we set $y = 0$, we will get the *x*-intercept. If we set $x = 0$, we will get the *y*-intercept. Connecting both the *x* and *y* intercepts gives us the graph of the line.

Examples (continued)

7. Graph $4x + 3y = 12$ by the *x-y* intercept method.

When we set $y = 0$, we get the *x*-intercept:

$$4x + 3y = 12$$
$$4x + 3(0) = 12$$
$$4x = 12$$
$$x = 3$$

The *x*-intercept is $(3, 0)$.

When we set $x = 0$, we get the *y*-intercept:

$$4x + 3y = 12$$
$$4(0) + 3y = 12$$
$$3y = 12$$
$$y = \frac{12}{3}$$
$$y = 4$$

The *y*-intercept is $(0, 4)$.

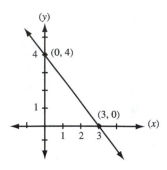

Instead of doing all of the arithmetic, however, all we really need to do is look at the equation. In the equation $4x + 3y = 12$, put your finger over the $3y$ term. What is left is $4x = 12$, which you can do in your mind, $x = 3$. The graph will cross the x-axis at 3. Similarly, put your finger over the $4x$ term. What is left is $3y = 12$, which you also can do in your mind, $y = 4$. The graph will cross the y-axis at 4.

8. Graph $2x + 5y = 10$ by the x-y intercept method.

If we try the cover-and-graph approach, we see that with a finger over $5y$, $2x = 10$ or $x = 2$. The x-intercept is 2. With a finger over $2x$, $5y = 10$ or $y = 2$. The y-intercept is 2. Plot the two points and draw the graph.

Graphing by the complete x-y intercept method, we have

When we set $y = 0$, we get the x-intercept:

$$2x + 5y = 10$$
$$2x + 5(0) = 10$$
$$2x = 10$$
$$x = 5$$

The x-intercept is $(5, 0)$.

When we set $x = 0$, we get the y-intercept:

$$2x + 5y = 10$$
$$2(0) + 5y = 10$$
$$5y = 10$$
$$y = 2$$

The y-intercept is $(0, 2)$.

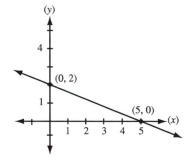

If an equation is in slope-intercept form, graph by the slope-intercept method. If an equation is in standard form, graph by the x-y intercept method. If graphing by the x-y intercept method produces fractions that are too difficult to graph, solve the equation for y and graph by the slope-intercept method.

➤ **Linear Equations, solutions of**   **Linear equations** are first-degree equations in one variable. When simplified, they can be solved for the variable.

In order to solve linear equations, the concepts of positive and negative numbers must first be completely clear. See **Positive and Negative Numbers**.

In this article, we will develop a four-step method for solving linear equations. Each example will be solved with this approach. The method applies the following steps:

1. Multiply into the parentheses (or brackets or braces) first. This will leave a long chain of terms, so to speak.

2. Move all the variable terms to one side of the equal sign, the number terms to the other, and change the signs of the terms that are moved.

3. Combine the variable terms into one term and the numbers into one number.

4. Divide or multiply as indicated.

In simplified form, we could say that, to solve linear equations, we multiply into the parentheses, then move, combine, and divide (or multiply).

Examples

1. Solve the equation $n + 4 = 10$ for n.

Since there aren't any parentheses, we skip the first step. Next, we move the $+4$ to the right-hand side of the $=$ and change it to -4. Then we combine and skip the last step:

$$n + 4 = 10$$
$$n = 10 - 4$$
$$n = 6$$

Simply, to solve a linear equation means we are looking for a number (or letter) that will satisfy the given conditions. If we are asked to solve the equation $n + 4 = 10$, we are being asked to find a number that added to 4 will equal 10. The number of course is 6 since $6 + 4 = 10$. As simple as this may seem, we can use this example to help develop a method with which to solve linear equations.

Notice, in the equation $n + 4 = 10$, if we subtract 4 from both sides of the equation, we get the desired answer:

$$n + 4 = 10$$
$$n + 4 - 4 = 10 - 4$$
$$n + 0 = 6$$
$$n = 6$$

We use opposite operations to solve the equation. Since the 4 of $n + 4$ is positive, we subtract 4 from both sides of the equation. This allows the left side of the equation to simplify to n and the right side to simplify to 6.

L

Subtracting 4 from both sides of the equation is cumbersome. We already know that the opposite operation on the left side of the equation is going to cancel. So, why write it? We just need to do the right side of the equation. This is why we make up the rule: pick up the $+4$, move it to the other side of the $=$, and change its sign to a -4:

$$n + 4 = 10$$
$$n = 10 - 4$$
$$n = 6$$

In order to complete problems like this, the concept of positive and negative numbers must be clear.

See **Positive and Negative Numbers**.

2. Solve the equation $n - 7 = 2$ for n.

$$n - 7 = 2$$
$$n = 2 + 7$$
$$n = 9$$

Move the -7 to the right side of the $=$ and change it to a $+7$.

Comparing this to the method of opposite operations, we have

$$n - 7 = 2$$
$$n - 7 + 7 = 2 + 7$$
$$n + 0 = 9$$
$$n = 9$$

3. Solve $n - 5 = -8$.

$$n - 5 = -8$$
$$n = -8 + 5$$
$$n = -3$$

Move the -5 and change it to a $+5$.

To combine $-8 + 5$, see **Positive and Negative Numbers**.

4. Solve $6n - 4n = 12$.

There are no parentheses, so we skip the first step. The variable and number terms are on their respective sides, so skip the second step. We must combine $6n - 4n$ and divide by 2:

$$6n - 4n = 12$$
$$2n = 12$$
$$\frac{2n}{2} = \frac{12}{2}$$
$$n = 6$$

Combine, then divide by 2 not by $2n$.

The equation $6n - 4n = 12$ means that we are asked to find a number such that 6 times the number subtracted by 4 times the number will equal 12. We know that 6 is the number because $6 \bullet 6 - 4 \bullet 6 = 36 - 24 = 12$.

L

The term $6n$ means 6 times the number n. We write it as $6n$ instead of $6xn$ because the x is quite often used as a variable. With x as a variable, the equation $6n - 4n = 12$ could have been written as $6x - 4x = 12$. We would then solve the equation for x:

$$6x - 4x = 12$$
$$2x = 12$$
$$\frac{2x}{2} = \frac{12}{2}$$
$$x = 6$$

Regardless of the letter used for the variable, the answer is 6.

In the solution of the equation when we get to $2n = 12$, we are being asked to find a number such that 2 times the number is equal to 12. Using opposite operations, we see that if we divide both sides of the equation by 2, we get the desired answer:

$$2n = 12$$
$$\frac{2n}{2} = \frac{12}{2} \qquad \text{Divide by 2 not } 2n.$$
$$n = 6$$

The last step of an equation is usually divide (or multiply). If the equation is a product, i.e., $2n = 12$, then we divide. If the equation is a quotient, i.e., $\frac{n}{3} = 5$ or $\frac{1}{3}n = 5$, then we multiply.

To solve the equation $\frac{n}{3} = 5$, we use opposite operations. Since n is divided by 3, we do the opposite, we multiply by 3:

$$\frac{n}{3} = 5 \qquad \text{or} \qquad \frac{1}{3}n = 5$$
$$3 \bullet \frac{n}{3} = 5 \bullet 3 \qquad\qquad 3 \bullet \frac{1}{3}n = 5 \bullet 3$$
$$n = 15 \qquad \text{Check: } \frac{15}{3} = 5. \qquad 1n = 15$$
$$n = 15 \qquad \text{Since } 1n \text{ means 1 times } n,$$
$$\text{1}n \text{ is the same as } n.$$

We could also have multiplied by the reciprocal, $\frac{3}{1}$:

$$\frac{1}{3}n = 5$$
$$\frac{3}{1} \bullet \frac{1}{3}n = 5 \bullet \frac{3}{1}$$
$$\frac{3}{3}n = \frac{15}{1}$$
$$n = 15 \qquad \text{We write } n \text{ instead of } 1n.$$

5. Solve $8n - 4 = 2n + 14$.

Since there aren't any parentheses, we skip the first step. By the second step, we will move the -4 and the $2n$ and change their signs, then combine and divide:

$$8n - 4 = 2n + 14$$
$$8n - 2n = 14 + 4 \qquad \text{$2n$ becomes $-2n$, and -4 becomes 4 when we move.}$$
$$6n = 18$$
$$\frac{6n}{6} = \frac{18}{6}$$
$$n = 3$$

When terms are moved, it doesn't matter on which side of each other they are placed:

$$8n - 4 = 2n + 14$$
$$-2n + 8n = 4 + 14$$
$$6n = 18$$
$$\frac{6n}{6} = \frac{18}{6}$$
$$n = 3$$

It can be seen with this problem why it is best to drop the usage of opposite operations. With opposite operations, we have

$$8n - 4 = 2n + 14$$
$$8n - 2n - 4 + 4 = 2n - 2n + 14 + 4$$
$$6n = 18$$
$$\text{etc.}$$

Sometimes students are taught to place opposite operations underneath the equation:

$$
\begin{array}{rcl}
8n - 4 & = & 2n + 14 \\
-2n + 4 & & -2n + 4 \\
\hline
6n + 0 & = & 0 + 18 \\
6n & = & 18 \\
& \text{etc.} &
\end{array}
$$

This, too, becomes cumbersome, especially in large problems.

6. Solve $4(2n - 1) = 2(n + 7)$.

We first need to multiply the 4 and the 2 inside the parentheses, then move, combine, and divide:

$$4(2n - 1) = 2(n + 7)$$
$$8n - 4 = 2n + 14$$

The rest of this problem is the same as example 5.

L

To multiply inside the parentheses, we are using the concept of the distributive property. See **Distributive Property**.

When multiplying $4(2n - 1)$, the 4 multiplies both the terms inside the parentheses. The result is $8n - 4$. Similarly, $2(n + 7)$ yields $2n + 14$.

7. Solve $-2(3n - 4) + 12 = -2n$.

We first multiply into the parentheses, then move, combine, and divide:

$$-2(3n - 4) + 12 = -2n$$
$$-6n + 8 + 12 = -2n$$
$$-6n + 2n = -8 - 12 \qquad \text{The } -2n \text{ moves over as a } +2n \text{ while } +8$$
$$-4n = -20 \qquad\qquad \text{and } +12 \text{ move over as } -8 \text{ and } -12.$$
$$\frac{-4n}{-4} = \frac{-20}{-4}$$
$$n = 5$$

To combine $-6n + 2n$, we use the following rule: if the signs are different, subtract and keep the sign of the larger number. Therefore, $-6n + 2n$ yields $-4n$. To combine $-8 - 12$, we use the other rule: if the signs are the same, add and keep the same sign. Therefore, $-8 - 12$ yields -20. See **Positive and Negative Numbers**.

When dividing $\frac{-4}{-4}$ or $\frac{-20}{-4}$, we use the first rule for multiplication and division of numbers: if the signs are alike, the answer is positive. See **Positive and Negative Numbers**.

As proficiency develops, combining steps is encouraged, also, doing steps mentally. We could have solved this problem more simply:

$$-2(3n - 4) + 12 = -2n$$
$$-6n + 8 + 12 = -2n$$
$$-4n = -20$$
$$n = 5$$

8. Solve $\frac{1}{5}(30n - 15) - \frac{1}{8}(24n + 16) = n + 5$.

We use all four steps to solve this equation:

$$\frac{1}{5}(30n - 15) - \frac{1}{8}(24n + 16) = n + 5$$
$$\frac{30}{5}n - \frac{15}{5} - \frac{24}{8}n + \frac{16}{8} = n + 5$$
$$6n - 3 - 3n - 2 = n + 5$$
$$6n - 3n - n = 5 + 3 + 2$$
$$2n = 10$$
$$\frac{2n}{2} = \frac{10}{2}$$
$$n = 5$$

L

Before we started this problem, we could have cleared the equation of any fractions. See **Rational Equations**. Sometimes this is advantageous. Think forward in the problem a step or two. If the fractions are multiplied into the equation, will they reduce? If so, then take that approach. If not, then clearing the equation of fractions might be beneficial. In our example they reduced.

9. Solve $6[4(3n - 5) - (7n - 6)] = -(9n + 20)$.

Some problems have a – in front of the parentheses. This really is a -1. The expression $-(7n - 6)$ is actually $-1(7n - 6)$. The 1 is usually not written. In either case the $-$, or -1, is distributed inside the parentheses just as any other number. Hence, $-(7n - 6)$ or $-1(7n - 6)$ both yield $-7n + 6$:

$$6\big[4(3n - 5) - (7n - 6)\big] = -(9n + 20)$$
$$6[12n - 20 - 7n + 6] = -9n - 20$$
$$6[5n - 14] = -9n - 20$$
$$30n - 84 = -9n - 20$$
$$30n + 9n = 84 - 20$$
$$39n = 64$$
$$\frac{39n}{39} = \frac{64}{39}$$
$$n = 1\frac{25}{39}$$

We could multiply into the parentheses and then into the brackets at $6[12n - 20 - 7n + 6]$. However, it is easier to simplify the terms inside the brackets and then multiply at $6[5n - 14]$. Either way will work.

We multiplied inside the parentheses first because of the order of operations. See **Order of Operations**.

➤ **Linear Inequalities**   See **Inequalities, linear solutions of**; **Inequalities, graph on a number line**; or **Inequalities, graph in the plane**.

➤ **Linear Interpolation**   See **Interpolation, linear**.

➤ **Linear Pair**   A **linear pair** is composed of two adjacent angles that have two of their rays as opposite rays. $\angle 1$ and $\angle 2$ form a linear pair because they are adjacent angles and because two of their rays, $\overrightarrow{OA}$ and $\overrightarrow{OB}$, are opposite rays. Obviously the angles of a linear pair are supplementary, but that is not part of their definition. This must either be proved or postulated.

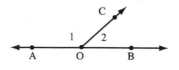

➤ **Linear System of Equations**   See **Systems of Linear Equations**.

➤ **Line, equation of**   See **Linear Equation, finding the equation of a line**.

➤ **Line, graph of**   See **Linear Equations, graphs of**.

➤ **Locus**   The **locus** of an equation is the graph of an equation. The curve that contains the set of points whose coordinates satisfy a given equation is called the locus of the equation.

➤ **Logarithm**   A **logarithm** is an exponent. In common or Briggsian logarithms, the logarithm is the exponent to which 10 must be raised to yield a given number. In natural or Napierian logarithms, the logarithm is the exponent to which e must be raised to yield a given number.

If the given number is 25, then in either case, respectively,

$$10^{1.3979} \approx 25 \qquad \text{where 1.3979 is the logarithm of 25, written } \log_{10} 25 = 1.3979$$

or $\quad e^{3.2189} \approx 25 \qquad$ where 3.2189 is the logarithm of 25, written $\log_e 25 = \ln 25 = 3.2189$

We use ln instead of writing $\log_e$.

For a clearer understanding of what a logarithm is, see **Logarithm Table (chart), how to use**. Also see **Common Logarithms**, **Natural Logarithms**, or articles listed under **Logarithm**.

➤ **Logarithm, base of**   In the equation $y = \log_b x$, b is called the **base of the logarithm**. It is referred to as a base because of its position in exponential form. The exponential form of the equation $y = \log_b x$ is written $b^y = x$. In this form, b is clearly the base and y is the exponent to which b must be raised to yield x.

In logarithmic form, the equation $y = \log_b x$ is read "y is the logarithm to the base b of x." Even though y is now called a logarithm, it is still the exponent to which the base b must be raised to yield x.

The base of a logarithm can be any positive real number greater than one. The most commonly used bases are 10 and e. The 10 is the base used in common logarithms and the e is the base used in natural logarithms. See **Common Logarithms** or **Natural Logarithms**.

Sometimes it is advantageous to change from one base to another. Whether the change is from base 10 to base e, or any other base change, a formula has been specifically developed. See **Change of Base Formula**.

➤ **Logarithm, common or Briggsian**   See **Common Logarithms**.

➤ **Logarithmic Equations**   In this article, we will solve equations that have logarithms in the equations. To solve equations using logarithms, see **Exponential Equation, solution of**.

There are generally two methods used to solve logarithmic equations. In one method, we simplify both sides of the equation until we get $\log_b p = \log_b q$, then we can set $p = q$ and solve for the variable. In the other, we simplify the equation until we get the form $\log_b x = y$, then we convert to an exponential equation and solve for the variable.

Examples

1. Solve $\log_7 x^3 - \log_7 x = 2$, with $x \neq 0$.

In this and the following examples, we will refer to the properties of logarithms as either property 1, 2, or 3. See **Logarithms, properties of (rules of)**.

L

To solve this equation, we need to use property 2 in reverse order, $\log_b p - \log_b q = \log_b \frac{p}{q}$. The reason we were given the restriction $x \neq 0$ is so there is no division by zero.

$$\log_7 x^3 - \log_7 x = 2$$

$$\log_7 \frac{x^3}{x} = 2 \qquad \text{Property 2. } (x \neq 0 \text{ restricts division by zero.})$$

$$\log_7 x^2 = 2$$

$$7^2 = x^2 \qquad \text{See \textbf{Logarithmic Form to Exponential Form, converting from}.}$$

$$x^2 = 49$$

$$\sqrt{x^2} = \sqrt{49} \qquad \text{See \textbf{Quadratic Equations, solutions of}.}$$

$$x = 7$$

The solution to the equation is $x = 7$. Although the equation $x^2 = 49$ has two solutions, $x = 7$ and $x = -7$, we use the positive value only. We cannot calculate the logarithms of negative numbers. To see this, substitute $x = -7$ in $\log_7 x$. If the substitution is valid, we should have $\log_7(-7) = a$. Changing to exponential form, we have $7^a = -7$. But no value of a can change 7^a to -7. Therefore, x cannot be -7.

2. Solve the equation $\log_7 16 + 2\log_7(x - 2) = 2\log_7 2(x + 1)$.

Watch for extraneous solutions.

$$\log_7 16 + 2\log_7(x - 2) = 2\log_7 2(x + 1)$$

$$\log_7 16 + \log_7(x - 2)^2 = \log_7[2(x + 1)]^2 \qquad \text{Property 3 on both sides.}$$

$$\log_7 16(x - 2)^2 = \log_7 4(x + 1)^2 \qquad \text{Property 2 on the left side.}$$

Now that we have $\log_7 p = \log_7 q$, we can set $p = q$:

$$16(x - 2)^2 = 4(x + 1)^2$$

$$16(x - 2)(x - 2) = 4(x + 1)(x + 1)$$

$$16(x^2 - 4x + 4) = 4(x^2 + 2x + 1) \qquad \text{See \textbf{FOIL}.}$$

$$16x^2 - 64x + 64 = 4x^2 + 8x + 4$$

$$12x^2 - 72x + 60 = 0$$

$$x^2 - 6x + 5 = 0 \qquad \text{Divide through by 12. See \textbf{Quadratic Equations,}}$$

$$(x - 1)(x - 5) = 0 \qquad \text{\textbf{solutions of}.}$$

$$x - 1 = 0 \qquad \text{or} \qquad x - 5 = 0$$

$$x = 1 \qquad\qquad\qquad x = 5$$

The answer $x = 1$ is an extraneous solution since substitution in $\log_7(x - 2)$ yields $\log_7(1 - 2)$ or $\log_7(-1)$, a negative logarithm. Logarithms are defined for positive values only. Suppose $\log_7(-1) = x$. In exponential form, we would have $7^x = -1$. No value of x can render 7 as a negative. Hence, $x = 1$ is not a solution of the equation.

On the other hand, the answer $x = 5$ causes no problems. Substitute it into the equation to verify that it is a viable solution.

3. Solve $2 + \log_2(2x - 7) = 2\log_2(x - 2)$.

At first glance, it does not appear that we can get the equation into the form $\log_2 p = \log_2 q$. So, we can try the second approach described at the beginning of this article:

$$2 + \log_2(2x - 7) = 2\log_2(x - 2)$$
$$\log_2(2x - 7) - 2\log_2(x - 2) = -2 \qquad \text{Put all logs on the same side.}$$
$$\log_2(2x - 7) - \log_2(x - 2)^2 = -2 \qquad \text{Property 3.}$$
$$\log_2 \frac{2x - 7}{(x - 2)^2} = -2 \qquad \text{Property 2.}$$

Now we can change from logarithmic form to exponential form:

$$2^{-2} = \frac{2x - 7}{(x - 2)^2} \qquad \text{See \textbf{Logarithmic Form to Exponential Form,}}$$
$$\textbf{converting from.}$$
$$\frac{1}{2^2} = \frac{2x - 7}{(x - 2)^2}$$
$$\frac{1}{4} = \frac{2x - 7}{(x - 2)^2}$$
$$(x - 2)^2 = 4(2x - 7) \qquad \text{Cross multiplication.}$$
$$x^2 - 4x + 4 = 8x - 28 \qquad \text{Left-hand side is } (x - 2)(x - 2). \text{ See \textbf{FOIL}.}$$
$$x^2 - 12x + 32 = 0$$
$$(x - 4)(x - 8) = 0 \qquad \text{See \textbf{Quadratic Equations, solutions of}.}$$
$$x - 4 = 0 \quad \text{or} \quad x - 8 = 0$$
$$x = 4 \qquad\qquad x = 8$$

By substitution, we can see that both answers are viable answers.

We could have solved this equation by the first method if we had recognized that $2 = \log_2 4$. The equivalence of this equation in $2^2 = 4$. By substitution, we have

$$2 + \log^2(2x - 7) = 2\log_2(x - 2)$$
$$\log_2 4 + \log_2(2x - 7) = 2\log_2(x - 2) \qquad \text{Substitute } 2 = \log_2 4.$$
$$\log_2 4(2x - 7) = \log_2(x - 2)^2 \qquad \text{Property 1 on the left and property 3 on the right.}$$

Now we have the form $\log_2 p = \log_2 q$, so

$$4(2x - 7) = (x - 2)^2$$

which is the same as line 8 of the previous solution. Hence, we have the solutions $x = 4$ and $x = 8$.

The examples in this article were for $\log_7$ and $\log_2$. The same approach can be used for $\log_{10}$, $\log_e$ (which is ln), or logarithms to any other base.

➤ **Logarithmic Form**   The equation $y = \log_b x$ is in **logarithmic form** while the equation $b^y = x$ is in exponential form.

➤ **Logarithmic Form to Exponential Form, converting from**   In general, the equation $y = \log_b x$ is equivalent to the equation $b^y = x$. The first equation is said to be in **logarithmic form** and the second equation is said to be in **exponential form**.

To convert from logarithmic form to exponential form, we need to observe where the y, b, and x are in each case:

$$y = \log_b x \qquad \text{is equivalent to } b^y = x$$
$$\text{or} \qquad \log_b x = y \qquad \text{is equivalent to } b^y = x$$

Similarly, to convert from exponential form to logarithmic form, the letters are reversed. The most notable fact in this reversal is that the exponent in exponential form is the logarithm in logarithmic form.

Examples

1. Convert $\log_2 8 = 3$ to exponential form.

$$\log_2 8 = 3 \text{ is equivalent to } 2^3 = 8$$

Also, $3 = \log_2 8$ is equivalent to $2^3 = 8$.

2. Convert $3^4 = 81$ to logarithmic form.

The 3 is the base and the 4 is the logarithm. Hence,

$$3^4 = 81 \text{ is equivalent to } 4 = \log_3 81$$

➤ **Logarithmic Function**   A **logarithmic function** is an equation defined by $y = \log_b x$ where $b > 0$, $b \neq 1$, x is a real number with $x > 0$, and y is a real number. The equations $y = \log_2 x$, $y = \log_{1/4} x$, and $y = -6 - 4 \log_{.75} x$ are logarithmic functions.

Examples

1. Graph $y = \log_2 x$ and state the domain and range.

At first, it might seem that we could graph $y = \log_2 x$ by the charting method. However, if we chose values for x, how could we calculate the values for y? We don't have a chart for $\log_2$. We have $\log_{10}$ charts and $\log_e$, or ln, charts. But we don't have a $\log_2$ chart. We could use the change of base formula to calculate each x. But that would be cumbersome.

Simply, we can graph $y = \log_2 x$ by thinking of it in exponential form, $2^y = x$. See **Logarithmic Form to Exponential Form, converting from**. Since these forms are equivalent, we need only graph $2^y = x$ by the charting method. In this form, we choose the y values and calculate the x values.

L

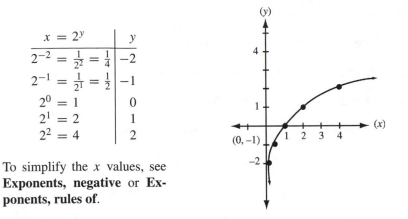

$$
\begin{array}{c|c}
x = 2^y & y \\
\hline
2^{-2} = \frac{1}{2^2} = \frac{1}{4} & -2 \\
2^{-1} = \frac{1}{2^1} = \frac{1}{2} & -1 \\
2^0 = 1 & 0 \\
2^1 = 2 & 1 \\
2^2 = 4 & 2 \\
\end{array}
$$

To simplify the x values, see **Exponents, negative** or **Exponents, rules of**.

Looking at the equation $x = 2^y$, we can see that any y we choose will not cause a violation of any definitions. Therefore, the range of y is all the real numbers. In set notation, the range is $\{y \mid -\infty < y < \infty\}$. In interval notation, the range is $(-\infty, \infty)$. Also, looking at the graph, we can see that the curve will increase indefinitely to the right and is asymptotic on the left. Hence, again, the range is all the real numbers. Finally, looking at the chart, we can see that we can choose y's from $-\infty$ to $+\infty$.

If we allow the y's to range from $-\infty$ to $+\infty$, we see that the x, in $x = 2^y$, will approach 0 or $+\infty$. Hence, the domain of x is all x's greater than 0. In set notation, the domain is $\{x \mid x > 0\}$. In interval notation, the domain is $(0, \infty)$. Also, looking at the graph or the chart, we can see that the x's, and hence the domain, are greater than 0.

See **Domain and Range, Set Notation,** or **Interval Notation**. To graph the domain and range on a number line, see **Inequalities, graph on a number line**.

2. Graph $y = \log_{1/2} x$ and state the domain and range.

Changing to exponential form and choosing values for y, we have

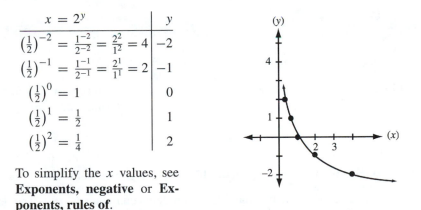

$$
\begin{array}{c|c}
x = 2^y & y \\
\hline
\left(\frac{1}{2}\right)^{-2} = \frac{1^{-2}}{2^{-2}} = \frac{2^2}{1^2} = 4 & -2 \\
\left(\frac{1}{2}\right)^{-1} = \frac{1^{-1}}{2^{-1}} = \frac{2^1}{1^1} = 2 & -1 \\
\left(\frac{1}{2}\right)^0 = 1 & 0 \\
\left(\frac{1}{2}\right)^1 = \frac{1}{2} & 1 \\
\left(\frac{1}{2}\right)^2 = \frac{1}{4} & 2 \\
\end{array}
$$

To simplify the x values, see **Exponents, negative** or **Exponents, rules of**.

Similar to the explanations in example 1, we see that the domain is all x's greater than 0 and the range is all y's from $-\infty$ to $+\infty$.

L

➤ **Logarithm, natural or Napierian**   See **Natural Logarithms**.

➤ **Logarithm Rules**   See **Logarithms, properties of (rules of)**.

➤ **Logarithms, change of base**   See **Change of Base Formula**.

➤ **Logarithms, characteristic of**   See **Characteristic**.

➤ **Logarithms, equations of**   See **Logarithmic Equations**.

➤ **Logarithms, graphs of**   See **Logarithmic Function**.

➤ **Logarithms, interpolation of**   See **Interpolation (linear)**.

➤ **Logarithms, mantissa of**   See **Mantissa** or **Logarithm Table (chart), how to use**.

➤ **Logarithms, properties of (rules of)**   There are three basic **properties of logarithms**. They are very similar to, as well as derived from, the rules for exponents. See **Exponents, rules of**.

Properties of Logarithms

1. $\log_b pq = \log_b p + \log_b q$.

    The log of a product is the sum of the logs.

2. $\log_b \dfrac{p}{q} = \log_b p - \log_b q$.

    The log of a quotient is the difference of the logs.

3. $\log_b p^n = n \log_b p$.

    It is useful to be familiar with three other equations: $\log_b b = 1$, $\log_b 1 = 0$, and $b^{\log_b p} = p$.

Examples

1. Show that $\log_b b = 1$, $\log_b 1 = 0$, and $b^{\log_b p} = p$.

    a. Let $\log_b b = x$. In exponential form, the equivalent equation is $b^x = b$. Since the exponent of b is 1, we have $b^x = b^1$ which means $x = 1$. Hence, by substitution,

    $$\log_b b = x \text{ means } \log_b b = 1$$

    See **Logarithmic Form to Exponential Form, converting from**.

    b. Let $\log_b 1 = x$. In exponential form, the equivalent equation is $b^x = 1$. The only way b^x can be equal to 1 is if $x = 0$, then $b^0 = 1$. See **Exponents, rules of**, special case. Hence, by substitution,

    $$\log_b 1 = x \text{ means } \log_b 1 = 0$$

c. If we convert the equation $\log_b p = \log_b p$ to exponential form, we have

$$b^{\log_b p} = p$$

See **Logarithmic Form to Exponential Form, converting from**. Also see **Exponential Equation, solution of**.

2. Write in expanded form $\log_7 \sqrt[3]{\frac{xy^2}{2z}}$.

$$\log_7 \sqrt[3]{\frac{xy^2}{2z}} = \log_7 \left(\frac{xy^2}{2z}\right)^{1/3}$$

See **Exponents, fractional or rational**.

$$= \frac{1}{3}\log_7 \frac{xy^2}{2z}$$

Property 3.

$$= \frac{1}{3}\left(\log_7 xy^2 - \log_7 2z\right)$$

Property 2.

$$= \frac{1}{3}\left[\log_7 x + \log_7 y^2 - (\log_7 2 + \log_7 2z)\right]$$

Property 1. Use parentheses when there is a negative.

$$= \frac{1}{3}\left(\log_7 x + 2\log_7 y - \log_7 2 - \log_7 z\right)$$

Property 3.

A problem like this is given for practice on applying the properties of logarithms.

3. Write $\log_a(x^2 - 2) + 3\log_a \frac{1}{x+2} - 5\log_a x$ as a logarithm of one expression.

$$\log_a(x^2 - 2) + 3\log_a \frac{1}{x+2} - 5\log_a x$$

$$= \log_a(x^2 - 2) + \log_a\left(\frac{1}{x+2}\right)^3 - \log_a x^5$$

Property 3.

$$= \log_a(x^2 - 2)\left[\frac{1}{(x+2)^3}\right] - \log_a x^5$$

Property 1 and the power-to-power rule.

$$= \log_a \frac{x^2 - 2}{(x+2)^3} - \log_a x^5$$

See **Exponents, rules of**.

$$= \log_a \frac{\frac{x^2-2}{(x+2)^3}}{x^5}$$

Property 2.

$$= \log_a \frac{x^2 - 2}{x^5(x+2)^3}$$

See **Fractions, complex**.

4. Expand $d = \frac{1}{2}gt^2$ using logarithms to the base b. Then solve for $\log_b t$.

L

Taking the logarithm of both sides, we have

$$\log_b d = \log_b \frac{1}{2} g t^2$$

$$\log_b d = \log_b \frac{1}{2} + \log_b g + \log_b t^2 \qquad \text{Property 1.}$$

$$\log_b d = \log_b 1 - \log_b 2 + \log_b g + 2 \log_b t \qquad \text{Properties 2 and 3.}$$
$$\log_b d = 0 - \log_b 2 + \log_b g + 2 \log_b t$$
$$\log_b d = -\log_b 2 + \log_b g + 2 \log_b t$$
$$\log_b 2 + \log_b d - \log_b g = 2 \log_b t \qquad \text{Add } \log_b 2 \text{ and subtract } \log_b g \text{ on both sides.}$$
$$2 \log_b t = \log_b 2 + \log_b d - \log_b g \qquad \text{Symmetric property.}$$

$$\log_b t = \frac{\log_b 2 + \log_b d - \log_b g}{2} \qquad \text{Divide both sides by 2.}$$

$$\log_b t = \frac{1}{2}(\log_b 2 + \log_b d - \log_b g) \qquad \text{Factor out the } \tfrac{1}{2}.$$

Although all the examples were done with log, they could have been done with ln and rendered the same results. For example, using e in example 1, we have

a. $\log_e e = 1$ means $\ln e = 1$, since $\log_e$ is defined as ln. See **Natural Logarithms**.
b. $\log_e 1 = 0$ means $\ln 1 = 0$.
c. $e^{\log_e p} = p$ means $e^{\ln p} = p$.

In the rest of the examples, write ln instead of log.

➤ **Logarithm Table (chart), how to use**   This article will cover tables related to **common logarithms** and their antilogarithms. For information related to natural or Napierian logarithms, see **Natural Logarithms**.

To look up a logarithm in the chart means to find the mantissa that corresponds to the given number. To find $\log_{10} 3.58$ go down the number column to 35, across the top of the page to 8, and read the number .5539 at the intersection of their respective row and column. We write $\log_{10} 3.58 = .5539$. Similarly, $\log_{10} 9.77 = .9897$. Although these are approximations, we write $=$ instead of $\approx$.

To find the antilogarithm using the chart means to find the number that corresponds with a given mantissa. Finding the antilogarithm is the reverse process of finding the logarithm. To find antilog .5539, first locate .5539 in the table of mantissas. From there, read 35 at the extreme left of the row and 8 at the extreme top of the column. We write antilog .5539 $= 3.58$. Similarly, antilog .9897 $= 9.77$.

The majority of the entries in the logarithm table, log chart, are exponents. They are exponents to which 10 can be raised to yield a given number. The far left column, or number column, along with the numbers across the top of the chart, make up the given number. Since the numbers in the number column have two digits and the numbers across the top of the page have one, the given numbers are composed of three digits, four if we interpolate. The rest of the chart are exponents, called mantissas.

The number column ranges from 1.0 to 9.9 although the decimals are usually not placed in the chart. The numbers across the top of the page are part of the given number. They range from .00 to .09 although they are written 0 to 9. In this manner the number 3.58 is represented in the chart as 35 down the number column and 8 across the top.

The mantissas, or exponents, range from .0000 to .9996 although the decimals are not shown in the chart. These are the exponents to which 10 can be raised to yield a given number. If the given number is 3.58, then the corresponding mantissa is .5539. To locate .5539 in the chart, look right of 35 in the number column and down from 8 at the top of the page. If the given number is 2.56, the corresponding mantissa is .4082. For 4.90, the mantissa is .6902. For 7.13, the mantissa is .8531. For 9.77, the mantissa is .9899.

In each case, the mantissa we have found is the exponent to which 10 can be raised to yield the given number. Let's verify this by using the y^x button (or 10^x) in a scientific calculator. The mantissa that corresponds with 3.58 is .5539. This means $10^{.5539}$ must equal 3.58. Enter 10 in the calculator and press the y^x button. Now enter .5539 in the calculator and press y^x again, or $=$. The answer is 3.5801399, which rounds off to 3.58. Hence,

$$10^{.5539} \approx 3.58 \quad \text{or} \quad 10^{.4082} \approx 2.56 \quad \text{or} \quad 10^{.6902} \approx 4.90$$
$$\text{or} \quad 10^{.8531} \approx 7.13 \quad \text{or} \quad 10^{.9899} \approx 9.77$$

Logarithms were invented to simplify large calculations. On a lesser scale, in order to multiply 3.58 and 2.56, we could use the powers of 10, the mantissas, instead of direct multiplication:

$$(3.58)(2.56) \approx 10^{.5539} \bullet 10^{.4082}$$

$$\approx 10^{.5539+.4082}$$
When we multiply similar bases, we add the exponents. See **Exponents, rules of**.

$$\approx 10^{.9621}$$

$$\approx 9.1643148$$
Use the y^x button of a calculator.

By direct multiplication $(3.58)(2.56) = 9.1648$. The reason for the approximation is because the mantissas have been rounded to four decimal places.

Of course, calculators weren't around when logarithms were invented. The way $10^{.9621}$ would have been dealt with then would have been to find the mantissa .9621 in the chart and see with what number it corresponded.

If we try to locate .9621, we see that it is between .9619 and .9624. This means, that the number that corresponds with .9621 is between the numbers that correspond with .9619 and .9624. The number that corresponds with .9619 is 9.16, that's 91 down the number column and 6 across the top. The number that corresponds with .9624 is 9.17. Hence, the number that corresponds with .9621 is between 9.16 and 9.17. We have already seen that it is 9.1648, the product of 3.58 and 2.56. The method used in logarithms for closer approximation is called interpolation. See **Interpolation (linear)**.

Although in disguise, we have already discussed the topics of logarithms and antilogarithms. We did this by writing everything in exponential form, i.e., with 10 as a base and the mantissas as exponents of the 10. We wrote 3.58 as $10^{.5539}$. This meant that 10, when raised to a power of .5539, would yield 3.58. If we write this in logarithmic form, however, the number .5539 is called the logarithm of 3.58 to the base 10 and we write

$$\log_{10} 3.58 = .5539, \quad \text{read "log to the base 10 of 3.58."}$$

This means that we have yet another name for the number .5539. It is an exponent, a mantissa, and now a logarithm.

Since the chart has been developed with a base of 10, we write $\log_{10}$. Although there are generally not charts for other bases we could have $\log_5$, $\log_7$, and $\log_{\text{any base}}$. See **Logarithm, base of**.

L

Do not let the mystique of LOGARITHMS throw you. A logarithm is an exponent. It is an exponent to which 10 can be raised to yield a given number, just as the mantissa is.

So, why so many different names for the same thing? It's just so we can tell the difference between one type of number and another.

In the expression $10^{.5539}$ the .5539 is generally referred to as an exponent. However, if we call it a mantissa, we know it is one of the numbers listed in the table of mantissas. Further, in general, we can refer to it as a logarithm. The logarithm of 3.58 is .5539.

To look up a logarithm in the chart means to find the mantissa that corresponds to the given number. To find $\log_{10} 3.58$ go down the number column to 35, across the top of the page to 8, and read the number .5539 at the intersection of their respective row and column. We write $\log_{10} 3.58 = .5539$. Similarly, $\log_{10} 2.56 = .4082$, $\log_{10} 4.90 = .6902$, $\log_{10} 7.13 = .8531$, and $\log_{10} 9.77 = .9897$. Although these are approximations, we write $=$ instead of $\approx$ or $\doteq$.

The chart can also be used in reverse, finding the antilogarithm. If we are given the mantissa and then asked to find its corresponding number, we would be finding the antilog of the mantissa. If the mantissa is .5539, then antilog .5539 = 3.58.

To find antilog .5539, first locate .5539 in the table of mantissas. From there, read 35 at the extreme left of the row and 8 at the extreme top of the column. We write antilog .5539 = 3.58. Similarly, antilog .4082 = 2.56, antilog .6902 = 4.90, antilog .8531 = 7.13, and antilog .9897 = 9.77.

If the mantissa we are looking for is not in the table, it will be between two mantissas. We can then interpolate to find its corresponding number. See **Interpolation (linear)**.

Since the numbers in the chart range from 1.00 to 9.90 any number beyond this range must be put into scientific notation and rounded to two decimal places, three if interpolation is used. See **Scientific Notation, Interpolation (linear), Characteristic**, or **Common Logarithms**.

Examples

1. Find $\log_{10} 6.47$.

   Looking down the number column to 64, across the top of the page to 7, and locating the intersection of their respective row and column, we have

   $$\log_{10} 6.47 = .8109$$

2. Find $\log_{10} 647,000$.

   In scientific notation,

   $$647,000 = 6.47 \times 10^5$$
   $$\log_{10} 647,000 = \log_{10} 6.47 \times 10^5 = 5.8109$$

   The exponent of the 10 is the characteristic, 5. See **Characteristic**.

3. Find antilog .8109.

   Locating .8109 in the table of mantissas, we read 64 at the extreme left of the row and 7 at the extreme top of the column. Hence,

   $$\text{antilog} .8109 = 6.47$$

If we had been asked to find antilog 5.8109, we would first have written 6.47 and then moved the decimal point five places. See **Characteristic**.

➤ **Logic**   **Logic** is a study that is concerned with reasoning and argumentation. It deals with the principles and methods utilized in evaluating various statements. Some of the statements are hypotheses (plural of hypothesis) that are accepted as a premise. Others are conclusions that can be drawn from the hypotheses. The hypotheses and conclusions can be observed through rules that determine whether they are logically correct, or valid, or whether they are logically incorrect, or invalid.

In symbolic logic, symbols such as p and q are used to represent various statements. In general, the meaning of p and q is not important. It is the relationship between p and q that is important in the development of the principles of logic. Once the principles and methods have been developed, application of the meaning of the statements to particular subjects can be pursued.

The topics of logic that are covered in this book are listed below, along with their definitions. At the end of the list, selected examples in logic are given.

1. **Biconditional statement**—A statement that can be written in if and only if form, denoted by $p \leftrightarrow q$, is called a **biconditional statement** (read "if p then q, and if q then p" or "p, if and only if q" or "p is equivalent to q"). It is a combined statement composed of a conditional statement and its converse. For more, see **Biconditional (equivalent) Statement**.

2. **Conclusion**—In a conditional statement $p \rightarrow q$, the then-part (q) is called the **conclusion**. For more, see **Hypothesis and Conclusion**.

3. **Conditional Statement**—A statement written in if-then form is called a **conditional statement**. If we let p denote the if-part of the statement and q denote the then-part, then the expression $p \rightarrow q$ denotes the conditional statement (read "if p, then q" or "p implies q"). For more, see **Conditional Statement**.

4. **Conjunction**—The compound statement p and q, written $p \wedge q$, where p and q represent any statement, is called the **conjunction** of p and q. For more, see **Conjunction**. For concepts related to the truth values of a conjunction, see **Truth Table**.

5. **Contradiction**—A statement that is both true and false at the same time is a **contradiction**. If p is given as a fact, then the statement both p and $\sim p$ (read "not p") is a contradiction. The following is a contradiction: The number 4 is not the number 4. For more on $\sim p$, see **Negation**.

6. **Contrapositive**—The **contrapositive** of a conditional statement is formed when the if-part and the then-part of the inverse ($\sim p \rightarrow \sim q$) are interchanged. The contrapositive of $p \rightarrow q$ is $\sim p \rightarrow \sim q$. For more, see **Contrapositive**.

7. **Converse**—The **converse** of a conditional statement is formed by interchanging the if-part and the then-part. If the conditional statement is $p \rightarrow q$, then its converse is $q \rightarrow p$. For more, see **Converse**.

8. **Disjunction**—The compound statement p or q, written $p \vee q$, where p and q represent any statement, is called the **disjunction** of p and q. For more, see **Disjunction**. For concepts related to the truth values of a disjunction, see **Truth Table**.

9. **Equivalence (equivalent statement)**—If two statements have exactly the same entries in each row of the truth table, then they are logically **equivalent**. In this manner, $p \rightarrow q$ and $q \rightarrow p$. In other words, they are biconditional statements. See **Biconditional (equivalent) Statement** or **Truth Table**.

10. **If and only if statements**—An **if and only if statement** is a biconditional statement in which both the conditional statement and its converse are written in one statement. In a biconditional statement, either the conditional and its converse are both true, or both false. See **Biconditional (equivalent) Statement**.

11. **Inverse**—The **inverse** of a conditional statement is formed by negating both the if-part and the then-part. If the conditional statement is $p \rightarrow q$, then its inverse is $\sim p \rightarrow \sim q$. For more, see **Inverse**.

12. **Negation**—The **negation** of a statement is the opposite of an original statement. If we let p be the original statement, then its negation is written $\sim p$ (read "not p"). For more, see **Negation**.

For more on other articles related to logic, see **Set**, **Set Notation**, or **Venn Diagram**.

Examples

1. State the hypothesis and the conclusion of the following statement: Vertical angles are congruent.

   We first need to write the statement in if-then form. Then we can identify the hypothesis and the conclusion.

   If two angles are vertical angles, then they are congruent.

   Hypothesis: If two angles are vertical angles

   Conclusion: then they are congruent

2. Prove that the conditional statement $(p \wedge q) \rightarrow p$ is true for all truth values of p and q.

   To derive the truth table for $(p \wedge q) \rightarrow p$, we need to compare its entries with those for a conditional statement. For convenience, we have put the entries for the conditional statement, $p \rightarrow q$, in the table.

p	q	$p \wedge q$	$p \rightarrow q$	$(p \wedge q) \rightarrow p$
T	T	T	T	T
T	F	F	F	T
F	T	F	T	T
F	F	F	T	T

   However, it is simpler to remember that when the hypothesis is true and the conclusion is false, the conditional statement is false. Otherwise, it is true. Since none of the entries from column 3 to column 1 (in that order or direction) are from T to F, the conditional statement $(p \wedge q) \rightarrow p$ cannot be false. Therefore, all of the entries for $(p \wedge q) \rightarrow p$ must be T.

3. Write the converse, inverse, and contrapositive of $p \rightarrow q$.

Applying $p \rightarrow q$ to the definitions of a converse, an inverse, and a contrapositive, we have

$$\text{Converse: } q \rightarrow p \qquad \text{Inverse: } \sim p \rightarrow \sim q \qquad \text{Contrapositive: } \sim q \rightarrow \sim p$$

4. Write the converse, inverse, and contrapositive of $r \rightarrow \sim s$.

   Since the double negative $\sim(\sim s)$ is s, we have

$$\text{Converse: } \sim s \rightarrow r \qquad \text{Inverse: } \sim r \rightarrow \sim(\sim s) \qquad \text{Contrapositive: } \sim(\sim s) \rightarrow \sim r$$
$$\qquad\qquad \text{or} \qquad \sim r \rightarrow s \qquad\qquad\qquad\qquad \text{or} \qquad s \rightarrow \sim r$$

5. Write the converse, inverse, and contrapositive of the following statement: If two lines are perpendicular, then they meet to form a right angle.

   Converse: If two lines meet to form a right angle, then they are perpendicular.

   Inverse: If two lines are not perpendicular, then they do not meet to form a right angle.

   Contrapositive: If two lines do not meet to form a right angle, then they are not perpendicular.

6. Write the truth table for the converse, inverse, and contrapositive. See **Truth Table**.

➤ **Logical Equivalence**   See **Biconditional (equivalent) Statement** or **Truth Table**.

➤ **Long Division**   In arithmetic, see **Division of Whole Numbers**.

➤ **Long Division of Polynomials**   **Long division of polynomials** is similar to the division process in arithmetic.

In arithmetic, we do the following:

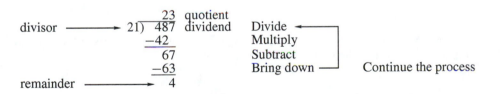

In long division of polynomials, we follow the same process. However, the divisor and the dividend must be in descending order of the exponents. If any term in the order is missing, it must be filled in with a 0. This is necessary because of place value. In arithmetic, the 2 and 3 of 203 are separated by the 0. In algebra, we do not write the 0 term in $2x^2 + 0x + 3$. We simply write $2x^2 + 3$. When we divide, the 0 term must be reinserted.

Examples

1. Using long division, divide $30x^3 + 15x^2 - 25x$ by $5x$.

L

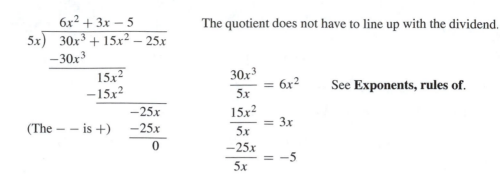

$$\frac{6x^2 + 3x - 5}{5x)\ 30x^3 + 15x^2 - 25x}$$

The quotient does not have to line up with the dividend.

$$\frac{30x^3}{5x} = 6x^2 \qquad \text{See \textbf{Exponents, rules of}.}$$

$$\frac{15x^2}{5x} = 3x$$

$$\frac{-25x}{5x} = -5$$

(The $--$ is $+$)

We can verify the answer by multiplying $5x(6x^2 + 3x - 5)$:

$$5x(6x^2 + 3x - 5) = 30x^3 + 15x^2 - 25x \qquad \text{See \textbf{Distributive Property}.}$$

2. Divide $6x^2 + 11x - 35$ by $3x - 5$.

$$\frac{2x}{3x - 5)\ 6x^2 + 11x - 35}$$
$$6x^2 - 10x$$

Divide the first term of the divisor into the first term of the dividend only, the $3x$ into the $6x^2$. So, $\frac{6x^2}{3x}$ is $2x$. Then multiply $2x(3x - 5)$ to get $6x^2 - 10x$.

When we subtract, we will have $-(6x^2 - 10x)$, which is $-6x^2 + 10x$:

$$\frac{2x}{3x - 5)\ 6x^2 + 11x - 35}$$
$$-(6x^2 - 10x)$$

becomes

$$\frac{2x}{3x - 5)\ 6x^2 + 11x - 35}$$
$$-6x^2 + 10x$$

In essence, when we subtract we change the signs of each term of the subtracted polynomial. We will put a circle around the sign to indicate that it has been changed.

$$\frac{2x}{3x - 5)\ 6x^2 + 11x - 35}$$
$$\ominus 6x^2 \oplus 10x$$

Once we change the signs, the subtraction in this problem becomes addition. We add $11x + 10x$.

Then we combine $6x^2$ and $-6x^2$, $11x$ and $10x$, and start the next division. We continue the process by dividing first term into first term, the $3x$ into $21x$. So $\frac{21x}{3x}$ is 7. The x's cancel. Then we multiply $7(3x - 5)$ and subtract (change the signs and combine):

$$\frac{2x + 7}{3x - 5)\ 6x^2 + 11x - 35}$$
$$\ominus 6x^2 \oplus 10x$$
$$21x - 35$$
$$\ominus 21x \oplus 35 \qquad \text{Change signs.}$$
$$0$$

3. Divide $8x^3 - 14x - 6$ by $2x - 3$.

Since we are missing the x^2 term in the descending order, we fill in that position with $0x^2$:

$$
\begin{array}{r}
4x^2 + 6x + 2 \\
2x - 3 \overline{)\ 8x^3 + 0x^2 - 14x - 6} \\
\ominus 8x^3 \oplus 12x^2 \\
\hline
12x^2 - 14x \\
\ominus 12x^2 \oplus 18x \\
\hline
4x - 6 \\
\ominus 4x \oplus 6 \\
\hline
0
\end{array}
$$

$$\frac{8x^3}{2x} = 4x^2$$

$$\frac{12x^2}{2x} = 6x$$

$$\frac{4x}{2x} = 2$$

We always divide the first term into the first term.

4. Divide $3y^4 - 4$ by $y^2 + 1$.

Both the divisor and divided are missing 0 terms:

$$
\begin{array}{r}
3y^2 - 3 \\
y^2 + 0y + 1 \overline{)\ 3y^4 + 0y^3 + 0y^2 + 0y - 4} \\
\ominus 3y^4 \ominus 0y^3 \ominus 3y^2 \\
\hline
-3y^2 + 0y - 4 \\
\oplus 3y^2 \oplus 0y \oplus 3 \\
\hline
-1
\end{array}
$$

$$\frac{3y^4}{y^2} = 3y^2$$

$$\frac{-3y^2}{y^2} = -3$$

Since there are three terms in the divisor, we bring down two terms from the dividend, the $0y$ and -4. The answer is $3y^2 - 3$ remainder -1 or $3y^2 - 3 - \frac{1}{y^2+1}$. Sometimes it is required to write the remainder as a fraction. To solve polynomial equations, see **Rational Roots Theorem**.

➤ **Lowest Common Denominator**   See **Common Denominator**.

➤ **Lowest Common Multiple**   See **Least Common Multiple**.

Mm

➤ **Magnitude of a Vector**   See **Vector**.

➤ **Major and Minor Arc**   In circle O, the arc from A to B is called the **minor arc** and is written $\overset{\frown}{AB}$. In circle P, the arc from X, through Z, and around to Y is called the **major arc**. It is written $\overset{\frown}{XYZ}$. Major arcs are named with three letters.

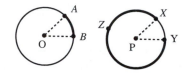

More formally, the minor arc $\overset{\frown}{AB}$ is composed of the points A and B and all the points on the circle between A and B. The major arc $\overset{\frown}{XYZ}$ is composed of the points X and Y and all the points on the circle exterior to $\angle XPY$.

The measure of a minor arc is equal to the measure of its central angle. The measure of a major arc is equal to 360 minus the measure of the central angle of its minor arc. See **Central Angle of a Circle**. Also see **Arc** or **Arc Length**.

➤ **Major Axis of an Ellipse**   The **major axis** of an ellipse is the longer axis of symmetry. The **minor axis** is the shorter axis of symmetry. See **Ellipse**.

➤ **Mantissa**   A mantissa is an exponent. In common logarithms, the mantissa is the exponent to which 10 can be raised to yield a given number. The logarithm table is a table of mantissas. See **Logarithm Table (chart), how to use**.

➤ **Math Symbols**   Throughout the centuries math symbols have evolved. They are used to perform various operations and, in general, are used to indicate concepts. A list of more commonly used math symbols is given below.

!	factorial
+	addition
−	subtraction
× or •	multiplication

÷ or ⟋	division
=	is equal to
≡	is identically equal to
≈ or ≐	is approximately equal to
≠	is not equal to
~	is similar to
≅	is congruent to
⊥	is perpendicular to
∥	is parallel to
±	plus or minus
<	is less than
>	is greater than
⩽	is less than or equal to
⩾	is greater than or equal to
Σ	summation
$|x|$	absolute value of x
$\sqrt{x}$	square root of x
$\sqrt[n]{x}$	nth root of x
x^n	x to the nth power
x_n	x subscript n, or x sub n
(x, y)	point with coordinates x and y, rectangular coordinates
(r, θ)	point with radius r and angle θ, polar coordinates
$f(x)$	function of x, read "f of x"
$f'(x)$ or $\frac{dy}{dx}$	first derivative, read "f prime of x" or "derivative of y with respect to x"
$f''(x)$ or $\frac{d^2y}{dx^2}$	second derivative, read "f double prime of x" or "second derivate of y with respect to x"
$n \to \infty$	n approaches infinity or increases without bound
$\int f(x)\,dx$	integral of $f(x)$
$f(x, y)$	function of x and y, read "f of x-y"
( ), [ ], { }	symbols of inclusion or grouping

➤ **Matrix**   A **matrix** is a rectangular array of numbers displayed between parentheses or brackets. The numbers, called entries or **elements**, are arranged in rows (horizontal entries) or columns (vertical entries) which determine the size or **dimension** of the matrix. If the matrix contains only one row, it is called a **row matrix**. If the matrix contains only one column, it is called a **column matrix**. A matrix that is composed of an equal amount of rows and columns is called a **square matrix**.

This article covers the foundations of matrices and matrix solution of systems of equations. For solutions of systems by using Cramer's Rule see **Cramer's Rule** or **Determinant**. To solve systems of equations by using augmented matrices, see **Matrix, augmented (matrix) solutions**. To solve systems of equations by the linear combination or addition method or by the substitution method, see **Systems of Linear Equations, solutions of**.

Properties of Matrices

1. **Equality of matrices:** Two matrices are equal if they have the same dimensions and their corresponding elements are equal. If $\begin{bmatrix} a & b \\ c & d \end{bmatrix} = \begin{bmatrix} 1 & 2 \\ 3 & 4 \end{bmatrix}$ then $a = 1$, $b = 2$, $c = 3$, and $d = 4$.

M

2. **Sum of two matrices:** Let A and B be two matrices with the same dimensions. Then the sum of A and B is a matrix $A + B$ with elements that are the sums of the corresponding elements of A and B. For example, let

$$A = \begin{bmatrix} a & b \\ c & d \end{bmatrix} \quad \text{and} \quad B = \begin{bmatrix} e & f \\ g & h \end{bmatrix}$$

Then

$$A + B = \begin{bmatrix} a+e & b+f \\ c+g & d+h \end{bmatrix}$$

3. **Additive identity matrix:** Let A and I be matrices with the same dimensions, then if $A + I = A$, I is called the additive identity matrix or zero matrix. For example, let $A = \begin{bmatrix} a & b \\ c & d \end{bmatrix}$ and $I = \begin{bmatrix} 0 & 0 \\ 0 & 0 \end{bmatrix}$. Then

$$A + I = \begin{bmatrix} a & b \\ c & d \end{bmatrix} + \begin{bmatrix} 0 & 0 \\ 0 & 0 \end{bmatrix} = \begin{bmatrix} a+0 & b+0 \\ c+0 & d+0 \end{bmatrix} = \begin{bmatrix} a & b \\ c & d \end{bmatrix} = A$$

4. **Additive inverse matrix:** Let A and $-A$ be matrices with the same dimensions. Then if $A + (-A) = I$, $-A$ is called the additive inverse matrix. For example, let

$$A = \begin{bmatrix} a & b \\ c & d \end{bmatrix} \quad \text{and} \quad -A = \begin{bmatrix} -a & -b \\ -c & -d \end{bmatrix}$$

Then

$$A + (-A) = \begin{bmatrix} a & b \\ c & d \end{bmatrix} + \begin{bmatrix} -a & -b \\ -c & -d \end{bmatrix} = \begin{bmatrix} a-a & b-b \\ c-c & d-d \end{bmatrix} = \begin{bmatrix} 0 & 0 \\ 0 & 0 \end{bmatrix} = I$$

5. **Commutative property of matrix addition:** Let A and B be any matrices with the same dimensions. Then $A + B = B + A$. For example, let

$$A = \begin{bmatrix} a & b \\ c & d \end{bmatrix} \quad \text{and} \quad B = \begin{bmatrix} e & f \\ g & h \end{bmatrix}$$

Then

$$A + B = \begin{bmatrix} a+e & b+f \\ c+g & d+h \end{bmatrix} \quad \text{and} \quad B + A = \begin{bmatrix} e+a & f+b \\ g+c & h+d \end{bmatrix} = \begin{bmatrix} a+e & b+f \\ c+g & d+h \end{bmatrix}$$

Therefore, $A + B = B + A$.

6. **Associative property of matrix addition:** Let A, B, and C be any matrices with the same dimensions. Then $(A + B) + C = A + (B + C)$. For example, let

$$A = \begin{bmatrix} a & b \\ c & d \end{bmatrix} \quad B = \begin{bmatrix} e & f \\ g & h \end{bmatrix} \quad C = \begin{bmatrix} i & j \\ k & l \end{bmatrix}$$

Then

$$(A + B) + C = \begin{bmatrix} a + e & b + f \\ c + g & d + h \end{bmatrix} + \begin{bmatrix} i & j \\ k & l \end{bmatrix} = \begin{bmatrix} a + e + i & b + f + j \\ c + g + k & d + h + l \end{bmatrix}$$

$$A + (B + C) = \begin{bmatrix} a & b \\ c & d \end{bmatrix} + \begin{bmatrix} e + i & f + j \\ g + k & h + l \end{bmatrix} = \begin{bmatrix} a + e + i & b + f + j \\ c + g + k & d + h + l \end{bmatrix}$$

Therefore, $(A + B) + C = A + (B + C)$.

7. **Product of a scalar (constant) and a matrix:** Let k be a scalar (any real number) and let A be a matrix of any dimension. Then the product of k and A is the matrix kA determined by multiplying each element of A by k. For example, let

$$A = \begin{bmatrix} a & b \\ c & d \end{bmatrix} \qquad \text{and} \qquad k = \text{a scalar}$$

Then

$$kA = k \begin{bmatrix} a & b \\ c & d \end{bmatrix} = \begin{bmatrix} ka & kb \\ kc & kd \end{bmatrix}$$

8. **Product of two matrices:** Let A be a matrix with dimensions $m \times n$ and let B be a matrix with dimensions $n \times r$. Then their product AB is a matrix with dimensions $m \times r$. Further, each row of A multiplies each column of B, the terms are combined, and the result is then entered in the corresponding location in the product matrix AB. For example, let

$$A = \begin{bmatrix} a & b \\ c & d \end{bmatrix} \qquad \text{and} \qquad B = \begin{bmatrix} e & f & g \\ h & i & j \end{bmatrix}$$

Then

$$AB = \begin{bmatrix} a & b \\ c & d \end{bmatrix} \begin{bmatrix} e & f & g \\ h & i & j \end{bmatrix}$$

$$= \begin{bmatrix} ae + bh & af + bi & ag + bj \\ ce + dh & cf + di & cg + dj \end{bmatrix} \qquad \begin{array}{l} ab \text{ multiplies } eh, \text{ then } fi, \text{ then } gi \\ cd \text{ multiplies } eh, \text{ then } fi, \text{ then } gi \end{array}$$

Because the dimensions of A must match up with the dimensions of B to yield the dimensions of AB, matrix multiplication is quite often not commutative. Although the product AB is defined, the product BA is not. In BA, the row with elements efg does not match up with the column $\begin{smallmatrix} a \\ b \end{smallmatrix}$.

9. **Multiplicative identity matrix:** Let A and I be square matrices with the same dimensions. Then if $A \bullet I = A$, I is called the multiplicative identity matrix. For example, let

$$A = \begin{bmatrix} a & b \\ c & d \end{bmatrix} \qquad \text{and} \qquad I = \begin{bmatrix} 1 & 0 \\ 0 & 1 \end{bmatrix}$$

Then

$$A \bullet I = \begin{bmatrix} a & b \\ c & d \end{bmatrix} \begin{bmatrix} 1 & 0 \\ 0 & 1 \end{bmatrix} = \begin{bmatrix} a \cdot 1 + b \cdot 0 & a \cdot 0 + b \cdot 1 \\ c \cdot 1 + d \cdot 0 & c \cdot 0 + d \cdot 1 \end{bmatrix} = \begin{bmatrix} a & b \\ c & d \end{bmatrix} = A$$

M

10. **Multiplicative inverse matrix:** Let A and A^{-1} be square matrices with the same dimensions. Then if $A \cdot A^{-1} = I$, A^{-1} is called the multiplicative inverse matrix. For example, let

$$A = \begin{bmatrix} a & b \\ c & d \end{bmatrix} \quad \text{and} \quad A^{-1} = \frac{1}{ad - bc} \begin{bmatrix} d & -b \\ -c & a \end{bmatrix}$$

Then

$$A \cdot A^{-1} = \begin{bmatrix} a & b \\ c & d \end{bmatrix} \left(\frac{1}{ad - bc} \right) \begin{bmatrix} d & -b \\ -c & a \end{bmatrix} = \frac{1}{ad - bc} \begin{bmatrix} ad - bc & -ab + ab \\ cd - cd & -bc + ad \end{bmatrix}$$

$$= \begin{bmatrix} 1 & 0 \\ 0 & 1 \end{bmatrix} = I$$

The reason for the complexity of A^{-1} can be seen from its derivation. If we let

$$A^{-1} = \begin{bmatrix} x & y \\ z & w \end{bmatrix} \text{ be the inverse of } A = \begin{bmatrix} a & b \\ c & d \end{bmatrix}$$

Then

$$A \cdot A^{-1} = \begin{bmatrix} a & b \\ c & d \end{bmatrix} \begin{bmatrix} x & y \\ z & w \end{bmatrix} = \begin{bmatrix} 1 & 0 \\ 0 & 1 \end{bmatrix}$$

Writing the equations that correspond with the product, we have (see properties 8 and 1)

$$ax + bz = 1$$
$$ay + bw = 0$$
$$cx + dz = 0$$
$$cy + dw = 1$$

Next, if we solve these equations simultaneously, we will determine the values of x, y, z, and w, the entries of A^{-1}. See **Systems of Linear Equations, solutions of**.

Solving equations 1 and 3:

$$
\begin{array}{lcl}
ax + bz = 1 & \rightarrow & adx + bdz = d \\
cx + dz = 0 & & -bcx - bdz = 0 \\
\hline
& & (ad - bc)x = d
\end{array}
$$

$$x = \frac{d}{ad - bc}$$

$$ax + bz = 1 \quad \rightarrow \quad acx + bcz = c$$
$$cx + dz = 0 \qquad\qquad \underline{-acx - adz = 0}$$
$$(bc - ad)z = c$$
$$z = \frac{c}{bc - ad}$$
$$z = \frac{-c}{ad - bc}$$

Solving equations 2 and 4:

$$ay + bw = 0 \quad \rightarrow \quad ady + bdw = 0$$
$$cy + dw = 1 \qquad\qquad \underline{-bcy - bdw = -b}$$
$$(ad - bc)y = -b$$
$$y = \frac{-b}{ad - bc}$$

$$ay + bw = 0 \quad \rightarrow \quad acy + bcw = 0$$
$$cy + dw = 1 \qquad\qquad \underline{-acy - adw = -a}$$
$$(bc - ad)w = -a$$
$$w = \frac{-a}{bc - ad}$$
$$w = \frac{a}{ad - bc}$$

Hence, we have the inverse A^{-1} where

$$A^{-1} = \begin{bmatrix} \frac{d}{ad-bc} & \frac{-b}{ad-bc} \\ \frac{-c}{ad-bc} & \frac{a}{ad-bc} \end{bmatrix} = \frac{1}{ad - bc} \begin{bmatrix} d & -b \\ -c & a \end{bmatrix}$$

Since $ad - bc$ is the value of the determinant of A (see **Determinant**), we see that A^{-1} is not defined if the value of the determinant of A is 0.

The inverse for a 3×3 matrix can be found by a different method. Simply stated, the inverse is

$$A^{-1}_{3\times3} = \frac{1}{D} \begin{bmatrix} \text{transpose matrix} \\ \text{of signed minors} \end{bmatrix}$$

where D is the determinant. The **transpose matrix of signed minors** is derived in two stages. The first is to replace each element of the given matrix by its signed minor. The second is to write the transpose of the new matrix. The resulting matrix is the transpose matrix of signed minors.

To replace each element of the given matrix by its signed minor, it is advantageous to use the following diagram:

$$\begin{vmatrix} + & - & + \\ - & + & - \\ + & - & + \end{vmatrix}$$

Each location in the diagram determines the positive or negative sign of its minor (see **Determinant**). The **minor** is the determinant that is formed when the row and column of an element are deleted. In the matrix $\begin{vmatrix} a & b & c \\ d & e & f \\ g & h & i \end{vmatrix}$, the signed minor of a is $+\begin{vmatrix} e & f \\ h & i \end{vmatrix}$, the signed minor of d is $-\begin{vmatrix} b & c \\ h & i \end{vmatrix}$, and the signed minor of f is $-\begin{vmatrix} a & b \\ g & h \end{vmatrix}$.

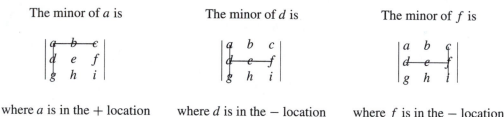

The minor of a is

where a is in the $+$ location in the diagram.

The minor of d is

where d is in the $-$ location in the diagram.

The minor of f is

where f is in the $-$ location in the diagram.

Writing the **transpose** of the new matrix is simple. Just interchange the rows and the columns. If the new matrix is $\begin{vmatrix} j & k & l \\ m & n & o \\ p & q & r \end{vmatrix}$, its transpose is $\begin{vmatrix} j & m & p \\ k & n & q \\ l & o & r \end{vmatrix}$.

To derive the inverse of a 3×3 matrix, see example 3.

11. **Associative property of matrix multiplication:** Let A, B, and C be any matrices with the same dimensions. Then if the products exist, $(A \cdot B) \cdot C = A \cdot (B \cdot C)$.

To solve systems of equations using matrices, we solve the equation $AX = C$ where A is the matrix of the coefficients of the variables, X is the matrix of the variables, and C is the matrix of the numerical values of the equations:

$$AX = C$$
$$A^{-1}AX = A^{-1}C \qquad \text{multiply by the inverse matrix}$$
$$IX = A^{-1}C \qquad \text{multiplicative inverse}$$
$$X = A^{-1}C \qquad \text{multiplicative identity}$$

For solutions of systems of equations using matrices, see examples 4 and 5.

Examples

1. Using the properties for equality of matrices and scalar multiplication, find x, y, and z if

$$6\begin{bmatrix} x & y-2 \\ 4 & z \end{bmatrix} = \begin{bmatrix} 36 & x \\ 24 & 3x+2y \end{bmatrix}.$$

By scalar multiplication, we can multiply 6 into the first matrix. Then, using the property for equality of matrices, we can equate each entry and solve the respective equations:

$$6\begin{bmatrix} x & y-2 \\ 4 & z \end{bmatrix} = \begin{bmatrix} 36 & x \\ 24 & 3x+2y \end{bmatrix} \qquad \begin{array}{l} 6x = 36, \text{ top left elements} \\ 6(y-2) = x, \text{ top right elements} \end{array}$$

$$\begin{bmatrix} 6x & 6(y-2) \\ 24 & 6z \end{bmatrix} = \begin{bmatrix} 36 & x \\ 24 & 3x+2y \end{bmatrix}$$

$24 = 24$, bottom left elements
$6z = 3x + 2y$, bottom right elements

From the first equation, we can find the value of x:

$$6x = 36$$
$$\frac{6x}{6} = \frac{36}{6}$$
$$x = 6$$

From the second equation, we can find the value of y:

$$6(y-2) = x$$
$$6y - 12 = 6 \qquad \text{Since } x = 6, \text{ we can substitute for } x.$$
$$6y = 6 + 12$$
$$\frac{6y}{6} = \frac{18}{6}$$
$$y = 3$$

From the fourth equation, we can find the value of z:

$$6z = 3x + 2y$$
$$6z = 3(6) + 2(3)$$
$$6z = 18 + 6$$
$$\frac{6z}{6} = \frac{24}{6}$$
$$z = 4$$

2. If $A = \begin{bmatrix} 2 & 3 \\ -1 & 4 \end{bmatrix}$ and $B = \begin{bmatrix} 0 & -4 \\ 5 & 6 \end{bmatrix}$, find $A + B$ and AB.

To find $A + B$ and AB, we can use the properties for the sum of two matrices and the product of two matrices:

$$A + B = \begin{bmatrix} 2 & 3 \\ -1 & 4 \end{bmatrix} + \begin{bmatrix} 0 & -4 \\ 5 & 6 \end{bmatrix} = \begin{bmatrix} 2+0 & 3-4 \\ -1+5 & 4+6 \end{bmatrix} = \begin{bmatrix} 2 & -1 \\ 4 & 10 \end{bmatrix}$$

$$AB = \begin{bmatrix} 2 & 3 \\ -1 & 4 \end{bmatrix} \begin{bmatrix} 0 & -4 \\ 5 & 6 \end{bmatrix} = \begin{bmatrix} 2(0)+3(5) & 2(-4)+3(6) \\ -1(0)+4(5) & -1(-4)+4(6) \end{bmatrix}$$

$$= \begin{bmatrix} 0+15 & -8+18 \\ 0+20 & 4+24 \end{bmatrix} = \begin{bmatrix} 15 & 10 \\ 20 & 28 \end{bmatrix}$$

3. Find the inverse of $A = \begin{bmatrix} 3 & -1 \\ 2 & 4 \end{bmatrix}$ and $B = \begin{bmatrix} 3 & 1 & 2 \\ 0 & -3 & -1 \\ 5 & -2 & 4 \end{bmatrix}$.

Equating A with $\begin{bmatrix} a & b \\ c & d \end{bmatrix}$ and using the formula for finding the inverse of a 2×2 matrix, we have (where $a = 3$, $b = -1$, $c = 2$, and $d = 4$)

$$A^{-1} = \frac{1}{ad - bc} \begin{bmatrix} d & -b \\ -c & a \end{bmatrix} = \frac{1}{3(4) - (-1)(2)} \begin{bmatrix} 4 & -(-1) \\ -2 & 3 \end{bmatrix} = \frac{1}{14} \begin{bmatrix} 4 & 1 \\ -2 & 3 \end{bmatrix}$$

To find the inverse of B, we need to find its matrix of signed minors, the transpose of the matrix of signed minors, the value of the determinant of B, and then substitute into the equation:

$$B_{3 \times 3}^{-1} = \frac{1}{D} \begin{bmatrix} \text{transpose matrix} \\ \text{of signed minors} \end{bmatrix}$$

The matrix of signed minors is composed of the minor associated with each element and the sign that corresponds with each minor. The minor is the determinant that is formed when the row and column of an element are deleted. The sign that corresponds with each minor is determined by the signs in the following diagram:

$$\begin{vmatrix} + & - & + \\ - & + & - \\ + & - & + \end{vmatrix}$$

See property 10 or **Determinant**.

$$\begin{bmatrix} +\begin{vmatrix} -3 & -1 \\ -2 & 4 \end{vmatrix} & -\begin{vmatrix} 0 & -1 \\ 5 & 4 \end{vmatrix} & +\begin{vmatrix} 0 & -3 \\ 5 & -2 \end{vmatrix} \\ -\begin{vmatrix} 1 & 2 \\ -2 & 4 \end{vmatrix} & +\begin{vmatrix} 3 & 2 \\ 5 & 4 \end{vmatrix} & -\begin{vmatrix} 3 & 1 \\ 5 & -2 \end{vmatrix} \\ +\begin{vmatrix} 1 & 2 \\ -3 & -1 \end{vmatrix} & -\begin{vmatrix} 3 & 2 \\ 0 & -1 \end{vmatrix} & +\begin{vmatrix} 3 & 1 \\ 0 & -3 \end{vmatrix} \end{bmatrix}$$

$$= \begin{bmatrix} +(-12 - 2) & -(0 - (-5)) & +(0 - (-15)) \\ -(4 - (-4)) & +(12 - 10) & -(-6 - 5) \\ +(-1 - (-6)) & -(-3 - 0) & +(-9 - 0) \end{bmatrix} = \begin{bmatrix} -14 & -5 & 15 \\ -8 & 2 & 11 \\ 5 & 3 & -9 \end{bmatrix}$$

Each minor is expanded by the following pattern (see **Determinant**):

$$\begin{vmatrix} a & b \\ c & d \end{vmatrix} = ad - bc$$

The transpose of the matrix of signed minors is formed by interchanging the rows and columns of the matrix of signed minors:

$$\begin{bmatrix} -14 & -8 & -5 \\ -5 & 2 & 3 \\ 15 & 11 & -9 \end{bmatrix}$$

The value of the determinant of B is

$$= +(-36) + (-5) + (0) - (-30) - (6) - (0)$$

$$= -41 + 30 - 6 = -17 \qquad \text{See } \textbf{Determinant.}$$

Substituting the transpose of the matrix of signed minors and the value of the determinant of B into the equation for the inverse of a 3×3 matrix, we have

$$B^{-1} = \frac{1}{D} \begin{bmatrix} \text{transpose matrix} \\ \text{of signed minors} \end{bmatrix} = -\frac{1}{17} \begin{bmatrix} -14 & -8 & 5 \\ -5 & 2 & 3 \\ 15 & 11 & -9 \end{bmatrix}$$

To verify that B^{-1} is the inverse of B, we can apply the concept of the multiplicative inverse, $B \cdot B^{-1} = I$:

$$\begin{bmatrix} 3 & 1 & 2 \\ 0 & -3 & -1 \\ 5 & -2 & 4 \end{bmatrix} \left(-\frac{1}{17}\right) \begin{bmatrix} -14 & -8 & 5 \\ -5 & 2 & 3 \\ 15 & 11 & -9 \end{bmatrix} = \begin{bmatrix} 1 & 0 & 0 \\ 0 & 1 & 0 \\ 0 & 0 & 1 \end{bmatrix}$$

4. Solve the following system of equations using matrices:

$$4x - 7y = -13$$
$$7x = 3y + 5$$

Before setting up the required matrices, each of the equations must first be in standard form, $Ax + By = C$. See **Standard Form of an Equation of a Line**. The first equation is already in standard form. Rewriting the second equation, we have $7x - 3y = 5$.

Using the system

$$4x - 7y = -13$$
$$7x - 3y = 5$$

we have the matrix of the coefficients $\begin{bmatrix} 4 & -7 \\ 7 & -3 \end{bmatrix}$, the matrix of the variables $\begin{bmatrix} x \\ y \end{bmatrix}$, and the matrix of the numerical values of the equations $\begin{bmatrix} -13 \\ 5 \end{bmatrix}$. To verify that the left sides of the equations are represented by the matrices $\begin{bmatrix} 4 & -7 \\ 7 & -3 \end{bmatrix}$ and $\begin{bmatrix} x \\ y \end{bmatrix}$, calculate their product. The result yields the expressions $4x - 7y$ and $7x - 3y$, respectively.

In order to solve the system, we need to multiply both sides of the matrix equation by the inverse of $\begin{bmatrix} 4 & -7 \\ 7 & -3 \end{bmatrix}$. Equating $\begin{bmatrix} 4 & -7 \\ 7 & -3 \end{bmatrix}$ with $\begin{bmatrix} a & b \\ c & d \end{bmatrix}$, we have the inverse:

$$\begin{bmatrix} 4 & -7 \\ 7 & -3 \end{bmatrix}^{-1} = \frac{1}{ad - bc} \begin{bmatrix} d & -b \\ -c & a \end{bmatrix} = \frac{1}{(4)(-3) - (-7)(7)} \begin{bmatrix} -3 & -(-7) \\ -(7) & 4 \end{bmatrix}$$

$$= \frac{1}{37} \begin{bmatrix} -3 & 7 \\ -7 & 4 \end{bmatrix} \qquad \text{See property 10, } \textbf{Multiplicative inverse matrix.}$$

Setting up the matrix equation and solving, we have

$$\begin{bmatrix} 4 & -7 \\ 7 & -3 \end{bmatrix} \bullet \begin{bmatrix} x \\ y \end{bmatrix} = \begin{bmatrix} -13 \\ 5 \end{bmatrix}$$

$$\frac{1}{37}\begin{bmatrix} -3 & 7 \\ -7 & 4 \end{bmatrix} \bullet \begin{bmatrix} 4 & -7 \\ 7 & -3 \end{bmatrix} \bullet \begin{bmatrix} x \\ y \end{bmatrix} = \frac{1}{37}\begin{bmatrix} -3 & 7 \\ -7 & 4 \end{bmatrix} \bullet \begin{bmatrix} -13 \\ 5 \end{bmatrix}$$

$$\frac{1}{37}\begin{bmatrix} 37 & 0 \\ 0 & 37 \end{bmatrix} \bullet \begin{bmatrix} x \\ y \end{bmatrix} = \frac{1}{37}\begin{bmatrix} 74 \\ 111 \end{bmatrix}$$

$$\begin{bmatrix} 1 & 0 \\ 0 & 1 \end{bmatrix} \bullet \begin{bmatrix} x \\ y \end{bmatrix} = \begin{bmatrix} 2 \\ 3 \end{bmatrix}$$

$$\begin{bmatrix} x \\ y \end{bmatrix} = \begin{bmatrix} 2 \\ 3 \end{bmatrix}$$

Therefore, the solution of the system is the point $(x, y) = (2, 3)$.

To verify the solution, solve the system by the substitution method or the linear combination method. See **Systems of Linear Equations, solutions of**. Keep in mind that the matrix equation

$$\begin{bmatrix} 4 & -7 \\ 7 & -3 \end{bmatrix} \bullet \begin{bmatrix} x \\ y \end{bmatrix} = \begin{bmatrix} -13 \\ 5 \end{bmatrix}$$

is equivalent to the system

$$4x - 7y = -13$$
$$7x - 3y = 5$$

This can be seen by multiplying the left side of the equation using the property for the product of two matrices and then setting the elements of the product equal to the matrix $\begin{bmatrix} -13 \\ 5 \end{bmatrix}$.

5. Solve the following system of equations using matrices:

$$2x + 8y - z = 17$$
$$5x + 4y + 3z = -6$$
$$6x - y - 5z = 7$$

Before we can solve the system, we need to calculate the value of the determinant of the coefficient matrix as well as its inverse matrix.

The value of the determinant is (see **Determinant**)

$$\begin{vmatrix} 2 & 8 & -1 \\ 5 & 4 & 3 \\ 6 & -1 & -5 \end{vmatrix} = +(-40) + (144) + (5) - (-24) - (-6) - (-200) = 339$$

The inverse matrix is (see property 10)

$$\begin{bmatrix} 2 & 8 & -1 \\ 5 & 4 & 3 \\ 6 & -1 & -5 \end{bmatrix} = \frac{1}{D} \begin{bmatrix} \text{transpose matrix} \\ \text{of signed minors} \end{bmatrix} = \frac{1}{339} \begin{bmatrix} -17 & 41 & 28 \\ 43 & -4 & -11 \\ -29 & 50 & -32 \end{bmatrix}$$

Now we can set up the matrix equation for the system, multiply both sides of the equation by the inverse matrix, and then solve the equation for the variable matrix:

$$\begin{bmatrix} 2 & 8 & -1 \\ 5 & 4 & 3 \\ 6 & -1 & -5 \end{bmatrix} \bullet \begin{bmatrix} x \\ y \\ z \end{bmatrix} = \begin{bmatrix} 17 \\ -6 \\ 7 \end{bmatrix}$$

$$\frac{1}{339} \begin{bmatrix} -17 & 41 & 28 \\ 43 & -4 & -11 \\ -29 & 50 & -32 \end{bmatrix} \bullet \begin{bmatrix} 2 & 8 & -1 \\ 5 & 4 & 3 \\ 6 & -1 & 5 \end{bmatrix} \bullet \begin{bmatrix} x \\ y \\ z \end{bmatrix} = \frac{1}{339} \begin{bmatrix} -17 & 41 & 28 \\ 43 & -4 & -11 \\ -29 & 50 & -32 \end{bmatrix} \bullet \begin{bmatrix} 17 \\ -6 \\ 7 \end{bmatrix}$$

$$\frac{1}{339} \begin{bmatrix} 339 & 0 & 0 \\ 0 & 339 & 0 \\ 0 & 0 & 339 \end{bmatrix} \bullet \begin{bmatrix} x \\ y \\ z \end{bmatrix} = \frac{1}{339} \begin{bmatrix} -339 \\ 678 \\ 1017 \end{bmatrix}$$

$$\begin{bmatrix} 1 & 0 & 0 \\ 0 & 1 & 0 \\ 0 & 0 & 1 \end{bmatrix} \bullet \begin{bmatrix} x \\ y \\ z \end{bmatrix} = \begin{bmatrix} -1 \\ 2 \\ -3 \end{bmatrix}$$

$$\begin{bmatrix} x \\ y \\ z \end{bmatrix} = \begin{bmatrix} -1 \\ 2 \\ -3 \end{bmatrix}$$

Therefore, the solution of the system is the point $(x, y, z) = (-1, 2, -3)$.

As with example 4, this problem too can be verified by the substitution method or the linear combination method. See **Systems of Linear Equations, solutions of**. A common mistake to avoid when calculating the inverse is remembering to apply the signs, $\begin{vmatrix} + & - & + \\ - & + & - \\ + & - & + \end{vmatrix}$, when deriving the matrix of signed minors.

➤ **Matrix, augmented (matrix) solution**   An **augmented matrix** is a matrix whose entries are composed of all of the elements of a system of equations, not including the variables. The augmented matrix of the system

$$\begin{array}{c} -x + 2y + 3z = 4 \\ 2x - y - z = 1 \\ 3x + 2y + 4z = 11 \end{array} \quad \text{is} \quad \begin{bmatrix} -1 & 2 & 3 & 4 \\ 2 & -1 & -1 & 1 \\ 3 & 2 & 4 & 11 \end{bmatrix}$$

To find the solution of an augmented matrix means that we want to find the solution of the original system of equations. In other words, we want to determine the values of the variables that satisfy the system. See **Systems of Linear Equations, solutions of**.

M

The values of the variables that satisfy a given system can be determined after the augmented matrix has been transformed into either of the following two forms:

$$\begin{bmatrix} a & b & c & d \\ 0 & f & g & h \\ 0 & 0 & k & l \end{bmatrix} \quad \text{or} \quad \begin{bmatrix} a & 0 & 0 & d \\ e & f & 0 & h \\ i & j & k & l \end{bmatrix}$$

As a system of equations, each matrix can then be written in the triangular form below (the form resembles a triangle):

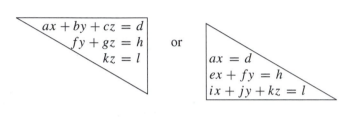

$$\begin{aligned} ax + by + cz &= d \\ fy + gz &= h \\ kz &= l \end{aligned} \quad \text{or} \quad \begin{aligned} ax &= d \\ ex + fy &= h \\ ix + jy + kz &= l \end{aligned}$$

In turn, either system can then be solved for the variables.

If a complete augmented matrix solution is derived, the augmented matrix must be transformed into the form $\begin{bmatrix} 1 & 0 & 0 & d \\ 0 & 1 & 0 & h \\ 0 & 0 & 1 & l \end{bmatrix}$ where $x = d$, $y = h$, and $z = l$.

There are three rules that are used to transform an augmented matrix into one of the preferred forms. How the rules should be applied, and when, will be covered in the examples. It is mostly through experience that a successful method can be developed.

Rules for Solving an Augmented Matrix

1. Interchange any two rows with each other.

2. Multiply or divide the elements of any row by any number (nonzero) that seems appropriate.

3. Combine each of the elements of a particular row with the corresponding elements of another particular row (not some with one row and some with another row).

Examples

1. Solve the following system using augmented matrices:

$$\begin{aligned} -x + 2y + 3z &= 4 \\ 2x - y - z &= 1 \\ 3x + 2y + 4z &= 11 \end{aligned}$$

There are many ways of starting the solution of a system using augmented matrices. One way is to get the leading term in any of the three rows of the augmented matrix to be a 1. Next, using this row, we can eliminate the first term in the other two rows. Then all we need to do is eliminate the second term in either of these rows. What remains will be an augmented matrix which can be rewritten as three equations in triangular form.

To get the leading term in row 1 to be a 1, we can multiply the row by -1:

$$-1(-1 \quad 2 \quad 3 \quad 4) = 1 \quad -2 \quad -3 \quad -4$$

Another way would be to combine row 2 with row 1:

$$\begin{matrix} -1 & 2 & 3 & 4 \\ 2 & -1 & -1 & 1 \end{matrix} = \begin{matrix} 1 & 1 & 2 & 5 \\ 2 & -1 & -1 & 1 \end{matrix}$$

Notice that after the rows have been combined, row 2 is written as it was previously. Or, if we chose, we could get a 1 in row 3 by multiplying row 2 by -1 and then combining with row 3. Using the first of the three choices, we have the first step of the solution.

Multiply row 1 by -1:

$$\begin{bmatrix} -1 & 2 & 3 & 4 \\ 2 & -1 & -1 & 1 \\ 3 & 2 & 4 & 11 \end{bmatrix} = \begin{bmatrix} 1 & -2 & -3 & -4 \\ 2 & -1 & -1 & 1 \\ 3 & 2 & 4 & 11 \end{bmatrix}$$

Since the 1 is in the first row, we can continue. If it wasn't, we could interchange rows until it was, and then continue.

Using the 1 in row 1, we can now eliminate the first term in rows 2 and 3, respectively. To eliminate the 2 in row 2, we can multiply row 1 by -2 and then combine with row 2. To eliminate the 3 in row 3, we can multiply row 1 by -3 and then combine with row 3. This beginning process is one that works well with augmented matrices.

Multiply row 1 by -2 and combine with row 2:

$$\begin{bmatrix} 1 & -2 & -3 & -4 \\ 2 & -1 & -1 & 1 \\ 3 & 2 & 4 & 11 \end{bmatrix} = \begin{bmatrix} 1 & -2 & -3 & -4 \\ 0 & 3 & 5 & 9 \\ 3 & 2 & 4 & 11 \end{bmatrix}$$

Multiply row 1 by -3 and combine with row 3:

$$\begin{bmatrix} 1 & -2 & -3 & -4 \\ 0 & 3 & 5 & 9 \\ 3 & 2 & 4 & 11 \end{bmatrix} = \begin{bmatrix} 1 & -2 & -3 & -4 \\ 0 & 3 & 5 & 9 \\ 0 & 8 & 13 & 23 \end{bmatrix}$$

Our next goal is to eliminate the second term in either the second or third row, the 3 or the 8 position. Depending on the problem, there are several approaches that can be taken. One that is very useful is to multiply the second and third rows by the second element in the third and second rows, respectively, making sure that the product leaves the second terms in the rows the same but opposite in sign. This is accomplished by making one of the numbers negative, if necessary. In our example, if we multiply row 2 by 8 and row 3 by -3, we will get second terms of 24 and -24, respectively. We can then combine row 2 with row 3, change the augmented matrix to three equations in triangular form, and then solve for the variables.

Multiply row 2 by 8 and row 3 by -3:

$$\begin{bmatrix} 1 & -2 & -3 & -4 \\ 0 & 3 & 5 & 9 \\ 0 & 8 & 13 & 23 \end{bmatrix} = \begin{bmatrix} 1 & -2 & -3 & -4 \\ 0 & 24 & 40 & 72 \\ 0 & -24 & -39 & -69 \end{bmatrix}$$

Combine row 2 with row 3:

$$\begin{bmatrix} 1 & -2 & 3 & -4 \\ 0 & 24 & 40 & 72 \\ 0 & -24 & -39 & -69 \end{bmatrix} = \begin{bmatrix} 1 & -2 & -3 & -4 \\ 0 & 24 & 40 & 72 \\ 0 & 0 & 1 & 3 \end{bmatrix}$$

Write the augmented matrix as three equations in triangular form:

$$x - 2y - 3z = -4$$
$$24y + 40z = 72$$
$$z = 3$$

Now we can solve for the variables.

$$\text{Solving for } z: \quad z = 3$$

$$
\begin{aligned}
\text{Solving for } y: \quad 24y + 40z &= 72 \\
24y + 40(3) &= 72 \\
24y + 120 &= 72 \\
24y &= -48 \\
y &= -2
\end{aligned}
$$

$$
\begin{aligned}
\text{Solving for } x: \quad x - 2y - 3z &= -4 \\
x - 2(-2) - 3(3) &= -4 \\
x + 4 - 9 &= -4 \\
x &= 1
\end{aligned}
$$

Hence, the solution of the system is the point $(x, y, z) = (1, -2, 3)$.

Sometimes eliminating the second term in the third row can be achieved by only multiplying or dividing the second row and then combining with the third row. For example, in the matrix

$$\begin{bmatrix} 1 & -2 & -1 & -6 \\ 0 & 2 & 1 & 5 \\ 0 & 4 & 3 & 13 \end{bmatrix}$$

if we multiply row 2 by -2 and combine with row 3, we can eliminate the second element.

Multiply row 2 by -2 and combine with row 3:

$$\begin{bmatrix} 1 & -2 & -1 & -6 \\ 0 & 2 & 1 & 5 \\ 0 & 4 & 3 & 13 \end{bmatrix} = \begin{bmatrix} 1 & -2 & -1 & -6 \\ 0 & 2 & 1 & 5 \\ 0 & 0 & 1 & 3 \end{bmatrix}$$

The system can then be solved from here.

Another approach would have been to divide row 3 by -2 and then combine with row 2, but this would have given us fractions. This will work but most people like to avoid fractions.

To see that there are other ways of solving the same system, let's solve the problem again. This time beginning by combining row 1 with row 2.

Combine row 1 with row 2:

$$\begin{bmatrix} -1 & 2 & 3 & 4 \\ 2 & -1 & -1 & 1 \\ 3 & 2 & 4 & 11 \end{bmatrix} = \begin{bmatrix} -1 & 2 & 3 & 4 \\ 1 & 1 & 2 & 5 \\ 3 & 2 & 4 & 11 \end{bmatrix}$$

We did this to obtain a 1 for the leading term in row 2. Next, we need to interchange row 2 with row 1 and then continue.

Interchange row 2 with row 1:

$$\begin{bmatrix} -1 & 2 & 3 & 4 \\ 1 & 1 & 2 & 5 \\ 3 & 2 & 4 & 11 \end{bmatrix} = \begin{bmatrix} 1 & 1 & 2 & 5 \\ -1 & 2 & 3 & 4 \\ 3 & 2 & 4 & 11 \end{bmatrix}$$

Combine row 1 with row 2:

$$\begin{bmatrix} 1 & 1 & 2 & 5 \\ -1 & 2 & 3 & 4 \\ 3 & 2 & 4 & 11 \end{bmatrix} = \begin{bmatrix} 1 & 1 & 2 & 5 \\ 0 & 3 & 5 & 9 \\ 3 & 2 & 4 & 11 \end{bmatrix}$$

Multiply row 1 by -3, combine with row 3:

$$\begin{bmatrix} 1 & 1 & 2 & 5 \\ 0 & 3 & 5 & 9 \\ 3 & 2 & 4 & 11 \end{bmatrix} = \begin{bmatrix} 1 & 1 & 2 & 5 \\ 0 & 3 & 5 & 9 \\ 0 & -1 & -2 & -4 \end{bmatrix}$$

Multiply row 3 by 3, combine with row 2:

$$\begin{bmatrix} 1 & 1 & 2 & 5 \\ 0 & 3 & 5 & 9 \\ 0 & -1 & -2 & -4 \end{bmatrix} = \begin{bmatrix} 1 & 1 & 2 & 5 \\ 0 & 0 & -1 & -3 \\ 0 & -1 & -2 & -4 \end{bmatrix}$$

Interchange row 2 with row 3:

$$\begin{bmatrix} 1 & 1 & 2 & 5 \\ 0 & 0 & -1 & -3 \\ 0 & -1 & -2 & -4 \end{bmatrix} = \begin{bmatrix} 1 & 1 & 2 & 5 \\ 0 & -1 & -2 & -4 \\ 0 & 0 & -1 & -3 \end{bmatrix}$$

Writing the equations in triangular form, we have the system

$$x + y + 2z = 5$$
$$-y - 2z = -4$$
$$-1z = -3$$

which has a solution of $(x, y, z) = (1, -2, 3)$.

Sometimes a complete augmented matrix solution is desired. If this is the case, continue with row operations until the matrix is transposed into the form $\begin{bmatrix} 1 & 0 & 0 & d \\ 0 & 1 & 0 & h \\ 0 & 0 & 1 & l \end{bmatrix}$.

Multiply row 2 by -1 and row 3 by -1:

$$\begin{bmatrix} 1 & 1 & 2 & 5 \\ 0 & -1 & -2 & -4 \\ 0 & 0 & -1 & -3 \end{bmatrix} = \begin{bmatrix} 1 & 1 & 2 & 5 \\ 0 & 1 & 2 & 4 \\ 0 & 0 & 1 & 3 \end{bmatrix}$$

Multiply row 2 by -1; combine with row 1:

$$\begin{bmatrix} 1 & 1 & 2 & 5 \\ 0 & 1 & 2 & 4 \\ 0 & 0 & 1 & 3 \end{bmatrix} = \begin{bmatrix} 1 & 0 & 0 & 1 \\ 0 & 1 & 2 & 4 \\ 0 & 0 & 1 & 3 \end{bmatrix}$$

Multiply row 3 by -2; combine with row 2:

$$\begin{bmatrix} 1 & 0 & 0 & 1 \\ 0 & 1 & 2 & 4 \\ 0 & 0 & 1 & 3 \end{bmatrix} = \begin{bmatrix} 1 & 0 & 0 & 1 \\ 0 & 1 & 0 & -2 \\ 0 & 0 & 1 & 3 \end{bmatrix}$$

Hence, in triangular form we have the equations

$$x = 1$$
$$y = -2$$
$$z = 3$$

which are the coordinates of the solution $(x, y, z) = (1, -2, 3)$.

2. Solve the following system using augmented matrices:

$$4x - 12y + 8z = 4$$
$$2x - 10y + 2z = -8$$
$$3x - 7y + 5z = 2$$

Divide row 1 by 4:

$$\begin{bmatrix} 4 & -12 & 8 & 4 \\ 2 & -10 & 2 & -8 \\ 3 & -7 & 5 & 2 \end{bmatrix} = \begin{bmatrix} 1 & -3 & 2 & 1 \\ 2 & -10 & 2 & -8 \\ 3 & -7 & 5 & 2 \end{bmatrix}$$

Multiply row 1 by -2; combine with row 2:

$$\begin{bmatrix} 1 & -3 & 2 & 1 \\ 2 & -10 & 2 & -8 \\ 3 & -7 & 5 & 2 \end{bmatrix} = \begin{bmatrix} 1 & -3 & 2 & 1 \\ 0 & -4 & -2 & -10 \\ 3 & -7 & 5 & 2 \end{bmatrix}$$

Multiply row 1 by -3; combine with row 3:

$$\begin{bmatrix} 1 & -3 & 2 & 1 \\ 0 & -4 & -2 & -10 \\ 3 & -7 & 5 & 2 \end{bmatrix} = \begin{bmatrix} 1 & -3 & 2 & 1 \\ 0 & -4 & -2 & -10 \\ 0 & 2 & -1 & -1 \end{bmatrix}$$

Multiply row 2 by $\frac{1}{2}$ or divide row 2 by 2; combine with row 3:

$$\begin{bmatrix} 1 & -3 & 2 & 1 \\ 0 & -4 & -2 & -10 \\ 0 & 2 & -1 & -1 \end{bmatrix} = \begin{bmatrix} 1 & -3 & 2 & 1 \\ 0 & -4 & -2 & -10 \\ 0 & 0 & -2 & -6 \end{bmatrix}$$

Writing the equations in triangular form, we have the system

$$x - 3y + 2z = 1$$
$$-4y - 2z = -10$$
$$-2z = -6$$

solving for z: $\qquad$ solving for y: $\qquad$ solving for x:

$$-2z = -6 \qquad -4y - 2z = -10 \qquad x - 3y + 2z = 1$$
$$z = 3 \qquad -4y - 2(3) = -10 \qquad x - 3(-1) + 2(3) = 1$$
$$-4y = -4 \qquad x + 3 = 1$$
$$y = 1 \qquad x = -2$$

Hence, the solution of the system is the point $(x, y, z) = (-2, 1, 3)$.

➤ Maximum and Minimum, concepts and problems related to $\quad$ Although there are several definitions of clarity concerning the maximum and minimum values of a function on an interval, it is the application of the concepts pertaining to maximum and minimum with which we are concerned. Usually the first two applications are curve sketching and optimization problems.

In general, if $f(x)$ is defined on an interval where a is any point in the interval, then we call $f(c)$ a maximum of $f(x)$ if $f(c) \geqslant f(x)$ for all x in the interval and we call $f(c)$ a minimum of $f(x)$ if $f(c) \leqslant f(x)$ for all x in the interval. Defined in this manner, the maximum and minimum are called the **extreme values** of the function. They are also referred to as **extrema** or the **absolute maximum** or **absolute minimum**.

If the extrema occur at the endpoints of the interval, they are referred to as **endpoint extrema.** If they occur at the interior points of the interval they are referred to as **relative extrema**.

In an open interval, if $f(c)$ is a maximum, then it is called a relative maximum. There could be other points outside the interval such that $f(c) < f(x)$.

In an open interval, if $f(c)$ is a minimum, then it is called a relative minimum. There could be other points outside the interval such that $f(c) > f(x)$.

The extrema of greatest interest are those that yield a derivative of zero or are relative extrema that are undefined. For example, these are extrema that occur in graphs such as $f(x) = -x^2$, $f(x) = x^2$, or $f(x) = |x|$. These types of extrema, whether absolute or relative, are referred to as **critical points**. The value of c that determines a critical point is referred to as a **critical number**.

As far as graphs of functions are concerned, if a tangent line to a graph has a slope, or derivative, that is equal to zero, this implies that it is parallel to the x-axis and hence determines an extremum. To determine whether the extremum is a maximum or a minimum would require more information. This leads to the first method of graphing functions, graphing by the first derivative method.

Graphing by the First Derivative Method

1. To determine the critical points, set the first derivative equal to zero and solve for x. These are the critical numbers of the graph, if any. Substitute any critical numbers into $f(x)$ to determine the y-values of the critical points. This will yield any critical points (x, y), if any.

2. To determine whether the critical points yield a maximum, minimum, or point of inflection (the point at which the graph changes from concave downward to concave upward), we need to choose x-values to the left and right of the critical points, substitute these into the first derivative, and then see whether the result is positive (which implies the graph is increasing as it approaches or leaves the critical point), negative (which implies the graph is decreasing as it approaches or leaves the critical point), or zero (which implies the graph is horizontal as it approaches or leaves the critical point). At this time, a sketch of the graph could be made.

3. If more accuracy in the graph is desired, a few more points of the graph can be determined. Do this by choosing x-values to the left and right of the critical points and then substituting these into $f(x)$ to determine the y-values of the points. Then plot the points (x, y) and sketch the graph. If the x- or y-intercepts can be calculated, they can be useful as well.

Another method of graphing can be determined by comparing the graphs of $f(x)$, $f'(x)$, and $f''(x)$. By looking at the three graphs, we can see that the x-values where the graph of the first derivative crosses the x-axis correspond with the maximum and minimum of $f(x)$, the x-value where the graph of the second derivative crosses the x-axis corresponds with the point of inflection of $f(x)$, and the y-values of the second derivative are negative when the original function opens concave downward and positive when the original function opens concave upward. This leads to the second method of graphing functions, graphing by the second derivative test.

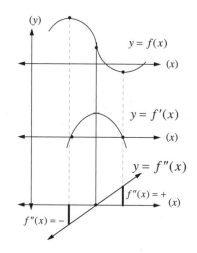

Graphing by the Second Derivative Test

1. Determine the critical points in the same manner as the first derivative method.

2. Find the point of inflection by setting the second derivative equal to zero and then solving for x. Substitute any x-values found into $f(x)$ to determine the y-values of the point(s) of inflection.

3. Substitute any critical numbers (the x-values found by setting the first derivative equal to zero) into the second derivative. If the result is positive, the original function opens upward at the critical point. If the result is negative, the original function opens downward at the critical point. If the result is zero, the critical point is a point of inflection.

4. If more accuracy in the graph is desired, find a few more points of the graph as was done in the first derivative test.

M

The methods described above can be used for graphing functions as well as solving optimization problems. As a result, optimization problems are sometimes referred to as **max-min** problems.

Examples

1. Graph $y = f(x) = x^3 - 6x^2 + 9x$ by applying the first derivative method.

The x-values of the critical points are found by setting the first derivative equal to zero and then solving for x:

$$f'(x) = 3x^2 - 12x + 9 = 0$$
$$x^2 - 4x + 3 = 0$$
$$(x-1)(x-3) = 0$$
$$x = 1 \qquad \text{and} \qquad x = 3$$

The y-values of the critical points are found by substituting $x = 1$ and $x = 3$ into $f(x)$:

$$y = f(1) = (1)^3 - 6(1)^2 + 9(1) = 4 \qquad y = f(3) = (3)^3 - 6(3)^2 + 9(3) = 0$$

Therefore the critical points are $(1, 4)$ and $(3, 0)$.

To determine the inclination of the graph on either side of the critical points, we can choose x-values of 0, 2, and 4 (0 and 2 are left and right of $(1, 4)$ while 2 and 4 are left and right of $(3, 0)$). Substituting into $f'(x)$, we have

$$f'(0) = 3(0)^2 - 12(0) + 9 = +, \qquad \text{increasing into } (1, 4)$$
$$f'(2) = 3(2)^2 - 12(2) + 9 = -, \qquad \text{decreasing out of } (1, 4) \text{ and into } (3, 0)$$
$$f'(4) = 3(4)^2 - 12(4) + 9 = +, \qquad \text{increasing out of } (3, 0)$$

Sketching the graph, we have

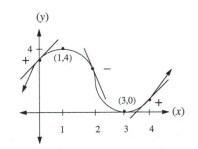

2. Graph $y = f(x) = x^3 - 6x^2 + 9x$ by applying the second derivative test.

The critical points are found just as in example 1. They are $(1, 4)$ and $(3, 0)$.

The x-value of the point of inflection is found by setting the second derivative equal to zero and solving for x:

$$f''(x) = 6x - 12 = 0$$
$$x = 2$$

The y-value of the point of inflection is found by substituting the x-value into $f(x)$.

$$y = f(2) = (2)^3 - 6(2)^2 + 9(2) = 2$$

Therefore the point of inflection is $(2, 2)$.

To test for concavity at the critical points, we test the critical numbers in the second derivative:

$$f''(1) = 6(1) - 12 = -, \qquad \text{opens concave downward at } (1, 4)$$
$$f''(3) = 6(3) - 12 = +, \qquad \text{opens concave upward at } (3, 0)$$

Sketching the graph, we have

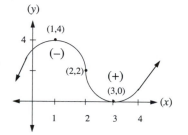

3. Find two positive numbers that have a sum of 30 while their product is as large as possible.

To solve an optimization problem, we need to derive an equation that involves the variable(s) to be optimized. If we let one of the numbers be x, then the other number is $30 - x$, and their product, which we can call P, is $P(x) = x(30 - x) = -x^2 + 30x$.

In order to find the maximum value of x, and choosing to use the second derivative test, we need to set the first derivative equal to zero, solve for x, and then test x in the second derivative:

$$P'(x) = -2x + 30 = 0$$
$$x = 15$$

$$P''(x) = -2, \qquad \text{which implies the graph of } P(x) \text{ is always concave downward.}$$

Hence, $x = 15$ yields a maximum. The two positive numbers are therefore the same, $x = 15$ and $30 - x = 30 - 15 = 15$.

4. Find the area of the largest rectangle that can be drawn with one side on the x-axis and the upper vertices on the curve $y = 12 - x^2$.

First, we need to graph $y = 12 - x^2$ and then derive the formula for the area of the rectangle.

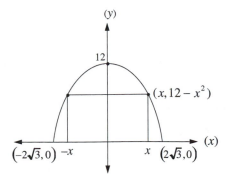

The base of the rectangle is $2x$ and the height is $12 - x^2$. If we let $A(x)$ represent the area of the rectangle, then $A(x) = 2x(12 - x^2) = -2x^3 + 24x$. Optimizing $A(x)$, we have

$$A'(x) = -6x^2 + 24 = 0$$
$$x^2 = 4$$
$$x = \pm 2$$

Testing $x = \pm 2$ in the second derivative, we have

$$A''(-2) = -12(-2) = +24,$$ which implies the graph is concave upward at $x = -2$; a minimum.

$$A''(2) = -12(2) = -24,$$ which implies the graph is concave downward at $x = 2$; a maximum.

Therefore, the area of the largest rectangle is

$$A(2) = -2(2)^3 + 24(2) = -16 + 48 = 32$$

➤ **Mean (average)**   The arithmetic mean, or just **mean**, is one of three commonly used averages or measures of the central tendency of data. The other two are median and mode.

The **mean** of a set of numbers is also known as the average of the numbers. It is found by adding the numbers and then dividing their sum by the number of members (frequency) in the set. In this manner, the central tendency of a set of numbers can be found.

Examples

1. Find the mean of the following test scores: 65, 82, 76, 98, 87, 100, 96, and 82.

   First, find the sum of the numbers in the list. Then, divide the sum by the number of members in the list.

   $$\text{Mean} = \frac{\text{Sum}}{\text{Number of members}} = \frac{\text{Sum}}{\text{Frequency}} = \frac{686}{8} = 85.75$$

2. Find the mean of the following list: $x - 14, x, x + 10, x - 5, x - 12, x + 9$.

$$\text{Mean} = \frac{\text{Sum}}{\text{Frequency}} = \frac{6x - 12}{6} = \frac{6(x - 2)}{6} = x - 2$$

For related articles, see **Median**, **Mode**, **Range (statistical definition)**, **Variance**, and **Standard Deviation**.

➤ **Mean Proportional** See **Geometric Mean**.

➤ **Means** In the proportion $\frac{a}{b} = \frac{c}{d}$, b and c are called the **means**. See **Proportion**.

➤ **Mean Value Theorem** The **mean value theorem** is mostly used in the proofs of other theorems. It does have some direct application in solving problems but its greatest merit is that it states that if $(a, f(a))$ and $(b, f(b))$ are two points of a continuous function $f(x)$, there is a tangent line to the curve, between the two points, that is parallel to the secant line through the points.

More formally, the **mean value theorem** states that if $f(x)$ is a continuous function on the closed interval $[a, b]$ and differentiable on the open interval (a, b), then there is at least one number c, between a and b, such that

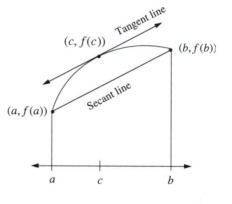

$$f'(c) = \frac{f(b) - f(a)}{b - a}$$

This equation states that the slope of the tangent line through the point $(c, f(c))$ is equal to the slope of the secant line through the points $(a, f(a))$ and $(b, f(b))$.

Example

1. If $f(x) = x^3$, find the value(s) of c, in the interval $(-2, 2)$ that satisfy the mean value theorem.

The derivative of $f(x)$ is $f'(x) = 3x^2$, which evaluated when $x = c$ is $f'(c) = 3c^2$. This is the left side of the equation of the mean value theorem. The right side of the equation is

$$\frac{f(b) - f(a)}{b - a} = \frac{(2)^3 - (-2)^3}{2 - (-2)} = 4$$

Setting the left and right sides of the equation equal to each other, we have

$$f'(c) = \frac{f(b) - f(a)}{b - a}$$
$$3c^2 = 4$$
$$c = \pm\sqrt{\frac{4}{3}} = \pm\frac{2}{\sqrt{3}} = \pm\frac{2\sqrt{3}}{3}$$

Therefore, there are two values of c, between -2 and 2, where the tangent line is parallel to the secant line through the points $(-2, -8)$ and $(2, 8)$.

➤ **Measure of an Arc**   See **Arc Measure**.

➤ **Measure of a Segment**   The **measure of a segment** is the distance between the endpoints of the segment as compared with a number line, or ruler. If the symbol for a segment is $\overline{AB}$, the symbol for its measure is AB. AB is the distance between A and B, whereas $\overline{AB}$ is a set of points. The segment $\overline{AB}$ is the set of points that include A, B, and all of the points between A and B.

➤ **Measure of an Angle**   In geometry, the **measure of an angle** is a real number between 0 and 180. It is determined by measuring the angle with a protractor. Since an angle is defined as the union of two noncollinear rays with the same endpoint (see **Angle, in geometry**), the measure of an angle determines the amount of separation of the two rays.

If the rays of $\angle BAC$ are separated by $39°$, we call the number 39 the measure of the angle BAC and we write $m\angle BAC = 39$, read "the measure of angle BAC is 39."

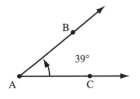

Since the angle is defined as a number, we do not need to include the word "degrees" or the symbol for degrees (°) when stating the measure of an angle. Hence, we say that $m\angle BAC = 39$, the measure of $\angle BAC$ is 39, or $\angle BAC$ has a measure of 39, even though $\angle BAC$ is a $39°$ angle. In the former example, the word "degree" is implied through the word "measure." In the last example, it must be stated.

In trigonometry, the **measure of an angle** is a real number measured in either degrees or radians of a circle. The measure determines the amount of separation of the initial and terminal sides of the angle. See **Angle, in trigonometry**.

➤ **Measure of the Angles of a Triangle**   See **Sum of the Measures of the Angles of a Triangle**.

➤ **Median**   The **median** is one of three commonly used averages or measures of the central tendency of data. The other two are mean and mode.

If a set of numbers is arranged in order of size, the **median** is the middle term if the number of elements in the set is odd, or, it is the average of the two middle terms if the number of elements in the set is even. The average of the two middle terms is found by adding them and dividing by 2.

Examples

1. Find the median of the following scores: 78, 86, 54, 99, 71, 68, 74, 89, 93.

   First arrange the scores in order:

$$54 \quad 68 \quad 71 \quad \underline{78} \quad 86 \quad 89 \quad 93 \quad 99$$

Since there is an odd number of terms in the list, there is only one middle term, 78. The number 78 is five terms in from either end of the list. Hence, 78 is the median.

2. Find the median of the following scores: 93, 62, 75, 82, 52, 67, 83, 96, 100, 71.

Arranging the scores in order, we have

$$52 \quad 62 \quad 67 \quad 71 \quad \underline{75 \quad 82} \quad 83 \quad 93 \quad 96 \quad 100$$

Since there is an even number of terms in the list, there are two middle terms, 75 and 82. In this case, the median is the average of the two middle terms. Hence,

$$\text{Median} = \frac{75 + 82}{2} = \frac{157}{2} = 78.5$$

The median can be a number that is not a term of the original data.

For related articles, see **Mean**, **Mode**, **Range (statistical definition)**, **Variance**, and **Standard Deviation**.

► **Midpoint**   The point that divides a segment into two equal parts is called the **midpoint**. It is equidistant from the endpoints of the segment. If the segment is $\overline{AB}$, then M is the midpoint of $\overline{AB}$ under the following stipulations: M must be between A and B, and $AM = MB$. In this sense, $AM = \frac{1}{2}AB$.

$$
\begin{array}{ccc}
A & M & B
\end{array}
$$
Midpoint

► **Midpoint Formula**   The **midpoint formula** is used to find the midpoint between any two points in the coordinate plane. Let $P(x_1, y_1)$ and $Q(x_2, y_2)$ be any two points in the coordinate plane. Then the midpoint M, between P and Q, is

$$M = \left(\frac{x_1 + x_2}{2}, \frac{y_1 + y_2}{2} \right)$$

The x-coordinate of the midpoint is determined by finding the average of the x-coordinates of the points. The y-coordinate is determined by finding the average of the y-coordinates of the points.

Examples

1. Find the midpoint of the segment between $P(-4, 7)$ and $Q(-2, 3)$.

Letting $(x_1, y_1) = (-4, 7)$ and $(x_2, y_2) = (-2, 3)$, we have

$$M\left(\frac{x_1 + x_2}{2}, \frac{y_1 + y_2}{2} \right) = M\left(\frac{-4 + (-2)}{2}, \frac{7 + 3}{2} \right) = M\left(\frac{-6}{2}, \frac{10}{2} \right) = M(-3, 5)$$

2. Find Q, the other endpoint of $\overline{PQ}$, if we are given $P(-4, 7)$ and the midpoint $M(-3, 5)$.

Letting P be represented by (x_1, y_1), with $(x_1, y_1) = (-4, 7)$ and letting Q be represented by (x_2, y_2), we have

$$M\left(\frac{x_1 + x_2}{2}, \frac{y_1 + y_2}{2}\right) = M(-3, 5)$$

Set the equation of the midpoint equal to the coordinates of the midpoint.

$$M\left(\frac{-4 + x_2}{2}, \frac{7 + y_2}{2}\right) = M(-3, 5)$$

$$\frac{-4 + x_2}{2} = -3 \quad \text{and} \quad \frac{7 + y_2}{2} = 5$$

If two points are equal, their corresponding coordinates are equal.

$$-4 + x_2 = -6 \qquad 7 + y_2 = 10$$
$$x_2 = -6 + 4 \qquad y_2 = 10 - 7$$
$$x_2 = -2 \qquad y_2 = 3$$

Hence, the point $Q(x_2, y_2)$ is $Q(-2, 3)$, just as it was in example 1.

➤ **Minor**   A **minor** is the determinant formed when the row and column of an entry of a given determinant are canceled. The given determinant must be of a size 3×3 or larger. The minor is one size less than the given determinant.

The sign of a minor is $(-1)^{i+j}$ where i is the number of the row of the entry and j is the number of the column of the entry. See below. Also see **Determinant** or **Cramer's Rule**.

Example

1. In the determinant below, find the minors and their signs for the entries 7, 2, and 3. Then evaluate each minor.

$$\begin{vmatrix} 7 & 2 & 3 \\ 4 & 8 & 9 \\ 6 & 5 & 1 \end{vmatrix}$$

The minor of 7 is $\begin{vmatrix} 8 & 9 \\ 5 & 1 \end{vmatrix}$. We get this by canceling the row (horizontal) and the column (vertical) of the entry 7. The row has entries 7, 2, and 3. The column has entries 7, 4, and 6. Once those have been canceled, what remains is the minor. The minor of 2 is $\begin{vmatrix} 4 & 9 \\ 6 & 1 \end{vmatrix}$. We cancel the row with 2 in it and the column with 2 in it. The minor of 3 is $\begin{vmatrix} 4 & 8 \\ 6 & 5 \end{vmatrix}$.

$$\begin{vmatrix} 7 & 2 & 3 \\ 4 & 8 & 9 \\ 6 & 5 & 1 \end{vmatrix} \qquad \begin{vmatrix} 7 & 2 & 3 \\ 4 & 8 & 9 \\ 6 & 5 & 1 \end{vmatrix} \qquad \begin{vmatrix} 7 & 2 & 3 \\ 4 & 8 & 9 \\ 6 & 5 & 1 \end{vmatrix}$$

The signs for each of the minors are derived from $(-1)^{i+j}$. The 7 is in row 1 and column 1, the 2 is in row 1 and column 2, and the 3 is in row 1 and column 3.

    a. For 7, we have $(-1)^{i+j} = (-1)^{1+1} = (-1)^2 = +1$

    b. For 2, we have $(-1)^{i+j} = (-1)^{1+2} = (-1)^3 = -1$

    c. For 3, we have $(-1)^{i+j} = (-1)^{1+3} = (-1)^4 = +1$

Hence, the minors, their signs, and the evaluation of the entries for 7, 2, and 3 are

$$\text{a. Entry 7, } (+1) \begin{vmatrix} 8 & 9 \\ 5 & 1 \end{vmatrix} = (+1)(8 - 45) = -37$$

$$\text{b. Entry 2, } (-1) \begin{vmatrix} 4 & 9 \\ 6 & 1 \end{vmatrix} = (-1)(4 - 54) = (-1)(-50) = 50$$

$$\text{c. Entry 3, } (+1) \begin{vmatrix} 4 & 8 \\ 6 & 5 \end{vmatrix} = (+1)(20 - 48) = -28$$

The determinants are of the form $\begin{vmatrix} a & b \\ c & d \end{vmatrix}$. They are expanded as follows:

$$\begin{vmatrix} a & b \\ c & d \end{vmatrix} = ad - bc \qquad \text{See \textbf{Determinant}.}$$

For a 3×3 determinant, the signs of each minor can be calculated or the following determinant can be memorized:

$$\begin{vmatrix} + & - & + \\ - & + & - \\ + & - & + \end{vmatrix}$$

The same can be done for 4×4 determinant or higher. To find the value of a 3×3 determinant using minors, see **Determinant**.

➤ **Minor Arc**   See **Major and Minor Arc**.

➤ **Minor Axis of an Ellipse**   See **Ellipse**.

➤ **Minus Numbers**   See **Positive and Negative Numbers**.

➤ **Minutes**   See **Degree-Minutes-Seconds**.

➤ **Mixed Decimal**   See **Decimal Numbers, definition**.

➤ **Mixed Number**   A **mixed number** is a number that is composed of a whole number and a fraction. Some examples are $3\frac{1}{2}$, $1\frac{3}{2}$, $10\frac{1}{6}$, and $100\frac{59}{61}$. When an improper fraction is simplified, or reduced, it becomes a mixed number. See **Fractions, improper**.

To change a mixed number to an improper fraction, we multiply the whole number and the denominator of the fraction, add the numerator, and place the resulting number over the denominator.

Examples

1. Change the mixed number $3\frac{1}{2}$ to an improper fraction.

In the mixed number $3\frac{1}{2}$, the 3 is the whole number, the 2 is the denominator, and the 1 is the numerator. To change $3\frac{1}{2}$ to an improper fraction, we multiply 3 and 2 (which is 6), add the 1 (now we have 7), and place that result over 2 (which gives us $\frac{7}{2}$). Hence,

$$3\frac{1}{2} = \frac{7}{2}$$

To verify this, reduce $\frac{7}{2}$. We divide 7 by 2 and write the remainder as a fraction:

$$\begin{array}{r} 3\frac{1}{2} \\ 2\overline{)7} \\ \underline{6} \\ 1 \end{array}$$

Hence, $\frac{7}{2} = 3\frac{1}{2}$.

See **Fractions, reducing**, examples 3 and 4.

2. Change $5\frac{2}{3}$ to an improper fraction.

We multiply 5 and 3, add the 2, and place that result over 3. Hence,

$$5\frac{2}{3} = \frac{17}{3}$$

We get 17 because the whole number (5) times the denominator (3) when added to the numerator (2) gives us 17. Then if we place 17 over the denominator (3), we get the improper fraction $\frac{17}{3}$.

➤ **Mode**   The **mode** is one of three commonly used averages or measures of central tendency. The other two are **mean** and **median**.

If a set of numbers is arranged in order of size, the **mode** is the number that occurs most often. The mode is the number that has the greater frequency, or occurrence.

There can be more than one mode. This can happen when two or more numbers have the same frequency.

Example

1. Find the mode of the following list of numbers: 1, 3, 5, 4, 8, 9, 5, 3, 3, 8, 7, 6, 7, 9, 8, 4, 8, 5.

First we arrange the numbers in order of size.

$$1 \quad 3 \quad 3 \quad 3 \quad 4 \quad 4 \quad 5 \quad 5 \quad 5 \quad 6 \quad 7 \quad 7 \quad 8 \quad 8 \quad 8 \quad 8 \quad 9 \quad 9$$

Since 8 occurs most often, the mode is 8. If there had been one more 3 and one more 5 in the list, we would have had three modes, 3, 5, and 8.

For related articles, see **Mean**, **Median**, **Range (statistical definition)**, **Variance**, or **Standard Deviation**.

➤ **Modulus of a Complex Number**   If $z = a + bi$ is a complex number, then its absolute value, called the **modulus**, is

$$|z| = |a + bi| = \sqrt{a^2 + b^2}$$

For a full explanation, see **Complex Number, absolute value**.

➤ **Monomial**   A **monomial** is a polynomial with one term. The terms 2, $-3x$, $8x^2y$, and $-5x^2y^3z$ are each monomials. A monomial is composed of a constant, a variable, or the product of a constant and one or more variables. For more, see **Polynomial**.

➤ **Monomial Factoring**   See **Factoring Common Monomials**.

➤ **Multiple**   A **multiple** is a product of a given whole number and another whole number. If 4 is the given whole number and it is multiplied by the whole numbers 0, 1, 2, 3, 4, and so on, then we generate the multiples of 4 which are 0, 4, 8, 12, 16, and so on. In this sense, 0 is the multiple of 4 and 0 ($4 \times 0 = 0$), 4 is the multiple of 4 and 1 ($4 \times 1 = 4$), 8 is the multiple of 4 and 2 ($4 \times 2 = 8$), 12 is the multiple of 4 and 3 ($4 \times 3 = 12$), and so on. Each of the multiples is a product of the given whole number, 4, and another whole number.

Other examples of multiples are

The multiples of 5: $0, 5, 10, 15, 20, \ldots$
The multiples of 8: $0, 8, 16, 24, 32, \ldots$

For related articles, see **Least Common Multiple** or **Common Denominator**.

➤ **Multiple, least common**   See **Least Common Multiple**.

➤ **Multiplicand**   See **Multiplication of Whole Numbers**.

➤ **Multiplication of Complex Numbers**   See **Complex Numbers**, example 5.

➤ **Multiplication of Decimals**   See **Decimal Numbers, multiplication of**.

➤ **Multiplication of Fractions**   See **Fractions, multiplication of**.

➤ **Multiplication of Polynomials**   See **Polynomials, multiplication of**.

➤ **Multiplication of Radicals**   See **Radicals, multiplication with**.

➤ **Multiplication of Rational Expressions**   See **Rational Expressions, multiplication and division of**.

➤ **Multiplication of Whole Numbers**   **Multiplication of whole numbers** is a process in which one number, the **multiplier**, multiplies another number, the **multiplicand**. The result is the product of the two numbers. The numbers in each step of the process are called **partial products**.

$$
\begin{array}{r}
\text{multiplicand} \\
\times\ \text{multiplier} \\
\hline
\text{partial product} \\
+\ \text{partial product} \\
\hline
\text{product}
\end{array}
$$

If the multiplier consists of one digit, multiply each digit of the multiplicand by this number, proceeding from right to left. See example 1.

If the multiplier contains more than one digit, form partial products by multiplying each digit of the multiplicand by each digit of the multiplier, proceeding from right to left. The right-hand digit of each partial product should line up with its corresponding digit in the multiplier. The answer is determined by adding the partial products. See example 2.

Examples

1. Multiply 486 by 3.

    Although not necessary, it is convenient to line up the ones digits of the multiplier and multiplicand:

    $$\begin{array}{r} {}^{1} \\ 486 \\ \times\ 3 \\ \hline 8 \end{array}$$

    Multiply 3 × 6 to get 18. Write units digit (8) in the product and carry the tens digit (1) to the next column.

    $$\begin{array}{r} {}^{21} \\ 486 \\ \times\ 3 \\ \hline 58 \end{array}$$

    Multiply 3 × 8 to get 24. Then add the 1 carried over from the previous step to get 25. Write the 5 in the product and carry the 2 to the next column.

    $$\begin{array}{r} {}^{21} \\ 486 \\ \times\ 3 \\ \hline 1458 \end{array}$$

    Multiply 3 × 4 to get 12. Then add the 2 carried over from the previous step to get 14. Write the 14 in the product.

    If 486 is multiplied by 3, the product is 1,458.

2. Multiply 486 by 53.

    $$\begin{array}{r} {}^{21} \\ 486 \\ \times\ 53 \\ \hline 1458 \end{array}$$

    Multiply each digit of 486 by 3 as in example 1. This gives the partial product 1458.

    $$\begin{array}{r} {}^{43} \\ 486 \\ \times\ 53 \\ \hline 1458 \\ 2430 \\ \hline 25758 \end{array}$$

    Multiply each digit of 486 by 5. This gives the partial product 2430. We line up the 0 of 2430 with the 5 of the partial product.

    We get the answer by adding the partial products. Place a comma in front of every third digit from the right.

    If 486 is multiplied by 53, the result is 25,758.

3. Find the product of 4,006 and 47.

First Partial Product	4 4006 × 47 28042	Second Partial Product	2 4006 × 47 28042 16024 188282

The product of 4,006 and 47 is 188,282.

4. Find the product of 82,050 and 7,041.

```
    82050
  ×  7041
    82050
   328200
   00000        →    Some people prefer to skip this step.
   574350            As long as the 0 of 574350 remains under the 7 of the
  577714050          multiplier, skipping the step is fine.
```

Answer: 577,714,050

➤ Multiplication Table

×	1	2	3	4	5	6	7	8	9	10	11	12
1	1	2	3	4	5	6	7	8	9	10	11	12
2	2	4	6	8	10	12	14	16	18	20	22	24
3	3	6	9	12	15	18	21	24	27	30	33	36
4	4	8	12	16	20	24	28	32	36	40	44	48
5	5	10	15	20	25	30	35	40	45	50	55	60
6	6	12	18	24	30	36	42	48	54	60	66	72
7	7	14	21	28	35	42	49	56	63	70	77	84
8	8	16	24	32	40	48	56	64	72	80	88	96
9	9	18	27	36	45	54	63	72	81	90	99	108
10	10	20	30	40	50	60	70	80	90	100	110	120
11	11	22	33	44	55	66	77	88	99	110	121	132
12	12	24	36	48	60	72	84	96	108	120	132	144

➤ **Multiplicative Identity**   The **multiplicative identity** is one of the properties of the real number system. It states that, for any real number a, there is a unique number 1 such that $a \bullet 1 = a$ and $1 \bullet a = a$. The number 1 is called the **multiplicative identity**. The equation $2x = 12$ can be solved by using the multiplicative identity.

$$2x = 12$$
$$\frac{2x}{2} = \frac{12}{2}$$
$$\underbrace{1 \bullet x} = 6$$
$$x = 6 \qquad \text{By using the multiplicative identity } 1 \bullet x = x.$$

See **Properties of the Real Number System**.

➤ **Multiplicative Inverse**   The **multiplicative inverse** is one of the properties of the real number system. It states that, for any real number $a \neq 0$, there is a real number $\frac{1}{a}$ (also written a^{-1}) such that $a \bullet \frac{1}{a} = 1$ and $\frac{1}{a} \bullet a = 1$. The multiplicative inverse of a is also called the **reciprocal** of a.
The equation $2x = 12$ can be solved by using the multiplicative inverse.

$$2x = 12$$
$$\underbrace{\frac{2x}{2}} = \frac{12}{2} \qquad \text{Since } \frac{2}{2} \text{ is the same as } 2 \bullet \frac{1}{2}, \text{ the multiplicative}$$
$$\text{inverse states that } 2 \bullet \frac{1}{2} = 1. \text{ Hence, } \frac{2}{2} = 1.$$
$$1 \bullet x = 6$$
$$x = 6$$

See **Properties of the Real Number System**.

➤ **Multiplicity**   When there is a repeated answer, or root, in an equation, we say the root has a **multiplicity** of 2, or 3, or etc. In the equation $2x(x - 3)(x - 3)(x + 4) = 0$, we can see the solution will have a repeated root for $(x - 3)(x - 3)$:

$$2x(x - 3)(x - 3)(x + 4) = 0$$

$2x = 0$	$x - 3 = 0$	$x - 3 = 0$	$x + 4 = 0$
$x = 0$	$x = 3$	$x = 3$	$x = -4$

We either say "3 is a double root" or "3 is a root of multiplicity 2." If the answer had been a triple root, we would have said that it was a root of multiplicity 3, etc.
To solve the equation, see **Factor Theorem**.

➤ **Multiplying Polynomials**   See **Polynomials, multiplication of**.

Nn

➤ **Napierian Logarithms**   See **Natural Logarithms**.

➤ **Natural Logarithms**   Logarithms with a base of e are called **natural** or **Napierian logarithms**. Instead of writing $\log_e$ we write ln.

Natural logarithms are extremely important in calculus. Without them there would not be a function with the derivative of $\frac{1}{x}$. So, what function does have a derivative of $\frac{1}{x}$? To find out, we need to consider the derivative of $y = \log_b x$. By definition,

$$\frac{d \log_b x}{dx} = \lim_{h \to 0} \frac{\log_b(x+h) - \log_b x}{h}$$ See **Derivative**.

$$= \lim_{h \to 0} \frac{\log_b \frac{x+h}{x}}{h}$$ See property 2 below.

$$= \lim_{h \to 0} \frac{1}{h} \log_b \left(1 + \frac{h}{x}\right)$$

$$= \lim_{h \to 0} \frac{1}{x} \cdot \frac{x}{h} \log_b \left(1 + \frac{h}{x}\right)$$ Multiply by 1, where $1 = \frac{x}{x}$.
$$\frac{1}{h} = 1 \cdot \frac{1}{h} = \frac{x}{x} \cdot \frac{1}{h} = \frac{1}{x} \cdot \frac{x}{h}$$

$$= \lim_{h \to 0} \frac{1}{x} \log_b \left(1 + \frac{h}{x}\right)^{x/h}$$ See property 3 below.

$$= \frac{1}{x} \lim_{h \to 0} \log_b \left(1 + \frac{h}{x}\right)^{x/h}$$ Since $\frac{1}{x}$ does not have an h in it, it will not be affected by the limit.

In calculus it can be shown that $\lim\limits_{h \to 0} \left(1 + \frac{h}{x}\right)^{x/h} = e$. Therefore, if we let $h \to 0$ we have

$$\frac{d \log_b x}{dx} = \frac{1}{x} \log_b e$$

The $\log_b e$ would stay with the derivative unless we define another logarithm. If we let $b = e$, then $\log_e e = 1$. See **Logarithms, properties of (rules of)**, example 1. Hence, it is convenient to use the logarithmic function with a base of e. Therefore, we define the natural logarithm as $\log_e$ and we write it ln. The derivative then becomes

$$\frac{d \ln x}{dx} = \frac{1}{x}$$

So, what has a derivative of $\frac{1}{x}$? The ln x does.

Although natural logarithms are very important in calculus, they are less important for computations. Common logarithms are more convenient for numerical work.

The properties or rules for natural logarithms are similar to those of common logarithms.

1. $\ln pq = \ln p + \ln q$.

2. $\ln \dfrac{p}{q} = \ln p - \ln q$.

3. $\ln(p)^n = n \ln p$.

It is also useful to be familiar with four other equations:

$$\ln e = 1$$
$$\ln 1 = 0$$
$$e^{\ln p} = p$$
$$\ln 10 = 2.3026$$

As with common logarithms, numbers that are not within the range (1.0 to 9.99) of the natural logarithm chart can be brought within range by using scientific notation. However, unlike common logarithms, natural logarithms do not have a characteristic. Any number to the left of the decimal point is part of the exponent. Hence, positive as well as negative exponents are dealt with in the same manner. The negative exponents do not have to be dealt with separately as in common logarithms, i.e., in common logarithms -4.3010 is treated as $6.3010 - 10$.

Examples

1. Calculate the value of ln 2,430.

First, we write 2,430 in scientific notation:

$$2{,}430 = 2.43 \times 10^3$$

Substituting, we have

$$
\begin{aligned}
\ln 2{,}430 &= \ln 2.43 \times 10^3 \\
&= \ln 2.43 + \ln 10^3 && \text{Property 1.} \\
&= \ln 2.43 + 3 \ln 10 && \text{Property 3.} \\
&= .8879 + 3(2.3026) && \text{Substitute } \ln 10 = 2.3026. \\
&= .8879 + 6.9078 \\
&= 7.7957
\end{aligned}
$$

To verify this, we can change $2{,}430 = 7.7957$ to exponential form and check it with a calculator. Since ln is $\log_e$, $\log_e 2{,}430 = 7.7957$ is $e^{7.7957} = 2{,}430$ in exponential form. See **Logarithmic Form to Exponential Form, converting from**. To calculate $e^{7.7957}$, we will use $e \approx 2.71828$. Enter 2.71828 in the calculator and press the y^x button. Next, enter 7.7957 and press y^x again, or $=$:

$$2.71828^{7.7957} = 2{,}430.1172 \approx 2{,}430$$

2. Calculate the value of ln .00682.

In scientific notation $.00682 = 6.82 \times 10^{-3}$, so

$$
\begin{aligned}
\ln .00682 &= \ln 6.82 \times 10^{-3} \\
&= \ln 6.82 + \ln 10^{-3} && \text{Property 1.} \\
&= \ln 6.82 - 3 \ln 10 && \text{Property 3.} \\
&= 1.9199 - 3(2.3026) && \text{Substitute } \ln 10 = 2.3026. \\
&= 1.9199 - 6.9078 \\
&= -4.9879 && \text{It is incorrect to write this as } 6.9879 - 10.
\end{aligned}
$$

The -4 of -4.9879 is not a characteristic as it would be in common logarithms. It is part of the exponent -4.9879. This is the exponent to which e can be raised to yield .00682. To verify this, we can change $\ln .00682 = -4.9879$ to $e^{-4.9879} = .00682$ and check it with a calculator. Using the y^x button as we did in example 1, we have

$$e^{-4.9879} = 2.71828^{-4.9879} = .00682$$

3. Find A if $\ln A = 7.7957$.

To find A, we need to take the antilog$_e$ of both sides of the equation:

$$
\begin{aligned}
\ln A &= 7.7957 \\
\text{antilog}_e \ln A &= \text{antilog}_e 7.7957 \\
A &= \text{antilog}_e 7.7957
\end{aligned}
$$

Since 7.7957 is outside the range of the chart, which is .000 to 2.3016, we need to successively subtract ln 10 until it is within the range of the chart. Since $\ln 10 = 2.3026$, this means that we need to successively subtract 2.3026 from 7.7957 until we are within the range of the chart:

$$
\begin{array}{ccc}
7.7957 & 5.4931 & 3.1905 \\
-2.3026 & -2.3026 & -2.3026 \\
\hline
5.4931 & 3.1905 & .8879
\end{array}
$$

Having subtracted three times means we subtracted $3(2.3026) = 6.9078$ from 7.7957. Since $3(2.3026) = 3 \ln 10$, we have

$$3 \ln 10 = \ln 10^3 = \ln 1000 = 6.9078$$

Now we can continue to find A:

$$A = \text{antilog}_e\, 7.7957 = \text{antilog}_e(.8879 + 6.9078)$$

Since .8879 is in the range of the chart, $.8879 = \ln 2.43$. If it were not exact, we would interpolate. Also we have $6.9078 = \ln 1000$, hence

$$
\begin{aligned}
&= \text{antilog}_e(\ln 2.43 + \ln 1000) \\
&= \text{antilog}_e \ln(2.43)(1000) \qquad && \text{Property 1.} \\
&= (2.43)(1000) \qquad && \text{Antilog}_e \text{ and ln cancel.} \\
&= 2,430
\end{aligned}
$$

4. Find A if $\ln A = -4.9879$.

To get -4.9879 within the range of the chart, we add 2.3026 to -4.9879 three times:

-4.9879	-2.6853	$-.3827$
$+2.3026$	$+2.3026$	$+2.3026$
-2.6853	$-.3827$	$+1.9199$

Adding three times means we added 6.9078 to -4.9879 to get 1.9199. With $6.9078 = \ln 1000$ and $1.9199 = \ln 6.82$, we have

$$
\begin{aligned}
\ln A &= -4.9879 \\
\text{antilog}_e \ln A &= \text{antilog}_e(-4.9879) \\
A &= \text{antilog}_e(1.9199 - 6.9078) \\
&= \text{antilog}_e(\ln 6.82 - \ln 1000) \\
&= \text{antilog}_e \ln \frac{6.82}{1000} \qquad && \text{Property 2.} \\
&= \frac{6.82}{1000} \\
&= .00682
\end{aligned}
$$

Paramount to all the examples is the equation $\ln 10 = 2.3026$. This is calculated by using the change of base formula:

$$\log_a N = \frac{\log_b N}{\log_b a}, \qquad \log_e 10 = \frac{\log_{10} 10}{\log_{10} e} = \frac{1}{.43429} = 2.3026$$

See **Change of the Base Formula**. Also see **Common Logarithms, Interpolation, linear, Antilogarithm, Logarithm Table (chart), how to use,** and **Logarithms, properties of rules of.**

▶ **Natural Numbers** **Natural numbers** are the counting numbers. These are the numbers that are used in everyday counting. They start with 1 and do not contain 0, any negative numbers, or any other type of number, i.e., fractions, decimals, etc.

$$n = \{1, 2, 3, \ldots\}$$

For more, see **Real Number System.**

➤ **Nature of Roots**   See **Discriminant**.

➤ **Negation**   The **negation** of a statement is the opposite of an original statement. If the original statement is true, the negation is false. If the original statement is false, the negation is true.

Let p be an original statement, then its negation is written

$$\sim p \qquad \text{read "not } p\text{."}$$

If the original statement says "all right angles have a measure of 90," then the negation says "not all right angles have a measure of 90." The original is true and the negation is false. See **Logic**.

➤ **Negative Angle**   See **Angle, in trigonometry**.

➤ **Negative Angle Identities**   See **Trigonometry, even and odd functions (functions or identities of negative angles)**.

➤ **Negative Exponent**   See **Exponents, negative**.

➤ **Negative Functions in Trigonometry**   See **Even and Odd Functions, in trigonometry**.

➤ **Negative Integers**   The **negative integers** are composed of the negative counting numbers. See **Integers**.

➤ **Negative Numbers**   See **Positive and Negative Numbers**.

➤ **Non-Euclidean Geometry**   **Non-Euclidean geometry** differs from Euclidean geometry in regards to the acceptance of the parallel postulate, postulate 5. If the fifth postulate is accepted as Euclid intended, then the geometry is Euclidean. If the fifth postulate is stated from other points of view, then the geometries that ensue are non-Euclidean.

The following postulates are quoted from *The Thirteen Books of Euclid's Elements*, Dover, New York, p. 154.

Euclid's Postulates

1. To draw a straight line from any point to any point.

2. To produce a finite straight line continuously in a straight line.

3. To describe a circle with any center and distance.

4. That all right angles are equal to one another.

5. That, if a straight line falling on two straight lines make the interior angles on the same side less than two right angles, the two straight lines, if produced indefinitely, meet on that side on which are the angles less than the two right angles.

Because of its length, the fifth postulate apprears more as a theorem then a postulate. From the time of Euclid on, many tried to prove the postulate as a theorem. It was from these efforts that an awareness of non-Euclidean geometry developed.

➤ **Nonlinear System of Equations**   See **Systems of Nonlinear Equations, solutions of**.

➤ **Nonrepeating Decimals**   A **nonrepeating decimal** is a decimal number that does not continually repeat a single digit or two or more digits. In other words, it is not a decimal number that repeats.

   If the nonrepeating decimal number terminates, it is a rational number. If the nonrepeating decimal does not terminate, it is an irrational number. See **Rational Number, definition, Irrational Numbers**, or **Decimal Numbers, changing a repeating decimal to a common fraction**.

➤ **Number in Front of a Letter**   See **Coefficient**.

➤ **Number Line**   A **number line** is a correspondence between the set of points of a line and the set of real numbers. Choose two points O and P on a straight line. We define O as the **origin** and assign the numbers 0 and 1 to O and P, respectively. With OP as a unit, we mark off equal units on both sides of O. The points marked off on the right sides of O are assigned the numbers 1, 2, 3, and so on. The points marked off on the left side of O are assigned -1, -2, -3, and so on. The remaining points on the line are placed into a one-to-one correspondence with the remaining numbers of the real numbers. In this manner, each point of the line is assigned a unique real number, resulting in a number line.

➤ **Number Line, distance on**   See **Distance on a Number Line**.

➤ **Number Line in Trigonometry**   See **Trigonometry, graphs of the six functions**, end of example 5.

➤ **Numbers**   The following numbers are listed by article under each name: **Complex Numbers, Counting Numbers, Decimal Numbers, Fraction, Integers, Irrational Numbers, Mixed Number, Perfect Number, Prime Numbers, Rational Number, definition, Real Number**, and **Whole Numbers**.

➤ **Number Sequence**   See **Numerical Sequence**.

➤ **Numerator**   In fractions, the **numerator** is the number above the fraction bar. The **denominator** is the number below the fraction bar. See **Fraction**.

➤ **Numerical Coefficient**   See **Coefficient**.

➤ **Numerical Sequence**   A **numerical sequence** is a list of numbers that has been ordered (arranged according to size). Each of the following ordered lists is a numerical sequence:

1. $1, 3, 5, \ldots$

2. $2, 4, 6, \ldots$

3. $1, \dfrac{1}{3}, \dfrac{1}{9}, \dfrac{1}{27}, \ldots$

4. $2, 2, 2, \ldots$

   For more, see **Sequence**.

O

➤ **Obtuse Angle**   An **obtuse angle** is an angle with a measure greater than 90 and less than 180. In the figure, $\angle BAC$ (read "angle BAC"), is an obtuse angle.

➤ **Obtuse Triangle**   A triangle with an obtuse angle as one of its angles is an **obtuse triangle**.

➤ **Odd Functions**   See **Even and Odd Functions**.

➤ **One-to-One Correspondence**   See **Number Line**.

➤ **One-to-One Function**   A function is said to be **one-to-one** if for every x in the domain there is a unique y in the range. Each x yields a different y. This means that two different x's cannot have the same y value. Hence, if a and b are in the same domain of the function and $f(a) \neq f(b)$, then the function is **one-to-one**. If $f(a) = f(b)$, then the function is not one-to-one.

Another way to tell if a function is one-to-one is to do the horizontal line test. Draw a horizontal line through any part of the graph. If it does not intersect the graph in more than one point, the graph (function) is one-to-one. If the horizontal line does intersect the graph in more than one point, the graph (function) is not one-to-one.

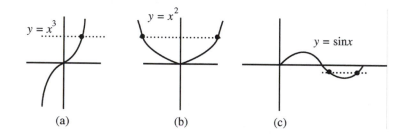

The function in graph (a) is one-to-one. The functions in graphs (b) and (c) are not one-to-one.

➤ **Order, ascending**   See **Ascending and Descending Order**.

➤ **Order, descending**   See **Ascending and Descending Order**.

➤ **Ordered Pair**   Each point in the coordinate plane is denoted by the **ordered pair** (x, y) where x and y are real numbers. The x and y are called **coordinates**. The x-coordinate is called the **abscissa** and the y-coordinate is called the **ordinate**. See **Coordinate Plane**.

An **ordered pair** can also be considered as a solution to a given system of equations. It is the point of intersection of the system. See **Systems of Linear Equations**.

➤ **Ordered Triple**   Each point in space is denoted by the **ordered triple** (x, y, z) where x, y, and z are real numbers. An ordered triple can also be the solution to a given system of equations.

➤ **Order of Operations**   Because of the many different symbols in algebra, there is an order that must be followed in the application of each, the **order of operations**.

1. If there are symbols of inclusion such as braces, brackets, or parentheses, do the operations within them first. If symbols of inclusion are within each other, do the operations in the innermost set first.

2. The first operation to be applied is exponents, working from left to right.

3. Then perform any multiplications or divisions, working from left to right.

4. Then perform any additions or subtractions, working from left to right.

Examples

1. Simplify $23 - 2 \times 3^2$.

$$\begin{aligned} 23 - 2 \times 3^2 &= 23 - 2 \times 9 \\ &= 23 - 18 \\ &= 5 \end{aligned}$$

Exponent first, not $23 - 2$, or 2×3.
Multiply next, not $23 - 2$.
Subtract last.

2. Simplify $6[4 + (8 - 5)] + 3(7 - 2)^2$.

$$\begin{aligned} 6[4 + (8 - 5)] + 3(7 - 2)^2 &= 6[4 + 3] + 3(5)^2 \\ &= 6[7] + 3 \cdot 25 \\ &= 42 + 75 \\ &= 117 \end{aligned}$$

Parentheses first.
Brackets and exponent next.
Then multiply.
Add last.

3. Simplify $6 - 2(3x - 4)^2$.

$$6 - 2(3x - 4)^2 = 6 - 2(3x - 4)(3x - 4)$$

Do not subtract $6 - 2$. Do not multiply -2 into the parentheses. Do not improperly square $(3x - 4)^2$ as $9x^2 + 16$ or $9x^2 - 16$.

$$\begin{aligned} &= 6 - 2(9x^2 - 24x + 16) \\ &= 6 - 18x^2 + 48x - 32 \\ &= -18x^2 + 48x - 26 \end{aligned}$$

Do not subtract $6 - 2$, do FOIL only. See **FOIL**.
Multiply -2 into the parentheses.
Combine 6 and -32.

➤ **Ordinal Number**   An **ordinal number** is a number that gives the order or succession as first, second, third, and so on (also written 1st, 2nd, 3rd, . . .). Ordinal numbers are distinguished from cardinal numbers. See **Cardinal Number**.

➤ **Ordinate**   The **ordinate** is the y-coordinate of a point. The **abscissa** is the x-coordinate of a point. Together they form an **ordered pair**. See **Coordinate Plane**.

➤ **Origin**   The **origin** is the point of intersection of the x-axis and y-axis. It is the point where the number 0 on the x-axis and the number 0 on the y-axis coincide. Therefore, the coordinates of the origin are $(0, 0)$. See **Coordinate Plane**.

O

Pp

> **Parabola**   A **parabola** is the set of all points in the plane such that its distance from a fixed point and a fixed line are equal. The fixed point is called the **focus** and the fixed line is called the **directrix**.

The line through the focus and perpendicular to the directrix is called the **principal axis**. The point where the principal axis and the parabola intersect is called the **vertex**. The vertex is midway between the focus and the **directrix**.

The principal axis is also called the **axis of symmetry**. Unless the graph has been rotated, the axis of symmetry passes through the vertex and is parallel or coincident with either the x- or y-axes.

The equation and related information of a parabola depends on the orientation of the axis of symmetry.

Vertical Axis of Symmetry

1. Standard form of the equation of a parabola, $y = a(x - h)^2 + k$.

2. Equation of the axis of symmetry, $x = h$.

3. Coordinates of vertex, (h, k).

4. Coordinates of focus, $\left(h, k + \frac{1}{4a}\right)$.

5. Equation of the directrix,
   $y = k - \frac{1}{4a}$.

6. The direction of the opening of the parabola is upward if $a > 0$ and downward if $a < 0$.

Graph of $y = a(x-h)^2 + k$:

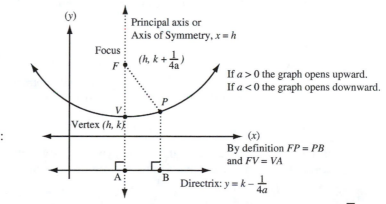

Principal axis or Axis of Symmetry, $x = h$

$\left(h, k + \frac{1}{4a}\right)$

If $a > 0$ the graph opens upward.
If $a < 0$ the graph opens downward.

Vertex (h, k)

By definition $FP = PB$
and $FV = VA$

Directrix: $y = k - \frac{1}{4a}$

317

Horizontal Axis of Symmetry

1. Standard form of the equation of the parabola, $x = a(y - k)^2 + h$.

2. Equation of the axis of symmetry, $y = k$.

3. Coordinates of the vertex, (h, k).

4. Coordinates of the focus, $\left(h + \frac{1}{4a}, k\right)$.

5. Equation of the directrix, $x = h - \frac{1}{4a}$.

6. The direction of the opening of the parabola is right if $a > 0$ and left if $a < 0$.

Graph of $x = a(x - k)^2 + h$:

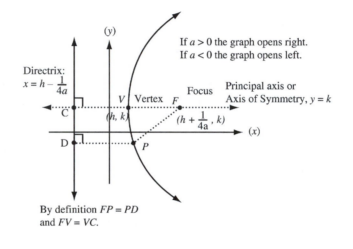

If $a > 0$ the graph opens right.
If $a < 0$ the graph opens left.

By definition $FP = PD$ and $FV = VC$.

Examples

1. Write the equation $y - 5 = x^2 + 8x$ in standard form. Then find the equation of the axis of symmetry, coordinates of the vertex, coordinates of the focus, equation of the directrix, and the direction of the opening. Finally, graph the equation.

   To write the equation in standard form, we need to solve the equation for y and then complete the square:

$$\begin{aligned}
y - 5 &= x^2 + 8x \\
y &= x^2 + 8x + 5 \\
y &= x^2 + 8x + 16 - 16 + 5 \\
y &= (x + 4)(x + 4) - 16 + 5 \\
y &= (x + 4)^2 - 11
\end{aligned}$$

   Take half the coefficient of x, square it, then add and subtract the result. See **Completing the Square**.

   Since the equation solves for y, the axis of symmetry is vertical. Comparing $y = (x + 4)^2 - 11$ to $y = a(x - h)^2 + k$, we see that h is -4. Hence, the equation of the axis of symmetry is $x = h$ or $x = -4$.

Since $h = -4$ and $k = -11$, the vertex (h, k) is $(-4, -11)$. Also comparing $y = (x + 4)^2 - 11$ with $y = a(x - h)^2 + k$, we see that the leading coefficient a is 1. Hence, the focus $\left(h, k + \frac{1}{4a}\right)$ is $\left(-4, -11 + \frac{1}{4(1)}\right) = \left(-4, -11 + \frac{1}{4}\right) = \left(-4, -10\frac{3}{4}\right)$.

From the equation for the directrix, we have

$$y = k - \frac{1}{4a} = -11 - \frac{1}{4(1)}$$

$$= -11 - \frac{1}{4} = -11\frac{1}{4}, \qquad \text{the equation of the directrix.}$$

Since $a = 1$ is greater than 0, the direction of the opening of the parabola is upward.

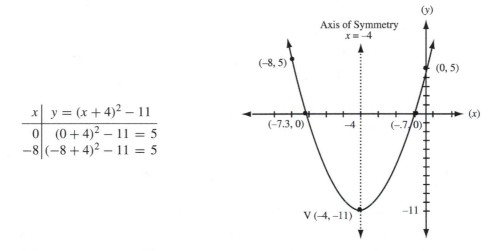

x	$y = (x + 4)^2 - 11$
0	$(0 + 4)^2 - 11 = 5$
-8	$(-8 + 4)^2 - 11 = 5$

To get more points of the graph, choose x-values and calculate the y-values, then plot the points i.e., $(0, 5)$ and $(-8, 5)$. See the chart.

Another way to get more points of the graph is to find the x-intercepts. These are the points where the graph crosses the x-axis. When the graph crosses the x-axis, the values of y are 0. Hence, setting $y = 0$ in $y = (x + 4)^2 - 11$, we have

$$(x + 4)^2 - 11 = 0$$
$$(x + 4)^2 = 11$$
$$\sqrt{(x + 4)^2} = \pm\sqrt{11} \qquad \text{See } \textbf{Quadratic Equations, solutions of.}$$
$$x = -4 \pm \sqrt{11}$$

$$\begin{array}{lll} x = -4 + \sqrt{11} & & x = -4 - \sqrt{11} \\ x \approx -4 + 3.3 & \text{or} & x \approx -4 - 3.3 \\ x \approx -.7 & & x \approx -7.3 \end{array}$$

The x-intercepts are located at $(-7.3, 0)$ and $(-.7, 0)$ but write them as $(-4 - \sqrt{11}, 0)$ and $(-4 + \sqrt{11}, 0)$.

Yet another way to get more points of the graph is to use the latus rectum. See **Latus Rectum**.

2. Find the equation of the parabola with focus $(4, 3)$ and directrix $x = -2$.

Since the directrix is solved for x, we know the axis of symmetry is horizontal. On the x-axis, the distance from 4 to -2 is $|4 - (-2)| = 6$. See **Distance on a Number Line**. Since the vertex is midway between the focus and the directrix, it is $\frac{6}{2} = 3$ units from each. Hence, the coordinates of the vertex are $V(1, 3)$. This gives us the values of h and k since $V(1, 3) = (h, k)$.

Substituting x and h into the equation of the directrix, we can find the value of a:

$$x = h - \frac{1}{4a}$$

$$-2 = 1 - \frac{1}{4a}$$

$$-3 = -\frac{1}{4a}$$

$$12a = 1$$

$$a = \frac{1}{12}$$

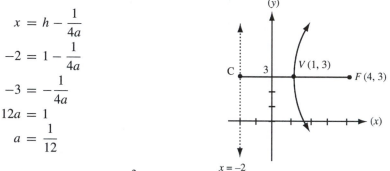

Substituting a, h, and k into $y = a(x - h)^2 + k$, we have the equation of the parabola $y = \frac{1}{12}(x - 1)^2 + 3$.

Similarly, the equation of the parabola could have been found by using $h + \frac{1}{4a}$ from the formula for the focus. Since $F(4, 3) = \left(h + \frac{1}{4a}, k\right)$, $h + \frac{1}{4a} = 4$. Substituting $h = 1$ into the equation and solving for a, we can find the equation of the parabola.

3. Find the equation of the parabola with vertex $(1, 3)$ and focus $(4, 3)$.

Since the y-coordinate of the vertex and the focus is the same, we know the axis of symmetry is horizontal (see the graph of example 2). The distance from the focus to the vertex is $|4 - 1| = 3$. See **Distance on a Number Line**. Therefore, since $FV = VC$, the equation of the directrix is $x = -2$. See **Horizontal and Vertical Lines**. With focus $(4, 3)$ and directrix $x = -2$, we can now find the equation of the parabola as we did in example 2.

4. Find the equation of the parabola with vertex $(1, 3)$ and directrix $x = -2$.

Since the directrix is vertical, the axis of symmetry is horizontal. See **Horizontal and Vertical Lines**.

The distance from the vertex to the directrix is $|1 - (-2)| = 3$. See **Distance on a Number Line**. Therefore, since $FV = VC$, the coordinate of the focus is $(4, 3)$. See the graph of example 2. With focus $(4, 3)$ and directrix $x = -2$, we can now find the equation of the parabola as we did in example 2.

▶ **Parallel Lines (geometric)**   **Parallel lines** are lines that are in the same plane but do not intersect. If two parallel lines are cut (crossed) by a transversal (a line intersecting two other lines), then the alternate interior angles are congruent, the alternate exterior angles are congruent, the corresponding angles are congruent, and the interior angles on the same side of the transversal are supplementary.

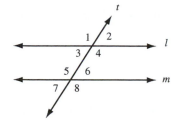

1. Alternate interior angles: $\angle 3 \cong \angle 6$, $\angle 5 \cong \angle 4$.

2. Alternate exterior angles: $\angle 1 \cong \angle 8$, $\angle 7 \cong \angle 2$.

3. Corresponding angles: $\angle 1 \cong \angle 5$, $\angle 2 \cong \angle 6$, $\angle 7 = \angle 3$, and $\angle 8 = \angle 4$.

4. Interior angles on the same side of the transversal: $\angle 4$ and $\angle 6$ are supplementary, $\angle 3$ and $\angle 5$ are supplementary.

To prove the lines are parallel, the converse of each of these must be proven true. The symbol for parallel is ∥. For more concerning the slopes of parallel lines, see **Parallel Lines (slopes with)**.

➤ **Parallel Lines (slopes with)**   In the coordinate plane, two lines are said to be **parallel lines** if their slopes are the same. Conversely, if the slopes of two lines are the same, the lines are said to be parallel.
    Let l_1 and l_2 represent two lines. If $l_1 \parallel l_2$, then $m_1 = m_2$. Conversely, if l_1 and l_2 represent two lines where $m_1 = m_2$, then $l_1 \parallel l_2$.

Examples

1. What is the slope of any line parallel to $y = 2x + 1$?

   The slope of any line parallel to $y = 2x + 1$ has the same slope as $y = 2x + 1$. From $y = mx + b$, we know the slope is m. So, the slope of $y = 2x + 1$ is $m = 2$, which is the same as the slope of any parallel line to $y = 2x + 1$. Any of those parallel lines also have a slope of $m = 2$.

2. Determine whether the lines $12x + 3y = 6$ and $8x + 2y = 10$ are parallel.

   If the two lines are parallel, then their slopes must be equal. We need to first determine their slopes. Solving each equation for y (to use $y = mx + b$ to identify m), we have

$$
\begin{array}{ll}
12x + 3y = 6 & \qquad 8x + 2y = 10 \\
3y = -12x + 6 & \qquad 2y = -8x + 10 \\
\dfrac{3y}{3} = \dfrac{-12x}{3} + \dfrac{6}{3} & \qquad \dfrac{2y}{2} = \dfrac{-8x}{2} + \dfrac{10}{2} \\
y = -4x + 2 & \qquad y = -4x + 5 \\
\text{Hence,} \quad m_1 = -4. & \qquad \text{Hence,} \quad m_2 = -4.
\end{array}
$$

   See **Slope-Intercept Form of an Equation of a Line**.

   Since $m_1 = m_2 = -4$, the slopes of the lines are the same. Hence, the two lines are parallel.

Another way to find m_1 and m_2 is to use $m = \frac{-A}{B}$. If an equation is in standard form $(Ax + By = C)$, then $m = \frac{-A}{B}$. Hence,

$$m_1 = -\frac{A}{B} = -\frac{12}{3} = -4 \qquad \text{and} \qquad m_2 = -\frac{A}{B} = -\frac{8}{2} = -4$$

3. Find the equation of the line through $(-8, 6)$ parallel to $y = 2x + 1$.

The slope of the line parallel to $y = 2x + 1$ is the same as the slope of $y = 2x + 1$, $m = 2$. Using the point $(-8, 6)$ and the slope $m = 2$, we can find the equation of the parallel line. We will do this by the slope-intercept method. The other method is the point-slope method. See **Linear Equation, finding the equation of a line**.

Substituting $(x, y) = (-8, 6)$ and $m = 2$ into $y = mx + b$, we have

$$y = mx + b$$
$$6 = 2(-8) + b$$
$$6 = -16 + b$$
$$6 + 16 = b$$
$$22 = b$$

This gives us the equation $y = 2x + 22$. Hence, the equation of the line through $(-8, 6)$ parallel to $y = 2x + 1$ is the line $y = 2x + 22$.

▶ **Parallelogram**   A **parallelogram** is a quadrilateral with both pairs of opposite sides parallel. Some of its properties are: the opposite sides are congruent, the opposite angles are congruent, any two consecutive angles are supplementary, each diagonal divides the parallelogram into two congruent triangles, the diagonals bisect each other, and the sum of the angles of the four vertices is 360°. The properties of a parallelogram also pertain to a rectangle, a rhombus, and a square.

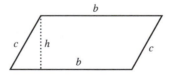

The **area** of a parallelogram can be found by multiplying the base and the altitude, $A = bh$. The **perimeter** of a parallelogram can be found by adding the lengths of all four sides, $P = b + c + b + c$ or $P = 2b + 2c$.

Examples

1. If the base of a parallelogram is 9 inches, its altitude is 5 inches, and its shorted side is 7 inches, find the area and perimeter.

Substituting into the equations for area and perimeter, we have

$$A = bh = (9)(5) = 45 \text{ square inches}$$
$$P = 2b + 2c = 2(9) + 2(7) = 18 + 14 = 32 \text{ inches}$$

2. If the area of a parallelogram is 45 square inches and its altitude is 5 inches, what is the length of its base?

To find the base b, we substitute A and h into $A = bh$ and solve for b:

$$A = bh$$
$$45 = b \bullet 5$$
$$\frac{45}{5} = \frac{b \bullet 5}{5}$$
$$9 = b$$
$$b = 9 \text{ inches}$$

For help in solving equations, see **Linear Equations, solutions of**.

➤ **Parentheses**   **Parentheses** are one of three sets of symbols used for grouping. **Brackets** [ ] and **braces** { } are the other two. The order of hierarchy is braces, brackets, and then parentheses.

> *Example*

1. Simplify $\{24 - 2[10 - (9 - 6)]\}$.

The order of hierarchy dictates that braces must be done before brackets and brackets before parentheses. In practice, however, this means to work from the inside out. We cannot combine the $24 - 2$ until we have calculated the $10 - (9 - 6)$ first. And to do that we must first calculate the numbers inside the parentheses.

$$\{24 - 2[10 - (9 - 6)]\} = \{24 - 2[10 - 3]\} = \{24 - 2[7]\} = \{24 - 14\} = 10$$

The most common mistake made is trying to combine the $24 - 2$ first. The multiplication takes precedence over the subtraction. See **Order of Operations**.

➤ **Partition**   There are two definitions for a partition: one for a **regular partition** and the other for a **general partition**. The most common usage of a partition is in calculus when summing rectangles under a curve.

Regular Partition

A **regular partition** is a subdivision of an interval $[a, b]$, where $[a, b]$ is divided into n subdivisions, each of width $\Delta x = \frac{b-a}{n}$.

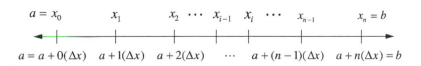

A regular partition can also be defined by a sequence x_n, where n is a nonnegative integer. We still have $\Delta x = \frac{b-a}{n}$ where $x_i - x_{i-1} = \Delta x$ for any ith subinterval:

$$a = x_0 < x_1 < x_2 < \cdots < x_{i-1} < x_i < \cdots < x_{n-1} < x_n = b$$

In a regular partition, the lengths of each subinterval are equal.

General Partition

A **general partition** is defined in relationship to the norm of the partition, where the **norm** is the width of the largest subinterval and is denoted by $\|\Delta\|$. If each of the subintervals is of equal length, then the partition is regular and we have

$$\|\Delta\| = \Delta x = \frac{b - a}{n}$$

If the lengths of the subintervals are unequal, then a **general partition** is defined in relationship to n where

$$\frac{b - a}{\|\Delta\|} \leqslant n$$

From the fraction, we can see that as the norm of the partition approaches zero, the number of subintervals approaches infinity.

Example

1. Draw a regular partition on the interval $[a, b] = [2, 6]$ where $n = 8$.

   First, we need to find Δx where $a = 2$, $b = 6$, and $n = 8$:

   $$\Delta x = \frac{b - a}{n} = \frac{6 - 2}{8} = \frac{4}{8} = \frac{1}{2}$$

   Then we can draw the regular partition between 2 and 6 with subintervals each of length $\frac{1}{2}$.

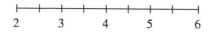

➤ **Pascal's Triangle**   **Pascal's triangle** is most often used to find the coefficients in the expansion of $(a + b)^n$ for any nonnegative integer n (meaning n is any of the numbers 0, 1, 2, 3, ...). Each row in the triangle is the sum of the entries in the previous row.

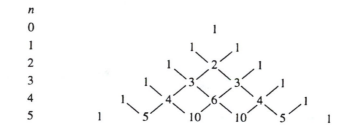

The triangle can be seen in the expansion of $(a + b)^n$.

$$1$$
$$a + b$$
$$a^2 + 2ab + b^2$$
$$a^3 + 3a^2b + 3ab^2 + b^3$$
$$a^4 + 4a^3b + 6a^2b^2 + 4ab^3 + b^4$$

See **Binomial Coefficient** or **Binomial Theorem**.

➤ **Percent**   **Percent** is a term used to represent hundredths. It is based on the Latin per centum or per hundred. If we consider 29 parts out of 100, we can refer to this as $\frac{29}{100}$ as a fraction, .29 as a decimal, or 29% as a percent where the % sign represents parts per hundred.

To have all of something is referred to as having 100% of it. This means we have 100 out of 100 parts where $\frac{100}{100} = 1$, or 100%, which represents the whole.

To have more than 100% can also be considered. If a pitcher contains enough water to fill a glass of water, then the pitcher contains 100% of the water needed to fill the glass. If the pitcher contains enough water to fill two glasses with water, then the pitcher contains 200% of the water necessary to fill the glass. It contains 200 parts out of 100 or 100 more than is needed for a whole glass.

➤ **Percent, changing from a decimal to a percent and vice versa**   To **change a decimal number to a percent**, move the decimal point two places to the right. To **change a percent to a decimal**, move the decimal point two places to the left.

Examples

1. Write .06, .45, .125, and 1.36 as percents.

   Moving the decimal two places to the right, we have

   a. $.06 = 6\%$
   b. $.45 = 45\%$
   c. $.125 = 12.5\%$ or $12\frac{1}{2}\%$
   d. $1.36 = 136\%$

2. Write 6%, 45%, $12\frac{1}{2}\%$, and 136% as decimals

   Moving the decimal two places to the left, we have

   a. $6\% = .06$
   b. $45\% = .45$
   c. $12\frac{1}{2}\% = 12.5\% = .125$, since $\frac{1}{2} = .5$, i.e., $2)\overline{1.0}$ $\frac{1.0}{0}$ with quotient .5
   d. $136\% = 1.36$

See **Decimal Numbers, changed to a fraction**; **Decimal Numbers, changed to a percentage**; or **Fractions, change from fraction to decimal to percent**.

➤ **Percent, finding a number when a percent of it is known**   To find a number when a percent of it is known, change the percent to a decimal and divide the given number by the decimal.

Examples

1. 20% of what number is 48?

   We change 20% to a decimal, 20% = .20, and divide 48 by this value:

$$
\begin{array}{r}
240. \\
.20\!\!\diagup\!\overline{)48.00}\diagdown \\
\underline{40} \\
80 \\
\underline{80} \\
0
\end{array}
$$

2. 48 is 20% of what number?

   This problem is the same as example 1, it's just written in reverse order. Still, change 20% to a decimal and then divide 48 by this value.

➤ **Percent, finding what percent one number is of another**   To find what percent one number is of another, make a fraction with the number in the numerator and the percentage of the other in the denominator. Then convert the fraction to a percent.

Examples

1. 3 is what percent of 5?

   Writing this as a fraction and then converting to a percent, we have

$$
\frac{3}{5} = .6 = 60\% \qquad \text{since} \qquad
\begin{array}{r}
.6 \\
5\overline{)3.0} \\
\underline{3.0} \\
0
\end{array}
$$

2. What percent of 5 is 3?

   This is the same as example 1, except in reverse order. Still, write the problem as a fraction, $\frac{3}{5}$, and then convert to a percent.

➤ **Percent, multiplication with or finding a percent of a number**   To multiply with percent, change the percent to a decimal and then multiply.

Examples

1. Multiply 346 by 28%.

First, we change 28% to a decimal, 28% = .28, and then we multiply:

$$
\begin{array}{r}
346 \\
\times\ .28 \\
\hline
2768 \\
692 \\
\hline
96.88
\end{array}
$$

2. Finding 28% of 346.

Since "of" means to multiply, this is the same problem as example 1. Change 28% to a decimal and then multiply. This process is called **finding a percent of a number**.

➤ **Perfect Number**   A **perfect number** is a number that is equal to the sum of its factors, including 1, but not itself. The number 6 is a perfect number since $6 = 1 + 2 + 3$. Other perfect numbers are 28, 496, and 8128.

➤ **Perfect Square Trinomial**   A **perfect square trinomial** is the result when a binomial is squared. If the binomial $a + b$ is squared, we have

$$
\begin{aligned}
(a + b)^2 &= (a + b)(a + b) \qquad \text{See \textbf{FOIL}.} \\
&= a^2 + ab + ab + b^2 \\
&= a^2 + 2ab + b^2
\end{aligned}
$$

If the binomial $a - b$ is squared, we have

$$
(a - b)^2 = (a - b)(a - b) = a^2 - ab - ab + b^2 = a^2 - 2ab + b^2
$$

Notice that in either case the first and last terms of the trinomial are the squares of the terms of the binomial and the center term is twice their product. The center sign of the trinomial is determined by the sign of the binomial. If the sign in the binomial is $+$, the center sign of the trinomial is $+$. If the sign in the binomial is $-$, the center sign of the trinomial is $-$.

Examples

1. Square the binomial $2x + 5$.

Applying the square of the binomial $a + b$, we have

$$
\begin{aligned}
(a + b)^2 &= a^2 + 2ab + b^2 \\
(2x + 5)^2 &= (2x)^2 + 2(2x \bullet 5) + (5)^2 = 4x^2 + 20x + 25
\end{aligned}
$$

In relation to $2x + 5$, the trinomial $4x^2 + 20x + 25$ is the result found by squaring the first term, doubling the product of the two terms, and squaring the last term. Since the sign in the binomial is $+$, the sign in the trinomial is $+$.

Another way to square $2x + 5$ is to multiply by FOIL. See **FOIL**.

$$(2x + 5)^2 = (2x + 5)(2x + 5) = 4x^2 + 10x + 10x + 25 = 4x^2 + 20x + 25$$

The *most common* mistake made in squaring a binomial is squaring the first and last terms only:

Wrong: $(2x + 5)^2 = 4x^2 + 25$

2. Simplify $(5x^3 - 2y^2)^2$.

Applying the square of the binomial $a - b$, we have

$$(a - b)^2 = a^2 - 2ab + b^2$$
$$(5x^3 - 2y^2)^2 = (5x^3)^2 - 2(5x^3 \bullet 2y^2) + (2y^2)^2 \qquad \text{Let } a = 5x^3 \text{ and } b = 2y^2 \ (\text{not} -2y^2).$$
$$= 25x^6 - 20x^3y^2 + 4y^4$$

By the FOIL method, we have

$$(5x^3 - 2y^2)^2 = (5x^3 - 2y^2)(5x^3 - 2y^2)$$
$$= 25x^6 - 10x^3y^2 - 10x^3y^2 + 4y^4 = 25x^6 - 20x^3y^2 + 4y^4$$

► Perimeter of a Polygon

The **perimeter of a polygon** is the distance around the polygon. In other words, the perimeter is the sum of the lengths of the sides of the polygon.

Examples

1. Find the perimeter of a rectangle with a length of 24 feet and width of 11 feet.

Each long side is 24 ft, and each short side is 11 ft. Therefore, the perimeter is

$$P = 24 + 24 + 11 + 11 = 70 \text{ ft}$$

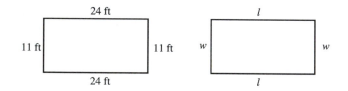

The distance around the rectangle is 70 ft. If we consider a rectangle with length l and width w, the perimeter is

$$P = l + l + w + w = 2l + 2w$$

To use the formula for a perimeter in our example, we need to substitute $l = 24$ and $w = 11$ into the equation.

$$P = 2l + 2w = 2(24) + 2(11) = 48 + 22 = 70 \text{ ft}$$

Finding the perimeter of a parallelogram is similar to the method for finding the perimeter of a rectangle.

2. Find the perimeter of a regular hexagon with a side of length 8 m.

   Since the hexagon is regular, each of the sides has a length of 8m. Hence, the perimeter is

$$p = 8 + 8 + 8 + 8 + 8 + 8$$
$$= 6(8) = 48 \text{ m}$$

8 m

➤ **Period**   Formally, the **period** of a function f is the smallest positive constant p of a periodic function. If f is a periodic function, then $f(x + p) = f(x)$ and p is called the period of the function f. See **Periodic Function**.

Less formally, the **period** of a function is the value at which the graph of the function starts to repeat itself. The graph of $y = f(x) = \sin x$ starts to repeat when $x = 360°$, or 2π in radians. Therefore, the period of $y = f(x) = \sin x$ is $p = 360°$, or $p = 2\pi$ in radians.

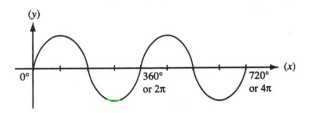

We can verify this by choosing any x in the domain of f and substituting it in $f(x + p) = f(x)$. Suppose we choose $x = 30°$. Then

$$f(x + p) = f(30° + 360°) = f(390°) = \sin 390° = .5$$

This is the same as $f(x)$ since $f(x) = f(30°) = \sin 30° = .5$. Hence, $f(x + p) = f(x)$ for any x we wish to choose in the domain of f.

Now let's look at the graph of $y = f(x) = \sin 2x$.

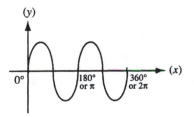

We see that the graph starts to repeat itself at $x = 180°$ or $x = \pi$. Therefore, the period of $y = f(x) = \sin 2x$ is $p = 180°$ or $p = \pi$.

If we would continue to look at graphs of $y = \sin 3x$ and $y = \sin 4x$, we would see that they would each start to repeat at $x = 120°$ or $x = 90°$, respectively. Hence, the period of $y = \sin 3x$ is $p = 120°$ or $p = \frac{2\pi}{3}$ and the period of $y = \sin 4x$ is $p = 90°$ or $p = \frac{\pi}{2}$.

From the above, we can see, in general, that the period p of the function $y = f(x) = \sin bx$ is

$$p = \frac{360°}{|b|} \text{ in degrees} \quad \text{or} \quad p = \frac{2\pi}{|b|} \text{ in radians}$$

This is also true for $y = \cos bx$, $y = \csc bx$, and $y = \sec bx$. The absolute value is used to keep the period positive.

The tangent and cotangent functions are different. For $y = f(x) = \tan bx$ and $y = f(x) = \cot bx$ the period of each is

$$p = \frac{180°}{|b|} \text{ in degrees} \quad \text{or} \quad p = \frac{\pi}{|b|} \text{ in radians}$$

This is because their graphs repeat in half the interval. See **Trigonometry, graphs of the six functions**.

Examples

1. Find the period of $y = f(x) = \cos 3x$ in both degrees and radians.

From the equations $p = \frac{360°}{|b|}$ and $p = \frac{2\pi}{|b|}$ with $b = 3$, we have

$$p = \frac{360°}{3} = 120° \text{ in degrees} \quad \text{and} \quad p = \frac{2\pi}{3} \text{ in radians}$$

Just as a matter of clarification, let's convert $\frac{2\pi}{3}$ to degrees:

$$\frac{2\pi}{3} = \frac{2(180)}{3} = \frac{360}{3} = 120°, \text{ the period in degrees}$$

The periods for $y = \csc 3x$ and $y = \sec 3x$ are found by the same process.

2. Find the period of $y = f(x) = -5 - 3\cot\left(-\frac{1}{2}x\right)$.

In comparison to $y = \cot bx$, we are concerned only with the "$-\frac{1}{2}x$" where $b = -\frac{1}{2}$. Hence, the period is

$$p = \frac{180}{\left|-\frac{1}{2}\right|} = 360° \text{ in degrees} \quad \text{or} \quad p = \frac{\pi}{\left|-\frac{1}{2}\right|} = 2\pi \text{ in radians.}$$

▶ **Periodic Function**   A function is said to be a **periodic function** if there exists a real number p such that

$$f(x + p) = f(x)$$

for all x in the domain of f. The number p is called the **period** of the function. It is the lowest number for which the function is periodic.

The function $y = f(x) = \sin x$ is a periodic function because for any x that we choose (for any x in the domain of f), we see that $f(x + p) = f(x)$. Since the period for $y = \sin x$ is $p = 360°$, let us choose any x in the domain of f, say $x = 45°$, and substitute these values in $f(x + p) = f(x)$. Substituting, we have

$$f(x + p) = f(45° + 360°) = f(405°) = \sin 405° = .7071$$

and

$$f(x) = f(45°) = \sin 45° = .7071$$

Hence, $f(x + p) = f(x)$ which shows that the function $y = f(x) = \sin x$ is periodic when $x = 45°$. A similar demonstration will hold for any x we choose in the domain of f, whether in degrees or radians.

We can use the concept of a periodic function to evaluate the functions of trigonometry.

Example

1. Evaluate $\sin \frac{9\pi}{4}$.

   Since $y = \sin x$ has a period of 2π, we want to rewrite $\frac{9\pi}{4}$ as a sum with 2π:

   $$\frac{9\pi}{4} = \frac{\pi}{4} + \frac{8\pi}{4} = \frac{\pi}{4} + 2\pi$$

   Now, using $f(x + p) = f(x)$, we have

   $$\sin \frac{9\pi}{4} = \sin\left(\frac{\pi}{4} + 2\pi\right) = \sin \frac{\pi}{4} = .7071$$

   This example is the same as the one given in our explanatory demonstration. The only difference is one is given in radians whereas the other is given in degrees.

   $$\sin 405° = \sin(45° + 360°) = .7071$$

➤ **Permutation**   A **permutation** is an arrangement of a set of elements in a specific order. If n objects (elements) are taken r at a time, then the number of permutations is

$$_nP_r = \frac{n!}{(n - r)!} \qquad \text{where } n \text{ and } r \text{ are whole numbers}$$

The term $n!$ is read "n factorial" where $n! = 1 \bullet 2 \bullet 3 \bullet \cdots$ or

$$n! = n \bullet (n - 1) \bullet (n - 2) \bullet \cdots \bullet 3 \bullet 2 \bullet 1 \qquad \text{See \textbf{Factorial}.}$$

Other symbols used to represent a permutation are $P(n, r)$ and P_r^n.

There are two basic types of permutations: **linear permutations** and **circular permutations**. If a linear permutation (elements in a line) has all of its elements considered at once, where the elements are all different, then the number of permutations is represented by

$$_n P_n = \frac{n!}{(n - n)!} = \frac{n!}{0!} = \frac{n!}{1} = n!$$

where 0! is defined to be 1. If a linear permutation is considered where some of the elements are the same, then the number of permutations can be found by evaluating the expression

$$\frac{n!}{p!q!r!\cdots}$$

where there are p things alike, q things alike, r things alike, and so on.

The difference between a circular permutation and a linear permutation is that in a circular permutation some of the arrangements are repeated. If there are n different elements arranged in a circle, then n of the arrangements (for each different arrangement) are repeated. The number of permutations can be found by evaluating the expression

$$\frac{n!}{n} = \frac{n(n - 1)(n - 2) \bullet \cdots \bullet 3 \bullet 2 \bullet 1}{n} = (n - 1)!$$

The reason n arrangements (for each different arrangement) are repeated has to do with the relative positions of the elements. If all of the elements are shifted either clockwise or counterclockwise around a circle (for example one place), the arrangement of the elements is not disturbed. This is not so with a linear permutation. If the elements in the list $a\ b\ c\ d\ e\ f$ are shifted left or right (for example one place), the new lists have different arrangements, either $b\ c\ d\ e\ f\ a$ or $f\ a\ b\ c\ d\ e$.

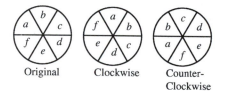

Original   Clockwise   Counter-
                         Clockwise

In a circle the arrangements remain in the same order when they are shifted.

In some circular permutation problems, the arrangement of elements does not have a definite top or bottom, such as a charm bracelet or a key ring. Problems of this type have arrangements on one side that are duplicated on the other side. In such cases, there are only half the amount of circular permutations or

$$\frac{(n - 1)!}{2}$$

For example, suppose a charm bracelet without a clasp has five charms. If we are asked to determine how many ways the charms can be arranged on the bracelet, we need to consider the fact that some of the arrangements on one side are the same as when the bracelet is turned over. Hence, the number of arrangements are

$$\frac{(n - 1)!}{2} = \frac{(5 - 1)!}{2} = \frac{4!}{2} = \frac{24}{2} = 12$$

Examples

1. A club has 14 members. In how many ways can a president, secretary, and treasurer be chosen?

Since there are three positions to be filled, we want to determine how many ways 14 people can fill the positions. We want to find the number of permutations of 14 things taken three at a time. From the equation for n objects taken r at a time, we have

$$_nP_r = {_{14}P_3} = \frac{14!}{(14-3)!} = \frac{14!}{11!} = \frac{14 \bullet 13 \bullet 12 \bullet 11 \bullet 10 \bullet \cdots \bullet 2 \bullet 1}{11 \bullet 10 \bullet \cdots \bullet 2 \bullet 1}$$

$$= 14 \bullet 13 \bullet 12 = 2,184$$

2. In how many ways can six different books be arranged on a bookshelf?

Since the books are considered all at one time, the number of permutations is 6!:

$$_nP_r = {_6P_6} = \frac{6!}{(6-6)!} = \frac{6!}{1!} = 6! = 6 \bullet 5 \bullet 4 \bullet 3 \bullet 2 \bullet 1 = 720$$

3. In how many different ways can five English books, three math books, and four history books be arranged, by subject, on a bookshelf?

This is a problem within a problem. We are required to find the number of permutations of the subjects as well as the number of permutations within each subject. Since there are three subjects, the number of their permutations is 3!.

Further, the number of permutations in each subject is 5!, 3!, and 4!, respectively. Hence, we have the total number of ways of arranging the books

$$= 3! \bullet 5! \bullet 3! \bullet 4! = 6 \bullet 120 \bullet 6 \bullet 24 = 103,680$$

This follows from the **counting principle**. See **Counting Principle**. If two or more events can occur in p and q different ways, then the number of ways both events can occur is determined by their product $p \bullet q$. If there are more than two events, then the number is determined by their product.

4. How many permutations are there for the letters in the following words:

TABLE, MAXIMUM, and MISSISSIPPI?

a. There are five letters in the word TABLE, with no repeats. Therefore, the number of permutations of the letters are 5! = 120.

b. There are 7 letters in the word MAXIMUM where the letter M is repeated 3 times. Hence, the number of permutations of the letters is

$$\frac{7!}{3!} = \frac{7 \bullet 6 \bullet 5 \bullet 4 \bullet 3 \bullet 2 \bullet 1}{3 \bullet 2 \bullet 1} = 7 \bullet 6 \bullet 5 \bullet 4 = 840$$

c. There are 11 letters in the word MISSISSIPPI, where the letter I is repeated 4 times, the letter S is repeated 4 times, and the letter P is repeated 2 times. Hence, the number of permutations of the letters is

$$\frac{11!}{4!4!2!} = \frac{39{,}916{,}800}{24 \bullet 24 \bullet 2} = \frac{39{,}916{,}800}{1{,}152} = 34{,}650$$

5. How many different seating arrangements are there for six people seated at a round table?

Since the number of permutations for a circle is represented by $(n - 1)!$, the number of seating arrangements is

$$(6 - 1)! = 5! = 120$$

For more, see **Combination** or **Counting Principle**.

➤ **Perpendicular (geometric)**   Two lines, segments, or rays are said to be **perpendicular** if they form a right angle when they intersect. Conversely, if two lines form a right angle when they intersect, then they are perpendicular.

Another form of this definition states that two lines are perpendicular if they form congruent adjacent angles. The symbol for perpendicular is $\perp$. For more, concerning the slopes of perpendicular lines, see **Perpendicular Lines, slopes of**.

➤ **Perpendicular Bisector**   The **perpendicular bisector** of a segment is a line, ray, or segment in the plane that is perpendicular to the segment at its midpoint.

If M is the midpoint of $\overline{AB}$ and $\angle 1$ and $\angle 2$ are right angles, then $\overleftrightarrow{CM} \perp \overline{AB}$ and $\overleftrightarrow{CM}$ passes through the midpoint of $\overline{AB}$. Hence, $\overleftrightarrow{CM}$ is the perpendicular bisector of $\overline{AB}$.

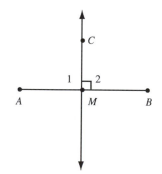

In reference to the symbols above, $\overleftrightarrow{CM}$ and $\overline{AB}$, the arrow ($\leftrightarrow$) and the bar ($-$) mean line and segment respectively.

➤ **Perpendicular Lines, slopes of**   The following applies to lines in the coordinate plane that are not horizontal or vertical. Two lines are said to be **perpendicular lines** if their slopes are negative reciprocals. Conversely, if the slopes of two lines are negative reciprocals, the lines are said to be perpendicular.

Let l_1 and l_2 represent two lines. If $l_1 \perp l_2$, then $m_1 = \frac{-1}{m_2}$. Conversely, if l_1 and l_2 represent two lines where $m_1 = \frac{-1}{m_2}$, $l_1 \perp l_2$.

Examples

1. What is the slope of any line perpendicular to $y = 2x + 1$?

   The slope of any line perpendicular to $y = 2x + 1$ is the negative reciprocal of the slope of $y = 2x + 1$. From $y = mx + b$, we know the slope of $y = 2x + 1$ is $m = 2$. Hence, the slope of any perpendicular line is $-\frac{1}{m} = -\frac{1}{2}$.

2. Determine whether the lines $2x - 3y = 15$ and $3x - 2y = -8$ are perpendicular.

   If the two lines are perpendicular, then their slopes must be negative reciprocals. So, we first must determine their slopes. Solving each equation for y (use $y = mx + b$ to identify m), we have

$$2x + 3y = 15 \qquad\qquad 3x - 2y = -8$$
$$3y = -2x + 15 \qquad\qquad -2y = -3x - 8$$
$$\frac{3y}{3} = \frac{-2x}{3} + \frac{15}{3} \qquad\qquad \frac{-2y}{-2} = \frac{-3x}{-2} - \frac{8}{-2}$$
$$y = -\frac{2}{3}x + 5 \qquad\qquad y = \frac{3}{2}x + 4$$
$$\text{Hence,} \quad m_1 = -\frac{2}{3}. \qquad\qquad \text{Hence,} \quad m_2 = \frac{3}{2}.$$

   See **Slope-Intercept Form of an Equation of a Line**.

   Substituting m_1 and m_2 into $m_1 = -\frac{1}{m_2}$, we have

$$m_1 = -\frac{1}{m_2}$$
$$-\frac{2}{3} = \frac{-1}{\frac{3}{2}}$$
$$-\frac{2}{3} = -\frac{2}{3}$$

   Since the results are the same, we know that the two slopes are negative reciprocals of one another. Therefore, the lines are perpendicular.

   Another way to find m_1 and m_2 is to use $m = -\frac{A}{B}$. If an equation is in standard form ($Ax + By = C$), then $m = -\frac{A}{B}$. Hence,

$$m_1 = -\frac{A}{B} = -\frac{2}{3} \qquad \text{and} \qquad m_2 = -\frac{A}{B} = -\frac{3}{-2} = \frac{3}{2}$$

3. Find the equation of the line through $(-8, 6)$ perpendicular to $y = 2x + 1$.

   The slope of the line perpendicular to $y = 2x + 1$ is the negative reciprocal of the slope of $y = 2x + 1$.

   The slope of $y = 2x + 1$ is $m = 2$. So, the slope of the perpendicular line is $-\frac{1}{m} = -\frac{1}{2}$.

   Using the point $(-8, 6)$ and the slope $-\frac{1}{2}$, we can find the equation of the perpendicular line. Let's do this by the slope-intercept method. The other method is the point-slope method. See **Linear Equation, finding the equation of a line**.

Substituting $(x, y) = (-8, 6)$ and $m = -\frac{1}{2}$ (don't confuse with the slope of $y = 2x + 1$, $m = 2$) into $y = mx + b$, we have

$$y = mx + b$$
$$6 = -\frac{1}{2}(-8) + b$$
$$6 = 4 + b$$
$$2 = b$$

which gives us the equation $y = -\frac{1}{2}x + 2$.

Horizontal and vertical lines are dealt with differently. The slope of a horizontal line is $m = 0$ and the slope of a vertical line is defined as no slope or ∞. Hence, if one line has a slope $m = 0$ and another line has no slope, the lines are perpendicular. See **Horizontal and Vertical Lines**.

▶ **Phase Shift**   In trigonometry, the **phase shift** of a function is a number by which the graph of the function has been shifted either left or right. In the equation $y = A \sin(Bx - C)$, the C creates a shift to the left if C is negative and a shift to the right if C is positive. Do not confuse the sign of C with the " $-$ " sign of $Bx - C$. If $C = -2$, then $Bx - C$ is $Bx - (-2) = Bx + 2$. If $C = 2$, then $Bx - C = Bx - 2$.

The number C is not necessarily the phase shift. The phase shift is the number $\frac{C}{B}$, which is determined by looking at the left and right endpoints of the interval on which $y = \sin x$ is graphed. The left endpoint occurs when $x = 0$ and the right endpoint occurs when $x = 2\pi$. See **Trigonometry, graphs of the six functions**.

In relation to $y = A \sin(Bx - C)$, this means that the left endpoint occurs at $Bx - C = 0$ or $x = \frac{C}{B}$ and the right endpoint occurs at $Bx - C = 2\pi$ or $x = \frac{2\pi + C}{B} = \frac{C}{B} + \frac{2\pi}{B}$. Remember that $\frac{2\pi}{B}$ is the formula for the period. See **Period**. Hence, the right endpoint is the phase shift and period combined. Here and in the rest of the article if it is required to have degrees instead of radians, convert the radians to degrees by replacing π with $180°$. See **Converting Radians to Degrees**. For example: $\pi = 180°$, $\frac{\pi}{2} = \frac{180°}{2} = 90°$, and $\frac{\pi}{4} = \frac{180°}{4} = 45°$.

After the left and right endpoints of the interval have been determined, the graphing techniques described in **Trigonometry, graphs of the six functions** can be applied. This allows us the means to determine the five key points needed to graph one period of the function.

Examples

1. Graph one period of $y = 2 \sin\left(x - \frac{\pi}{2}\right)$.

   Although the phase shift is $\frac{C}{B} = \frac{\pi/2}{1} = \frac{\pi}{2}$, what we are really concerned with is the five key points. We first need to find the left and right endpoints of the graphing interval:

Left endpoint	**Right endpoint**
(set $Bx - C = 0$)	(set $Bx - C = 2\pi$)
$x - \dfrac{\pi}{2} = 0$	$x - \dfrac{\pi}{2} = 2\pi$
$x = \dfrac{\pi}{2}$	$x = 2\pi + \dfrac{\pi}{2}$
	$x = 2\frac{1}{2}\pi$

Using the period $P = \frac{2\pi}{|B|} = \frac{2\pi}{1} = 2\pi$, we can find the increment of the graphing interval, $\frac{period}{4} = \frac{2\pi}{4} = \frac{2\pi}{4} = \frac{\pi}{2}$. Then we can find the remaining x-values of the key points. We do this by successively adding $\frac{\pi}{2}$ from the left endpoint:

1st x-value: $\qquad \frac{\pi}{2}$, left endpoint

2nd x-value: $\qquad \frac{\pi}{2} + \frac{\pi}{2} = \frac{2\pi}{2} = \pi$

3rd x-value: $\qquad \pi + \frac{\pi}{2} = 1\frac{1}{2}\pi$, middle

4th x-value: $\qquad 1\frac{1}{2}\pi + \frac{\pi}{2} = 2\pi$

5th x-value: $\qquad 2\frac{1}{2}\pi$, right endpoint

Since the sine graph starts low and ends low the y-values follow the pattern $0, A, 0, -A$, and 0. Hence, the key points are $\left(\frac{\pi}{2}, 0\right)$, $(\pi, 2)$, $\left(1\frac{1}{2}\pi, 0\right)$, $(2\pi, -2)$, and $\left(2\frac{1}{2}\pi, 0\right)$.

If it is required to extend the graph, continue the pattern of the key points.

Notice that the graph of $y = 2\sin x$ has been shifted to the right $\frac{\pi}{2}$ units. The phase shift of $y = 2\sin\left(x - \frac{\pi}{2}\right)$ is $\frac{C}{B} = \frac{\pi}{2}$.

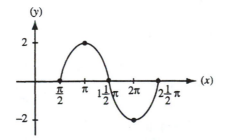

If it is preferred to graph the function in degrees rather than radians, set $Bx - C$ equal to $0°$ and $360°$, respectively. Then calculate the key points. They are $(90°, 0)$, $(180°, 2)$, $(270°, 0)$, $(360°, -2)$ and $(450°, 0)$. The phase shift is $90°$.

Graphing the phase shift for cosine, cosecant, and secant is similar to that of the sine. Determine the graphing interval, find the key points, and then sketch the graph. If the graph involves asymptotes, use the values of the graphing interval to determine the asymptotes.

2. Graph one period of $y = 2\sec\left(x + \frac{\pi}{4}\right)$.

First, we need to write the equation in the form $y = A\sec(Bx - C)$, which is $y = 2\sec\left(x - \left(\frac{-\pi}{4}\right)\right)$. Now we can determine the phase shift $\frac{C}{B} = \frac{-\pi/4}{1} = \frac{-\pi}{4}$, the period $\frac{2\pi}{|B|} = \frac{2\pi}{1} = 2\pi$, and the increment of the graphing interval $\frac{period}{4} = \frac{2\pi}{4} = \frac{\pi}{2}$. With this information, we can calculate the left and right endpoints of the graphing interval as well as its intermediate values:

Left endpoint	Right endpoint
$x + \dfrac{\pi}{4} = 0$	$x + \dfrac{\pi}{4} = 2\pi$
$x = \dfrac{-\pi}{4}$	$x = 2\pi - \dfrac{\pi}{4} = \dfrac{7\pi}{4}$

The graphing interval is found by successively adding the increment of the graphing interval from the left endpoint:

1st x-value: $\quad -\dfrac{\pi}{4} = -45°,\ \text{left endpoint}$

2nd x-value: $\quad -\dfrac{\pi}{4} + \dfrac{\pi}{2} = \dfrac{\pi}{4} = 45°$

3rd x-value: $\quad \dfrac{\pi}{4} + \dfrac{\pi}{2} = \dfrac{3\pi}{4} = 135°,\ \text{middle}$

4th x-value: $\quad \dfrac{3\pi}{4} + \dfrac{\pi}{2} = \dfrac{5\pi}{4} = 225°$

5th x-value: $\quad \dfrac{5\pi}{4} + \dfrac{\pi}{2} = \dfrac{7\pi}{4} = 315°,\ \text{right endpoint}$

The asymptotes for the secant occur at the second, and third x-values and the key points occur at the first, third, and fifth x-values. The y-values for the key points follow the pattern A, $-A$, and A. Therefore, the keypoints are $\left(\frac{-\pi}{4}, 2\right)$, $\left(\frac{3\pi}{4}, -2\right)$, and $\left(\frac{7\pi}{4}, 2\right)$. See **Trigonometry, graphs of the six functions**.

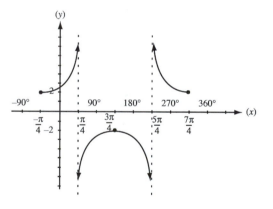

The method for graphing the shift of the tangent and the cotangent is slightly different from the other four functions. The period for $y = \tan x$ has a left asymptote at $-\frac{\pi}{2}$ and a right asymptote at $\frac{\pi}{2}$. When finding the left and right endpoints of the graphing interval, set $Bx - C$ equal to $-\frac{\pi}{2}$ and $\frac{\pi}{2}$, respectively. The period for $y = \cot x$ has a left asymptote at 0 and a right asymptote at π. When finding the left and right endpoints of its graphing interval, set $Bx - C$ equal to 0 and π, respectively. Remember, in either case, the graphing intervals of $y = \tan x$ and $y = \cot x$ are half those of the other four functions.

▶ **Place Value**   In the decimal numeration system, the position of each member of a number determines the power of 10 it uses. Each position is called a **place value** and is 10 times the value of the next place to the right of it.

In the number 73, the place occupied by the 7 represents 7×10 or 70. In the number 753 the place occupied by the 7 represents 7×100 or 700. In either case the value of the 7 is determined by the location of its place in the number. The place value of the 7 is determined by its location in the number. A list of place values is given below.

| trillions | , | hundred billions | ten billions | billions | , | hundred millions | ten millions | millions | , | hundred thousands | ten thousands | thousands | , | hundreds | tens | ones or units | . and | tenths | hundredths | thousandths | ten-thousandths | hundred-thousandths | millionths |

To the left of the decimal point, we separate each third entry with a comma. No such method is applied to the right of the decimal point.

In the number 400.25, the decimal point is referred to as "*and*." We say "four hundred *and* twenty-five hundredths." The most common mistake made is to refer to the number 425 as "four hundred and twenty-five." Correctly, we say "four hundred twenty-five."

➤ **Plane**   A **plane** is a set of points on a flat surface that extends indefinitely. It is endless and has no boundaries. As soon as we try to draw a plane, we limit it by giving it boundaries. These boundaries may be used to describe the flat nature of the surface, but should not be confused with any idea that a plane has edges. It does not. It extends outward as a flat surface indefinitely.

A plane is usually drawn as a four-sided figure. It is named by using three points in the plane, such as plane ABC, or it is simply given a name, such as plane P.

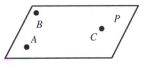

The points of a plane can be represented by the ordered pair (x, y). See **Coordinate Plane**.

➤ **Plane Geometry**   **Plane geometry** or Euclidean geometry is a branch of mathematics that deals with the properties related to points, lines, and planes. It is distinguished from other geometries by the parallel postulate. In plane geometry, there is exactly one parallel to a given line, considering that the parallel goes through a point not on the given line. Non-Euclidean geometry differs in this regard.

➤ **Plus and Minus Numbers**   See **Positive and Negative Numbers**.

➤ **Plus and Minus Signs in Trigonometry**   See **Trigonometry, signs of the six functions**.

➤ **Plus or Minus ($\pm$)**   The concept of $\pm$, read "plus or minus," is developed from solutions of equations involving exponents and radicals. See **Root, *n*th**.

➤ **Point-Slope Method of Finding the Equation of a Line**   See **Linear Equation, finding the equation of a line**.

P

➤ **Polar Axis**   In the polar coordinate system, the **polar axis** is an initial ray with its endpoint at the pole (origin). By comparison with rectangular coordinates, the polar axis is coincident with the positive *x*-axis. See **Polar Coordinate System**.

➤ **Polar Coordinates**   In the polar coordinate system, points in the plane are assigned **polar coordinates** denoted by (r, θ). The coordinate r is a **directed distance** and the coordinate θ is a **directed angle**. See **Polar Coordinate System**.

➤ **Polar Coordinate System**   The **polar coordinate system** is another means of locating points in the plane. The more common method is the rectangular coordinate system. See **Coordinate Plane**.

In the polar coordinate system, the origin is called the **pole** and the initial ray, emanating from the pole, is called the **polar axis**. Also emanating from the pole is a **directed distance** r which forms a **directed angle** θ with the polar axis, measured counterclockwise.

In this manner, each point in the plane can be assigned a pair of numbers (r, θ) which are called the **polar coordinates** of the plane P.

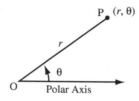

To facilitate the location of points, a grid of concentric circles with the pole as center is intersected by **radial lines** that pass through the pole.

Although each point in rectangular coordinates is unique, this is not so in polar coordinates. The coordinates (r, θ) and $(-r, \theta + \pi)$ locate the same point because r is a directed distance. For conversion from rectangular coordinates to polar coordinates or polar coordinates to rectangular coordinates, see examples 2 and 3.

Examples

1. Plot the points $(r, \theta) = \left(2, \frac{\pi}{6}\right)$, $(r, \theta) = \left(3, \frac{7\pi}{4}\right)$, $(r, \theta) = \left(3, -\frac{\pi}{4}\right)$, and $(r, \theta) = \left(-2, \frac{\pi}{6}\right)$.

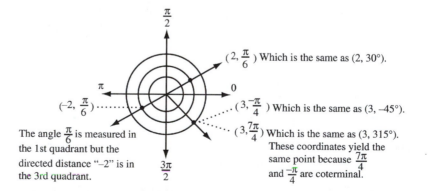

The relationship between polar and rectangular coordinates can be seen from the following graph.

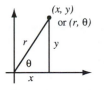

We can calculate r in terms of x and y since $r^2 = x^2 + y^2$ or $r = \sqrt{x^2 + y^2}$. We can determine θ from x and y since $\tan\theta = \frac{y}{x}$ or $\theta = \tan^{-1}\frac{y}{x}$.

Using r and θ, we can calculate x since $\cos\theta = \frac{x}{r}$ or $x = r\cos\theta$. Also, using r and θ, we can calculate y since $\sin\theta = \frac{y}{r}$ or $y = r\sin\theta$.

To convert from rectangular coordinates to polar coordinates, we use the first two pieces of information. To convert from polar coordinates to rectangular coordinates, we use the second two pieces of information.

2. Convert the rectangular coordinates $(-1, -1)$ to polar coordinates.

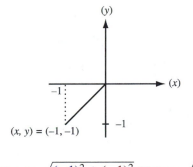

From $r = \sqrt{x^2 + y^2}$, we have $r = \sqrt{(-1)^2 + (-1)^2}$ or $r = \sqrt{1+1} = \sqrt{2}$. To find θ, we use $\tan\theta = \frac{y}{x} = \frac{-1}{-1} = 1$. Hence, $\theta = \tan^{-1}$ or $\theta = \frac{\pi}{4}$ or $\frac{5\pi}{4}$. Since $(-1, -1)$ is in third quadrant, $\theta = \frac{5\pi}{4}$. Therefore, the point $(-1, -1)$ converts to $(r, \theta) = \left(\sqrt{2}, \frac{5\pi}{4}\right)$ in polar coordinates.

3. Convert the polar coordinates $\left(\sqrt{2}, \frac{5\pi}{4}\right)$ to rectangular coordinates.

The x-value of the coordinates is:

$$x = r\cos\theta = \sqrt{2}\cos\frac{5\pi}{4}$$
$$= \sqrt{2}\left(\frac{-1}{\sqrt{2}}\right) = -1$$

The y-value of the coordinates is:

$$x = r\sin\theta = \sqrt{2}\sin\frac{5\pi}{4}$$
$$= \sqrt{2}\left(\frac{-1}{\sqrt{2}}\right) = -1$$

Therefore, the point $\left(\sqrt{2}, \frac{5\pi}{4}\right)$ converts to $(x, y) = (-1, -1)$ in rectangular coordinates.

➤ **Polar Form**   **Polar form** is another term for the trigonometric form of a complex number. See **Complex Number, trigonometric form**.

➤ **Pole**   In the polar coordinate system, the **pole** is the center or origin of the system. See **Polar Coordinate System**.

➤ **Polygon**   A **polygon** is a closed plane figure with edges, or sides, which are segments. If all the sides are equal in length and all the angles are equal in measure, the polygon is called a **regular polygon**. Some common examples of a polygon are a triangle (3 sides), a quadrilateral (4 sides), a pentagon (5 sides), a hexagon (6 sides), and an octagon (8 sides).

A polygon that does not collapse in on itself is called a **convex polygon**. In a convex polygon, no line containing any of its sides will contain a point inside the polygon. The following figure is not a convex polygon:

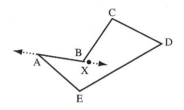

The figure is not a convex polygon because there are points of $\overleftrightarrow{AB}$ (like X) that are inside the figure. The diagonal of a polygon is a segment, other than an edge, that connects two vertices. A triangle, therefore, doesn't have diagonals.

➤ **Polygon, area of a regular**   See **Area of a Regular Polygon**.

➤ **Polygonal Region**   A **polygonal region** consists of the edges of a polygon and all the points interior to the polygon.

➤ **Polynomial**   A **polynomial** is either a monomial by itself or the sum or difference of two or more monomials. A **monomial** is composed of a constant, a variable, or the product (not sum or difference) of one or more variables. The exponents of the variables are nonnegative integers.

The following monomials are one-term polynomials:

$$12, \quad x^2, \quad -7a^2b^3, \quad -35xyz^2$$

The **degree** of a monomial is the sum of the exponents of its variables. The degree of a nonzero constant term is 0, i.e., 7 is equivalent to $7x^0$, therefore, its degree is 0. If the polynomial consists only of the term 0, then it has no degree. In the examples, the degree of 12 is 0, the degree of x^2 is 2, the degree of $-7a^2b^3$ is 5, and the degree of $-35xyz^2$ is 4 (x and y each have exponents of 1).

The sum or difference of two or more monomials is also a polynomial. If the polynomial is composed of two monomials, or **terms**, it is called a **binomial**. If the polynomial is composed of three terms, it is called a **trinomial**. If it is composed of more than three terms, it is simply referred to as a **polynomial**.

The following are examples of polynomials:

1. $6x^2 - 3$ (binomial)

2. $-4x^3 - 5x + 3$ (trinomial)

3. $7x^3 - 2x^2y + xy^2 - y^3$ (polynomial)

The following are not polynomials:

1. $\sqrt{2x^2 - 1} + 4x$

2. $3x^{-2} + 8$

3. $\dfrac{3}{x^2} + 8$

Each term of a polynomial contains a numerical factor called the **coefficient**. The polynomial $7x^3y - 2x^2 - x + 3$ has coefficients of 7, −2, −1, and 3, respectively. The term 3 is called a coefficient because it is the numerical factor of the term $3x^0$. See **Exponents, rules of**, special case.

The degree of a polynomial is the same as the monomial of highest degree. The polynomial $5xy^3 + 7x^2y^3 - 2x$ is of degree 5 because that is the degree of the monomial $7x^2y^3$. Usually polynomials are written in descending order of the variable or, if there is more than one variable, descending order of the dominant variable (usually alphabetical order). The following polynomials are written in descending order of x:

1. $3x^4 - 7x^3 - 12x - 2$

2. $6x^4 - 4x^3y + 5x^2y^2 - 2xy^3 + 4y^4$

In the second case the x's are in **descending order** and the y's are in **ascending order**. See **Ascending and Descending Order**. The term $6x^4$ can be considered to be $6x^4y^0$ and the term $4y^4$ can be considered to be $4x^0y^4$. Remember, variables with no exponent actually have an exponent of 1, i.e., $x = x^1$ or $y = y^1$. See **Exponents, rules of**.

Examples

In the following polynomials state the degree, identify the coefficients, classify by the number of terms, and then arrange in descending order of the exponents.

1. $7 - 2x^2 + 3x^4$

    Degree is 4
    Coefficients are 7, −2, 3
    Classified as a trinomial
    $3x^4 - 2x^2 + 7$

2. $-24x^2y^2 + 4x^3$

    Degree is 4
    Coefficients are −24, 4
    Classified as a binomial
    $4x^3 - 24x^2y^2$

3. $-8 + 5x - 2x^3 + 9x^2$

    Degree is 3
    Coefficients are −8, 5, −2, 9
    Classified as a polynomial
    $9x^2 - 2x^3 + 5x - 8$

For more, see **Polynomials, sums and differences** or **Polynomials, multiplication of**.

► **Polynomial Equations**   This article deals with solutions of equations that are of fourth degree and higher. For solutions of second degree equations, see **Quadratic Equations, solutions of**. For solutions of third-degree equations, see **Cubic Equations, solutions of** or **Rational Roots Theorem**.

To solve a **polynomial equation**, the equation must be set equal to zero, factored, each factor set equal to zero, and then the solution of each factor must be determined. See **Product Theorem of Zero**. The reason why polynomial equations have multiple answers is discussed in **Zeros of a Function or Polynomial**.

Examples

1. Solve $x^4 = x^2 + 12$ for real values of x.

We need to set the equation equal to zero, factor it, set each factor equal to zero, and then solve each factor.

$$x^4 = x^2 + 12$$
$$x^4 - x^2 - 12 = 0$$
$$(x^2 + 3)(x^2 - 4) = 0$$
$$(x^2 + 3)(x + 2)(x - 2) = 0$$

$$x^2 + 3 = 0 \quad \text{or} \quad x + 2 = 0 \quad \text{or} \quad x - 2 = 0$$
$$\varnothing \qquad\qquad\qquad x = -2 \qquad\qquad\qquad x = 2$$

The equation $x^2 + 3 = 0$ does not have a real solution. Therefore, the solutions of the equation $x^4 = x^2 + 12$ are $x = -2$ and $x = 2$. If complex solutions were allowed, then $x^2 + 3 = 0$ would give two more solutions of $x = \pm\sqrt{3}\,i$. See **Complex Numbers**, example 6.

2. Solve $x^5 = 10x^3 - 9x$.

Setting the equation equal to 0 and solving, we have

$$x^5 = 10x^3 - 9x$$
$$x^5 - 10x^3 + 9x = 0$$
$$x(x^4 - 10x^2 + 9) = 0$$
$$x(x^2 - 1)(x^2 - 9) = 0$$
$$x(x + 1)(x - 1)(x + 3)(x - 3) = 0$$

$$x = 0 \quad \text{or} \quad x + 1 = 0 \quad \text{or} \quad x - 1 = 0$$
$$x = -1 \qquad\qquad\qquad x = 1$$

$$\text{or} \quad x + 3 = 0 \quad \text{or} \quad x - 3 = 0$$
$$x = -3 \qquad\qquad\qquad x = 3$$

Hence, there are five solutions.

3. Solve $2x^6 = 100x^4 - 98x^2$.

Setting the equation equal to 0 and solving, we have

$$2x^6 = 100x^4 - 98x^2$$
$$2x^6 - 100x^4 + 98x^2 = 0$$
$$2x^2(x^4 - 50x^2 + 49) = 0$$
$$2x^2(x^2 - 1)(x^2 - 49) = 0$$
$$2x^2(x + 1)(x - 1)(x + 7)(x - 7) = 0$$

$$2x^2 = 0 \quad \text{or} \quad x + 1 = 0 \quad \text{or} \quad x - 1 = 0$$
$$x = 0 \qquad\qquad\qquad x = -1 \qquad\qquad\qquad x = 1$$

$$\text{or} \quad x + 7 = 0 \quad \text{or} \quad x - 7 = 0$$
$$x = -7 \qquad\qquad\qquad x = 7$$

Hence, there are five solutions. The reason there are not six solutions is due to the double root when $2x^2 = 0$. See **Roots of Multiplicity**.

➤ **Polynomial Functions**   A **polynomial function** of degree n is a function defined by the equation

$$f(x) = a_n x^n + a_{n-1} x^{n-1} + \cdots + a_1 x + a_0$$

where n is a nonnegative integer and the leading coefficient a_n is not equal to 0.

If $n = 1$, then $f(x) = a_1 x + a_0$. This is a linear function or a polynomial function of degree 1. The slope of the equation is a_1.

If $n = 2$, then $f(x) = a_2 x^2 + a_1 x + a_0$. This is a quadratic function or polynomial function of degree 2.

➤ **Polynomial Inequalities**   For more information related to quadratic inequalities, see **Inequalities, quadratic solutions of**. There are, however, quadratic inequalities in this article as well.

There are two methods for solving **polynomial inequalities**, strictly by definition or by test intervals. We will consider the latter. The intervals we will look at will be written in interval notation rather than inequality notation. See **Interval Notation**. For a conversion from one type to the other, see example 1.

To solve polynomial inequalities by using test intervals, the following steps may be applied:

1. Add or subtract terms until the polynomial is on the left side of the inequality sign and 0 is on the right.

2. Determine the key numbers of the inequality. These are the numbers that will yield a 0 in the inequality.

3. Graph the key numbers on a number line.

4. From the number line determine the intervals to be tested and choose a number from each.

5. Substitute each test number into the polynomial to determine the sign of the polynomial in that interval.

6. Determine the solution from the collective information. Remember to consider whether the inequality is $<$ or $>$ or $\leqslant$ or $\geqslant$.

Examples

1. Solve $x^2 \leqslant -8x - 12$.

   We first put all terms on the left side of the inequality sign:

   $$x^2 \leqslant -8x - 12$$
   $$x^2 + 8x + 12 \leqslant 0$$

   Factoring, we can determine the key numbers.

   $$x^2 + 8x + 12 \leqslant 0$$
   $$(x + 6)(x + 2) \leqslant 0$$

The key numbers are -6 and -2 since these yield 0 in the inequality. Graphing these on a number line, we have

$$\begin{array}{ccc} (-\infty, -6) & (-6, -2) & (-2, -\infty) \end{array}$$

$$\xleftarrow{\qquad\qquad\qquad\qquad\qquad} (x)$$

$$\begin{array}{ccccc} x < -6 & -6 & -6 < x < -2 & -2 & x > -2 \end{array}$$

The intervals to be tested are $(-\infty, -6)$, $(-6, -2)$ and $(-2, \infty)$. See **Interval Notation**. If it is required to write the intervals in inequality notation, they are $x < -6$, $-6 < x < -2$, and $x > -2$. We need to choose test numbers from each interval, say -7, -3, and 0, respectively. Testing these, we have

Interval	Test Number	$x + 6$	$x + 2$	$(x + 6)(x + 2)$
$(-\infty, -6)$, $x < -6$	-7	$-$	$-$	$+$
$(-6, -2)$, $-6 < x < -2$	-3	$+$	$-$	$-$
$(-2, \infty)$, $x > -2$	0	$+$	$+$	$+$

Since $x^2 + 8x + 12 \leqslant 0$, we want intervals that are $\leqslant 0$, or negative. The only interval that satisfies this requirement is $(-6, -2)$. We must also test for $x^2 + 8x + 12 = 0$. This occurs when $x = -6$ and $x = -2$. Hence, the solution is $[-6, -2]$. As an inequality, the solution is $-6 \leqslant x \leqslant -2$.

2. Solve $x^2 > 7x - 10$.

Setting the inequality > 0 and factoring, we have

$$x^2 > 7x - 10$$
$$x^2 - 7x + 10 > 0$$
$$(x - 2)(x - 5) > 0$$

Hence, the key numbers are 2 and 5.

$$\begin{array}{ccc} (-\infty, 2) & (2, 5) & (5, \infty) \end{array}$$

$$\xleftarrow{\qquad\qquad\qquad\qquad\qquad} (x) \qquad \text{Choose 0, 3, and 6 to test.}$$

$$\begin{array}{ccccc} x < 2 & 2 & 2 < x < 5 & 5 & x > 5 \end{array}$$

Interval	Test Number	$x - 2$	$x - 5$	$(x + 2)(x - 5)$
$(-\infty, 2)$, $x < 2$	0	$-$	$-$	$+$
$(2, 5)$, $2 < x < 5$	3	$+$	$-$	$-$
$(5, \infty)$, $x > 5$	6	$+$	$+$	$+$

The intervals that are $\geqslant 0$ are $(-\infty, 2)$ and $(5, \infty)$ or $x < 2$ and $x > 5$ in inequalities. Since the inequality $x^2 - 7x + 10 > 0$ is strictly > 0, the key numbers are not part of the solution. Hence, the solution is $(-\infty, 2) \cup (5, \infty)$. As an inequality, the solution is $\{x \mid x < 2\} \cup \{x \mid x > 5\}$.

3. Solve $x^2(x - 3) \geqslant 25x - 75$.

Setting the inequality $\geqslant 0$ and factoring, we have

$$x^2(x - 3) \geqslant 25x - 75$$
$$x^3 - 3x^2 - 25x + 75 \geqslant 0$$
$$x^2(x - 3) - 25(x - 3) \geqslant 0$$
$$(x - 3)(x^2 - 25) \geqslant 0 \qquad \text{See \textbf{Factoring by Grouping (4-term polynomials)}.}$$
$$(x - 3)(x + 5)(x - 5) \geqslant 0$$

Hence, the key numbers are -5, 3, and 5.

Choose -6, 0, 4, and 6 to test.

Interval	Test Number	$x - 3$	$x + 5$	$x - 5$	$(x - 3)(x + 5)(x - 5)$
$(-\infty, 5), x < -5$	-6	$-$	$-$	$-$	$-$
$(-5, 3), -5 < x < 3$	0	$-$	$+$	$-$	$+$
$(3, 5), 3 < x < 5$	4	$+$	$+$	$-$	$-$
$(5, \infty), x > 5$	6	$+$	$+$	$+$	$+$

The intervals that are $\geqslant 0$ are $(-5, 3)$ and $(5, \infty)$ or $-5 < x < 3$ and $x > 5$ in inequalities. Also the key numbers satisfy $(x - 3)(x + 5)(x - 5) = 0$, so they are part of the solution. Hence, the solution is $[-5, 3] \cup [5, \infty]$. As an inequality, the solution is $\{x \mid -5 \leqslant x \leqslant 3\} \cup \{x \mid x \geqslant 5\}$.

4. Solve $\frac{3x+1}{x-1} < \frac{2}{x-4} + 2$.

Setting the inequality < 0, we have

$$\frac{3x + 1}{x - 1} - \frac{2}{x - 4} - 2 < 0$$

Since we do not know whether $(x - 1)(x - 4)$ is positive or negative, we cannot clear the fractions of their denominators. In other words, we cannot multiply through the inequality by $(x - 1)(x - 4)$. Instead, we need to write the entire left-hand side of the inequality as a single fraction, factor the numerator, and then consider the key values that yield a zero.

$$\frac{(3x + 1)(x - 4) - 2(x - 1) - 2(x - 1)(x - 4)}{(x - 1)(x - 4)} < 0$$

$$\frac{x^2 - 3x - 10}{(x - 1)(x - 4)} < 0$$

$$\frac{(x + 2)(x - 5)}{(x - 1)(x - 4)} < 0$$

Hence, the key numbers are $-2, 5, 1,$ and 4.

We choose the numbers $-3, 0, 2, 4\frac{1}{2},$ and 6 to test.

Interval	Test Number	$(x + 2)(x - 5)$	$(x - 1)(x - 4)$	$\frac{(x+2)(x-5)}{(x-1)(x-4)}$
$(-\infty, -2), x < -2$	-3	$+$	$+$	$+$
$(-2, 1), -2 < x < 1$	0	$-$	$+$	$-$
$(1, 4), 1 < x < 4$	2	$-$	$-$	$+$
$(4, 5), 4 < x < 5$	$4\frac{1}{2}$	$-$	$+$	$-$
$(5, \infty), x > 5$	6	$+$	$+$	$+$

The intervals that are < 0 are $(-2, 1)$ and $(4, 5)$ or $-2 < x < 1$ and $4 < x < 5$ in inequalities. Since the inequality is strictly < 0, we do not need to consider the zeros of the inequality, -2 and 5. Also, since the inequality is undefined at 1 and 4, we exclude these values from the solution as well. Hence, the solution is $(-2, 1) \cup (4, 5)$. As an inequality, the solution is $\{x \mid -2 < x < 1\} \cup \{x \mid 4 < x < 5\}$.

➤ **Polynomial Multiplication**   See **Polynomials, multiplication of**.

➤ **Polynomials, combining sums and differences.**   This is the addition and subtraction of polynomials. See **Polynomials, sums and differences.**

➤ **Polynomials, division**   See **Long Division of Polynomials** or **Synthetic Division**.

➤ **Polynomials, multiplication of**   In the process of **multiplication of polynomials**, each term of one polynomial multiplies each term in the adjacent polynomial. The terms are combined and then multiplied into the next adjacent polynomial. When all the polynomials have been multiplied, the terms are simplified and arranged in descending order of the exponents of the variable.

The process of multiplying a monomial with a binomial, or trinomial, or multiterm polynomial is an application of the distributive property. The monomial multiplies each of the terms in the parenthetical expression.

The process for multiplying two binomials is the FOIL method (multiply the terms in the order <u>F</u>irst, <u>O</u>uter, <u>I</u>nner, and <u>L</u>ast). See **FOIL**.

Examples

In the following examples, multiply the polynomials, combine like terms (see **Combining Like Terms**), and then leave the answers in descending order of the exponents of the variables. See **Polynomial**. Also see **Polynomials, sums and differences**.

1. Multiply $6x(-3 + 2x^2)$.

$$6x\left(-3+2x^2\right) = -18x + 12x^3 \qquad \text{Distributive property.}$$
$$= 12x^3 - 18x$$

2. Multiply $(3x + 2)(4x - 7)$.

$$(3x + 2)(4x - 7) = 12x^2 - 21x + 8x - 14 \qquad \text{By FOIL.}$$
$$= 12x^2 - 13x - 14$$

3. Multiply $6x(x - 3)(4x - 7)$.

$$6x(x - 3)(4x - 7) = \left(6x^2 - 18x\right)(4x - 7) \qquad \text{Distributive property.}$$
$$= 24x^3 - 42x^2 - 72x^2 + 126x \qquad \text{FOIL.}$$
$$= 24x^3 - 114x^2 + 126x$$

An alternate method to this problem is to multiply $(x - 3)$ and $(4x - 7)$, combine like terms, and then multiply $6x$ into that result.

4. Multiply $(x - 3)(x^2 + 2x + 1)$.

First x multiplies each term in $(x^2 + 2x + 1)$ and then -3 multiplies each term in $(x^2 + 2x + 1)$.

$$(x - 3)\left(x^2 + 2x + 1\right) = x^3 + 2x^2 + x - 3x^2 - 6x - 3$$
$$= x^3 - x^2 - 5x - 3$$

5. Simplify $(3a^2 - 2a - 5)^2$.

$$\left(3a^2 - 2a - 5\right)^2 = \left(3a^2 - 2a - 5\right)\left(3a^2 - 2a - 5\right)$$
$$= 9a^4 - 6a^3 - 15a^2 - 6a^3 + 4a^2 + 10a - 15a^2 + 10a + 25$$
$$= 9a^4 - 12a^3 - 26a^2 + 20a + 25$$

6. Multiply $(y - 3)(y + 4)(y - 6)$.

$$(y - 3)(y + 4)(y - 6) = \left(y^2 + 4y - 3y - 12\right)(y - 6) \qquad \text{Multiply } (y - 3)(y + 4).$$
$$= \left(y^2 + y - 12\right)(y - 6)$$
$$= y^3 - 6y^2 + y^2 - 6y - 12y + 72$$
$$= y^3 - 5y^2 - 18y + 72$$

➤ **Polynomials, sums and differences**   To find the **sum of two or more polynomials**, combine like terms and then leave the answer in descending order of the exponents of the variable. To find the **difference of two or more polynomials**, multiply any negative terms into its adjacent polynomial, combine like terms, and then leave the answer in descending order of the exponents of the variable. This process applies the distributive property. See **Distributive Property**.

Examples

Leave all answers in descending order of the exponents of the variable.

1. Simplify $(4x^2 - 3x + 2) + (5x^2 - 2x + 6)$.

$$\left(4x^2 - 3x + 2\right) + \left(5x^2 - 2x + 6\right) = 4x^2 + 5x^2 - 3x - 2x + 2 + 6 = 9x^2 - 5x + 8$$

2. Simplify $(x^2 - 3x) - 2x(-8x^2 + 5)$.

$$\left(x^2 - 3x\right) - 2x\left(-8x^2 + 5\right) = x^2 - 3x + 16x^3 - 10x \qquad \text{Distributive property.}$$
$$= 16x^3 + x^2 - 13x$$

3. Subtract $m^3 - 6m^2$ from $6m^4 + m^3 + 5m - 3$.

   To subtract B from A means that we write $A - B$.

   To subtract $m^3 - 6m^2$ from $6m^4 + m^3 + 5m - 3$ means that we write

$$\left(6m^4 + m^3 + 5m - 3\right) - \left(m^3 - 6m^2\right) = 6m^4 + m^3 + 5m - 3 - m^3 + 6m^2$$
$$= 6m^4 + 6m^2 + 5m - 3$$

4. Subtract $-3a^3 + 4a^2 - 3$ from $5a^3 + 2$.

$$\left(5a^3 + 2\right) - \left(-3a^3 + 4a^2 - 3\right) = 5a^3 + 2 + 3a^3 - 4a^2 + 3$$
$$= 8a^3 - 4a^2 + 5$$

➤ **Positive and Negative Numbers**   There are two sets of rules for **positive and negative numbers**. One set of rules is used for combining numbers (adding and subtracting) and the other set is used for multiplying and dividing numbers. The rules for multiplying and dividing are also used to deal with double signs. Although the rules in this article are developed using integers, they apply to all algebraic calculations. See the list at the end of the last set of examples.

Rules for Combining Numbers

(Memorize Them)

I.   If the signs are the same, add and keep the same sign.

II. If the signs are different, subtract and keep the sign of the larger number.

Examples:

$$\begin{aligned} +8 + 2 &= +10 \\ -8 - 2 &= -10 \end{aligned} \Big\} \quad \text{Rule I}$$

$$\begin{aligned} +8 - 2 &= +6 \\ -8 + 2 &= -6 \end{aligned} \Big\} \quad \text{Rule II}$$

Each of these examples can be thought of in terms of money:

$+8 + 2 = +10$ If you have 8 dollars and then someone gives you 2 dollars more, then you have 10 dollars.

$-8 - 2 = -10$ If you owe 8 dollars and then you owe 2 dollars more, then you owe a total of 10 dollars.

$+8 - 2 = +6$ If you have 8 dollars and then give 2 dollars to a friend, then you have 6 dollars left.

$-8 + 2 = -6$ If you owe 8 dollars and pay off 2 dollars, then you still owe 6 dollars.

The rules for combining numbers must be memorized. Then, whenever there is a doubt concerning how to proceed with a problem, reciting the rule will describe exactly what to do.

Examples

1. Combine: $+9 + 2$.

   Since the signs are the same, $+$ and $+$, we add and keep the same sign, $+9 + 2 = +11$.

2. Combine: $-9 - 2$.

   Since the signs are the same, $-$ and $-$, we add and keep the same sign, $-9 - 2 = -11$.

3. Combine: $+9 - 2$.

   Since the signs are different, $+$ and $-$, we subtract and keep the sign of the larger number, $+9 - 2 = +7$.

4. Combine: $-9 + 2$.

   Since the signs are different, $-$ and $+$, we subtract and keep the sign of the larger number, $-9 + 2 = -7$.

Rules for Multiplication and Division of Numbers

(Memorize Them)

A. If the signs are alike, the answer is positive.

B. If the signs are not alike, the answer is negative.

Multiplication:

$$\left.\begin{array}{l} (+3)(+5) = +15 \\ (-3)(-5) = +15 \end{array}\right\} \quad \text{Rule A}$$

$$\left.\begin{array}{l} (+3)(-5) = -15 \\ (-3)(+5) = -15 \end{array}\right\} \quad \text{Rule B}$$

Division:

$$\left\{ \frac{+10}{+2} = +5, \quad \frac{-10}{-2} = +5 \right.$$

$$\left\{ \frac{+10}{-2} = -5, \quad \frac{-10}{+2} = -5 \right.$$

P

The rules for multiplication and division of numbers do not have simple, commonsense, explanations. Whereas the rules for combining numbers can be thought of in terms of money or temperature, the rules for multiplication and division are either postulated (accepted without proof) or developed through proof. As a result, it is best to just memorize the rules. Their understanding can come later.

The rules for multiplication and division of numbers must be memorized. Then whenever there is a doubt concerning how to proceed with a problem, reciting the rule will describe exactly what to do.

Examples

1. Multiply $+6$ and $+2$.

   Since the signs are alike, $+$ and $+$, the answer is positive.

   $$(+6)(+2) = +12$$

2. Multiply -6 and -2.

   Since the signs are alike, $-$ and $-$, the answer is positive.

   $$(-6)(-2) = +12$$

3. Multiply $+6$ and -2.

   Since the signs are not alike, $+$ and $-$, the answer is negative.

   $$(+6)(-2) = -12$$

4. Multiply -6 and $+2$.

   Since the signs are not alike, $-$ and $+$, the answer is negative.

   $$(-6)(+2) = -12$$

5. Divide $+12$ by $+3$.

   $$\frac{+12}{+3} = +4 \qquad \text{alike, } +.$$

6. Divide -12 by -3.

   $$\frac{-12}{-3} = +4 \qquad \text{alike, } +.$$

7. Divide $+12$ by -3.

   $$\frac{+2}{-3} = -4 \qquad \text{not alike, } -.$$

8. Divide -12 by $+3$.

   $$\frac{-12}{+3} = -4 \qquad \text{not alike, } -.$$

Of course, the most common mistakes made when multiplying or dividing numbers is to confuse the rules for multiplying and dividing with those for combining numbers. These mistakes are completely avoidable if the rules for each are clearly memorized.

Double Sign

Sometimes double signs occur in calculations. They also are taught in some classes. If they are causing problems, they are easy to deal with. Just get rid of them by applying the rules for multiplication and division of numbers.

$$\left. \begin{array}{l} +{}^+6 = +6 \\ -{}^-6 = +6 \end{array} \right\} \quad \text{Rule A} \quad \text{alike, } +$$

$$\left. \begin{array}{l} +{}^-6 = -6 \\ -{}^+6 = -6 \end{array} \right\} \quad \text{Rule B} \quad \text{not alike, } -$$

Examples

1. $12 + {}^+4 = 12 + 4 = 16$ In each case, we get rid of the double signs by applying the rules for multiplication and division of numbers.

2. $12 - {}^-4 = 12 + 4 = 16$

3. $12 + {}^-4 = 12 - 4 = 8$

4. $12 - {}^+4 = 12 - 4 = 8$

Before continuing, the rules for combining numbers and the rules for multiplication and division of numbers must be clear. Also, the method for dealing with double signs must be understood.

Examples

In the following examples, a number without a sign in front of it is considered to be positive, i.e., 8 is $+8$.

1. Simplify $8 - 2 + 6 - 3$.

   If we simplify from left to right, we have

   $$8 - 2 + 6 - 3 = 6 + 6 - 3 = 12 - 3 = 9$$

   Some people prefer to combine the $+$ terms, combine the $-$ terms, and then combine those:

   $$8 - 2 + 6 - 3 = 14 - 5 = 9$$

2. Simplify $-9 + 4 - 12 - 7 + 10$.

From left to right	Combining $+$ and $-$ terms
$-9 + 4 - 12 - 7 + 10$	$-9 + 4 - 12 - 7 + 10$
$= -5 - 12 - 7 + 10$	$= 14 - 28 = -14$
$= -17 - 7 + 10$	
$= -24 + 10 = -14$	

P

3. Simplify $(-6)(-2) - 10 + (3)(-4)$.

$(-6)(-2) - 10 + (3)(-4)$
$= 12 - 10 + {}^-12$　　　Combine $12 - 10$ and eliminate the double
$= 2 - 12 = -10$　　　signs, $+{}^-12$ is -12 (not alike, $-$).

4. Simplify $\frac{-12}{-4} - \frac{6}{-2} - (-2)(-5)$.

$$\frac{-12}{-4} - \frac{6}{-2} - (-2)(-5)$$
$$= 3 - {}^-3 - {}^+10$$
$$= 3 + 3 - 10$$
$$= 6 - 10 = -4$$

$\frac{6}{-2}$ is -3.
Eliminate double signs.

For an application to algebra, see **Combining Like Terms**; **Polynomials, sums and differences**; or **Polynomials, multiplication of**.

➤ **Positive Numbers**　　　See **Positive and Negative Numbers**.

➤ **Postulate**　　　A **postulate**, or **axiom**, is a statement that is accepted without proof. It is assumed to be true and used to develop other concepts, or theorems.

In the past, there used to be a distinction between a postulate and axiom. Today, they are mostly used synonymously. A distinction between the terms is only important to a scrutinizing view of mathematics. For more, see **Property** or **Proof in Geometry**.

➤ **Power**　　　The term **power** is used synonymously with the term **exponent**. It is a number (or letter) that is written as a superscript to a base. For more, see **Exponent**.

➤ **Prime**　　　In calculus, the symbol $'$ is used to denote the derivative of a function. If $y = f(x)$, then the derivative is written as y' (read "y **prime**") or $f'(x)$ (read "f prime of x").

In geometry, the symbol $'$ is used to delineate one point from another. Whereas one point might be called A and another point called B, the two points could have been called P and P' (read "P prime"), respectively.

➤ **Prime Factorization**　　　The **prime factorization** of a number is the result yielded when a number has been completely factored into primes (prime numbers).

Examples

1. Find the prime factorization of 24.

First, we write 24 as the product of any two of its factors:

$$24 = 4 \bullet 6$$

Next, we write each of the factors as a product of any two of their factors. We continue this process until only prime numbers remain:

$$24 = 4 \bullet 6$$
$$= 2 \bullet 2 \bullet 2 \bullet 3 \qquad \text{Since } 4 = 2 \bullet 2 \text{ and } 6 = 2 \bullet 3.$$
$$= 2^3 \bullet 3 \qquad \qquad \text{See \textbf{Exponent}.}$$

Suppose, instead, we had first factored 24 as $24 = 3 \bullet 8$. Then

$$24 = 3 \bullet 8$$
$$= 3 \bullet 2 \bullet 4 \qquad \text{Since } 8 = 2 \bullet 4.$$
$$= 3 \bullet 2 \bullet 2 \bullet 2 \qquad \text{Since } 4 = 2 \bullet 2.$$
$$= 2^3 \bullet 3$$

2. Find the prime factorization of 60.

$$60 = 2 \bullet 30 \qquad \text{or} \qquad 60 = 4 \bullet 15$$
$$= 2 \bullet 2 \bullet 15 \qquad \qquad \qquad = 2 \bullet 2 \bullet 3 \bullet 5$$
$$= 2 \bullet 2 \bullet 3 \bullet 5 \qquad \qquad \quad = 2^2 \bullet 3 \bullet 5$$
$$= 2^2 \bullet 3 \bullet 5$$

➤ **Prime Numbers**   A **prime number** is a counting number that is divisible only by itself and 1. The first 10 prime numbers are

$$2,\ 3,\ 5,\ 7,\ 11,\ 13,\ 17,\ 19,\ 23,\ 29$$

Numbers other than 0 or 1 that can be written as a product of two or more smaller whole numbers are called **composite numbers**. For more, see **Composite Number** or **Prime Factorization**.

➤ **Principal Axis**   The **principal axis** is one of the axes (plural of axis) considered in conic sections. See **Parabola**, **Ellipse**, or **Hyperbola**.

➤ **Principal nth Root**   The **principal nth root** of a radical is the positive root of the radical. The square root of 25 is 5 and we write $\sqrt{25} = 5$. We do not consider the $\sqrt{25}$ to be ±5 (read "plus or minus 5"). This is considered in equations only. See **Roots, nth**.
    Similarly, $\sqrt[4]{16} = 2$, not ±2, and $\sqrt[6]{64} = 2$, not ±2. This is the case for even values of the radical index. Odd values of the radical index do not need to be considered since their roots do not involve the plus or minus situation. Also see **Radical**.

➤ **Prism**   To find the **lateral area** of a right prism (the outside area, not including the areas of the top and bottom), we use the formula

$$L = Ph$$

where P is the perimeter of the base and h is the height or altitude.
    For an example, see **Area of the Lateral Surface of a Right Prism (and total area)**.

h = height
base

To find the **total area** of a right prism (lateral area and area of the top and bottom), we use the formula

$$\text{T.A.} = Ph + 2B$$

where Ph is the lateral area and $2B$ is the combined areas of the top and bottom. To find B, if B is a regular polygon (all sides are equal), see **Area of a Regular Polygon**. If B is not a regular polygon, divide it into smaller polygons of which their respective areas can be calculated, then combine their areas. For an example of finding the total area, see **Area of the Lateral Surface of a Right Prism (and total area)**.

To find the volume of a right prism, we use the formula

$$V = Bh$$

where B is the area of the base.

➤ **Probability**   In relation to favorable outcomes and total possibilities, **probability** is the ratio of the two.

$$\text{Probability} = \frac{\text{Favorable outcomes}}{\text{Total possibilities}}$$

From the fraction, it can be seen that if there are no favorable outcomes, then the probability is 0. If all of the total possibilities are favorable outcomes, then the probability is 1.

Consider a bag containing 15 marbles. If 5 of the marbles are green, then the probability of selecting a green marble is five out of fifteen. As a ratio this is $\frac{5}{15} = \frac{1}{3}$. Hence, we see that there is an equivalence between 15 possibilities with 5 favorable outcomes and 3 possibilities with 1 favorable outcome. In either case, the probability is the same. Notice that $\frac{1}{3}$ is between 0 and 1, which are the bounds of a probability.

➤ **Product**   In multiplication, the **product** is the answer after the numbers have been multiplied. The number with that we multiply is called the **multiplier**. The number that we multiply is called the **multiplicand**. The result is the **product**.

$$\begin{array}{r} \text{multiplicand} \\ \times \text{ multiplier} \\ \hline \text{product} \end{array}$$ or   (multiplicand)(multiplier) = product
Parentheses next to each other indicate multiplication.

The multiplicand and the multiplier are the **factors** of the product.

➤ **Product Theorem of Zero**   The **product theorem of zero** is used to solve equations. In general, it states that if A and B are real numbers, then $A \bullet B = 0$ if and only if $A = 0$ or $B = 0$, or both A and B equal 0.

This follows from common sense. How can two numbers multiply each other to yield 0? If $A \bullet B = 0$, then one of them would have to be 0. Otherwise, $A \bullet B$ would yield something other than 0. The only way $A \bullet B$ can equal 0 is if either A is 0, B is 0, or both A and B are 0.

Suppose $(x - 4)(x + 3) = 0$. The only way the product can yield 0 is if $x - 4$ is 0 or $x + 3$ is 0, or both are 0. If $x - 4 = 0$, then $x = 4$ is the value of x that will yield 0. If $x + 3 = 0$, then $x = -3$ is the value of x that will yield 0. As can be seen, both $x = 4$ and $x = -3$ are the values of x that will yield a 0. Hence, they are the solutions to the equation $(x - 4)(x + 3) = 0$.

The product theorem of zero can be used to solve equations that are of second degree or higher. The only qualification is that the equation is factorable. To apply the theorem, the equation must be set equal to zero, factored, each factor set equal to zero, and then each factor solved for the variable if possible.

Example

1. Solve $x^2 = x + 12$ for x.

    To solve the equation, we need to set the equation equal to zero, factor the equation, set each factor equal to zero, and then solve each factor.

    $$x^2 = x + 12$$
    $$x^2 - x - 12 = 0 \qquad \text{Subtract } x \text{ and 12 from both sides.}$$
    $$(x - 4)(x + 3) = 0 \qquad \text{Factor the trinomial.}$$
    $$x - 4 = 0 \quad \text{or} \quad x + 3 = 0 \qquad \text{Set the factors equal to zero.}$$
    $$x = 4 \qquad\qquad x = -3 \qquad \text{Solve each factor.}$$

    The values $x = 4$ and $x = -3$ are the solutions to the equation. Check them in the original equation; they will balance.

➤ **Product to Sum or Sum to Product Identities**   In trigonometry, it is sometimes useful to write a product of two trigonometric functions as a sum, or to write a sum of two trigonometric functions as a product.

Product to Sum

$$\sin\theta \sin\phi = \frac{1}{2}\left[\cos(\theta - \phi) - \cos(\theta + \phi)\right]$$

$$\cos\theta \cos\phi = \frac{1}{2}\left[\cos(\theta - \phi) + \cos(\theta + \phi)\right]$$

$$\sin\theta \cos\phi = \frac{1}{2}\left[\sin(\theta - \phi) + \sin(\theta + \phi)\right]$$

$$\cos\theta \sin\phi = \frac{1}{2}\left[\sin(\theta - \phi) - \sin(\theta + \phi)\right]$$

Sum to Product

$$\sin\theta + \sin\phi = 2\sin\left(\frac{\theta + \phi}{2}\right)\cos\left(\frac{\theta - \phi}{2}\right)$$

$$\sin\theta - \sin\phi = 2\cos\left(\frac{\theta + \phi}{2}\right)\sin\left(\frac{\theta - \phi}{2}\right)$$

$$\cos\theta + \cos\phi = 2\cos\left(\frac{\theta + \phi}{2}\right)\cos\left(\frac{\theta - \phi}{2}\right)$$

$$\cos\theta - \cos\phi = -2\sin\left(\frac{\theta + \phi}{2}\right)\sin\left(\frac{\theta - \phi}{2}\right)$$

P

The product to sum identities can be derived by adding the corresponding members of the sum and difference identities. See **Sum and Difference Formulas** or **Trigonometry, identities**.

$$\sin\theta\cos\phi + \cos\theta\sin\phi = \sin(\theta + \phi)$$
$$\underline{\sin\theta\cos\phi - \cos\theta\sin\phi = \sin(\theta - \phi)}$$

$$2\sin\theta\cos\phi + 0 = \sin(\theta + \phi) + \sin(\theta - \phi) \qquad \text{Combine the two equations.}$$

$$\frac{2\sin\theta\cos\phi}{2} = \frac{\sin(\theta + \phi) + \sin(\theta - \phi)}{2}$$

$$\sin\theta\cos\phi = \frac{1}{2}\left[\sin(\theta + \phi) + \sin(\theta - \phi)\right]$$

$$\sin\theta\cos\phi = \frac{1}{2}\left[\sin(\theta - \phi) + \sin(\theta + \phi)\right] \qquad \text{Commutative property.}$$

This is the third product to sum identity. Similarly, we can derive the remaining three identities. To derive the sum to product identities, we let $A = \theta + \phi$ and $B = \theta - \phi$. Combining $A + B$ and $A - B$, solving for θ and ϕ, respectively, and then substituting into the third product to sum identity, we have

$$A + B = \theta + \phi + \theta - \phi \qquad\qquad A - B = \theta + \phi - (\theta - \phi)$$
$$A + B = 2\theta \qquad\qquad\qquad\qquad A - B = \theta + \phi - \theta + \phi$$
$$\theta = \frac{A + B}{2} \qquad\qquad\qquad\qquad A - B = 2\phi$$
$$\phi = \frac{A - B}{2}$$

$$\sin\theta\cos\phi = \frac{1}{2}\left[\sin(\theta - \phi) + \sin(\theta + \phi)\right]$$

$$\sin\left(\frac{A + B}{2}\right)\cos\left(\frac{A - B}{2}\right) = \frac{1}{2}\left[\sin\left(\frac{A + B}{2} - \frac{A - B}{2}\right) + \sin\left(\frac{A + B}{2} + \frac{A - B}{2}\right)\right]$$

$$\sin\left(\frac{A + B}{2}\right)\cos\left(\frac{A - B}{2}\right) = \frac{1}{2}\left[\sin\left(\frac{A + B - A + B}{2}\right) + \sin\left(\frac{A + B + A - B}{2}\right)\right]$$

$$\sin\left(\frac{A + B}{2}\right)\cos\left(\frac{A - B}{2}\right) = \frac{1}{2}(\sin B + \sin A) \qquad \text{Now eliminate } \frac{1}{2} \text{ by multiplying both sides}$$
$$\text{by 2.}$$

$$2\sin\left(\frac{A + B}{2}\right)\cos\left(\frac{A - B}{2}\right) = \sin B + \sin A$$

$$\sin A + \sin B = 2\sin\left(\frac{A + B}{2}\right)\cos\left(\frac{A - B}{2}\right)$$

This is the first sum to product identity. In a similar fashion the remaining three identities can be derived.

Examples

1. Express $\sin 45° \cos 30°$ as a sum.

   Letting $\theta = 45°$, $\phi = 30°$ and substituting into the third product to sum identity, we have

$$\sin\theta\cos\phi = \frac{1}{2}\big(\sin(\theta-\phi)+\sin(\theta+\phi)\big)$$

$$\sin 45°\cos 30° = \frac{1}{2}\big(\sin(45°-30°)+\sin(45°+30°)\big)$$

$$\sin 45°\cos 30° = \frac{1}{2}(\sin 15° + \sin 75°)$$

If the difference of the angles, $\theta-\phi$, yields a negative angle, simplify the term by applying the concepts of even and odd functions. See **Trigonometry, even and odd function (functions or identities of negative angles)**.

To verify the solution, we can look up the value of each function in a chart or calculator and then simplify:

$$\sin 45°\cos 30° = \frac{1}{2}(\sin 15° + \sin 75°)$$

$$(.7071)(.8660) = \frac{1}{2}(.2588 + .9659)$$

$$.6123 = \frac{1}{2}(1.2247)$$

$$.6123 = .6123$$

2. Express $\sin 15° + \sin 75°$ as a product.

Letting $\theta = 15°$, $\phi = 75°$ and substituting into the first sum to product identity, we have

$$\sin\theta + \sin\phi = 2\sin\left(\frac{\theta+\phi}{2}\right)\cos\left(\frac{\theta-\phi}{2}\right)$$

$$\sin 15° + \sin 75° = 2\sin\left(\frac{15°+75°}{2}\right)\cos\left(\frac{15°-75°}{2}\right)$$

$$\sin 15° + \sin 75° = 2\sin\left(\frac{90°}{2}\right)\cos\left(\frac{-60°}{2}\right)$$

$$\sin 15° + \sin 75° = 2\sin 45°\cos(-30°)$$

$$\sin 15° + \sin 75° = 2\sin 45°\cos 30° \qquad \text{Since } \cos(-30°) = \cos 30°.$$
$$\text{See \textbf{Trigonometry, even and odd function}}$$
$$\text{\textbf{(functions of identities of negative angles)}}$$

We can verify our answer by multiplying through the equation by 2. This gives us the same equation we had in example 1.

▶ **Proof in Geometry**   **Proof** is a process used to show that a particular statement follows logically from other accepted statements. Once the statement is proved, it becomes a **theorem** and can be used to prove other statements. In this manner, all the assertions made are a logical consequence of definitions, postulates (or axioms or properties), or previously proved theorems.

In this article, we will concern ourselves with **proof in geometry**. The demonstration can be carried out either in paragraph form or in two-column deductive form. We will discuss two-column deductive proof. This type of proof allows for statements to be made in one column and supporting reasons listed in the other.

Recall that a *conditional* statement is a statement that can be written in *if-then* form. See **Conditional Statement**. The *if-part* of the statement is called the *hypothesis* and the *then-part* is called the *conclusion*. In a two-column deductive proof, the hypothesis is called the Given and the conclusion is called the Prove. *Our goal is to start with the Given and then develop assertions that will lead to the conclusion stated in the Prove.*

Obviously, it's the development of assertions that causes problems. The other area of difficulty is knowing the reason for each assertion. This is easy to take care of, memorize all essential definitions, postulates, and theorems.

Memorization is all important to being successful with geometric proof. It not only allows for an understanding of the reason behind each assertion but, more importantly, allows for the development of the assertions. No matter what text you are using, memorize all pertinent information.

Remember that even though many of the theorems proved might appear mundane, it is the knowledge of proof that is important. In time, the theorems become more meaningful. When this stage is reached, a great appreciation for deductive proof, as well as mathematics in general, unfolds. Even more, if you stick with it, clarity of the universe. Ask Einstein.

Examples

1. Given: $m\angle A = 28$; $m\angle B = 62$.

Prove: Angles A and B are complementary.

To do this proof, the definition of complementary angles must be known: two angles whose measures have a sum of 90.

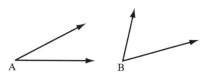

Statements	Reasons
1. $m\angle A = 28$; $m\angle B = 62$	1. Given.
2. $m\angle A + m\angle B = 90$	2. Addition property (If $a = b$ and $c = d$, then $a + c = b + d$).
3. Angles A and B are complementary.	3. Definition of complementary angles.

In order to prove angles A and B are complementary, we first have to show the angles have a sum of 90. We know we have to first show this because the definition of complementary angles says so.

This is what gives us the idea of adding the numbers in the Given. But how do we add numbers? We must first know the addition property. It states that if $a = b$ and $c = d$ then $a + c = b + d$. We can't do the problem if we don't know this property. Obviously, memorization is important.

2. Given: $AB = CD$.

Prove: $AC = BD$.

To do this proof, the definition of betweenness of points must be known: Y is between X and Z if X, Y, and Z are distinct collinear points and $XY + YZ = XZ$.

$$\underset{X}{\bullet} \rule{2cm}{0.4pt} \underset{Y}{\bullet} \underset{Z}{\bullet}$$

Statements	Reasons
1. $AB = CD$	1. Given.
2. $AC + CD = AD$	2. Definition of betweenness of points.
$AB + BD = AD$	
3. $AC + CD = AB + BD$	3. Substitution of 2a into 2b, since they both equal AD.
4. $AC + CD = CD + BD$	4. Substitution of given into 3.
5. $AC + CD = BD + CD$	5. Commutative property of addition.
6. $AC = BD$	6. Addition property (actually the converse of the addition property: If $a + c = b + c$, then $a = c$).

To a beginning student, it is usually not clear how to start this type of problem. As a suggestion, just choose any two sums and see if they can be hooked together. As an example: can $AB + BC = AC$ be hooked together with $BC + CD = BD$? Since we are given that $AB = CD$, we need to solve each equation for AB and CD, respectively, then set them equal to each other and solve. The result is $AC = BD$.

Statements	Reasons
1. $AB = CD$	1. Given.
2. $AB + BC = AC$	2. Definition of betweenness of points.
$BC + CD = BD$	
3. $AB = AC - BC$	3. Subtraction property (subtract BC from both sides of 2a and 2b).
$CD = BD - BC$	
4. $AC - BC = BD - BC$	4. Substitute 3 into 1.
5. $AC = BD$	5. Converse of the addition property (If $a + (-c) = b + (-c)$, then $a = b$).

3. Given: $m\angle AOB = m\angle COD$.

   Prove: $m\angle AOC = m\angle BOD$.

   Notice the similarity between this example and example 2.

   To do this proof, the definition of betweenness of rays must be known: $\overrightarrow{AY}$ is between $\overrightarrow{AX}$ and $\overrightarrow{AZ}$ if $\overrightarrow{AX}$, $\overrightarrow{AY}$, and $\overrightarrow{AZ}$ are coplanar and $m\angle XAY + m\angle YAZ = m\angle XAZ$.

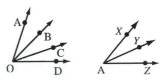

Statements	Reasons
1. $m\angle AOB = m\angle COD$	1. Given.
2. $m\angle AOB + m\angle BOC = m\angle AOC$ $m\angle BOC + m\angle COD = m\angle BOD$	2. Definition of betweenness of rays.
3. $m\angle AOB = m\angle AOC - m\angle BOC$ $m\angle COD = m\angle BOD - m\angle BOC$	3. Subtraction property (subtract $m\angle BOC$ from both sides of 2a and 2b).
4. $m\angle AOC - m\angle BOC =$ $m\angle BOD - m\angle BOC$	4. Substitute 3 into 1.
5. $m\angle AOC = m\angle BOD$	5. Converse of the addition property (If $a + (-c) = b + (-c)$, then $a = b$).

4. Given: $\angle 1 \cong \angle 3$.

   Prove: $\angle 1 \cong \angle 2$.

   To do this proof, the vertical angle theorem
   must be known: vertical angles are congruent.

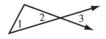

Statements	Reasons
1. $\angle 1 \cong \angle 3$	1. Given.
2. $\angle 3 \cong \angle 2$	2. Vertical angle theorem.
3. $\angle 1 \cong \angle 2$	3. Transitive property of congruence (If $\angle a \cong \angle b$ and $\angle b \cong \angle c$, then $\angle a \cong \angle c$).

Notice that step 2 was written $\angle 3 \cong \angle 2$ and not $\angle 2 \cong \angle 3$. Both are correct but $\angle 3 \cong \angle 2$ matches up with step 1. The transitive property could not work for $\angle 2 \cong \angle 3$.

5. Given: $\overline{AC} \cong \overline{BC}$.

   Prove: $\angle 3 \cong \angle 1$.

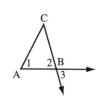

Statements	Reasons
1. $\overline{AC} \cong \overline{BC}$	1. Given.
2. $\angle 1 \cong \angle 2$	2. If two sides of a triangle are congruent, then the angles opposite are congruent.
3. $\angle 2 \cong \angle 3$	3. Vertical angle theorem. See example 4.
4. $\angle 1 \cong \angle 3$	4. Transitive property of congruence (If $\angle a \cong \angle b$ and $\angle b \cong \angle c$, then $\angle a \cong \angle c$).
5. $\angle 3 \cong \angle 1$	5. Symmetric property of congruence (If $\angle a \cong \angle b$, then $\angle b \cong \angle a$).

Notice the similarity between examples 4 and 5.

6. If two sides of a quadrilateral are congruent and parallel, the quadrilateral is a parallelogram.

We first need to write the "if" part of the statement as the Given and the "then" part as the Prove.

Given: $\overline{AB} \cong \overline{CD}$, $\overline{AB} \parallel \overline{CD}$.

Prove: Quadrilateral $ABCD$ is a parallelogram.

Statements	Reasons
1. $\overline{AB} \cong \overline{CD}$, $\overline{AB} \parallel \overline{CD}$	1. Given.
2. Draw $\overline{AC}$	2. Between every two points there is a unique line.
3. $\overline{AC} \cong \overline{CA}$	3. Reflexive property of congruence ($\angle a \cong \angle a$, correspondence is ACD with CAB).
4. $\angle 3 \cong \angle 4$	4. If two parallel lines are cut by a transversal, then the alternate interior angles are congruent.
5. $\triangle ACD \cong \triangle CAB$	5. SAS (If the side ($\overline{AB}$), angle ($\angle 3$) and another side ($\overline{AC}$) of one triangle are congruent to the side (CD), angle ($\angle 4$) and another side ($\overline{CA}$) of a second triangle, then the triangles are congruent).
6. $\overline{AD} \cong \overline{CB}$	6. Corresponding parts of congruent triangles are congruent (abbreviated CPCTC).
7. Quad $ABCD$ is a $\square$	7. If both pairs of opposite sides of a quadrilateral are congruent, then the quadrilateral is a parallelogram. (We were given $\overline{AB} \cong \overline{CD}$ and we proved $\overline{AD} \cong \overline{CB}$.)

► **Proper Fraction**   A **proper fraction** is a fraction whose numerator is less than the denominator. As a result, proper fractions are always less than one. The following are examples of proper fractions:

$$\frac{1}{2}, \frac{3}{4}, \frac{99}{100}, \text{ and } \frac{1}{1,000}$$

► **Properties of the Real Number System**   The **properties of the real number system** are based on the representation of the real numbers on the real number line. See **Real Number** or **Real Number Line**. Once the real numbers have been placed into a one-to-one correspondence with the points on a line, a property of order is established.

Property (or Axiom) of Comparison

If a and b are real numbers, then either

$$a < b, a = b, \text{ or } a > b$$

If three or more points on a number line are considered, we have a further comparison of order.

Transitive Property (or Axiom) of Order

If a, b, and c are real numbers, then either of the following hold:

1. If $a < b$ and $b < c$, then $a < c$.

2. If $a > b$ and $b > c$, then $a > c$.

The pair of inequalities $a < b$ and $b < c$ are quite often combined into the triple inequality $a < b < c$. This implies that b is between a and c.

When an order has been established, the sum and product of two real numbers can be defined. The **sum** of a and b is the result of adding a and b and is written $a + b$. Each of the numbers a and b is called a **term**. The **product** of a and b is the result of multiplying a and b and is written $a \times b = a \bullet b = a(b) = (a)b = (a)(b) = ab$. Each of the numbers a and b is called a **factor**.

When two real numbers are either added or multiplied, the result is a real number. This is stated in the next property, The Property of Closure.

Property (or Axiom) of Closure for Addition and Multiplication

If a and b are real numbers, then their sum or their product are each equal to a unique (one and only one) real number.

The equality of the real numbers a and b is determined by properties as well.

Properties (or Axioms) of Equality

Let a, b, and c be real numbers.

1. Reflexive property of equality: $a = a$.

2. Symmetric property of equality: If $a = b$, then $b = a$.

3. Transitive property of equality: If $a = b$ and $b = c$, then $a = c$.

Sometimes *substitution* is included as a property, although it can be proved as a theorem.

Let a, b, c, and d be real numbers. Then, for addition, the substitution property states that if $a = b$ and $a + c = d$, then $b + c = d$. For multiplication, it states that if $a = b$ and $ac = d$, then $bc = d$.

Addition and multiplication of real numbers follow distinct properties. Including closure, we have the following:

Properties (or Axioms) of the Real Numbers

Let a, b, and c denote real numbers.

1. Closure property of addition and multiplication: For all numbers $a, b \in R$, $a + b \in R$ and $ab \in R$. This simply means that the sum and product of two real numbers are also real numbers.

2. Commutative property of addition and multiplication: $a + b = b + a$ and $ab = ba$.

3. Associative property of addition and multiplication: $a + (b + c) = (a + b) + c$ and $a(bc) = (ab)c$.

4. Additive identity: there exists a unique real number 0 such $a + 0 = a$.

5. Multiplicative identity: there exists a unique real number 1 such that $a \bullet 1 = a$.

6. Additive inverse: for each real number a, there exists a real number $-a$ such that $a + (-a) = 0$.

7. Multiplicative inverse: for each real number a, $a \neq 0$, there exists a real number $\frac{1}{a}$ such that $a\frac{1}{a} = 1$.

8. Distributive property: $a(b+c) = ab + ac$, called "left distributive" and $(b+c)a = ab + ac$, called "right distributive."

Subtraction is defined in reference to a sum. If a and b are real numbers, then $a - b = a + (-b)$. Division is defined in reference to multiplication. If a and b are real numbers, then $a \div b = \frac{a}{b} = a \bullet \frac{1}{b}$. These lead to the following properties:

Properties (Theorems) of Equations

Let a, b, and c denote real numbers.

1. Addition property: if $a = b$, then $a + c = b + c$.

2. Subtraction property: if $a = b$, then $a - c = b - c$.

3. Multiplication property: if $a = b$, then $a \bullet c = b \bullet c$.

4. Division property: if $a = b$, then $\frac{a}{c} = \frac{b}{c}$, where $c \neq 0$.

In addition to the properties of equations, we have the following:

Properties (Theorems) of Absolute Value

Let a and b denote real numbers.

1. $|a| \geqslant 0$.

2. $a \leqslant |a|$ and $-a \leqslant |a|$.

3. $|-a| = |a|$.

4. $|ab| = |a| \, |b|$.

5. $\left| \dfrac{a}{b} \right| = \dfrac{|a|}{|b|}$ with $b \neq 0$.

6. $|a + b| \leqslant |a| + |b|$, called the **triangle inequality**.

► **Property**   A **property** is a statement. Some properties are accepted without proof, as an axiom. Yet other properties are proven, as a theorem. A property is usually assumed, or sometimes proven, to be true (logical) and is used to develop other concepts called laws, propositions, or theorems. These, in turn, are used to develop even more complicated theorems. In this manner, the most intricate theorems are derived from a small list of properties. For a list of real number properties, see **Properties of the Real Number System**. For properties related to matrices, see **Matrix**.

➤ **Proportion**   If a, b, c, and d are real numbers, then the equation $\frac{a}{b} = \frac{c}{d}$, where $b \neq 0$ and $d \neq 0$, is called a **proportion**. If $\frac{a}{b} = \frac{c}{d}$, then by cross multiplication $ad = bc$.

Written as a ratio, the equation $\frac{a}{b} = \frac{c}{d}$ is $a \div b = c \div d$, read "a is to b as c is to d." See **Ratio**. The terms a and d are called the **extremes** and the terms b and c are called the **means**. Hence, in a proportion, the product of the extremes is equal to the product of the means.

The proportion $\frac{a}{b} = \frac{c}{d}$ can be solved for any one of the letters a, b, c, or d. For examples see **Cross Multiplication**. For an application of proportion, see **Direct Variation**, **Combined Variation**, **Inverse Variation**, or **Joint Variation**. Also see **Similar Triangles**.

➤ **Proposition**   A **proposition** is a theorem. It is a statement that can be proved to be logically sound by other theorems or postulates. See **Theorem**. Also see **Proof in Geometry**.

➤ **Pure Imaginary Number**   In a complex number $a + bi$, the bi term is called a **pure imaginary number**. See **Complex Numbers** or **Complex Plane**.

➤ **Pyramid**   To find the **lateral area** of a regular pyramid (sides of the base are equal and all edges are equal), we use the formula

$$L = \frac{1}{2} s P$$

where s is the slant height and P is the perimeter of the base.

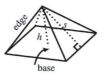

For an example, see **Area of the Lateral Surface of a Regular Pyramid (and total area)**.

To find the **total area** of a regular pyramid (lateral area and the area of the base), we use the formula

$$\text{T.A.} = \frac{1}{2} s P + B$$

where $\frac{1}{2} s P$ is the lateral area and B is the area of the base. For the area of the base, see **Area of a Regular Polygon**. For an example, see **Area of the Lateral Surface of a Regular Pyramid (and total area)**.

To find the **volume** of a regular pyramid, we use the formula

$$V = \frac{1}{3} B h$$

where B is the area of the base (see **Area of a Regular Polygon**) and h is the altitude or height of the pyramid.

> **Pythagorean Identities**   In trigonometry, the identities that are derived from the Pythagorean theorem are known as the **Pythagorean identities**.

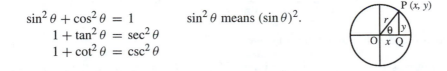

$$\sin^2 \theta + \cos^2 \theta = 1 \qquad \sin^2 \theta \text{ means } (\sin \theta)^2.$$
$$1 + \tan^2 \theta = \sec^2 \theta$$
$$1 + \cot^2 \theta = \csc^2 \theta$$

In $\triangle OPQ$ consider the sine and cosine, where $\sin \theta = \frac{y}{r}$ and $\cos \theta = \frac{x}{r}$. Solving for y and x, we have $y = r \sin \theta$ and $x = r \cos \theta$. Substituting x and y into the Pythagorean theorem, we get the first of the three Pythagorean identities:

$$x^2 + y^2 = r^2 \qquad \text{See \textbf{Pythagorean Theorem}.}$$
$$(r \cos \theta)^2 + (r \sin \theta)^2 = r^2$$
$$r^2 \cos^2 \theta + r^2 \sin^2 \theta = r^2 \qquad \text{Since } (\cos \theta)^2 \text{ is } \cos^2 \theta.$$
$$\frac{r^2 \cos^2 \theta}{r^2} + \frac{r^2 \sin^2 \theta}{r^2} = \frac{r^2}{r^2}$$
$$\sin^2 \theta + \cos^2 \theta = 1$$

To derive the other two identities, divide through the first identity by $\cos^2 \theta$ and then $\sin^2 \theta$, respectively:

$$\sin^2 \theta + \cos^2 \theta = 1 \qquad\qquad \sin^2 \theta + \cos^2 \theta = 1$$
$$\frac{\sin^2 \theta}{\cos^2 \theta} + \frac{\cos^2 \theta}{\cos^2 \theta} = \frac{1}{\cos^2 \theta} \qquad\qquad \frac{\sin^2 \theta}{\sin^2 \theta} + \frac{\cos^2 \theta}{\sin^2 \theta} = \frac{1}{\sin^2 \theta}$$
$$\tan^2 \theta + 1 = \sec^2 \theta \qquad\qquad 1 + \cot^2 \theta = \csc^2 \theta$$
$$1 + \tan^2 \theta = \sec^2 \theta$$

The identities for $\frac{\sin^2 \theta}{\cos^2 \theta} = \tan^2 \theta$, $\frac{\cos^2 \theta}{\sin^2 \theta} = \cot^2 \theta$, $\frac{1}{\cos^2 \theta} = \sec^2 \theta$, and $\frac{1}{\sin^2 \theta} = \csc^2 \theta$ are covered in the articles **Quotient Identities**, **Reciprocal Identities**, and **Trigonometry, identities**.

Examples

1. Write $y = \frac{\cos x}{\sin x - 1}$ in terms of $\sin x$.

   We need to write $\cos x$ in terms of $\sin x$. We can do this by solving the first Pythagorean identity for $\cos x$:

$$\sin x + \cos^2 x = 1$$
$$\cos^2 x = 1 - \sin^2 x$$
$$\sqrt{\cos^2 x} = \sqrt{1 - \sin^2 x}$$
$$\cos x = \sqrt{1 - \sin^2 x}$$

Substituting into the given equation, we have

$$y = \frac{\cos x}{\sin x - 1} = \frac{\sqrt{1 - \sin^2 x}}{\sin x - 1}$$ which is in terms of $\sin x$

2. Express $\sec^2 x + \tan^2 x$ in terms of $\cos x$.

Substituting $\sec^2 x = \frac{1}{\cos^2 x}$ and $\tan^2 x = \frac{\sin^2 x}{\cos^2 x}$ (see **Reciprocal Identities** or **Quotient Identities**), we have

$$\begin{aligned}
\sec^2 x + \tan^2 x &= \frac{1}{\cos^2 x} + \frac{\sin^2 x}{\cos^2 x} \\
&= \frac{1 + \sin^2 x}{\cos^2 x} \qquad \text{Common denominator.} \\
&= \frac{1 + (1 - \cos^2 x)}{\cos^2 x} \qquad \text{Substitute } \sin^2 x = 1 - \cos^2 x. \\
&= \frac{2 - \cos^2 x}{\cos^2 x} \qquad \text{which is in terms of } \cos x.
\end{aligned}$$

The substitution $\sin^2 x = 1 - \cos^2 x$ is derived by solving the first Pythagorean identity for $\sin^2 x$:

$$\begin{aligned}
\sin^2 x + \cos^2 x &= 1 \\
\sin^2 x &= 1 - \cos^2 x
\end{aligned}$$

3. Prove the identity $\sin^2 x - \cos^2 x = 2 \sin^2 x - 1$.

We need to prove that one side of the identity is equivalent to the other (see **Trigonometry, identities**). Working on the left side, we can substitute $\cos^2 x = 1 - \sin^2 x$ from the first Pythagorean identity:

$$\begin{aligned}
\sin^2 x - \cos^2 x &= 2 \sin^2 x - 1 \\
\sin^2 x - (1 - \sin^2 x) &= 2 \sin^2 x - 1 \qquad \text{Substitute } \cos^2 x = 1 - \sin^2 x. \\
\sin^2 x - 1 + \sin^2 x &= 2 \sin^2 x - 1 \\
2 \sin^2 x - 1 &= 2 \sin^2 x - 1
\end{aligned}$$

➤ **Pythagorean Theorem**   The **Pythagorean theorem** states that in any right angle triangle, the square of the length of the hypotenuse is equal to the sum of the squares of the lengths of the other two sides (legs). In $\triangle ABC$, let a, b, and c represent the lengths of sides $\overline{BC}$, $\overline{AC}$, and $\overline{AB}$, respectively. Then, by the Pythagorean theorem, we have

$$c^2 = a^2 + b^2$$

Since a^2 means $a \bullet a$, the a^2 term can be thought of as representing the area of the square on the side $\overline{BC}$. Similarly, b^2 and c^2 represent the area of the squares on sides $\overline{AC}$ and $\overline{AB}$, respectively. Therefore, the Pythagorean theorem can also be thought of as stating that "in any right triangle, the area of the square on the hypotenuse is equal to the sum of the areas of the squares on the legs."

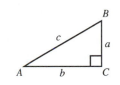

The **converse** to the Pythagorean theorem is also a useful theorem. It states that "if the square of the length of one side of a triangle is equal to the sum of the squares of the lengths of the other two sides, then the triangle is a right triangle." See **Converse**.

One obvious consequence of the converse is that it can be used to prove a triangle is a right triangle. Another way to do this is to use slopes. See **Perpendicular Lines, slopes of**.

For application of the Pythagorean theorem in the plane, see **Distance Formula**.

Examples

1. In right $\triangle ABC$, if $a = 6$ and $b = 8$, find c.

   Substituting into the formula, we have

   $$c^2 = a^2 + b^2$$
   $$c^2 = 6^2 + 8^2$$
   $$c^2 = 36 + 64$$
   $$c^2 = 100$$
   $$\sqrt{c^2} = \sqrt{100} \qquad \text{See \textbf{Radicals, equations with}.}$$
   $$c = 10$$

2. In right $\triangle ABC$, if $b = 10$ and $c = 6\sqrt{5}$, find a. Substituting into the formula, we have

   $$c^2 = a^2 + b^2$$
   $$\left(6\sqrt{5}\right)^2 = a^2 + 10^2$$
   $$36 \bullet 5 = a^2 + 100$$
   $$180 = a^2 + 100$$
   $$180 - 100 = a^2$$
   $$80 = a^2$$
   $$\sqrt{a^2} = \sqrt{80} \qquad \text{Symmetric property and } \sqrt{} \text{ both sides.}$$
   $$a = \sqrt{16 \bullet 5} \qquad \text{Reduce the square root.}$$
   $$a = 4\sqrt{5}$$

3. Prove that the triangle with sides 3, 4, and 5 is a right triangle.

   If the triangle, is a right triangle, then it should satisfy the equation $c^2 = a^2 + b^2$. Substituting, we have

$$c^2 = a^2 + b^2$$
$$5^2 = 3^2 + 4^2 \qquad \text{Substituting the longest side in for } c.$$
$$25 = 9 + 16$$
$$25 = 25$$

Since the equation remains balanced, the triangle with sides 3, 4, and 5 is a right triangle.

P

Qq

➤ Quadrant

In the coordinate plane, the x- and y-axes (plural of axis) separate the plane into four sections. Each section is called a **quadrant**. The quadrants are numbered counterclockwise: quadrant I, quadrant II, quadrant III, and quadrant IV. For more, see **Coordinate Plane**.

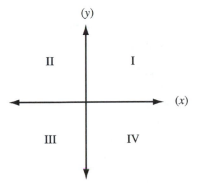

➤ Quadratic Equation

A **quadratic equation** is an equation of the form $ax^2 + bx + c = 0$, where a, b, and c are real numbers with $a \neq 0$. If a were 0, the equation would be $bx + c = 0$, which is a linear equation, i.e., $mx + b = 0$.

To solve a quadratic equation, see **Quadratic Equations, solutions of**. To graph a quadratic equation, see **Quadratic Equations, graphs of**.

➤ Quadratic Equations, graphs of

The **graph of a quadratic equation** is a parabola. To see this, we can graph $y = x^2$ by the charting method. Choosing x values from -2 to 2, we have

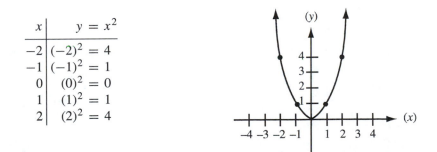

x	$y = x^2$
-2	$(-2)^2 = 4$
-1	$(-1)^2 = 1$
0	$(0)^2 = 0$
1	$(1)^2 = 1$
2	$(2)^2 = 4$

This is the graph of the quadratic equation $y = a^2x + bx + c$ where $a = 1$, $b = 0$, and $c = 0$. If $a > 0$ (positive), the graph opens upward. If $a < 0$ (negative), the graph opens downward.

Graphing by the charting method can be cumbersome, especially if the graph has been moved away from the origin. To graph equations of this type, we can apply the **axis of symmetry method** or the **completing the square method**.

Axis of Symmetry Method

1. Calculate the coordinates of the vertex.

   a. x-value of the vertex.

   To find the x-value of the vertex, we use the equation of the axis of symmetry, $x = -\frac{b}{2a}$.

   b. y-value of the vertex.

   To find the y-value of the vertex, we substitute the x-value of the axis of symmetry into the original equation for y.

2. Calculate more points of the graph.

   a. x-intercept.

   We can find two more points of the graph by setting the original equation equal to 0 and solving for x. We solve the equation either by factoring or by the quadratic formula. See **Quadratic Equations, solutions of** or **Quadratic Formula**.

   If the equation doesn't have a solution, i.e., if $b^2 - 4ac < 0$ (is negative), then apply step b. See **Discriminant**.

   b. Find two or more function values.

   To find more points by using the function, we choose x-values to the left and right of the vertex, calculate those function values, and plot the points. See **Function**.

3. Plot the vertex and more points, then sketch the curve of the graph.

The axis of symmetry, $x = \frac{-b}{2a}$, is derived from the equation of a parabola $y = a(x - h)^2 + k$, where the vertex is (h, k) and the axis of symmetry is $x = h$. See **Parabola**. If we complete the square on $y = ax^2 + bx + c$ and compare with $y = a(x - h)^2 + k$, we see that the axis of symmetry is $x = \frac{-b}{2a}$. See **Completing the Square**. This is the x-value of the vertex. The y-value of the vertex is $c - \frac{b^2}{4a}$:

$$y = ax^2 + bx + c = a\left(x^2 + \frac{b}{a}x\right) + c$$

$$= a\left(x^2 + \frac{b}{a}x + \frac{b^2}{4a^2} - \frac{b^2}{4a^2}\right) + c \qquad \frac{b}{a} \div 2 = \frac{b}{2a}, \text{ squared is } \frac{b^2}{4a^2}, \text{ then add and}$$

$$= a\left(x^2 + \frac{b}{a}x + \frac{b^2}{4a^2}\right) + c - \frac{ab^2}{4a^2} \qquad \text{subtract } \frac{b^2}{4a^2}. \ a\left(-\frac{b^2}{4a^2}\right) = \frac{-ab^2}{4a^2}$$

$$= a\left(x + \frac{b}{2a}x\right) + c - \frac{b^2}{4a}$$

Comparing this last equation with $y = a(x - h)^2 + k$, we see that the vertex (h, k) is

$$v(h, k) = \left(-\frac{b}{2a}, c - \frac{b^2}{4a}\right)$$

where $h = x = -\frac{b}{2a}$ is the axis of symmetry.

Completing the Square Method

When a quadratic equation is solved by completing the square, the coordinates of the vertex and x-intercepts are present in the solution. The vertex is under the $\sqrt{}$ signs and any real solutions (meaning real numbers) to the equation are the x-intercepts.

Vertex

If we set the equation $y = x^2 - 6x + 5$ equal to 0 and solve by completing the square, the square root portion of the solution is $\sqrt{(x-3)^2} = \pm\sqrt{4}$. From this, we have the vertex (h, k) which is $(3, -4)$. The coordinates are opposite in sign from the numbers under the square root signs.

If the equation to be graphed has a coefficient on the x^2 term, i.e., $2x^2$, the y-value of the vertex must be multiplied by this value. This is a direct result of the division that takes place early on when completing the square. To complete the square, the coefficient of the x^2 term is divided through the equation. When determining the y-value of the vertex, the number under its square root must be multiplied by this value. Nothing is done with the x-value.

x-intercepts

The solutions to the equation $x^2 - 6x + 5 = 0$ are $x = 1$ and $x = 5$. From these, we have the x-intercepts $(1, 0)$ and $(5, 0)$. The real solutions of a quadratic equation are the x-intercepts.

An alternate form that is sometimes used is to complete the square on $y = ax^2 + bx + c$ to yield the form

$$y = a(x - h)^2 + k$$

In this form, the vertex is (h, k) and the axis of symmetry is $x = h$. To get the x-intercepts, set the equation equal to 0 and solve for x. See example 2.

Examples

1. Graph the equation $y = 2x^2 - 12x + 10$ by the axis of symmetry method.

   The x-value of the vertex is

   $$x = \frac{-b}{2a} = \frac{-(-12)}{2(2)} = 3$$

   The y-value of the vertex is

   $$y = 2x^2 - 12x + 10 = 2(3)^2 - 12(3) + 10 \qquad \text{Substitute } x = 3.$$
   $$= 18 - 36 + 10 = -8$$

Hence, we have the coordinates of the vertex $(h, k) = (3, -8)$.

We can get the x-intercepts by setting the original equation equal to 0 and solving for x:

$$y = 2x^2 - 12x + 10$$
$$2x^2 - 12x + 10 = 0$$
$$2(x^2 - 6x + 5) = 0$$
$$2(x - 1)(x - 5) = 0$$
$$x - 1 = 0 \quad \text{or} \quad x - 5 = 0$$
$$x = 1 \qquad\qquad x = 5$$

Hence, the x-intercepts are $(1, 0)$ and $(5, 0)$.

Using the coordinates $(3, -8)$, $(1, 0)$, and $(5, 0)$, we can now sketch the graph of $y = 2x^2 - 12x + 10$.

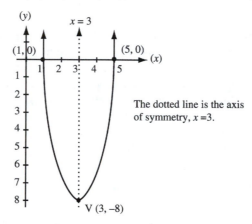

The dotted line is the axis of symmetry, $x = 3$.

2. Graph $y = 2x^2 - 12x + 10$ by completing the square.

We set the equation equal to 0 and solve by completing the square. See **Completing the Square**.

$$2x^2 - 12x + 10 = 0$$
$$2x^2 - 12x = -10 \qquad \text{Add 10 to both sides.}$$
$$x^2 - 6x = -5 \qquad \text{Divide through the equation by 2.}$$
$$x^2 - 6x + 9 = -5 + 9 \qquad 6 \div 2 = 3, 3^2 = 9, \text{ add 9 to both sides.}$$
$$(x - 3)(x - 3) = 4 \qquad \text{Factor.}$$
$$(x - 3)^2 = 4 \qquad \text{Write as the square of a binomial.}$$
$$\sqrt{(x - 3)^2} = \pm\sqrt{4} \qquad \text{Take the square root of both sides.}$$
$$x - 3 = \pm 2$$
$$x = 3 \pm 2$$
$$x = 3 + 2 \quad \text{or} \quad x = 3 - 2$$
$$x = 5 \qquad\qquad x = 1$$

The vertex is under the $\sqrt{}$ signs. The x-value is 3 and the y-value is $-4 \bullet 2 = -8$. We multiply by the "2" of $2x^2$. The x-intercepts are $(5, 0)$ and $(1, 0)$. The sketch of the graph is the same as example 1.

An alternate method of graphing the equation is to rewrite it in the form $y = a(x - h)^2 + k$:

$$y = 2x^2 - 12x + 10$$
$$= 2(x^2 - 6x) + 10 \qquad \text{Factor the coefficient of } x^2 \text{ (the 2) from } 2x^2 \text{ and } -12x.$$
$$= 2(x^2 - 6x + 9 - 9) + 10 \qquad (6 \div 2)^2 = 9, \text{ add and substract 9.}$$
$$= 2(x^2 - 6x + 9) - 18 + 10 \qquad \text{Multiply } -9 \text{ by 2 but leave the } +9 \text{ in the parentheses.}$$
$$= 2(x - 3)(x - 3) - 8$$
$$= 2(x - 3)^2 - 8$$

Written in the form $y = a(x - h)^2 + k$, we have

$$y = 2(x - 3)^2 + (-8)$$

where the vertex (h, k) is $(3, -8)$ and the axis of symmetry $x = h$ is $x = 3$. To find the x-intercepts, set the equation equal to 0 and solve for x. To find the y-intercepts, substitute 0 for the x-values and solve for y. Or, find more points of the graph by choosing x-values to the left and right of the vertex, then calculating their corresponding y-values.

3. Find the vertex and x-intercepts for $y = 3x^2 - 12x - 36$ by the axis of symmetry method and by completing the square.

Axis of symmetry method

x-value of the vertex:

$$x = -\frac{b}{2a} = -\frac{-12}{2(3)} = 2$$

y-value of the vertex:

$$y = 3(2)^2 - 12(2) - 36 = -48$$

Vertex is $(2, -48)$.

x-intercepts:
Set equation $= 0$ and solve:

$$3x^2 - 12x - 36 = 0$$
$$3(x^2 - 4x - 12) = 0$$
$$3(x + 2)(x - 6) = 0$$

$$x = -2 \qquad \text{or} \qquad x = 6$$

The x-intercepts are $(-2, 0)$ and $(6, 0)$.

Completing the square method

$$3x^2 - 12x - 36 = 0$$
$$3x^2 - 12x = 36$$
$$x^2 - 4x = 12$$
$$x^2 - 4x + 4 = 12 + 4$$
$$(x - 2)(x - 2) = 16$$
$$(x - 2)^2 = 16$$
$$\sqrt{(x - 2)^2} = \pm\sqrt{16} \qquad \text{Multiply 16 by 3 to get}$$
$$x - 2 = \pm 4 \qquad\qquad \text{a vertex of } (2, -48).$$
$$x = 2 \pm 4$$

$$x = 2 + 4 \qquad \text{or} \qquad x = 2 - 4$$
$$x = 6 \qquad\qquad\qquad x = -2$$

The x-intercepts are $(-2, 0)$ and $(6, 0)$.

4. Graph $y = x^2 - 4x + 7$.

By the axis of symmetry method, the x-value of the vertex is

$$x = \frac{-b}{2a} = \frac{-(-4)}{2(1)} = 2$$

and the y-value of the vertex is

$$y = 2^2 - 4(2) + 7 = 3$$

Hence, the vertex is $(2, 3)$.

Setting the equation equal to 0, we see that it will not factor and does not have a real solution when solved by the quadratic formula. This happens because $b^2 - 4ac < 0$. The solutions are complex. All of this implies that the graph does not cross the x-axis. There are no x-intercepts.

To find more points of the graph, we need to choose x-values to the left and right of the vertex, substitute those into the original equation to find their corresponding y-values, then plot the points.

An x-value to the left of $(2, 3)$, i.e., to the left of 2, is 0. An x-value to the right of $(2, 3)$ is 4. Their corresponding y-values are

for 0, $y = 0^2 - 4(0) + 7 = 7$, which gives the point $(0, 7)$.
for 4, $y = 4^2 - 4(4) + 7 = 7$, which gives the point $(4, 7)$.

Hence, with the vertex $(2, 3)$ and two other points, $(0, 7)$ and $(4, 7)$, we can sketch the graph.

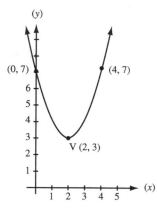

For more, see **Parabola**.

➤ **Quadratic Equations, solutions of**   Quadratic equations are solved by factoring, by the quadratic formula, or by completing the square. For the latter two, see **Quadratic Formula** or **Completing the Square**. Some of the equations that can be solved by factoring can also be solved using square roots. See example 3.

To solve a quadratic equation by factoring, we apply the following steps:

1. Set the equation equal to 0 by adding or subtracting terms from both sides of the equation.

2. Factor the polynomial, usually a binomial or a trinomial. See **Factoring the Difference of Two Squares**, **Factoring Trinomials**, or **FOIL**.

3. Set each factor equal to 0.

4. Solve each factor.

The process works because of the following convention or property:

If $A \bullet B = 0$, then either $A = 0$, $B = 0$, or both A and B equal 0. In order for two terms to multiply each other and yield 0, one of them must be 0, or possibly both. See **Product Theorem of Zero**.

In the solution of a quadratic equation, the A and B are usually binomials. As an example, let $A = x + 5$ and $B = x - 3$. Then

$$A \bullet B = 0$$
$$\text{or} \quad (x + 5)(x - 3) = 0 \quad \text{By substituting for } A \text{ and } B.$$

Since we have a product that equals 0, either the $x + 5$ is 0 or the $x - 3$ is 0, or both. Treating both of them as 0, since we don't know which might be, we have

$$x + 5 = 0 \quad \text{or} \quad x - 3 = 0$$
$$x = -5 \quad\quad\quad x = 3$$

Actually both of these are the answer. A quadratic equation can have no answers, one answer, or two answers depending on whether the graph of the equation does not pass through the x-axis, touches the x-axis at one point, or goes through the x-axis at two points, respectively. Since we have two answers, $x = -5$ and $x = 3$, the graph of the equation passes through these points on the x-axis. See **Quadratic Equations, graphs of**.

If we multiply $x + 5$ and $x - 3$, we get the original quadratic equation:

$$(x + 5)(x - 3) = x^2 - 3x + 5x - 15 = x^2 + 2x - 15. \quad \text{See **FOIL**.}$$

The answers, $x = -5$ and $x = 3$, are the solutions to the equation $x^2 + 2x - 15 = 0$.

In order to solve the equation $x^2 + 2x - 15 = 0$, we must first factor the equation. See **Factoring Trinomials**. Then we set each factor equal to 0 and solve for x. This is one of the reasons why factoring is so important.

Examples

1. Solve the equation $x^2 = -2x + 15$ by factoring.

   We first must set the equation equal to 0. We do this by adding $2x$ to both sides and subtracting 15 from both sides. Then we factor the trinomial, set each factor equal to 0, and solve each factor:

$$x^2 = -2x + 15$$
$$x^2 + 2x - 15 = 0 \quad\quad\quad \text{Set } = 0.$$
$$(x + 5)(x - 3) = 0 \quad\quad\quad \text{Factor. See **Factoring Trinomials**.}$$
$$x + 5 = 0 \quad \text{or} \quad x - 3 = 0 \quad\quad \text{Solve each factor.}$$
$$x = -5 \quad\quad\quad x = 3$$

There are two solutions to the equation, $x = -5$ and $x = 3$. To see if the answers are correct, we can substitute them into the original equation and simplify. If both sides remain balanced, the solution is correct.

Check for $x = -5$ $\qquad$ Check for $x = 3$

$$x^2 = -2x + 15 \qquad\qquad x^2 = -2x + 15$$
$$(-5)^2 = -2(-5) + 15 \qquad (3)^2 = -2(3) + 15$$
$$25 = 10 + 15 \qquad\qquad 9 = -6 + 15$$
$$25 = 25 \qquad\qquad\qquad 9 = 9$$

Hence, both answers are correct.

2. Solve $3x^2 = 4 + 11x$ by factoring.

$$3x^2 = 4 + 11x$$
$$3x^2 - 11x - 4 = 0$$
$$(3x + 1)(x - 4) = 0 \qquad \text{See \textbf{Factoring Trinomials}.}$$

$$3x + 1 = 0 \qquad \text{or} \qquad x - 4 = 0$$
$$3x = -1 \qquad\qquad\qquad x = 4$$
$$x = -\frac{1}{3}$$

The solutions are $x = -\frac{1}{3}$ and $x = 4$.

3. Solve $x^2 - 25 = 0$ by factoring.

$$x^2 - 25 = 0$$
$$(x + 5)(x - 5) = 0$$

$$x + 5 = 0 \qquad \text{or} \qquad x - 5 = 0$$
$$x = -5 \qquad\qquad\qquad x = 5$$

The solutions are $x = -5$ and $x = 5$.

Another way to solve this problem is to use square roots. We do this by isolating the x^2 term, taking the square root of both sides, and then solving for x:

$$x^2 - 25 = 0$$
$$x^2 = 25 \qquad \text{Isolate the } x^2 \text{ term.}$$
$$\sqrt{x^2} = \pm\sqrt{25} \qquad \text{Square root both sides.}$$
$$x = \pm 5$$

$$x = 5 \qquad \text{or} \qquad x = -5 \qquad \text{Solve for } x.$$

Since $\sqrt{x^2}$ is 25 for either $x = 5$ or $x = -5$, we use $\pm$ when solving the equation. It is placed in front of the $\sqrt{}$ sign when the square root of both sides is taken.

As another example, solve $4x^2 - 49 = 0$ using square roots.

$$4x^2 - 49 = 0$$
$$4x^2 = 49 \qquad \text{Isolate the } x^2 \text{ term.}$$

$$\sqrt{4x^2} = \pm\sqrt{49}$$
$$2x = \pm 7$$
$$x = \pm\frac{7}{2}$$

Square both sides. Remember to place $\pm$ in front of the $\sqrt{\ }$.
Now solve for x.

$$x = \frac{7}{2} \qquad \text{or} \qquad x = -\frac{7}{2}$$

4. Solve $4x^2 - 25 = 0$ by factoring.

$$4x^2 - 25 = 0$$
$$(2x + 5)(2x - 5) = 0$$

See **Factoring the Difference of Two Squares**.

$$2x + 5 = 0 \qquad \text{or} \qquad 2x - 5 = 0$$
$$2x = -5 \qquad\qquad 2x = 5$$
$$x = -\frac{5}{2} \qquad\qquad x = \frac{5}{2}$$

5. Solve $3x^2 = -6x + 45$.

$$3x^2 = -6x + 45$$
$$3x^2 + 6x - 45 = 0$$
$$3(x^2 + 2x - 15) = 0$$
$$3(x + 5)(x - 3) = 0$$

See **Factoring Common Monomials**.
Use the idea that if $3 \bullet A \bullet B = 0$, then either
$A = 0$ or $B = 0$. We do nothing with the 3.

$$x + 5 = 0 \qquad \text{or} \qquad x - 3 = 0$$
$$x = -5 \qquad\qquad x = 3$$

Notice the similarity between this problem and example 1. The only difference is the factor 3. Otherwise, the solutions of each are similar.

6. Solve $2x^4 + 50 = 52x^2$.

$$2x^4 + 50 = 52x^2$$
$$2x^4 - 52x^2 + 50 = 0$$
$$2(x^4 - 26x^2 + 25) = 0$$
$$2(x^2 - 1)(x^2 - 25) = 0$$

See **Factoring Trinomials**.

$$x^2 - 1 = 0 \qquad \text{or} \qquad x^2 - 25 = 0 \qquad \text{Each binomial has solutions.}$$
$$(x + 1)(x - 1) = 0 \qquad\qquad (x + 5)(x - 5) = 0$$

$$x + 1 = 0 \quad \text{or} \quad x - 1 = 0 \qquad x + 5 = 0 \quad \text{or} \quad x - 5 = 0$$
$$x = -1 \qquad\qquad x = 1 \qquad\qquad x = -5 \qquad\qquad x = 5$$

There are four answers to the equation because the graph of a fourth-degree polynomial can cross the x-axis, at most, four times.

7. Solve $-3x^2 = 10x - x^3$.

$$-3x^2 = 10x - x^3$$
$$x^3 - 3x^2 - 10x = 0$$
$$x(x^2 - 3x - 10) = 0 \qquad \text{Factor the common monomial } x.$$
$$x(x + 2)(x - 5) = 0 \qquad \text{Use the idea that if } A \bullet B \bullet C = 0, \text{ then either}$$
$$A = 0, B = 0, \text{ or } C = 0.$$

$$x = 0 \qquad \text{or} \qquad x + 2 = 0 \qquad \text{or} \qquad x - 5 = 0$$
$$x = -2 \qquad\qquad\qquad x = 5$$

There are three solutions to this equation, $x = 0$, $x = -2$, and $x = 5$. The graph crosses the x-axis at these points. A third-degree polynomial can have, at most, three solutions.

➤ **Quadratic Formula** The **quadratic formula** is used to solve quadratic equations of the form $ax^2 + bx + c = 0$. If the equation $ax^2 + bx + c = 0$ is solved for x by completing the square, the answer is the quadratic formula:

$$x = \frac{-b \pm \sqrt{b^2 - 4ac}}{2a}$$

In application, the quadratic formula is usually used to solve quadratic equations that do not factor easily. To apply the formula to this type of equation, substitute the a, b, and c from the given equation into the formula, then simplify.

To see the quadratic formula derived from $ax^2 + bx + c = 0$, see **Completing the Square**, example 5.

Examples

1. Solve $3x^2 = 7x - 2$ by using the quadratic formula.

The equation must first be set equal to 0. Then the a, b, and c can be substituted into the formula:

$$3x^2 = 7x - 2$$
$$3x^2 - 7x + 2 = 0$$

Substituting $a = 3$, $b = -7$, and $c = 2$ into the formula, we have

$$x = \frac{-b \pm \sqrt{b^2 - 4ac}}{2a} = \frac{-(-7) \pm \sqrt{(-7)^2 - 4(3)(2)}}{2(3)}$$
$$= \frac{7 \pm \sqrt{49 - 24}}{6} = \frac{7 \pm \sqrt{25}}{6} = \frac{7 \pm 5}{6}$$

$$x = \frac{7 + 5}{6} = \frac{12}{6} \qquad \text{or} \qquad x = \frac{7 - 5}{6} = \frac{2}{6}$$
$$x = 2 \qquad\qquad\qquad\qquad x = \frac{1}{3}$$

The solutions to $3x^2 = 7x - 2$ are $x = 2$ and $x = \frac{1}{3}$. To see if these are correct, substitute them into the original equation. If both sides remain balanced, the solutions are correct.

$$\text{Check for } x = 2 \qquad\qquad \text{Check for } x = \tfrac{1}{3}$$

$$3x^2 = 7x - 2 \qquad\qquad 3x^2 = 7x - 2$$

$$3(2)^2 = 7(2) - 2 \qquad\qquad 3\left(\frac{1}{3}\right)^2 = 7\left(\frac{1}{3}\right) - 2$$

$$25 = 10 + 15 \qquad\qquad \frac{3}{9} = \left(\frac{7}{3}\right) - 2$$

$$12 = 12 \qquad\qquad\qquad \frac{1}{3} = \frac{1}{3}$$

Since each is balanced, the solutions are correct.

2. Solve $8x = -1 - 4x^2$ by using the quadratic formula.

$$8x = -1 - 4x^2$$

$$4x^2 + 8x + 1 = 0$$

Substituting $a = 4$, $b = 8$, and $c = 1$ into the formula, we have

$$x = \frac{-8 \pm \sqrt{(8)^2 - 4(4)(1)}}{2(4)}$$

$$= \frac{-8 \pm \sqrt{64 - 16}}{8}$$

$$= \frac{-8 \pm \sqrt{48}}{8} \qquad \sqrt{48} = \sqrt{16 \bullet 3} = 4\sqrt{3}.$$

$$= \frac{-8 \pm 4\sqrt{3}}{8}$$

$$= \frac{4(-2 \pm \sqrt{3})}{8} \qquad \text{Factor out 4.}$$

$$= \frac{-2 \pm \sqrt{3}}{2}$$

The solutions are $x = \frac{-2+\sqrt{3}}{2}$ and $x = \frac{-2-\sqrt{3}}{2}$.

If the number inside the square root sign is negative, write the result as a complex number (only if required). For example, if the answer to an equation is $x = \frac{6 \pm \sqrt{-5}}{7}$, write the answer in the complex number form $x = \frac{6 \pm i\sqrt{5}}{7}$. See **Complex Numbers**.

➤ **Quadratic Inequalities**   See **Inequalities, quadratic solutions of** or **Polynomial Inequalities**.

➤ **Quadrilateral**   A **quadrilateral** is a four-sided polygon. If the quadrilateral is convex, the sum of the measures of the four angles is 360. Five quadrilaterals with special qualities are the parallelogram, rectangle, rhombus, square, and trapezoid. See articles under these names.

➤ **Quotient**   For quotients in algebra, see **Rational Expressions, multiplication and division of**.

In arithmetic, the **quotient** is the answer found when dividing the dividend by the divisor. In long division, the quotient is written above the dividend. See **Division of Whole Numbers**.

$$\begin{array}{r} \text{quotient} \qquad\qquad\qquad\qquad \\ \hline \text{divisor)} \quad \text{divident} \qquad\qquad\qquad \\ - \text{ product (of quotient and divisor)} \\ \hline \text{remainder (if any)} \end{array}$$

➤ **Quotient Identities**   The **quotient identities** are a pair of identities in trigonometry that relate the tangent and cotangent functions with the sine and cosine functions:

$$\tan A = \frac{\sin A}{\cos A} \qquad \text{and} \qquad \cot A = \frac{\cos A}{\sin A}$$

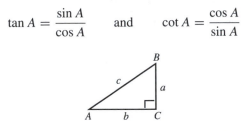

The **quotient identities** can be derived by looking at a right $\triangle ABC$. From the triangle, it can be seen that $\sin A = \frac{a}{c}$, $\cos A = \frac{b}{c}$, and $\tan A = \frac{a}{b}$. Solving the first two equations for a and b, we have $a = c \sin A$ and $b = c \cos A$. Substituting a and b into the third equation, we have

$$\tan A = \frac{a}{b} = \frac{c \sin A}{c \cos A} = \frac{\sin A}{\cos A}$$

Similarly, substituting a and b into $\cot A = \frac{a}{b}$, we have

$$\cot A = \frac{b}{a} = \frac{c \cos A}{c \sin A} = \frac{\cos A}{\sin A}$$

For an application of the quotient identities, see **Reciprocal Identities**. Also see **Trigonometry, identities**.

Q

R r

> **Radian**   An angle of rotation of a circle has a measure of 1 **radian** if it intercepts an arc with the same length as the radius.

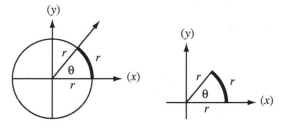

We say that the measure of the angle θ is 1 radian. As we will see shortly, 1 radian $= 57.3°$.

Another way to think of radians is to consider the circumference of a unit circle (a circle with a radius of 1). From the equation of the circumference of a circle, we have

$$C = 2\pi r = 2\pi(1) = 2\pi$$

The circumference around a unit circle is 2π. If we consider half the circumference, the distance is π, or half of π is $\frac{\pi}{2}$, the length of $\frac{1}{4}$ of the circumference.

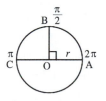

Now consider the angle AOB. Since $m\angle AOB = 90$, we see that the arc length $\frac{\pi}{2}$ and $90°$ describe the same angle, $\angle AOB$. Similarly, since the straight angle $\angle AOC$ has a measure of 180, we see that the arc length π and $180°$ describe the same angle, $\angle AOC$. Hence, we have two ways of describing the separation of the rays of an angle. We can say the rays are separated by degrees or by arc length. The separation by arc length is what is referred to as radians.

In this manner we can say that the rays $\overrightarrow{OA}$ and $\overrightarrow{OC}$ are separated by $180°$ or by the arc $\overset{\frown}{ABC}$. The arc $\overset{\frown}{ABC}$ has an arc length of π. We call this π radians and write

$$\pi^R = 180°$$

To know what 1 radian is, we can divide both sides of this equation by π:

$$\pi^R = 180°$$
$$\frac{\pi^R}{\pi} = \frac{180°}{\pi}$$
$$1^R = \frac{180°}{\pi}$$
$$1^R = \frac{180°}{3.14} \qquad \text{Substituting } \pi = 3.14.$$
$$1^R = 57.3°$$

To convert from one system to the other, see **Converting Radians to Degrees** or **Converting Degrees to Radians**.

➤ **Radian Conversion to Degrees**   See **Converting Radians to Degrees**.

➤ **Radical**   A **radical** indicates the root of a number and is written $\sqrt[n]{a}$, read "the nth root of a." The n is called the **radical index** and is a natural number greater than one. The a is called the **radicand** and is a real number with the following restrictions:

1. If n is even, then $a \geqslant 0$ (avoid negative roots).

2. If n is odd, then a is any real number.

The radical $\sqrt[n]{a}$ is the principal, or positive, nth root of a. The concept of $\pm$ (read "plus or minus") is developed from solutions of equations involving exponents and radicals. See **Roots, nth**.

When $n = 2$, the radical is called a square root. For a square root, the radical index is usually omitted. We write $\sqrt{5}$ instead of $\sqrt[2]{5}$.

The square root of a given number is a number that, when multiplied by itself, will yield the given number:

$$\sqrt{25} = 5 \qquad \text{because } 5 \bullet 5 = 25$$
$$\sqrt{49} = 7 \qquad \text{because } 7 \bullet 7 = 49$$
$$\sqrt{144} = 12 \qquad \text{because } 12 \bullet 12 = 144$$

When $n = 3$, the radical is called a **cube root**. The cube root of a given number is a number that, when multiplied by itself three times, will yield the given number:

$$\sqrt[3]{8} = 2 \qquad \text{because } 2 \bullet 2 \bullet 2 = 8$$
$$\sqrt[3]{27} = 3 \qquad \text{because } 3 \bullet 3 \bullet 3 = 27$$
$$\sqrt[3]{1000} = 10 \qquad \text{because } 10 \bullet 10 \bullet 10 = 1000$$

When $n = 4$ or higher, we refer to the root as the fourth root, the fifth root, etc. We use ordinal numbers.

$$\sqrt[4]{81} = 3 \qquad \text{because } 3 \bullet 3 \bullet 3 \bullet 3 = 81$$
$$\sqrt[5]{32} = 2 \qquad \text{because } 2 \bullet 2 \bullet 2 \bullet 2 \bullet 2 = 32$$

For more, see **Root, *n*th** or the various articles under **radical**.

➤ **Radical Equation**   See **Radicals, equations with**.

➤ **Radical Form**   An expression such as $\sqrt[3]{7}$ is said to be in **radical form** while the equivalent expression $7^{1/3}$ is said to be in fractional exponent form. In general, if *n* is a natural number with $n > 1$ and *x* is a real number, then

$$\sqrt[n]{x} = x^{1/n}$$

except when *n* is even and $x < 0$. See **Complex Numbers**.
    Also, in general,

$$\sqrt[n]{x^m} = \left(\sqrt[n]{x}\right)^m = x^{m/n}$$

where $\frac{m}{n}$ is a rational number in lowest terms, *n* is a natural number with $n > 1$, and *x* is a real number with $x \neq 0$. This last stipulation prevents division by 0. Division by 0 could occur if *m* is negative when *x* is 0.

Examples

1. Change $\sqrt{5}$, $\sqrt[3]{7}$, and $\sqrt[4]{x}$ from radical form to fractional exponents.

$$\sqrt{5} = 5^{1/2}, \qquad \sqrt[3]{7} = 7^{1/3}, \qquad \sqrt[4]{x} = x^{1/4}$$

   Notice that the denominator is the root (root index) of the radical.

2. Change $\sqrt[3]{x^2}$ and $\sqrt[5]{x^3}$ to fractional exponents.

$$\sqrt[3]{x^2} = x^{2/3}, \qquad \sqrt[5]{x^3} = x^{3/5}$$

   Notice that the numerator is the power (or exponent) of *x* and the denominator is the root (root index) of the radical.

3. Change $2^{1/5}$ and $x^{3/4}$ from fractional exponents to radical form.

$$2^{1/5} = \sqrt[5]{2}, \qquad x^{3/4} = \sqrt[4]{x^3}$$

4. Write $5x^{1/3}$ and $(5x)^{1/3}$ in radical form.

$$5x^{1/3} = 5\sqrt[3]{x}, \qquad (5x)^{1/3} = \sqrt[3]{5x}$$

   For more, see **Exponents, fractional or rational** or **Radicals, simplifying (reducing)**.

➤ **Radicals, combining**   Radicals are combined in the same manner as combining like terms. See **combining Like Terms**. Nothing should be done inside the radical sign when the terms are combined. Combine the numerical coefficients only.

R

Examples

1. Simplify $3\sqrt{7} + 8\sqrt{7}$.

$$3\sqrt{7} + 8\sqrt{7} = 11\sqrt{7}$$

Add the 3 and 8 only, not the 7 and the 7. Think of this as you would $3x + 8x$ combine to yield $11x$. You don't do anything with the x. Similarly, you don't do anything with the $\sqrt{7}$. Leave it alone. The following is the most common mistake:

Mistake: $3\sqrt{7} + 8\sqrt{7} = 11\sqrt{14}$.

2. Simplify $3\sqrt{5} + 8\sqrt{6} - 8\sqrt{5} - 4\sqrt{6}$.

$$\begin{aligned} 3\sqrt{5} + 8\sqrt{6} - 8\sqrt{5} - 4\sqrt{6} &= 3\sqrt{5} - 8\sqrt{5} + 8\sqrt{6} - 4\sqrt{6} &&\text{Regroup.} \\ &= -5\sqrt{5} + 4\sqrt{6} &&\text{Combine.} \end{aligned}$$

To regroup the like terms, the concepts of the commutative and associative properties are applied. See **Combining Like Terms**.

3. Simplify $3\sqrt{18} + 4\sqrt{50} - 2\sqrt{32} - \sqrt{8}$.

$$\begin{aligned} 3\sqrt{18} + 4\sqrt{50} - 2\sqrt{32} - \sqrt{8} &= 3\sqrt{9\bullet2} + 4\sqrt{25\bullet2} - 2\sqrt{16\bullet2} - \sqrt{4\bullet2} \\ &= 9\sqrt{2} + 20\sqrt{2} - 8\sqrt{2} - 2\sqrt{2} = 19\sqrt{2} \end{aligned}$$

To simplify the radicals, i.e., $\sqrt{18} = \sqrt{9\bullet2} = 3\sqrt{2}$, see **Radicals, simplifying (reducing)**.

4. Simplify $(\sqrt{6} - 2\sqrt{3})(\sqrt{6} - \sqrt{3})$.

$$\begin{aligned} (\sqrt{6} - 2\sqrt{3})(\sqrt{6} - \sqrt{3}) &= \sqrt{36} - \sqrt{18} - 2\sqrt{18} + 2\sqrt{9} &&\text{See **FOIL**.} \\ &= 6 - 3\sqrt{18} + 6 = 12 - 3\sqrt{18} \\ &= 12 - 3\sqrt{9\bullet2} = 12 - 9\sqrt{2} \end{aligned}$$

5. Simplify $6\sqrt[3]{3} + \sqrt[3]{24} + \sqrt[3]{81}$.

$$\begin{aligned} 6\sqrt[3]{3} + \sqrt[3]{24} + \sqrt[3]{81} &= 6\sqrt[3]{3} + \sqrt[3]{8\bullet3} + \sqrt[3]{27\bullet3} &&\text{See **Radicals, simplifying (reducing)**.} \\ &= 6\sqrt[3]{3} + 2\sqrt[3]{3} + 3\sqrt[3]{3} = 11\sqrt[3]{3} \end{aligned}$$

For more, see the articles under **Radical**.

➤ **Radicals, division with**   If x and y are real numbers with $y \neq 0$ and n is a natural number greater than 1, then

$$\sqrt[n]{\frac{x}{y}} = \frac{\sqrt[n]{x}}{\sqrt[n]{y}} \qquad \text{provided both sides of the equation are defined.}$$

The reason y must be chosen with $y \neq 0$ is so there will not be a division by 0. The radical index n must be greater than 1 to consider a square root or higher roots. Finally, both sides of the equation must be defined in order to avoid even roots of negative numbers. If $n = 4$ when $x = -16$ then $\sqrt[4]{-16}$ is not defined.

Examples

1. Simplify $\sqrt{\frac{125}{16}}$.

   For square roots, we don't show the radical index. We write $\sqrt{\ }$ instead of $\sqrt[2]{\ }$.

   $$\sqrt{\frac{125}{16}} = \frac{\sqrt{125}}{\sqrt{16}} = \frac{\sqrt{25 \bullet 5}}{4} = \frac{\sqrt{25}\sqrt{5}}{4} = \frac{5\sqrt{5}}{4}$$

   See **Radicals, simplifying (reducing)**.

2. Simplify $\frac{\sqrt{21}}{\sqrt{7}}$.

   Since $\sqrt{21}$ and $\sqrt{7}$ cannot each simplify, write them as one radical, then simplify.

   $$\frac{\sqrt{21}}{\sqrt{7}} = \sqrt{\frac{21}{7}} = \sqrt{3}$$

3. Simplify $\frac{\sqrt{45x^9}}{\sqrt{5x^3}}$ for $x > 0$.

   The reason x must be chosen with $x > 0$ is so there will not be a division by 0. Also, so we don't have to consider $\sqrt[n]{x^n} = |x|$ when n is an even number. For this, see **Radicals, simplifying (reducing)**.

   $$\frac{\sqrt{45x^9}}{\sqrt{5x^3}} = \sqrt{\frac{45x^9}{5x^3}} = \sqrt{9x^6} = \sqrt{9}\sqrt{x^6} = 3x^3$$

   To simplify $\frac{x^9}{x^3}$, we use the rule $\frac{a^m}{a^n} = a^{m-n}$:

   $$\frac{x^9}{x^3} = x^{9-3} = x^6$$

   See **Exponents, rules of**.

   To simplify $\sqrt{x^6}$, we use the rule $\sqrt{ab} = \sqrt{a}\sqrt{b}$:

   $$\sqrt{x^6} = \sqrt{x^2 \bullet x^2 \bullet x^2} = \sqrt{x^2} \bullet \sqrt{x^2} \bullet \sqrt{x^2} = x \bullet x \bullet x = x^3. \quad \text{See } \textbf{Radicals, rules for}.$$

   A shorter way to simplify $\sqrt{x^6}$ is to divide 6 by 2, where 2 represents the square root, i.e., $\sqrt[2]{x^6}$. Hence, $6 \div 2 = 3$, so $\sqrt[2]{x^6} = x^3$. See **Radicals, simplifying (reducing)**.

   Since $x > 0$, we do not need to consider the rule $\sqrt{x^2} = |x|$. For examples involving this case, see **Radicals, simplifying (reducing)**.

4. Simplify $\sqrt[3]{\frac{16x^{12}y^2}{z^9}}$.

   $$\sqrt[3]{\frac{16x^{12}y^2}{z^9}} = \frac{\sqrt[3]{16x^{12}y^2}}{\sqrt[3]{z^9}} = \frac{2x^4\sqrt[3]{2y^2}}{z^3}$$

R

Individually, we have the following roots:

$$\sqrt[3]{16} = \sqrt[3]{8 \bullet 2} = \sqrt[3]{8}\sqrt[3]{2} = 2\sqrt[3]{2}$$
$$\sqrt[3]{x^{12}} = \sqrt[3]{x^3 \bullet x^3 \bullet x^3 \bullet x^3} = \sqrt[3]{x^3}\sqrt[3]{x^3}\sqrt[3]{x^3}\sqrt[3]{x^3} = x \bullet x \bullet x \bullet x = x^4$$

Essentially, $12 \div 3 = 4$, so $\sqrt[3]{x^{12}} = x^4$. If the problem had been $\sqrt[3]{x^{14}}$ the remainder would have been left inside the radical, i.e., $\sqrt[3]{x^{14}} = x^4\sqrt[3]{x^2}$. See **Radicals, simplifying (reducing)**.

To rationalize radicals such as $\frac{3}{\sqrt{5}}$ and $\frac{1}{3+\sqrt{7}}$, see **Rationalizing the Denominator**.

➤ **Radicals, equations with**   Equations with a variable in the radicand (inside the radical sign) are called **radical equations**. The method for solving radical equations is to isolate the radical (get it by itself on one side of the $=$ sign), raise each side of the equation to a power equivalent to that of the radical index (square root is 2, cube root is 3, etc.), and then solve for the variable. If there is still a radical, repeat the process until all radicals with variables have been dealt with.

Since each side of the equation is raised to a power, there is a possibility that **extraneous solutions** (extra, usually incorrect solutions) have been introduced. As a result, each solution to the equation must be checked.

Examples

1. Solve $\sqrt{2x - 6} - 6 = 0$.

$$\sqrt{2x - 6} - 6 = 0$$
$$\sqrt{2x - 6} = 6 \qquad \text{Isolate } \sqrt{2x - 6} \text{ by adding 6 to both sides.}$$
$$\left(\sqrt{2x - 6}\right)^2 = (6)^2 \qquad \text{Square both sides since the radical index is 2.}$$
$$2x - 6 = 36 \qquad \text{Now, solve the equation for } x.$$
$$2x = 36 + 6$$
$$2x = 42$$
$$\frac{2x}{2} = \frac{42}{2}$$
$$x = 21$$

Check:
$$\sqrt{2x - 6} - 6 = 0$$
$$\sqrt{2(21) - 6} - 6 = 0$$
$$\sqrt{42 - 6} - 6 = 0$$
$$\sqrt{36} - 6 = 0$$
$$6 - 6 = 0 \qquad \text{Principal root only.}$$
$$0 = 0$$

Since $0 = 0$, $x = 21$ is a solution to the equation.

R

2. Solve $-2\sqrt{x+1} - x = -2$.

$$-2\sqrt{x+1} - x = -2$$
$$-2\sqrt{x+1} = x - 2 \qquad \text{Isolate the radical.}$$
$$\left(-2\sqrt{x+1}\right)^2 = (x-2)^2 \qquad \text{Square both sides.}$$
$$4(x+1) = (x-2)(x-2) \qquad \text{The } (\sqrt{x+1})^2 \text{ is } x+1.$$
$$4x + 4 = x^2 - 4x + 4 \qquad \text{Solve for } x. \text{ See } \textbf{Quadratic Equations,}$$
$$0 = x^2 - 8x \qquad \qquad \textbf{solutions of.}$$
$$0 = x(x-8)$$

$$x = 0 \qquad \text{or} \qquad x - 8 = 0$$
$$x = 8$$

Check:

For $x = 0$, we have

$$-2\sqrt{0+1} - 0 = -2$$
$$-2\sqrt{1} = -2$$
$$-2 = -2$$

For $x = 8$, we have

$$-2\sqrt{8+1} - 8 = -2$$
$$-2\sqrt{9} - 8 = -2$$
$$-6 - 8 = -2$$
$$-14 = -2$$

Since $-2 = -2$, $x = 0$ is a solution. Since -14 is not equal to -2, $x = 8$ is not a solution. It is an extraneous solution. See **Extraneous Solutions (roots)**.

3. Solve $\sqrt{3x - 2} + \sqrt{2x - 2} - 1 = 0$.

Since there are two radicals, we need to choose one to isolate. Either is fine. Let's isolate the $\sqrt{3x - 2}$.

$$\sqrt{3x - 2} + \sqrt{2x - 2} - 1 = 0$$
$$\sqrt{3x - 2} = 1 - \sqrt{2x - 2} \qquad \text{Isolate either radical.}$$
$$\left(\sqrt{3x - 2}\right)^2 = \left(1 - \sqrt{2x - 2}\right)^2 \qquad \text{Square both sides.}$$
$$3x - 2 = \left(1 - \sqrt{2x - 2}\right)\left(1 - \sqrt{2x - 2}\right) \qquad \text{Multiply by FOIL.}$$
$$3x - 2 = 1 - 2\sqrt{2x - 2} + \sqrt{(2x - 2)^2}$$
$$3x - 2 = 1 - 2\sqrt{2x - 2} + 2x - 2$$
$$2\sqrt{2x - 2} = -x + 1 \qquad \text{Isolate the radical.}$$
$$\left(2\sqrt{2x - 2}\right)^2 = (-x + 1)^2 \qquad \text{Square both sides.}$$
$$4(2x - 2) = (-x + 1)(-x + 1) \qquad \text{Multiply by FOIL.}$$
$$8x - 8 = x^2 - 2x + 1 \qquad \text{Solve for } x.$$
$$0 = x^2 - 10x + 9$$
$$0 = (x - 9)(x - 1)$$

$$x - 9 = 0 \qquad \text{or} \qquad x - 1 = 0$$
$$x = 9 \qquad \qquad \qquad x = 1$$

R

Check:

For $x = 9$, we have

$$\sqrt{3(9) - 2} + \sqrt{2(9) - 2} - 1 = 0$$
$$\sqrt{25} + \sqrt{16} - 1 = 0$$
$$5 + 4 - 1 = 0$$
$$8 = 0$$

For $x = 1$, we have

$$\sqrt{3(1) - 2} + \sqrt{2(1) - 2} - 1 = 0$$
$$\sqrt{1} + \sqrt{0} - 1 = 0$$
$$1 - 1 = 0$$
$$0 = 0$$

Since 8 is not equal to 0, $x = 9$ is an extraneous solution. Since $0 = 0$, $x = 1$ is a solution.

4. Solve $4 + \sqrt[3]{3x - 1} = 6$.

$$4 + \sqrt[3]{3x - 1} = 6$$
$$\sqrt[3]{3x - 1} = 6 - 4$$
$$\left(\sqrt[3]{3x - 1}\right)^3 = (2)^3$$
$$3x - 1 = 8$$
$$3x = 9$$
$$x = 3$$

Check:

$$4 + \sqrt[3]{3x - 1} = 6$$
$$4 + \sqrt[3]{3(3) - 1} = 6$$
$$4 + \sqrt[3]{9 - 1} = 6$$
$$4 + \sqrt[3]{8} = 6$$
$$4 + 2 = 6$$
$$6 = 6$$

Since $6 = 6$, $x = 3$ is a solution.

➤ **Radicals, fractional exponents of**   See **Exponents, fractional or rational**.

➤ **Radicals, multiplication with**   If x and y are real numbers and n is a natural number greater than 1, then

$$\sqrt[n]{xy} = \sqrt[n]{x} \bullet \sqrt[n]{y}$$ provided both sides of the equation are defined.

The radical index n must be greater than 1 to consider a square root, i.e., $n = 2$, or higher roots. Both sides of the equation must be defined in order to avoid even roots of negative numbers. If $n = 4$ when $x = -16$, then $\sqrt[4]{-16}$ is not defined.

Examples

1. Multiply $\sqrt{3} \bullet \sqrt{5}$ and $4\sqrt{6} \bullet 2\sqrt{3}$.

$$\sqrt{3} \bullet \sqrt{5} = \sqrt{3 \bullet 5} = \sqrt{15}$$
$$4\sqrt{6} \bullet 2\sqrt{3} = 8\sqrt{6 \bullet 3} = 8\sqrt{18} = 8\sqrt{9 \bullet 2} = 8\sqrt{9} \bullet \sqrt{2} = 24\sqrt{2}$$

2. Simplify $\sqrt{36x^2}$ if $x \geqslant 0$.

$$\sqrt{36x^2} = \sqrt{36} \bullet \sqrt{x^2} = 6 \bullet x = 6x$$

The reason x is chosen with $x \geqslant 0$ is to avoid negative values of x. When x is allowed to be negative, say -2, then we would have

$$\sqrt{36x^2} = 6x$$
$$\sqrt{36(-2)^2} = 6(-2)$$

$$\sqrt{36 \bullet 4} = -12$$
$$\sqrt{144} = -12$$
$$12 = -12$$

which is not true.

From another viewpoint, if x is allowed to be negative, then $\sqrt{x^2} = |x|$. Therefore, if x is any real number, $\sqrt{36x^2} = 6|x|$. See **Radicals, rules for**.

3. Simplify $\sqrt{x^2y} \bullet \sqrt{xy^2}$ for positive values of x and y.

$$\sqrt{x^2y} \bullet \sqrt{xy^2} = \sqrt{x^3y^3} = xy\sqrt{xy}$$

Individually, the $\sqrt{x^3}$, as well as the $\sqrt{y^3}$, simplifies as follows:

$$\sqrt{x^3} = \sqrt{x^2x} = \sqrt{x^2} \bullet \sqrt{x} = x\sqrt{x}$$

4. Use the rule for multiplication of radicals to simplify $\sqrt{18}$.

$$\sqrt{18} = \sqrt{9 \bullet 2} = \sqrt{9} \bullet \sqrt{2} = 3\sqrt{2}$$

We chose the factors of 18 to be 9 and 2 because we know the square root of 9. We would not choose 6 and 3 because they both do not have perfect square roots.

5. Simplify $\sqrt[4]{16x^4y^5}$ for positive values of x and y.

$$\sqrt[4]{16x^4y^5} = \sqrt[4]{16} \bullet \sqrt[4]{x^4} \bullet \sqrt[4]{y^5} = 2xy \bullet \sqrt[4]{y}$$

Individually, we have

$$\sqrt[4]{16} = 2 \qquad \text{Since } 2 \bullet 2 \bullet 2 \bullet 2 = 16.$$
$$\sqrt[4]{x^4} = x \qquad \text{Since } x \text{ is positive, we do not use } \sqrt[4]{x^4} = |x|.$$
$$\sqrt[4]{y^5} = \sqrt[4]{y^4y} = \sqrt[4]{y^4} \bullet \sqrt[4]{y} = y \bullet \sqrt[4]{y}$$

For more, see **Radicals, simplifying (reducing)**, example 4; **Radicals, rules for**; or **Radicals, combining**.

➤ **Radicals, rules for**   If x and y are real numbers and n is a natural number greater than 1, then

1. $\sqrt[n]{xy} = \sqrt[n]{x}\sqrt[n]{y}$ Provided both sides of the equation are defined. See below.

2. $\sqrt[n]{\dfrac{x}{y}} = \dfrac{\sqrt[n]{x}}{\sqrt[n]{y}}$ With $y \neq 0$, and both sides of the equation defined. See below.

3. $\left(\sqrt[n]{x}\right)^n = x$ For $x \geq 0$ when n is even and for all x when n is odd.

4. $\sqrt[n]{x^n} = |x|$ When n is even and x is negative.

5. $\sqrt[n]{x^n} = x$ When n is odd and x is any real number.

Rules 1 and 2 are not defined for even roots of negative numbers. For example, if $n = 4$ when $x = -16$, then $\sqrt[4]{-16}$ is not defined. Or, if $n = 2$ when $x = -25$, then $\sqrt{-25}$ is not defined.

Some texts dispense with the need for rule 4 by always choosing x with $x \geqslant 0$. However, if all real values of x are allowed, then when n is even we would have an even root equal to a negative number. This is not defined. Hence, we use the absolute value to keep the definition correct. For example:

1. $\sqrt{(-2)^2} = |-2| = 2$ This makes sense since $\sqrt{(-2)^2} = \sqrt{4} = 2$.
2. $\sqrt[4]{(-2)^4} = |-2| = 2$ Similarly, $\sqrt[4]{(-2)^4} = \sqrt[4]{16} = 2$.
3. $\sqrt[4]{x^4} = |x|$ Since we don't know whether x is negative or not, we need $|x|$.
4. $\sqrt{x^2} = |x|$

For examples of rules 1 and 2, see **Radicals, multiplication with** or **Radicals, division with**. For an example of rule 3, see **Radicals, equations with**. For an example of rule 4, see **Radicals, multiplication with**, example 1. For more on rules 4 and 5, see **Radicals, simplifying (reducing)**.

For square roots, the corresponding rules are

1. $\sqrt{xy} = \sqrt{x}\sqrt{y}$ for $x \geqslant 0$ and $y \geqslant 0$

2. $\sqrt{\dfrac{x}{y}} = \dfrac{\sqrt{x}}{\sqrt{y}}$ for $x \geqslant 0$ and $y > 0$

3. $\left(\sqrt{x}\right)^2 = x$ for $x \geqslant 0$

4. $\sqrt{x^2} = |x|$ for all x

➤ **Radicals, simplifying (reducing)**   To **simplify or reduce radicals**, the concept for multiplication of radicals is applied. In general, $\sqrt[n]{xy} = \sqrt[n]{x}\sqrt[n]{y}$ for real numbers x and y with n a natural number greater than 1. For the limitations on this rule, see **Radicals, rules for**.

Examples

1. Simplify $\sqrt{12}$.

To simplify $\sqrt{12}$, we need to think of the factors of 12 that are perfect square roots, numbers such as 4, 9, 16, 25, and so on. Since $12 = 4 \bullet 3$ where 4 is a perfect square root, we have

$$\sqrt{12} = \sqrt{4 \bullet 3} = \sqrt{4}\sqrt{3} = 2\sqrt{3}$$

We use $12 = 4 \bullet 3$ instead of $12 = 6 \bullet 2$ because neither of the numbers 6 and 2 is a perfect square root.

2. Simplify $\sqrt{75}$.

Using the factors 25 and 3, we have

$$\sqrt{75} = \sqrt{25 \bullet 3} = \sqrt{25}\sqrt{3} = 5\sqrt{3}$$

3. Simplify $\sqrt{16x^2}$ for $x \geqslant 0$.

$$\sqrt{16x^2} = \sqrt{16}\sqrt{x^2} = 4x$$

In elementary texts, the restriction $x \geqslant 0$ is placed on x so the possibility of negative values of x does not have to be considered. If the restriction is lifted, then

$$\sqrt{16x^2} = \sqrt{16}\sqrt{x^2} = 4|x|$$

This follows from the rule $\sqrt[n]{x^n} = |x|$ where n is an even number and x is any real number (possibly negative). See **Radicals, rules for.** Further examples are $\sqrt[4]{x^4} = |x|$ and $\sqrt[6]{x^6} = |x|$.

4. Simplify $\sqrt{72x^3y^4}$ for positive values of x and y.

If x and y are positive, we do not have to consider absolute values:

$$\sqrt{72x^3y^4} = \sqrt{72} \bullet \sqrt{x^3} \bullet \sqrt{y^4} = 6\sqrt{2} \bullet x\sqrt{x} \bullet y^2 = 6xy^2\sqrt{2x}$$

Individually, we have

$$\sqrt{72} = \sqrt{36 \bullet 2} = \sqrt{36}\sqrt{2} = 6\sqrt{2}$$
$$\sqrt{x^3} = \sqrt{x^2x} = \sqrt{x^2}\sqrt{x} = x\sqrt{x}$$
$$\sqrt{y^4} = \sqrt{y^2y^2} = \sqrt{y^2}\sqrt{y^2} = y \bullet y = y^2$$

An easy way to simplify roots of variables with powers is to divide the radical index ($\sqrt{}$ is 2, $\sqrt[3]{}$ is 3, etc.) into the power. The quotient is written outside of the radical sign and the remainder stays within:

1. $\sqrt{x^5} = x^2 \bullet \sqrt{x^1}$ since $5 \div 2 = 2$ remainder 1
2. $\sqrt[3]{x^5} = x\sqrt[3]{x^2}$ since $5 \div 3 = 1$ remainder 2
3. $\sqrt[4]{x^{11}} = x^2\sqrt[4]{x^3}$ since $11 \div 4 = 2$ remainder 3
4. $\sqrt{y^4} = y^2$ since $4 \div 2 = 2$ remainder 0

5. Simplify $\sqrt[4]{16x^5yz^4}$.

Using the method of example 4, we have

$$\sqrt[4]{16x^5yz^4} = \sqrt[4]{16}\sqrt[4]{x^5}\sqrt[4]{y}\sqrt[4]{z^4} = 2xz\sqrt[4]{xy}$$

6. Simplify $\sqrt[3]{54}$.

$$\sqrt[3]{54} = \sqrt[3]{27 \bullet 2} = \sqrt[3]{27}\sqrt[3]{2} = 3\sqrt[3]{2}$$

7. Simplify $\sqrt[3]{192x^5y^3}$.

Since $\sqrt[3]{192} = \sqrt[3]{64 \bullet 3} = \sqrt[3]{64}\sqrt[3]{3} = 4\sqrt[3]{3}$, we have

$$\sqrt[3]{192x^5y^3} = \sqrt[3]{192} \bullet \sqrt[3]{x^5} \bullet \sqrt[3]{y^3} = 4xy\sqrt[3]{3x^2}$$ using the method of example 4

If all the steps are put in, the problem simplifies as follows:

$$\sqrt[3]{192x^5y^3} = \sqrt[3]{192}\sqrt[3]{x^5}\sqrt[3]{y^3} = \sqrt[3]{64 \bullet 3}\sqrt[3]{x^3x^2}y = \sqrt[3]{64}\sqrt[3]{3}\sqrt[3]{x^3}\sqrt[3]{x^2}\,y$$
$$= 4\sqrt[3]{3}\,x\sqrt[3]{x^2}\,y = 4xy\sqrt[3]{3x^2}$$

➤ **Radicals, sums and differences**   See **Radicals, combining**.

➤ **Radicand**   The number inside a radical sign is called the **radicand**. In the terms $\sqrt{5}$ and $\sqrt[3]{x}$, the 5 and the x are each **radicands**.

➤ **Radius**   In a circle, a **radius** is a segment with one endpoint at the center of the circle and the other endpoint on the circle.

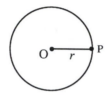

If O is the center of a circle and P is any point on the circle, then $\overline{OP}$ is a radius of the circle. Since $C = 2\pi r$ is the circumference of a circle (see **Circumference (of a circle)**), the equation of the radius is

$$r = \frac{C}{2\pi} \qquad \text{where } \pi \approx 3.14$$

Since the diameter of a circle passes through the center, the diameter is equal to twice the radius, $d = 2r$.

In a sphere, a **radius** is a segment with one endpoint at the center of the sphere and the other endpoint on the sphere.

In a regular polygon, the **radius** of the polygon is the radius of the circumscribed circle.

➤ **Range (algebraic definition)**   See also **Range (statistical definition)**. In reference to a set of points (x, y) where x and y are real numbers, the **range** is the set of all y-coordinates. The set of all x-coordinates is called the **domain**. See **Domain and Range**.

➤ **Range (statistical definition)**   See also **Range (algebraic definition)**.

In statistics, the **range** is the difference between the largest and smallest numbers in a set of data. Consider the following test scores: 48, 56, 69, 70, 78, 81, 88, 90, 93, and 97. The scores range from a low of 48 to a high of 97. The range is $97 - 48 = 49$. From the lowest to the highest scores, there is a range of 49 points.

To calculate the range, it is usually useful to arrange the data in order from smallest to largest. Then it is clear which numbers are going to be used in finding the range.

For related articles, see **Mean (average)**, **Median**, **Mode**, **Variance**, or **Standard Deviation**.

➤ **Ratio**   When two numbers are compared by division, the comparison is called a **ratio**. The comparison can not be made with zero since division by zero is not defined. If the ratio is expressed as a fraction, it is usually reduced to lowest terms. Hence, the ratio of 8 to 12 is $\frac{8}{12}$ or $\frac{2}{3}$. Therefore, the ratio of 8 to 12 is equivalent to the ratio of 2 to 3.

The ratio $\frac{a}{b}$ can also be written as $a : b$. This form is especially useful when comparing three or more numbers. The ratio $3 : 5 : 11$ means that the ratio of the first number to the second number is 3 to 5, the second number to the third number is 5 to 11, and the first number to the third number is 3 to 11.

For more, see **Proportion**, **Direct Variation**, **Combined Variation**, **Inverse Variation**, **Joint Variation**, or **Cross Multiplication**. Also see **Similar Triangles**.

➤ **Ratio and Proportion**   See **Proportion**.

➤ **Rational Equations**   **Rational equations** can be solved by multiplying both sides of the equation by the lowest common denominator. The lowest common denominator is the least common multiple of all the denominators.

If the lowest common denominator has a variable in it, extraneous solutions (usually false solutions) may have been introduced into the equation. As a result, each of the answers must be checked in the original equation. Also, if the denominator of the original equation has a variable in it, the answer must be checked. We cannot allow division by zero.

Examples

1. Solve $\frac{x}{4} - \frac{9}{2} = \frac{-x}{5}$.

We first need to determine the least common multiple of the denominators:

$$\text{multiples of } 4 \rightarrow 4, 8, 12, 16, \boxed{20}, 24, \ldots$$

$$\text{multiples of } 2 \rightarrow 2, 4, 6, 8, 10, 12, 14, 16, 18, \boxed{20}, 22, \ldots$$

$$\text{multiples of } 5 \rightarrow 5, 10, 15, \boxed{20}, 25, \ldots$$

By listing the multiples of each denominator, we see that the least common multiple, which is the lowest common denominator, is 20. See **Least Common Multiple** or **Common Denominator**.

Next, we multiply both sides of the equation by 20. Then we solve the equation. See **Linear Equations, solutions of**. We do not have to check the answer since the lowest common denominator does not have a variable in it.

$$\frac{x}{4} - \frac{9}{2} = \frac{-x}{5}$$

$$20\left(\frac{x}{4} - \frac{9}{2}\right) = 20\left(\frac{-x}{5}\right)$$

$$\frac{20x}{4} - \frac{180}{2} = \frac{-20x}{5} \qquad \text{See } \textbf{Distributive Property}.$$

$$5x - 90 = -4x$$

$$5x + 4x = 90$$

$$9x = 90$$

$$\frac{9x}{9} = \frac{90}{9}$$

$$x = 10$$

R

2. Solve $\frac{3}{2}x^2 = 2 - \frac{26}{3}x$.

The denominator of 2 is 1. The lowest common denominator of 2, 1, and 3 is 6. We need to multiply both sides of the equation by 6 and then solve the quadratic equation. See **Quadratic Equations, solutions of.**

$$\frac{3}{2}x^2 = 2 - \frac{26}{3}x$$

$$6\left(\frac{3}{2}x^2\right) = 6\left(\frac{2}{1} - \frac{26}{3}x\right)$$

$$\frac{18}{2}x^2 = \frac{12}{1} - \frac{156}{3}x$$

$$9x^2 = 12 - 52x$$

$$9x^2 + 52x - 12 = 0$$

$$(9x - 2)(x + 6) = 0 \qquad\qquad \text{Solve by factoring.}$$

$$9x - 2 = 0 \qquad \text{or} \qquad x + 6 = 0$$
$$9x = 2 \qquad\qquad\qquad x = -6$$
$$x = \frac{2}{9}$$

The solutions are $x = \frac{2}{9}$ and $x = -6$.

3. Solve $\frac{1}{x} - \frac{5}{2x} = \frac{-3}{10}$. Remember to check the answer.

The lowest common denominator for x, $2x$, and 10 is $10x$. Each of the denominators is contained in $10x$. Multiplying both sides of the equation by $10x$, we have

$$\frac{1}{x} - \frac{5}{2x} = \frac{-3}{10} \qquad\qquad \text{Check:}$$

$$10x\left(\frac{1}{x} - \frac{5}{2x}\right) = 10x\left(-\frac{3}{10}\right) \qquad \frac{1}{5} - \frac{5}{2(5)} = -\frac{3}{10}$$

$$\frac{10x}{x} - \frac{50x}{2x} = \frac{-30x}{10} \qquad\qquad \frac{1}{5} - \frac{1}{2} = -\frac{3}{10}$$

$$10 - 25 = -3x \qquad\qquad\qquad \frac{2 - 5}{2} = \frac{-3}{10}$$

$$-15 = -3x \qquad\qquad\qquad -\frac{3}{10} = -\frac{3}{10} \qquad \text{which checks.}$$

$$5 = x$$

$$x = 5$$

Since the check of $x = 5$ balances, $x = 5$ is the answer to the equation.

4. Solve $\frac{2}{x} + \frac{2}{x-3} = 1$. Check the answers.

The denominator of the 1 is 1, $\frac{1}{1}$. The lowest common denominator of x, $x-3$, and 1 is $x(x-3)$. Each of the denominators is contained in $x(x-3)$. This means that if each of the denominators is divided into $x(x-3)$, they will divide in evenly, i.e., $\frac{x(x-3)}{x} = x - 3$, $\frac{x(x-3)}{x-3} = x$, and $\frac{x(x-3)}{1} = x(x-3)$.

The common denominator cannot be $x - 3$ by itself. If this were the case, then x, from $\frac{2}{x}$, would have to divide evenly into $x - 3$. But it doesn't, i.e., $\frac{x-3}{x} = 1 - \frac{3}{x}$, not -3.

Multiplying both sides of the equation by $x(x - 3)$, we have

$$\frac{2}{x} + \frac{2}{x - 3} = 1$$

$$x(x - 3)\left(\frac{2}{x} + \frac{2}{x - 3}\right) = x(x - 3)\left(\frac{1}{1}\right)$$

$$\frac{2x(x - 3)}{x} + \frac{2x(x - 3)}{x - 3} = \frac{x(x - 3)}{1} \qquad \text{Distributive property.}$$

$$2(x - 3) + 2x = x(x - 3) \qquad \text{Multiply the terms and set the}$$
$$2x - 6 + 2x = x^2 - 3x \qquad \text{equation equal to zero.}$$
$$0 = x^2 - 7x + 6$$

$$x^2 - 7x + 6 = 0 \qquad \text{Symmetric property.}$$
$$(x - 1)(x - 6) = 0 \qquad \text{Solve by factoring.}$$

$$x - 1 = 0 \qquad \text{or} \qquad x - 6 = 0$$
$$x = 1 \qquad\qquad\qquad x = 6$$

We need to see if substituting either answer into the denominators of the original equation will yield a zero. Division by zero is not allowed. If $x = 1$ is substituted into the denominators x and $x - 3$ things are fine, no zeros. Also, substituting $x = 6$ causes no problems. Hence, both answers are fine, so $x = 1$ and $x = 6$ are the solutions to the equation.

5. Solve $\frac{16}{x^2 - 4x} + \frac{x+4}{x} = \frac{-3}{x-4}$. Remember to check the answer.

We first need to factor $x^2 - 4x$, so we can determine the lowest common denominator, i.e., $x^2 - 4x = x(x - 4)$. See **Factoring Common Monomials**. We do the same thing when finding the common denominator of fractions. For example, the common denominator of $\frac{7}{10} + \frac{1}{2} + \frac{1}{5}$ is 10, because the factors of $10 = 2 \bullet 5$ contain the other two denominators.

The lowest common denominator must contain each of the factors of all the denominators with no duplication. Since the denominators are $x(x - 4)$, x, and $x - 4$, respectively, the common denominator must have an x and an $x - 4$ in it. Hence, the lowest common denominator is $x(x - 4)$. Factoring and multiplying both sides of the equation by $x(x - 4)$, we have

$$\frac{16}{x^2 - 4x} + \frac{x + 4}{x} = \frac{-3}{x - 4}$$

$$\frac{16}{x(x - 4)} + \frac{x + 4}{x} = \frac{-3}{x - 4}$$

$$x(x - 4)\left(\frac{16}{x(x - 4)} + \frac{x + 4}{x}\right) = x(x - 4)\left(\frac{-3}{x - 4}\right)$$

$$\frac{16x(x - 4)}{x(x - 4)} + \frac{x(x - 4)(x + 4)}{x} = \frac{-3x(x - 4)}{x - 4}$$

$$16 + (x - 4)(x + 4) = -3x$$

$$16 + x^2 - 16 = -3x$$
$$x^2 = -3x$$
$$x^2 + 3x = 0$$
$$x(x + 3) = 0$$

$$x = 0 \qquad \text{or} \qquad x + 3 = 0$$
$$x = -3$$

Since the common denominator has a variable in it, we need to check the answer in the original equation. We do not need to do a full check since all we are interested in is any violations of definitions. Otherwise, the solution is fine.

We see that if $x = 0$ is substituted into the second fraction, $\frac{x+4}{x}$, we get a division by 0. Division by 0 is undefined. Hence, $x = 0$ is not a solution.

Substituting $x = -3$ into all three fractions causes no violations. Hence, $x = -3$ is a good solution. The solution of the equation has one answer, $x = -3$.

6. Solve $\frac{3x+1}{x^2-9} = \frac{1-5x}{x+3} - \frac{x+3}{3-x}$.

We first need to factor the $x^2 - 9$ and $3 - x$ so we can determine the lowest common denominator. In the $3 - x$ term, we are factoring out a $-$ sign:

$$3 - x = -(-3 + x) = -(x - 3)$$

For the denominators $(x + 3)(x - 3)$, $x + 3$ and $x - 3$, the lowest common denominator is the combination of the three denominators with no duplication, $(x + 3)(x - 3)$.

$$\frac{3x + 1}{x^2 - 9} = \frac{1 - 5x}{x + 3} - \frac{x + 3}{3 - x}$$

$$\frac{3x + 1}{(x + 3)(x - 3)} = \frac{1 - 5x}{x + 3} - \frac{x + 3}{-(x - 3)}$$

$$\frac{(x + 3)(x - 3)(3x + 1)}{(x + 3)(x - 3)} = (x + 3)(x + 3)\left(\frac{1 - 5x}{x + 3} + \frac{x + 3}{x - 3}\right) \qquad \text{The } -- \text{ is } +.$$

$$3x + 1 = \frac{(x + 3)(x - 3)(1 - 5x)}{x + 3} + \frac{(x + 3)(x - 3)(x + 3)}{x - 3}$$

$$3x + 1 = (x - 3)(1 - 5x) + (x + 3)(x + 3)$$

$$3x + 1 = x - 5x^2 - 3 + 15x + x^2 + 3x + 3x + 9$$

$$4x^2 - 19x - 5 = 0$$

$$(4x + 1)(x - 5) = 0$$

$$4x + 1 = 0 \qquad \text{or} \qquad x - 5 = 0$$
$$4x = -1 \qquad\qquad x = 5$$
$$x = \frac{-1}{4}$$

Checking both of the solutions in the original equation, we see that there are no violations of definitions. Those would be caused by a 3 and a -3. Hence, $x = \frac{-1}{4}$ and $x = 5$ are the solutions of the equation.

7. What is the lowest common denominator for the equation $\frac{x-2}{3x-9} + \frac{3}{x^2-6x+9} = \frac{x}{2x-6}$?

To find the lowest common denominator, we need to factor each of the denominators and then determine their combination:

$$\frac{x-2}{3(x-3)} + \frac{3}{(x-3)(x+3)} = \frac{x}{2(x-3)}$$

The lowest common denominator must have a 3, an $x - 3$, another $x - 3$, and a 2 in it. Hence, the lowest common denominator is $6(x-3)(x-3)$.

If you wish to continue solving the problem, the solution to the equation is $x = -6$ and $x = 5$.

➤ **Rational Exponents**   See **Exponents, fractional or rational**.

➤ **Rational Expression**   A **rational expression** is a fraction that contains a variable.

➤ **Rational Expressions, complex fractions of**   To simplify a **complex rational expression**, the concepts for simplifying complex fractions can be utilized. See **Fractions, complex**.

$$\frac{\frac{a}{b}}{\frac{c}{d}} = \frac{a}{b} \div \frac{c}{d} = \frac{a}{b} \bullet \frac{d}{c} = \frac{ad}{ac}$$

As a rule, we can say "flip the bottom fraction and multiply it with the top fraction." This utilizes the concept of multiplying by the reciprocal. The rule works only for fractions that are exactly in the form $\frac{\frac{a}{b}}{\frac{c}{d}}$. If the expression is not in this form, then the goal is to get it into this form.

R

Examples

1. Simplify $\dfrac{1-\frac{1}{x^2}}{1-\frac{1}{x}}$.

To get the fraction into the proper form, we first need to combine the terms in the numerator and the denominator. See **Rational Expressions, sums and differences**.

$$\frac{1-\frac{1}{x^2}}{1-\frac{1}{x}} = \frac{\frac{x^2-1}{x^2}}{\frac{x-1}{x}} = \frac{x^2-1}{x^2} \bullet \frac{x}{x-1}$$ Flip and multiply.

$$= \frac{x(x^2-1)}{x^2(x-1)}$$

To reduce the fraction, we factor, cancel, and multiply what's left. **Rational Expressions, simplifying (reducing)**.

$$= \frac{x(x+1)x-1}{x^2(x-1)}$$

$$= \frac{x+1}{x}$$

2. Simplify $\dfrac{\frac{3}{x-2}+\frac{2x}{x^2-4}}{\frac{4}{x-2}+\frac{5}{x+2}}$.

$$\dfrac{\frac{3}{x-2}+\frac{2x}{x^2-4}}{\frac{4}{x-2}+\frac{5}{x+2}} = \dfrac{\frac{3}{x-2}+\frac{2x}{(x+2)(x-2)}}{\frac{4}{x-2}+\frac{5}{x+2}} = \dfrac{\frac{3(x+2)+2x}{(x+2)(x-2)}}{\frac{4(x+2)+5(x-2)}{(x-2)(x+2)}}$$

$$= \dfrac{5x+6}{(x+2)(x-2)} \bullet \dfrac{(x-2)(x+2)}{9x-2} = \dfrac{5x+6}{9x-2}$$

See **Rational Expressions, simplifying (reducing)**.

► **Rational Expressions, multiplication and division of**   **Multiplication and division of rational expressions** is very similar to the method used to simplify rational expressions: factor, cancel, and multiply what's left. See **Rational Expressions, simplifying (reducing)**. In either, it is imperative that the ability to factor has been mastered. See various articles under **Factoring**.

As with division of fractions, division of rational expressions requires the second fraction to be turned around. We multiply by the reciprocal.

Examples

1. Multiply: $\dfrac{x^2-y^2}{y^2} \bullet \dfrac{y^3}{y-x}$.

$$\dfrac{x^2-y^2}{y^2} \bullet \dfrac{y^3}{y-x} = \dfrac{(x+y)(x-y)}{y^2} \bullet \dfrac{y^3}{-(x-y)} \qquad \text{Factor.}$$

$$= \dfrac{x+y}{1} \bullet \dfrac{y}{-1} \qquad \text{Cancel.}$$

$$= -y(x+y) \qquad \text{Multiply what's left.}$$

$$\text{or} \qquad = -xy - y^2$$

2. Multiply: $\dfrac{3x^2y^3z}{4xy^2} \bullet 6z$.

$$\dfrac{3x^2y^3z}{4xy^2} \bullet 6z = \dfrac{3xyz}{2} \bullet \dfrac{3z}{1} \qquad \text{There is nothing to factor so just cancel and multiply what's left.}$$

$$= \dfrac{9xyz^2}{2}$$

3. Divide: $\dfrac{x^3y}{z} \div \dfrac{xy}{z^2}$.

$$\dfrac{x^3y}{z} \div \dfrac{xy}{z^2} = \dfrac{x^3y}{z} \bullet \dfrac{z^2}{xy} \qquad \text{Turn the second fraction around (the reciprocal of } \tfrac{xy}{z^2}\text{).}$$

$$= \dfrac{x^2}{1} \bullet \dfrac{z}{1} \qquad \text{Cancel.}$$

$$= x^2z \qquad \text{Multiply what's left.}$$

4. Divide: $\frac{x^2-9}{x^2-2x-15} \div \frac{x^2-6x+9}{12-4x}$.

$$\frac{x^2-9}{x^2-2x-15} \div \frac{x^2-6x+9}{12-4x} = \frac{(x+3)(x-3)}{(x-5)(x+3)} \cdot \frac{-4(x-3)}{(x-3)(x-3)} \qquad \text{Factor and turn around.}$$

$$= \frac{1}{x-5} \cdot \frac{-4}{1} = \frac{-4}{x-5}$$

➤ **Rational Expressions, simplifying (reducing)**   **Simplifying rational expressions** is similar to reducing fractions. In general, the following applies: factor, cancel, and multiply what's left. To simplify rational expressions, it is imperative that the ability to factor has been mastered. See the various articles under **Factoring**.

Examples

1. Simplify $\frac{x^2-4}{3x-6}$.

$$\frac{x^2-4}{3x-6} = \frac{(x+2)(x-2)}{3(x-2)} = \frac{x+2}{3}$$

Each binomial is factored, the $x-2$ terms are canceled, and the product $\frac{x+2}{3}$ is the answer.

2. Simplify $\frac{3x-15}{10-2x}$.

The – as well as the 2 must be factored out of $10-2x$:

$$10-2x = -2(-5+x) = -2(x-5)$$

This will allow cancellation in the next step:

$$\frac{3x-15}{10-2x} = \frac{3(x-5)}{-2(x-5)} = \frac{-3}{2}$$

3. Simplify $\frac{-2-2a}{-10}$ and $\frac{-2-2\sqrt{5}}{-10}$.

$$\frac{-2-2a}{-10} = \frac{-2(1+a)}{-10} = \frac{1+a}{5} \qquad \frac{-2-2\sqrt{5}}{-10} = \frac{-2(1+\sqrt{5})}{-10} = \frac{1+\sqrt{5}}{5}$$

4. Simplify $\frac{x^3-27}{5x^2+15x+45}$.

$$\frac{x^3-27}{5x^2+15x+45} = \frac{(x-3)(x^2+3x+9)}{5(x^2+3x+9)} \qquad \text{See \textbf{Factoring the Sum and Difference of Two Cubes}.}$$

$$= \frac{x-3}{5} \qquad \text{See \textbf{Factoring Common Monomials}.}$$

Also see **Rational Expressions, multiplication and division of**.

➤ **Rational Expressions, sums and differences**   Finding the **sum and difference of rational expressions** is very similar to finding the sum and difference of fractions. Each expression must be changed to a common denominator, then combined, and then simplified or reduced.

The lowest common denominator of a rational expression is the least common multiple of their denominators. To find the lowest common denominator, we first factor each denominator. Then, we form the product of each factor of the denominators with no duplications. This is the lowest common denominator.

Examples

1. Combine $\frac{1}{2} + \frac{5}{6x^2} - \frac{1}{x}$.

   By observation, we can see that the lowest common denominator is $6x^2$. However, if we consider the factors of each, we have 2, $2 \bullet 3 \bullet x \bullet x$, and x. The product of these factors, with no duplications, is $2 \bullet 3 \bullet x \bullet x = 6x^2$.

   Now we need to change each denominator to $6x^2$. We multiply the numerator and denominator of $\frac{1}{2}$ by $3x^2$, leave $\frac{5}{6x^2}$ alone, and multiply the top and bottom of $\frac{1}{x}$ by $6x$:

   $$\frac{1}{2} + \frac{5}{6x^2} - \frac{1}{x} = \frac{1}{2} \bullet \frac{3x^2}{3x^2} + \frac{5}{6x^2} - \frac{1}{x} \bullet \frac{6x}{6x} = \frac{3x^2}{6x^2} + \frac{5}{6x^2} - \frac{6x}{6x^2}$$

   $$= \frac{3x^2 + 5 - 6x}{6x^2} = \frac{3x^2 - 6x + 5}{6x^2}$$

2. Combine $\frac{x}{6x-12y} - \frac{y}{3x-6y}$.

   Factoring each denominator, we have $6x - 12y = 6(x-2y) = 2 \bullet 3(x-2y)$ and $3x - 6y = 3(x-2y)$. The product of these factors, with no duplications, is $2 \bullet 3(x - 2y) = 6(x - 2y)$.

   $$\frac{x}{6x - 12y} - \frac{y}{3x - 6y} = \frac{x}{6(x - 2y)} - \frac{y}{3(x - 2y)}$$

   $$= \frac{x}{6(x - 2y)} - \frac{y}{3(x - 2y)} \bullet \frac{2}{2} \qquad \text{Multiply the top and bottom by 2 to get the common denominator } 6(x - 2y).$$

   $$= \frac{x}{6(x - 2y)} - \frac{2y}{6(x - 2y)}$$

   $$= \frac{x - 2y}{6(x - 2y)} = \frac{1}{6}$$

3. Combine $\frac{4x}{x^2-1} + \frac{2}{1-x} - \frac{2}{x}$.

   Factoring the denominator of the first fraction, we have $x^2 - 1 = (x+1)(x-1)$. The denominator of the second fraction is reversed so we need to factor out a negative, $1 - x = -(-1+x) = -(x-1)$. Hence, we see that the lowest common denominator must have an $x + 1$, an $x - 1$, and an x. The lowest common denominator is $x(x + 1)(x - 1)$.

   $$\frac{4x}{x^2 - 1} + \frac{2}{1 - x} - \frac{2}{x} = \frac{4x}{(x + 1)(x - 1)} + \frac{2}{-(x - 1)} - \frac{2}{x} = \frac{4x}{(x + 1)(x - 1)} - \frac{2}{x - 1} - \frac{2}{x}$$

R

$$= \frac{4x}{(x+1)(x-1)} \cdot \frac{x}{x} - \frac{2}{x-1} \cdot \frac{x(x+1)}{x(x+1)} - \frac{2}{x} \cdot \frac{(x+1)(x-1)}{(x+1)(x-1)}$$

$$= \frac{4x^2}{x(x+1)(x-1)} - \frac{2x(x+1)}{x(x+1)(x-1)} - \frac{2(x+1)(x-1)}{x(x+1)(x-1)}$$

$$= \frac{4x^2 - 2x^2 - 2x - 2(x^2-1)}{x(x+1)(x-1)} = \frac{2x^2 - 2x - 2x^2 + 2}{x(x+1)(x-1)}$$

$$= \frac{-2x+2}{x(x+1)(x-1)} = \frac{-2(x-1)}{x(x+1)(x-1)} \qquad \text{See \textbf{Rational Expressions},}$$
$$\text{simplifying (reducing).}$$

$$= \frac{-2}{x(x+1)}$$

➤ **Rationalizing the Denominator**   **Rationalizing the denominator** is a procedure used to rid fractions of denominators that have radicals in them. If the fraction has a single root in the denominator, we multiply both the denominator and the numerator by a root that will simplify with the root of the denominator. If the fraction has a denominator of the form $a + \sqrt{b}$ or $\sqrt{a} + \sqrt{b}$, we multiply both the denominator and the numerator by the **conjugate**. The conjugate of $a + \sqrt{b}$ is $a - \sqrt{b}$ and the conjugate of $\sqrt{a} + \sqrt{b}$ is $\sqrt{a} - \sqrt{b}$. See **Conjugate**. The same holds for denominators with negative signs.

To rationalize denominators of complex numbers, see **Complex Numbers**.

Examples

1. Rationalize the fractions $\frac{3}{\sqrt{5}}$ and $\frac{\sqrt{7}}{5\sqrt{3}}$.

   If the denominator is the square root of a number, we multiply both the denominator and the numerator by that square root. We utilize the property $\sqrt{a} \cdot \sqrt{b} = \sqrt{ab}$. See **Radicals, multiplication with**.

   a. $\dfrac{3}{\sqrt{5}} = \dfrac{3}{\sqrt{5}} \cdot \dfrac{\sqrt{5}}{\sqrt{5}} = \dfrac{3\sqrt{5}}{\sqrt{25}} = \dfrac{3\sqrt{5}}{5}$

   b. $\dfrac{\sqrt{7}}{5\sqrt{3}} = \dfrac{\sqrt{7}}{5\sqrt{3}} \cdot \dfrac{\sqrt{3}}{\sqrt{3}} = \dfrac{\sqrt{21}}{5\sqrt{9}} = \dfrac{\sqrt{21}}{5 \cdot 3} = \dfrac{\sqrt{21}}{15}$

2. Rationalize $\frac{1}{\sqrt[3]{2ab^4}}$ for a and b not equal to 0 (can't divide by 0).

   We need to multiply both the denominator and the numerator by a root that will simplify with $\sqrt[3]{2ab^4}$. Since $2 \cdot 4 = 8$ and we know the $\sqrt[3]{8}$, our root will contain $\sqrt[3]{4}$. Also, since $a \cdot a^2 = a^3$ and we know the $\sqrt[3]{a^3}$, our root will contain $\sqrt[3]{4a^2}$. Finally, since $b^4 \cdot b^2 = b^6$ and we know the $\sqrt[3]{b^6}$, the root we want to multiply by is $\sqrt[3]{4a^2b^2}$:

   $$\frac{1}{\sqrt[3]{2ab^4}} = \frac{1}{\sqrt[3]{2ab^4}} \cdot \frac{\sqrt[3]{4a^2b^2}}{\sqrt[3]{4a^2b^2}} = \frac{\sqrt[3]{4a^2b^2}}{\sqrt[3]{8a^3b^6}} = \frac{\sqrt[3]{4a^2b^2}}{2ab^2}$$

R

3. Rationalize $\frac{3}{\sqrt[4]{8ab^5}}$ for a and b greater than 0.

We need to think of numbers, or terms, that when multiplied with 8, a, and b^5 will yield numbers, or terms, of which we know the fourth root. If we multiply 8 by 2, a by a^3, and b^5 by b^3, we get 16, a^4, and b^8, each of which we know the fourth root. Hence, the root we want to multiply by is $\sqrt[4]{2a^3b^3}$:

$$\frac{3}{\sqrt[4]{8ab^5}} = \frac{3}{\sqrt[4]{8ab^5}} \bullet \frac{\sqrt[4]{2a^3b^3}}{\sqrt[4]{2a^3b^3}} = \frac{3\sqrt[4]{2a^3b^3}}{\sqrt[4]{16a^4b^8}} = \frac{3\sqrt[4]{2a^3b^3}}{2ab^2}$$

4. Rationalize $\frac{1}{1+\sqrt{2}}$.

We need to multiply both the denominator and the numerator by the conjugate $1 - \sqrt{2}$. Since $1 + \sqrt{2}$ and $1 - \sqrt{2}$ differ only by their signs, the inside and outside terms of their product will cancel. As a result, only the products of the first and last terms remain. These are squared terms and will cancel any square roots.

$$\frac{1}{1+\sqrt{2}} = \frac{1}{(1+\sqrt{2})} \bullet \frac{(1-\sqrt{2})}{(1-\sqrt{2})} = \frac{1-\sqrt{2}}{1-\sqrt{4}} = \frac{1-\sqrt{2}}{1-2} = \frac{1-\sqrt{2}}{-1} = -(1-\sqrt{2}) \text{ or } -1 + \sqrt{2}$$

To multiply $(1 + \sqrt{2})(1 - \sqrt{2})$, see **FOIL**.

5. Rationalize $\frac{\sqrt{5}+2}{3\sqrt{5}+9}$.

The conjugate for $3\sqrt{5} + 9$ is $3\sqrt{5} - 9$. Multiplying both the denominator and the numerator by the conjugate, we have

$$\frac{\sqrt{5}+2}{3\sqrt{5}+9} = \frac{(\sqrt{5}+2)}{(3\sqrt{5}+9)} \bullet \frac{(3\sqrt{5}-9)}{3\sqrt{5}-9} = \frac{3\sqrt{25}-9\sqrt{5}+6\sqrt{5}-18}{9\sqrt{25}-81}$$

$$= \frac{3 \bullet 5 - 3\sqrt{5} + 6\sqrt{5} - 18}{9 \bullet 5 - 81} = \frac{-3 - 3\sqrt{5}}{-36}$$

To simplify the fraction, we need to factor, cancel, and multiply what's left. See **Rational Expressions, simplifying (reducing)**, example 3.

$$\frac{-3-3\sqrt{5}}{-36} = \frac{-3(1+\sqrt{5})}{-36} = \frac{1+\sqrt{5}}{12}$$

6. Rationalize $\frac{\sqrt{x}}{\sqrt{x}-\sqrt{y}}$ for x and y greater than 0.

$$\frac{\sqrt{x}}{\sqrt{x}-\sqrt{y}} = \frac{\sqrt{x}}{(\sqrt{x}-\sqrt{y})} \bullet \frac{(\sqrt{x}+\sqrt{y})}{(\sqrt{x}+\sqrt{y})} = \frac{\sqrt{x^2}+\sqrt{xy}}{\sqrt{x^2}-\sqrt{y^2}} = \frac{x+\sqrt{xy}}{x-y}$$

➤ **Rationalizing the Numerator**   **Rationalizing the numerator** is a procedure used to rid fractions of numerators that have radicals in them. This process is sometimes useful in calculus. It follows the same method as rationalizing the denominator only in reverse order. See **Rationalizing the Denominator**.

➤ **Rational Number, definition**   A **rational number** is a fraction of the form $\frac{a}{b}$ where a and b are both integers and $b \neq 0$. Each of the numbers 21, -5, $-6\frac{3}{4}$, 7.35, and $8.333\ldots$ or $8.\overline{3}$ are rational numbers.

1. 21 is the fraction $\frac{21}{1}$.

2. -5 is the fraction $\frac{-5}{1}$.

3. $-6\frac{3}{4}$ is a mixed fraction which can be written as an improper fraction $\frac{-27}{4}$. The 27 comes from $6 \bullet 4 + 3$. See **Mixed Number**.

4. 7.35 is the fraction $7\frac{35}{100} = 7\frac{5}{20} = \frac{145}{20}$. See **Decimal Numbers, changed to a fraction** or **Mixed Number**.

5. $8.333\ldots$ or $8.\overline{3}$ is the fraction $8\frac{1}{3} = \frac{25}{3}$. See **Decimal Numbers, changing a repeating decimal to a common fraction**.

   The fractions $\frac{1}{\sqrt{2}}$, $\frac{\sqrt{3}}{\sqrt{7}}$, $\frac{\pi}{4}$, $\frac{e}{5}$, and $\frac{i}{6}$ are not rational numbers. Each contains numbers that are not integers.

   The set of rational numbers contains the following subsets: natural or counting numbers, whole numbers, and integers. See the articles under each. Also see **Subset**.

➤ **Rational Roots Theorem**   The **rational roots theorem** is used to solve polynomial equations with integer coefficients that have at least one rational solution. It states that if $f(x) = a_n x^n + a_{n-1} x^{n-1} + \cdots + a_1 x + a_0$ where n is a natural number with $n \geqslant 1$ and $a_n \neq 0$ and $\frac{p}{q}$ is a rational number in simplest form with $\frac{p}{q}$ a root of $f(x)$, then p is a factor of a_0 and q is a factor of a_n.

   In order to solve a polynomial equation, it must first be factored. This can be accomplished by using the solution $\frac{p}{q}$. Since $\frac{p}{q}$ is a solution of $f(x)$, it will divide $f(x)$ evenly with a remainder of 0. Hence, $f(x)$ is equal to the product of its factors, the divisor and the quotient. To be able to solve $f(x)$, we must first find one of its solutions.

   To find one of the solutions of $f(x)$, we look at the factors of the a_0 and a_n terms of the polynomial. Since $\frac{p}{q}$ is a solution of $f(x)$, where p is a factor of a_0 and q is a factor of a_n, the solution is in the factors of the a_0 and a_n terms. But which factors? To find out, we must divide the various $\frac{p}{q}$ possibilities until we get a remainder of 0. We do this by using synthetic division. See **Synthetic Division**. Once we get a remainder of 0, we can factor the polynomial, set each factor equal to 0, and solve each factor. See **Product Theorem of Zero**.

Examples

1. Find all rational zeros of $f(x) = 2x^3 - 5x^2 - 4x + 3$.

   To find all rational zeros of an equation means to find all the rational solutions of the equation when it has been set equal to 0. See **Zeros of a Function or Polynomial**. We first need to find a solution of $f(x)$, use it to factor $f(x)$, and then solve each factor.

   As the rational roots theorem states, the solutions to $f(x)$ are in the factors of the terms of $a_0 = 3$ and $a_n = 2$. Since p is a factor of 2, $p = \pm1, \pm2$. Since q is a factor of 3, $q = \pm1, \pm3$.

   The possible solutions of $f(x)$ are $\frac{p}{q} = \pm1, \pm2, \pm\frac{1}{3}$, and $\pm\frac{2}{3}$. There are eight possible solutions.

R

To determine which of the possible solutions is an actual solution of $f(x)$, we need to divide $f(x)$ by each. The first to give a remainder of 0 is a solution of $f(x)$. We can then use it to factor $f(x)$ and find the remaining solutions. Hence, using synthetic division (see **Synthetic Division**), we have

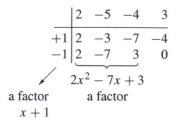

$$2x^2 - 7x + 3$$

a factor   a factor

$x + 1$

Since -1 yields, a remainder of 0, it is a solution of $f(x)$.

We can now solve the equation. Setting $f(x) = 0$, we have

$$f(x) = 2x^3 - 5x^2 - 4x + 3 = 0$$
$$(x + 1)(2x^2 + 7x + 3) = 0$$
$$(x + 1)(2x - 1)(x - 3) = 0 \qquad \text{See \textbf{Solutions of Equations}.}$$

$$\begin{array}{ccccc} x + 1 = 0 & \text{or} & 2x - 1 = 0 & \text{or} & x - 3 = 0 \\ x = -1 & & 2x = 1 & & x = 3 \\ & & x = \dfrac{1}{2} & & \end{array}$$

The rational zeros of $f(x)$ are $x = -1$, $x = \frac{1}{2}$, and $x = 3$.

The reason the rational roots theorem works can be seen by reversing the process. Notice where the solutions go when we move back up the problem. Prior to being solutions, the -1, $\frac{1}{2}$, and 3 were in the binomials set equal to 0. Before that, they were in a product of binomials set equal to 0, $(x + 1)(2x - 1)(x - 3) = 0$. In the next two steps, the product of the binomials turns into the original polynomial. Notice that the denominator of the $\frac{1}{2}$ has become the coefficient of the x^3 term. Also, the numerators $-1 = \frac{-1}{1}$ and $3 = \frac{3}{1}$ have become the last term of the polynomial. The q of $\frac{p}{q}$ is in the coefficient of x^3 and the p is in the last term. The solutions to the equation are in the coefficient q and p.

2. Find all rational zeros of $f(x) = x^3 - 5x^2 - 2x + 24$.

The coefficient of the x^3 term has factors $q = \pm 1$. The last term of the polynomial has factors $p = \pm 1, \pm 2, \pm 3, \pm 4, \pm 6, \pm 8, \pm 12, \pm 24$. All of the possible roots of $f(x)$ are $\frac{p}{q} = \frac{\pm 1}{\pm 1}$, $\frac{\pm 2}{\pm 1}, \frac{\pm 3}{\pm 1}, \frac{\pm 4}{\pm 1}, \frac{\pm 6}{\pm 1}, \frac{\pm 8}{\pm 1}, \frac{\pm 12}{\pm 1}, \frac{\pm 24}{\pm 1} = \pm 1, \pm 2, \pm 3, \pm 4, \pm 6, \pm 8, \pm 12, \pm 24$. There are 16 possible roots.

Looking for the first root of $f(x)$ by synthetic division, we have

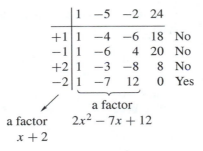

Since -2 yields a remainder of 0, we know that $x + 2$ and $x^2 - 7x + 12$ are factors of $f(x)$. Hence, we have

$$x^3 - 5x^2 - 2x + 24 = 0$$
$$(x + 2)(x^2 - 7x + 12) = 0$$
$$(x + 2)(x - 3)(x - 4) = 0$$

$$x + 2 = 0 \quad \text{or} \quad x - 3 = 0 \quad \text{or} \quad x - 4 = 0$$
$$x = -2 \qquad\qquad x = 3 \qquad\qquad x = 4$$

The rational zeros of $f(x)$ are $x = -2$, $x = 3$, and $x = 4$.

▶ **Ray**   A **ray** is part of a line. If A and B are any two distinct points on a line l, then the ray $\overrightarrow{AB}$ is that part of the line that contains point A and extends indefinitely through point B. The point A is called the **endpoint** of the ray (although it is at the beginning of the ray). The first letter in a ray symbol always names the endpoint of the ray.

▶ **Real Axis**   In the complex plane, the horizontal axis is called the **real axis**. See **Complex Plane**.

▶ **Real Number**   Any number that can be expressed in decimal form is a **real number**. The following are examples of real numbers: $9 = 9.000\ldots$, $-16 = -16.000\ldots$, $\frac{3}{4} = .75$, $-\frac{1}{2} = -.5$, $\frac{1}{3} = .333\ldots$, $\frac{2}{3} = .\overline{6}$, $\sqrt{3} = 1.732\ldots$, $\sqrt{3} = 1.732\ldots$, $\pi = 3.1415$, and $e = 2.718\ldots$.

The set of real numbers is composed of the **rational numbers** and the **irrational numbers**. For more, see **Real Number System**.

▶ **Real Number Line**   If the set of real numbers is placed into a one-to-one correspondence with the points on a line, the result is a **real number line**. In this manner, every real number corresponds with a point on the line and every point on the line corresponds with a real number.

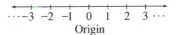

The number that corresponds with the point on the line is called the **coordinate** of the point. The point with a coordinate of 0 is called the **origin**. The points to the left of the origin have negative coordinates and the points to the right of the origin have positive coordinates.

Also see **Number Line, Distance on a Number Line**, or **Real Number System**.

➤ **Real Number System**   The **real number system** is composed of the set of real numbers, the properties pertaining to the set of real numbers, the operations of addition, subtraction, multiplication, and division, and the properties pertaining to those operations.

  The set of real numbers is composed of the **rational numbers** and the **irrational numbers**. Rational numbers are fractions of the form $\frac{a}{b}$ where a and b are integers with $b \neq 0$. Irrational numbers are decimal numbers that do not terminate and do not repeat. See **Rational Number, definition**, **Irrational Numbers**, **Properties of the Real Number System**, or **Real Number**.

➤ **Real Part**   In a complex number $a + bi$, the a term is referred to as the **real part** of the complex number. See **Complex Numbers**, or **Complex Plane**.

➤ **Real Root**   If the solution of an equation is an element of the real numbers, then the solution is called a **real root** of the equation.

➤ **Reals**   The term **reals** refers to the real numbers. If a number belongs to the reals, it is an element of the set of real numbers. See **Real Number**.

➤ **Reciprocal**   When two numbers multiply each other to yield 1, then one is said to be the multiplicative inverse or **reciprocal** of the other. There is no reciprocal for 0.

  The reciprocal of 7 is $\frac{1}{7}$ since $7 \bullet \frac{1}{7} = 1$. The reciprocal of $-\frac{3}{4}$ is $-\frac{4}{3}$ since $\left(-\frac{3}{4}\right) \bullet \left(-\frac{4}{3}\right) = 1$. The reciprocal of $1\frac{3}{5} = \frac{8}{5}$ is $\frac{5}{8}$ since $\frac{8}{5} \bullet \frac{5}{8} = 1$.

  The multiplicative inverse states that, for any real number $a \neq 0$, there is a unique (one and only one) real number $\frac{1}{a}$ such that $a \bullet \frac{1}{a} = 1$. Each of the numbers a and $\frac{1}{a}$ are reciprocals of one another. See **Multiplicative Inverse** or **Properties of the Real Number System**.

➤ **Reciprocal Identities**   In trigonometry, the pairs sine and cosecant, cosine and secant, and tangent and cotangent are reciprocal functions of one another. Combined, they form the set of identities known as the **reciprocal identities**. See **Trigonometry, identities**.

  In right $\triangle ABC$, we have

$$\sin A \bullet \csc A = 1 \qquad \cos A \bullet \sec A = 1 \qquad \tan A \bullet \cot A = 1$$
$$\sin A = \frac{1}{\csc A} \qquad \cos A = \frac{1}{\sec A} \qquad \tan A = \frac{1}{\cot A}$$
$$\csc A = \frac{1}{\sin A} \qquad \sec A = \frac{1}{\cos A} \qquad \cot A = \frac{1}{\tan A}$$

Some texts include the **quotient identities** as reciprocal identities:

$$\tan A = \frac{\sin A}{\cos A} \qquad \text{and} \qquad \cot A = \frac{\cos A}{\sin A}$$

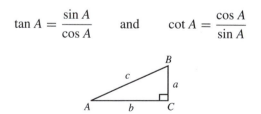

Why the reciprocal identities occur can be seen by looking at a right $\triangle ABC$. From the triangle, we see that $\sin A = \frac{a}{c}$. By algebra, we know that the reciprocal of $\frac{a}{c}$ is $\frac{c}{a}$. Hence, $\frac{a}{c} \bullet \frac{c}{a} = 1$. Since $\csc A = \frac{c}{a}$, we have $\frac{a}{c} \bullet \frac{c}{a} = \sin A \bullet \csc A = 1$ by substitution.

Similarly, $\cos A \bullet \sec A = 1$ since $\frac{b}{c} \bullet \frac{c}{b} = 1$ and $\tan A \bullet \cot A = 1$ since $\frac{a}{b} \bullet \frac{b}{a} = 1$. See **Reciprocal** or **Trigonometry, right triangle definition of**.

1. Prove $\tan A \bullet \csc A = \sec A$.

   It is easiest to work on the left side of the identity and change to sin and cos. See **Trigonometry, identities**. Since $\tan A = \frac{\sin A}{\cos A}$ and $\csc A = \frac{1}{\sin A}$, we have

$$\tan A \bullet \csc A = \sec A$$

$$\frac{\sin A}{\cos A} \bullet \frac{1}{\sin A} = \sec A \qquad \text{Substitute for } \tan A \text{ and } \csc A.$$

$$\frac{1}{\cos A} = \sec A$$

$$\sec A = \sec A \qquad \text{Substitute } \sec A \text{ for } \frac{1}{\cos A}.$$

2. Prove $2\cos A = \sin A \cot A + \frac{1}{\sec A}$.

   Changing to sin and cos and working on the right side of the identity only, we have

$$2\cos A = \sin A \cot A + \frac{1}{\sec A}$$

$$2\cos A = \sin A \bullet \frac{\cos A}{\sin A} + \frac{1}{\frac{1}{\cos A}} \qquad \text{Substitute for } \cot A \text{ and } \sec A.$$

$$2\cos A = \cos A + \cos A$$

$$2\cos A = 2\cos A$$

The reciprocal $\frac{1}{\frac{1}{\cos A}} = 1 \div \frac{1}{\cos A} = 1 \bullet \frac{\cos A}{1} = \cos A$.

➤ **Rectangle**   A **rectangle** is a parallelogram with four right angles. See **Parallelogram**. In addition to the properties of a parallelogram, a rectangle has the property that its diagonals are congruent. A special rectangle is a square. See **Square**.

   The **area of a rectangle** can be found by multiplying its length and width, $A = LW$. The perimeter of a rectangle can be found by adding the lengths of all four sides,

$$P = L + W + L + W = 2L + 2W$$

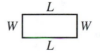

1. Find the area and the perimeter of a rectangle with length $L = 18$ inches and width $W = 12$ inches.

Substituting into the equations for area and perimeter, we have

$$A = LW = (18)(12) = 216 \text{ inches}$$
$$P = 2L + 2W = 2(18) + 2(12) = 36 + 24 = 60 \text{ inches}$$

➤ **Rectangular Coordinate System**   See **Coordinate Plane**.

➤ **Reduce**   **Reduce** is a term usually applied to fractions. To reduce a fraction means to express it in lowest terms. See **Fractions, reducing**.

➤ **Reducing Algebraic Fractions**   See **Rational Expressions, simplifying (reducing)**.

➤ **Reducing Fractions**   See **Fractions, reducing**.

➤ **Reducing Square Roots**   See **Radicals, simplifying (reducing)**.

➤ **Reference Angle**   See **Trigonometry, reference angle**.

➤ **Reflection**   See **Translation or Reflection**.

➤ **Reflexive Property**   The **reflexive property** is one of the three axioms of equality. The other two are the **symmetric** and **transitive properties**. See the articles under those names.
    The reflexive property states that if a is a real number then $a = a$. Also see **Property** or **Properties of the Real Number System**.

➤ **Regular Polygon**   A **regular polygon** is a polygon that is equilateral and equiangular. The first two regular polygons are an equilateral triangle and a square. For more, see **Polygon**, **Perimeter**, **of a Polygon**, or **Area of a Regular Polygon**.

➤ **Regular Prism**   A right prism with bases that are regular congruent polygons is a **regular prism**. For more, see **Prism** or **Area of the Lateral Surface of a Right Prism (and total area)**.

➤ **Regular Pyramid**   A **regular pyramid** has a base that is a regular polygon, lateral edges of equal length, and an altitude that passes through the center of the base. The length of the altitude of each face is called the **slant height** and the faces, called **lateral faces**, are congruent triangles. For more, see **Pyramid** or **Area of the Lateral Surface of a Regular Pyramid (and total area)**.

➤ **Related Rates**   In differential calculus, the derivative is introduced as the slope of a tangent line to the graph of a function $f(x)$. Since slope is defined as the change in y divided by the change in x, the fraction $\frac{\Delta y}{\Delta x}$ represents the average rate of change of y with respect to x. When y changes by an amount $\Delta y = f(x + \Delta x) - f(x)$, the x-value changes by an amount Δx. Hence, we have the formula for a derivative

$$\frac{dy}{dx} = \lim_{\Delta x \to 0} \frac{\Delta y}{\Delta x} = \lim_{\Delta x \to 0} \frac{f(x + \Delta x) - f(x)}{\Delta x}$$

which represents the rate of change of y with respect to x.
    Since the change in y is related to the change in x, problems that involve such an interplay are referred to as **related rate** problems. Of particular interest are problems that relate the change of a variable with respect to time.

Examples

1. A ladder 26 feet long leans against a wall, and its foot is being moved away from the wall at the rate of 4 ft/sec (4 feet for every 1 second). When the foot of the ladder is 10 feet from the wall, how fast is the top of the ladder moving down the wall?

The question being asked is, what is the rate of change of the distance the ladder moves down the wall with respect to time? In one second, the top of the ladder moves a certain amount of feet down the wall. In the next second, the top of the ladder moves down the wall some more. How many feet per second is the top of the ladder moving down the wall?

Notice that as the foot of the ladder moves away from the wall, the x-value gets larger while the y-value get smaller. The length of the ladder remains constant. What the problem is asking for is, how fast is the y-value changing with respect to time? This translates into how fast is the top of the ladder moving down the wall and is written as $\frac{dy}{dt}$. So, what we want to find is $\frac{dy}{dt}$.

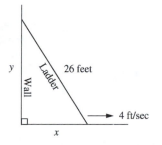

The main thing in a related rate problem is finding an equation that relates all of the rates. In the right triangle, we can see that we are relating x, y, and 26, which reminds us of the Pythagorean theorem. So, using the Pythagorean theorem and differentiating it with respect to time, we have

$$\frac{d}{dt}(x^2 + y^2) = \frac{d}{dt}(26)^2$$
$$2x\frac{dx}{dt} + 2y\frac{dy}{dt} = 0$$

where the $\frac{dx}{dt}$ and $\frac{dy}{dt}$ remain because they are variable. The right side of the equation is zero because 26 is a constant. Solving the equation for $\frac{dy}{dt}$, we have $\frac{dy}{dt} = \frac{-x}{y} \bullet \frac{dx}{dt}$.

Next, we need to see if we have all of the pertinent information pertaining to the equation for $\frac{dy}{dt}$. When the foot of the ladder is 10 feet from the wall means $x = 10$. The ladder is moved away from the wall at a rate of $4\frac{ft}{sec}$ means $\frac{dx}{dt}$. Further using $x = 10$ and $z = 26$ in the Pythagorean theorem, we see that $y = 24$. Therefore, we have everything we need to find $\frac{dy}{dt}$. Substituting, we have

$$\frac{dy}{dt} = \frac{-x}{y} \bullet \frac{dx}{dt} = \frac{-10}{24} \bullet 4 = \frac{-5}{3}$$

This means that the top of the ladder is moving down the wall at a rate of $\frac{5}{3}$ ft/sec ($\frac{5}{3}$ feet for every 1 second). The negative refers to down, as opposed to up (+).

2. Water runs into a conical tank at the rate of 3 ft^3 per minute. The height of the tank is 12 feet and the radius is 4 feet. When the water is 8 feet deep, how fast is the water level rising?

The water level is determined by y. If we want to know how fast the water level is rising, we want to know the change of y with respect to time. So, what we are looking for is $\frac{dy}{dt}$.

As far as the equation we will use in determining $\frac{dy}{dt}$ is concerned, we notice, by looking at the drawing, that there are two: one which is the volume equation for a cone and the other which involves similar triangles.

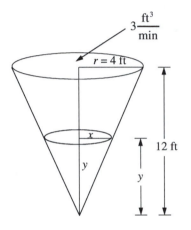

The volume equation involves all of the variables:

$$V = \frac{1}{3}\pi r^2 h = \frac{1}{3}\pi x^2 y$$

R

We need to differentiate it with respect to time and then solve for $\frac{dy}{dt}$. If we differentiated the equation now, the problem would be more cumbersome because it has the variables x and y in it. In the future, we will have to make a substitution for x anyway. Experience will teach you this. So, make the substitution now. This is done by using the equation for similar triangles:

$$\frac{x}{y} = \frac{12}{4} = 3$$
$$x = 3y$$

For practice, try making the substitution later.

Substituting x into the equation for volume, we have

$$V = \frac{1}{3}\pi x^2 y = \frac{1}{3}\pi(3y)^2 y = 3\pi y^3$$

Differentiating this equation with respect to time and then solving for $\frac{dy}{dt}$, we have

$$\frac{dV}{dt} = \frac{d}{dt}(3\pi y^3)$$

$$\frac{dV}{dt} = 9\pi y^2 \frac{dy}{dt}$$

So, $$\frac{dy}{dt} = \frac{\frac{dV}{dt}}{9\pi y^2}$$

We have all of the pertinent information pertaining to the right side of this equation. If we didn't, we would have to consider the drawing further to find a suitable substitution.

Since $\frac{dV}{dt}$ means the change of volume with respect to time, we know that $\frac{dV}{dt} = 3\frac{ft^3}{min}$. We are asked to find $\frac{dy}{dt}$ when the water level is 8 feet deep, so $y = 8$. Substituting, we have

$$\frac{dy}{dt} = \frac{3}{9\pi(8)^2} = \frac{1}{192\pi} \approx .0164 \frac{ft}{min}$$

➤ **Relation**   A **relation** is a set of ordered pairs. The set of all first coordinates of the ordered pairs is called the **domain** and the set of all second coordinates of the ordered pairs is called the **range**. See **Domain and Range**.

Visitors	1	0	1	0	2
Home	0	2	3	0	4

The set of scores for the first five innings of a baseball game determines a relation. If we let the visitors scores represent the domain and the home scores represent the range, then the relation of the scores is the set of ordered pairs $(1, 0)$, $(0, 2)$, $(1, 3)$, $(0, 0)$, and $(2, 4)$.

A relation can be determined by a rule of correspondence which assigns each element of the domain to an element of the range. The solution set of $y = \sqrt{x}$ determines a relation. If we choose x-values of 0, 1, and 4, the y-values of 0, ±1, and ±2 can be calculated.

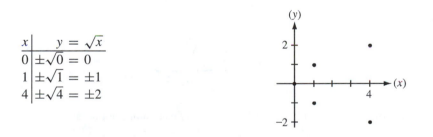

x	$y = \sqrt{x}$
0	$\pm\sqrt{0} = 0$
1	$\pm\sqrt{1} = \pm1$
4	$\pm\sqrt{4} = \pm2$

The relation determined by $y = \sqrt{x}$ is the set of ordered pairs $(0, 0)$, $(1, 1)$, $(1, -1)$, $(4, 2)$, and $(4, -2)$. Notice that each ordered pair does not necessarily have a different first coordinate. As a result, some points are vertically stacked one above the other.

Relations that have different first coordinates and have no points vertically stacked one above the other are in a category of their own. They are called functions. See **Function**.

➤ **Remainder**   In division, if the divisor does not divide completely into the dividend, the remaining portion is called the **remainder**. See **Division of Whole Numbers**. Also see **Difference**.

➤ **Remainder, in subtraction**   See **Difference**.

➤ **Remainder Theorem**   Let $f(x)$ be a polynomial. Then if $f(x)$ is divided by $x - r$, the remainder is $f(r)$. In essence, if we divide the polynomial $f(x)$ by $x - r$, the remainder is the same as evaluating the polynomial when $x = r$.

> *Example*

1. Let $f(x) = 3x^2 - 4x + 7$. Show that if $f(x)$ is divided by $x - 2$, the remainder is the same as $f(2)$.

   We first need to divide $3x^2 - 4x + 7$ by $x - 2$ using either long division or synthetic division. The remainder we get should be the same as the function value we get for $f(2)$.

   The remainder theorem says that we should use r, not $-r$. In our case, 2 not -2. If the binomial had been $x + 3$, we would have used -3 in the function, i.e., $f(-3)$. Doing the calculations side by side, we have

$$
\begin{array}{r}
3x + 2 \\
x - 2 \overline{)\ 3x^2 - 4x + 7} \\
\ominus 3x^2 \oplus 6x \\
\hline
2x + 7 \\
\ominus 2x \oplus 4 \\
\hline
11
\end{array}
\qquad
\begin{aligned}
f(x) &= 3x^2 - 4x + 7 \\
f(2) &= 3(2)^2 - 4(2) + 7 \\
&= 3 \bullet 4 - 8 + 7 \\
&= 12 - 1 = 11
\end{aligned}
$$

   Hence, we see that the remainder and the function value are the same.

   See **Long Division of Polynomials** or **Function Notation**. For examples using synthetic division, see **Synthetic Substitution**.

➤ **Remote Interior Angles**   In $\triangle ABC$ extend side $\overline{AC}$ to D. Then, in relationship to the exterior angle BCD, the **remote interior angles** of $\triangle ABC$ are $\angle A$ and $\angle B$.

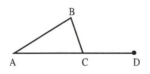

   The remote interior angles of a triangle are the angles that are not adjacent to an exterior angle of a triangle. Since there are six possible extensions to the three sides of $\triangle ABC$, the exterior angle must be determined first. Then the remote interior angles of the exterior angle can be determined.

   Of the many relationships between an exterior angle, its remote interior angles, and the triangle itself, two are the most outstanding. The exterior angle of a triangle has a measure that is greater than either of the measures of the remote interior angles. Also, the measure of the exterior angle is equal to the sum of the measure of the remote interior angles.

➤ **Repeating Decimals**   See **Decimal Numbers, repeating**.

➤ **Repeating Decimals, changing to a common fraction**   See **Decimal Numbers, changing a repeating decimal to a common fraction**.

➤ **Replacement Set**   A **replacement set** is the set of all numbers that a variable can represent. It is a set whose elements serve as replacements for the variable. The replacement set is also called the **domain**. See **Domain and Range**. The replacement set is also called the **universe** or **universe set**. See **Universal Set**.

➤ **Resultant of Vectors**   See **Vector**.

➤ **Rhombus**   A **rhombus** is a parallelogram with four congruent sides. The sides are equal in measure. See **Parallelogram**.

In addition to its attributes as a parallelogram, a rhombus has three other main qualities. The diagonals of a rhombus are perpendicular. The diagonals of a rhombus each bisect two angles of the rhombus. Also, a rhombus is a square.

To calculate the area or perimeter of a rhombus, see **Parallelogram**.

➤ **Right Angle**   A **right angle** is an angle that has a measure of 90. An angle with a measure greater than 90 is an **obtuse angle** and an angle with a measure less than 90 is an **acute angle**.

For a relationship to perpendicular, see **Perpendicular (geometric)** or **Perpendicular Lines, slopes of**.

➤ **Right Triangle**   A triangle with a right angle as one of its angles is called a **right triangle**. The side opposite the right angle is called the **hypotenuse**. The other two sides are called **legs**. The remaining angles are acute angles.

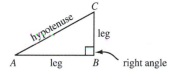

When referring to right $\triangle ABC$, the vertex of the right angle, B, is placed in the center. We do not say right $\triangle ACB$.

➤ **Rise and Run**   The **rise** is the y-value of the slope of a line and the **run** is the x-value. If $y = mx + b$ is the equation of a line, then

$$m = \frac{y}{x} = \frac{\text{rise}}{\text{run}}$$

For more, see **Slope**.

➤ **Root**   In reference to an equation, the **root** of an equation is the solution of the equation.

➤ **Root, cube**   See **Cube Root**.

➤ **Root Index**   In the radical $\sqrt[n]{a}$, the n is called the **root index** or radical index. See **Radical**.

➤ **Root, nth**   If a and b are real numbers and n is a natural number greater than one, then we have either of the following:

1. For even numbers n, if $a^n = b$ then $a = \sqrt[n]{b}$ or $a = -\sqrt[n]{b}$.

R

2. For odd numbers n, if $a^n = b$ then $a = \sqrt[n]{b}$.

The term $\sqrt[n]{b}$ is read "the nth root of b." When n is even, the solution of $a^n = b$ could have two answers, one positive and one negative. Instead of writing the two answers separately, we write $a = \pm\sqrt[n]{b}$, read "plus or minus nth root of b."

Examples

1. Solve $x^2 = 25$ by using roots.

   Since $n = 2$ is an even number, we have

   $$x^2 = 25$$
   $$\sqrt{x^2} = \pm\sqrt{25} \qquad \text{We solve the equation by taking the square root of both sides.}$$
   $$x = \pm 5$$

   The answers are $x = 5$ and $x = -5$.

   We can see the necessity of the $\pm$ if we solve the equation by factoring. See **Quadratic Equations, solutions of**.

   $$x^2 = 25$$
   $$x^2 - 25 = 0 \qquad \text{Set the equation equal to zero.}$$
   $$(x + 5)(x - 5) = 0 \qquad \text{Factor.}$$

   $$x + 5 = 0 \qquad \text{or} \qquad x - 5 = 0 \qquad \text{Solve each factor.}$$
   $$x = -5 \qquad\qquad\qquad x = 5$$

   Clearly, by this approach, there are two answers. If we only allowed the principal square root (positive root) when taking the $\sqrt{25}$ in the first approach, we would only have one answer. Therefore, as a rule we let $x = \pm 5$.

2. Solve $4x^2 - 49 = 0$ by using roots.

   $$4x^2 - 49 = 0$$
   $$4x^2 = 49$$
   $$x^2 = \frac{49}{4} \qquad \text{Isolate the } x^2 \text{ terms.}$$
   $$\sqrt{x^2} = \pm\sqrt{\frac{49}{4}} \qquad \text{Take the } \sqrt{\ } \text{ of both sides.}$$
   $$x = \pm\frac{7}{2} \qquad \text{Since } \sqrt{\frac{49}{4}} = \frac{\sqrt{49}}{\sqrt{4}} = \frac{7}{2}. \text{ See } \textbf{Radicals, division with.}$$

   The solutions are $x = \frac{7}{2}$ and $x = -\frac{7}{2}$.

3. Solve $8x^3 - 27 = 0$ by using roots.

$$8x^3 - 27 = 0$$
$$8x^3 = 27$$
$$x^3 = \frac{27}{8} \qquad \text{Isolate the } x^2 \text{ term.}$$
$$\sqrt[3]{x^3} = \sqrt[3]{\frac{27}{8}} \qquad \text{Take the } \sqrt[3]{\ } \text{ of both sides.}$$
$$x = \frac{3}{2} \qquad \text{Since } \sqrt[3]{\frac{27}{8}} = \frac{\sqrt[3]{27}}{\sqrt[3]{8}} = \frac{3}{2}. \text{ See } \textbf{Radicals, division with}.$$

Since $n = 3$ is an odd number, there is only one answer. The reason there is only one answer can be seen by solving $8x^2 - 27 = 0$ by factoring:

$$8x^3 - 27 = 0$$
$$(2x - 3)(4x^2 + 6x + 9) = 0$$

See **Factoring the Sum and Difference of Two Cubes**. Also see **Quadratic Equations, solutions of**.

$$2x - 3 = 0 \qquad \text{or} \qquad 4x^2 + 6x + 9 = 0$$
$$2x = 3 \qquad\qquad\qquad \varnothing \text{ (empty set)}$$
$$x = \frac{3}{2}$$

The first factor has a solution but the second does not. This can be seen by evaluating the discriminant $b^2 - 4ac$:

$$b^2 - 4ac = (6)^2 - 4(4)(9) = 36 - 144 = -108$$

Since the discriminant is negative, the solution is complex. See **Discriminant**.

➤ **Root, square**   See **Square Root**.

➤ **Roots of Multiplicity**   If an equation has a root (solution) that is repeated k times, then we say that the equation has a **root of multiplicity** k. For example, if the equation has two solutions the same, we say that it has a root of multiplicity 2, also called a **double root**. If the equation has three solutions the same, we say that it has a root of multiplicity 3.

If we solve the equation $x^2 - 2x + 1 = 0$, we will see that it has a root of multiplicity 2. See **Quadratic Equations, solutions of**.

$$x^2 - 2x + 1 = 0$$
$$(x - 1)(x - 1) = 0 \qquad\qquad \text{Factor.}$$

$$x - 1 = 0 \qquad \text{or} \qquad x - 1 = 0 \qquad \text{Solve each factor.}$$
$$x = 1 \qquad\qquad\qquad x = 1$$

Since the factors are the same, the solutions are the same. There is only one solution to the equation, $x = 1$. Therefore, $x^2 - 2x + 1 = 0$ has a root of multiplicity 2.

➤ **Rounding**   **Rounding** or **rounding off** is a method used to approximate numbers to a given place value. When rounding a number to a given place value, one of two rules must be followed:

1. If the digit to the right of the given place value is 5 or greater, then round up.

2. If the digit to the right of the given place is less than 5, then round down.

To **round up** means to raise the given digit by 1. The rest of the digits, to the right of the given digit, are replaced by zeros. The phrase rounding up implies that the entire number is raised in value.

To **round down** does not mean that the given digit is lowered by 1. In fact, the given digit remains the same. To round down, we leave the given digit alone while replacing the digits to the right of the given digit with zeros. In this manner, the entire number is lowered in value. It is rounded down.

Examples

1. Round to the nearest hundred: 68,476; 68,456; and 68,436.

   Since we are asked to round to the nearest hundred, the given place value to which we want to approximate each number is 4. In each number, we need to look at the digit to the right of 4, the 7, 5, and 3, respectively. In the first two cases, the 4 will be raised to 5. See rule 1. In the last case, the 4 remains the same. See rule 2.

68,476	68,456	68,436
↑	↑	↑
Since 7 is greater than 5,	Since 5 is equal to 5,	Since 3 is less than 5,
we round up.	we round up.	we round down.
= 68,500	= 68,500	= 68,400

   In the first two cases, the 4 is raised to a 5 and the remaining digits, to the right of 4, are replaced by zeros. In the last case, the 4 remains the same while the remaining digits, to the right of 4, are replaced by zeros.

2. Round to the nearest tenth: 6.984, 6.954, and 6.904.

   Since we are asked to round to the nearest tenth, the given place value to which we want to approximate each number is 9. In each number, we need to look at the digit to the right of 9, the 8, 5, and 0, respectively. In the first two cases, the 9 will be raised to a 10. The tens digit will be carried over to and added with the 6. The 0 will remain in the tenths place. In the, last case, the 9 remains the same.

6.984	6.954	6.904
↑	↑	↑
Since 8 is greater than 5,	Since 5 is equal to 5,	Since 0 is less than 5,
we round up.	we round up.	we round down.
= 7.000	= 7.000	= 6.900

For a list of the names of each position in a decimal number, see **Place Value**.

➤ **Rules for Exponents**   See **Exponents, rules of**.

➤ **Rules for Logarithms**   See **Logarithms, properties of (rules of)**.

➤ **Rules for Radicals**   See **Radicals, rules for**.

➤ **Run**   See **Rise and Run**. Also see **Slope**.

R

S s

➤ **Scalene Triangle**   A **scalene triangle** is a triangle in which no two sides are congruent (have equal measure).

➤ **Scientific Notation**   Sometimes very large or very small numbers are difficult to work with. In these cases, it is easier to carry out various operations if the numbers are changed to scientific notation. **Scientific notation** is a convenient way of writing numbers in the form $a \times 10^n$ where a is a number between 1 and 10 and n is an integer.

To write a number in scientific notation, we need to move the decimal point to the right of the first nonzero digit of the number and then indicate the movement by multiplying by a power of 10, i.e., 10^n. If we move the decimal to the left, the n is $+$ (positive). If we move the decimal to the right, the n is $-$ (negative).

> *Examples*

1. Write 46,580,000 and .0000371 in scientific notation.

   In the first number, we move the decimal point from its current location, behind the extreme right 0, to the right of the first nonzero digit, the 4. The new location of the decimal point is between the 4 and the 6. Then we multiply by 10^7, where the 7 indicates how many places we moved the decimal point to the left.

$$46,580,000. = 4.658 \times 10^7$$

   Move the decimal 7 places to the left.

   In the second number, we move the decimal point from its current location and place it to the right of the first nonzero digit, the 3. The new location of the decimal point is between the 3 and the 7. Then we multiply by 10^{-5}, where the -5 indicates how many places we moved the decimal point to the right.

$$.0000371 = 3.71 \times 10^{-5}$$

   Move the decimal 5 places to the right (-5).

2. Write 4.658×10^7 and 3.71×10^{-5} in decimal form.

What is required is to change the numbers from scientific notation back into decimal form. To reverse the process, we look at the power of 10. If it is + (positive), we will move the decimal point to the right. If it is − (negative), we will move the decimal point to the left. Remember that the movement changing out of scientific notation is reversed from the movement changing into scientific notation.

In the first number, since the 7 is +, we move the decimal point to the right.

$$4.658 \times 10^7 = 4.6580000 = 46{,}580{,}000$$

Move the decimal 7 places to the right.

In the second number, since the 5 is −, we move the decimal point to the left.

$$3.71 \times 10^{-5} = 00003.71 = .0000371$$

Move the decimal 5 places to the left (−5).

3. Multiply 46,580,000 and .0000371 using scientific notation.

We first need to write each number in scientific notation and then multiply. When multiplying the powers of 10, we apply the rules of exponents, i.e., $10^m \bullet 10^n = 10^{m+n}$. See **Exponents, rules of**.

$$
\begin{aligned}
(46{,}580{,}000)(.0000371) &= \left(4.658 \times 10^7\right)\left(3.71 \times 10^{-5}\right) \\
&= (4.658)(3.71) \times 10^7 \times 10^{-5} \\
&= 17.28118 \times 10^2 \qquad \text{In scientific notation.} \\
&= 1{,}728.118 \qquad \text{In decimal form.}
\end{aligned}
$$

4. Simplify $\frac{(560{,}000)(.0012)}{42{,}000{,}000}$ using scientific notation.

$$
\begin{aligned}
\frac{(560{,}000)(.0012)}{42{,}000{,}000} &= \frac{(5.6 \times 10^5)(1.2 \times 10^{-3})}{4.2 \times 10^7} \\
&= \frac{(5.6)(1.2)}{4.2} \times 10^5 \times 10^{-3} \times 10^{-7} \qquad \text{Move } 10^7 \text{ up as } 10^{-7}, \text{ or see below.} \\
&= 1.6 \times 10^{-5} \qquad \text{In scientific notation.} \\
&= .000016 \qquad \text{In decimal form.}
\end{aligned}
$$

If it is confusing to move the 10^7 from the denominator to the numerator and write it as 10^{-7}, deal with the powers of 10 another way:

$$\frac{(10^5)(10^{-3})}{10^7} = \frac{10^2}{10^7} = 10^{2-7} = 10^{-5}$$

See **Exponents, negative** or **Exponents, rules of**. If significant digits are to be considered, see **Significant Digit**.

➤ **Secant, in trigonometry**   The **secant** is one of the six functions of trigonometry. It should not be confused with a secant line. In trigonometry, the secant is defined as a fraction of two sides of a right triangle. It is a fraction of the **hypotenuse** to the **adjacent** side (reciprocal of cosine).

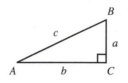

From angle A, $\sec A = \frac{c}{b}$. From angle B, $\sec B = \frac{c}{a}$. For more, see **Trigonometry, right Triangle definition of**.

➤ **Secant Line**   In relation to a curve, a **secant line** is any line that intersects the curve in two different points. It passes through the curve.

In geometry, if we consider a circle, a **secant line** is any line that contains a chord, including the diameter. In the figure, both $\overleftrightarrow{CD}$ and $\overleftrightarrow{AB}$ are secant lines.

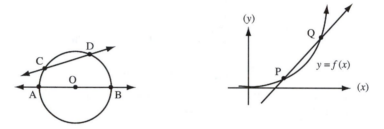

In calculus, when deriving a derivative by definition, the secant line to a curve is considered. If the curve is defined by $y = f(x)$, then the secant line to the curve through the points P and Q is the line $\overleftrightarrow{PQ}$. However, it is the slope of the segment $\overline{PQ}$ that is of interest. See **Derivative**, Derivative **by Definition**.

➤ **Seconds**   See **Degree-Minutes-Seconds**.

➤ **Sector of a Circle**   If two radii are drawn to the endpoints of an arc, then the region bounded by the radii and the arc is called a **sector** of the circle.

In the figure, the sector of the circle from O to A to B through C is denoted $O - \overparen{ACB}$. The sector from O to A to B through D is denoted $O - \overparen{ADB}$. See **Major and Minor Arc**.

The equations for the **area** and **perimeter** of a sector with an arc of measure m are

$$A = \frac{m}{360} \bullet \pi r^2 \qquad \text{for } m \text{ measured in degrees.}$$

$$P = 2r + s \qquad$$

where s is the arc length and $s = \frac{m}{180} \bullet \pi r$, if m is in degrees or $s = r\theta$ where θ is the measure of the central angle in radians. See **Arc Length**.

Examples

1. If the central angle of an arc is 60°, find the area and perimeter of its sector if the radius of the circle is 5 inches.

Since the measure of an arc is equal to the measure of its central angle, the measure of the arc is 60°. Hence, we have the area of the sector subtended by a 60° arc with a radius of 5 inches:

$$A = \frac{m}{360} \bullet \pi r^2 = \frac{60}{360} \bullet \pi (5)^2 = \frac{6.25\pi}{36} = \frac{25\pi}{6} \text{ in}^2$$
$$\approx \frac{25(3.14)}{6} \approx 13.08 \text{ in}^2$$

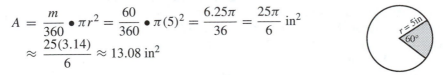

To find the perimeter, we first need to find the length of the arc. Since the arc is measured in degrees, we can use the first equation to find s:

$$s = \frac{m}{180} \bullet \pi r = \frac{60}{180} \bullet \pi (5) = \frac{5\pi}{3}$$

Hence, we have the perimeter

$$P = 2r + s = 2(5) + \frac{5\pi}{3} = 10 + \frac{5\pi}{3} \approx 10 + \frac{5(3.14)}{3} = 15.23 \text{ in}$$

➤ **Sector of a Circle, area of**   See **Sector of a Circle.**

➤ **Sector of a Circle, perimeter of**   See **Sector of a Circle.**

➤ **Segment**   A **segment** is the set of all points between two points A and B, and including A and B. The points A and B are called endpoints of the segment.

•———————————•
A B

The symbol for a segment is a bar (—) written over AB and we write $\overline{AB}$. Do not confuse $\overline{AB}$ with AB. $\overline{AB}$ is a segment whereas AB is the measure of the segment. See **Congruent Segments.**

➤ **Segment Addition**   See **Betweenness of Points.**

➤ **Segment Bisector**   See **Bisector of a Segment.**

➤ **Semicircle**
A **semicircle** is the set of all points of a circle lying on one side of the diameter and including the endpoints of the diameter. Arc $\overparen{ACB}$ is a semicircle of circle O and so is $\overparen{ADB}$.

▶ **Sequence**   See **Arithmetic Progression** or **Geometric Progression**.

▶ **Series**   See **Arithmetic Series** or **Geometric Series**.

▶ **Set**   A **set** is a collection of objects. Each of the objects is called a **member** or **element** of the set and is said to "belong to" or to be "contained in" the set. A set is enclosed by braces and the elements of the set are listed within the braces. We use the symbol $\in$ to indicate that an object is an element of a set and the symbol $\notin$ to indicate that an object is not an element of a set.

The set of natural numbers between 2 and 7 is $S = \{3, 4, 5, 6\}$ where S represents the set and the elements of the set are the numbers 3, 4, 5, and 6. We say that $3 \in S$, $4 \in S$, $5 \in S$, and $6 \in S$. The numbers 2 and 7 are not in S and we write $2 \notin S$ and $7 \notin S$.

Whether S is written $\{3, 4, 5, 6\}$ or $\{5, 3, 6, 4\}$ makes no difference. Sets are **equal sets** if they contain the same elements. The order of the elements does not matter.

If it is possible to count all the elements in a set, the set is said to be **finite**. The set S is finite. It has four elements. If it is not possible to count all the elements in a set, the set is said to be **infinite**. The set of natural numbers is infinite, $N = \{1, 2, 3, \ldots\}$. They cannot be counted.

A set that has no elements in it is called an **empty set** or **null set**. We denote an empty set by ϕ (pronounced fee) or $\{\ \}$. A common mistake is to say that $\{\phi\}$ or $\{0\}$ represent an empty set. These are incorrect. $\{\phi\}$ means the set with the Greek letter $\{\phi\}$ in it while $\{0\}$ means the set with the number 0 (zero) in it.

If every element of a set A is contained in another set B, we say that A is a subset of B and we write $A \subset B$. Consider the set of natural numbers $N = \{1, 2, 3, \ldots\}$ and the set of integers $Z = \{-\infty, \ldots, -2, -1, 0, 1, 2, \ldots, +\infty\}$. Since every element of N is contained in Z, N is a subset of Z and we write $N \subset Z$.

Since every element of the natural numbers is contained in the set of natural numbers, the set of natural numbers is a subset of itself. Such a subset is called an **improper subset**. All other subsets are called **proper subsets**.

For more, see **Union and Intersection**, **Solution Set**, **Subset**, **Set Notation**, or **Venn Diagram**. For a comparison of set notation and interval notation, see **Interval Notation**.

▶ **Set Notation**   When describing sets or concepts related to sets, various forms of notation or set notation are used. The following is a list of examples and explanations.

1. $x \in A$, means that x is an element of the set A.

2. $x \notin A$, means that x is not an element of the set A.

3. ϕ, phi is the symbol for the empty set, pronounced "fee."

4. $\{\ \}$, also a symbol for the empty set.

5. $A \subset B$, means that the set A is a subset (is contained in) the set B.

6. $A \not\subset B$, means that the set A is not a subset of set B.

7. $\{1, 2, 3, \ldots, 10\}$, means the finite set of numbers from 1 to 10.

8. $\{0, 1, 2, \ldots\}$, means the infinite set of numbers from 0 on up to positive infinity.

9. : or |, either a colon or a vertical bar means "such that."

10. $\forall$, means "for all."

11. $A \cup B$, means the union of set A and set B.

12. $A \cap B$, means the intersection of set A and set B.

13. $\overline{A}$, means the complement of the set A.

Examples

1. Consider the number a on a number line. Using set notation, describe the numbers to the right of a, to the left of a, and including a.

Numbers to the right of a are $x > a$. In set notation, this is written $\{x : x \in R, \ x > a\}$, read "the set of all x such that x is an element of the real numbers and x is greater than a."

Numbers to the right of a including a are $x \geqslant a$. In set notation, we write $\{x : x \in R, \ x \geqslant a\}$.

Numbers to the left of a are $x < a$. In set notation, we write $\{x \mid x \in R, \ x < a\}$, using the vertical bar for such that.

Numbers to the left of a and including a are $x \leqslant a$. In set notation, we write $\{x \mid x \in R, \ x \leqslant a\}$.

Numbers to the right of a, left of a, and including a, in other words the whole number line, are written as $\{x \mid x \in R, \ -\infty < x < \infty\}$.

2. Let a and b be two numbers on a real number line. Using set notation, describe the numbers to the left of a, between a and b including b, and to the right of b including b.

a. The numbers to the left of a are $\{x : x \in R, \ x < a\}$.

b. The numbers between a and b and including b are $\{x : x \in R, \ a < x \leqslant b\}$.

c. The numbers to the right of b and including b are $\{x \mid x \in R, \ x \geqslant b\}$.

3. If x is a real number, describe the domain of $y = \sqrt{x}$, using set notation.

The $\sqrt{x}$ is not defined for negative values of x. We cannot take the square root of a negative number. The square root is defined for 0 and any positive values. It is defined for $x \geqslant 0$. In set notation, we say that the domain of x is $\mathcal{D} = \{x \mid x \in R, \ x \geqslant 0\}$. See **Domain and Range**. For more, see **Union and Intersection**.

► **Set of Real Numbers**   The **set of real numbers** is the union of the set of **rational numbers** with the set of **irrational numbers**. Hence, the elements of the set of real numbers are numbers that can be expressed in decimal form. See **Rational Number (definition)**, **Irrational Numbers**, or **Real Number**.

► **Shift**   See **Phase Shift**.

➤ **Sigma Notation** **Sigma notation** is a compact form of writing a sum. The sum of the first n terms of the sequence $a_1, a_2, \ldots, a_n$ where n is a natural number is $a_1 + a_2 + a_3 + \bullet s + a_n$.

A simpler way of writing the sum using the summation sign $\sum$ (sigma) is

$$\sum_{k=1}^{n} a_k \qquad \text{where } k \text{ is a natural number}$$

Hence, we have $\sum_{k=1}^{n} a_k = a_1 + a_2 + a_3 + \bullet s + a_n$.

As an example find $\sum_{k=1}^{4} 3k$, read "the sum from k equal 1 to 4 of $3k$," where k is called the **index of summation**. Expanding the sum, we have

$$\sum_{k=1}^{4} 3k = 3(1) + 3(2) + 3(3) + 3(4) = 3 + 6 + 9 + 12 = 30$$

Examples

1. Find the sum of $\sum_{k=3}^{6}(k+2)^2$.

$$\sum_{k=3}^{6}(k+2)^2 = (3+2)^2 + (4+2)^2 + (5+2)^2 + (6+2)^2$$

$$= 5^2 + 6^2 + 7^2 + 8^2 = 25 + 36 + 49 + 64 = 174$$

2. Write, using sigma notation, $2 + 5 + 10 + 17$.

Since

$$2 + 5 + 10 + 17 = (1+1) + (4+1) + (9+1) + (16+1)$$
$$= (1^2 + 1) + (2^2 + 1) + (3^2 + 1) + (4^2 + 1)$$

we see that each term is generated by $k^2 + 1$. Hence, we have

$$2 + 5 + 10 + 17 = \sum_{k=1}^{4}(k^2 + 1)$$

➤ **Signed Numbers** See **Positive and Negative Numbers**.

➤ **Significant Digit** When a number is approximated, the digits that determine the accuracy of the number are called **significant digits**. If the number has two significant digits, it is said to have "two-digit accuracy." If the number has three significant digits, it is said to have "three-digit accuracy," and so on.

In general, we determine the amount of significant digits in a number by counting the digits from left to right, starting with the first nonzero digit. Zeros that appear between numerical digits are significant while zeros at the end of the number are not, unless otherwise specified. If there are no numerals before the decimal point, zeros between the decimal and the first numeral to the right are not significant. If there are numerals to the left of the decimal point, zeros to the right of the decimal point are significant. All the digits in the first factor of a number expressed in scientific notation are significant.

Example

Number	Number of Significant Digits
752	3
702	3
700	1
752.0	4
2,030	3, from 203
3.24	3
0.81	2
0.07	1
.07	1
496.1	4
.0038	2
42.150	5
7.004	4
8×10^{12}	1
8.0×10^{12}	2
4.921×10^{-5}	4

➤ **Signs in Trigonometry**   See **Trigonometry, signs of the six functions**.

➤ **Signs or Symbols in Math**   See **Math Symbols**.

➤ **Similar Polygons**   **Similar polygons** are polygons that have corresponding angles congruent and corresponding sides proportional. See **Congruent Angles** or **Proportion**. Also see **Polygon**.

Examples of similar polygons are: any two equilateral triangles, any two squares, and any two regular pentagons, or hexagons, etc. Any polygon is similar to itself and congruent polygons are similar to each other.

S

Example

1. If quad $ABCD \sim$ quad $WXYZ$ ($\sim$ means similar), find AB, AD, CD, $m\angle A$, and $m\angle B$.

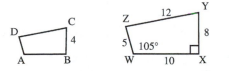

Since the quadrilaterals are similar, their corresponding angles are congruent. Therefore, $m\angle A = 105$ since $\angle A \cong \angle W$ ($\cong$ means congruent), $m\angle B = 90$ since $\angle B \cong \angle X$, and $m\angle X = 90$. If two segments are perpendicular, they meet to form a right angle, and right angles have a measure of 90.

Also, since the quadrilaterals are similar, their corresponding sides are proportional. Hence, we have

$$\frac{AB}{WX} = \frac{BC}{XY} \qquad \text{or} \qquad \frac{AD}{WZ} = \frac{BC}{XY} \qquad \text{or} \qquad \frac{CD}{YZ} = \frac{BC}{XY}$$

$$\frac{AB}{10} = \frac{4}{8} \qquad\qquad \frac{AD}{5} = \frac{4}{8} \qquad\qquad \frac{CD}{12} = \frac{4}{8}$$

$$AB = \frac{4 \bullet 10}{8} \qquad\qquad AD = \frac{5 \bullet 4}{8} \qquad\qquad CD = \frac{12 \bullet 4}{8}$$

$$AB = 5 \qquad\qquad AD = 2\frac{1}{2} \qquad\qquad CD = 6$$

See **Cross Multiplication**. Also see **Similar Triangles**.

➤ **Similar Terms**   See **Combining Like Terms**.

➤ **Similar Triangles**   Two triangles are **similar triangles** if the angles of one triangle are congruent to the angles of the other triangle. If $\angle A \cong \angle D$ ($\cong$ means congruent; see **Congruent Angles**), $\angle B \cong \angle E$, and $\angle C \cong \angle F$, then the two triangles are similar and we write $\triangle ABC \sim \triangle DEF$ read "triangle ABC is similar to triangle DEF."

For two triangles to be similar triangles means that they are similar in shape, not equal in size. Triangles that are equal in size are called congruent triangles. See **Congruent Triangles**.

One of the most important aspects of similar triangles is that their corresponding sides are proportional. If $\triangle ABC \sim \triangle DEF$, then

$$\frac{AB}{DE} = \frac{BC}{EF}, \qquad \frac{BC}{EF} = \frac{AC}{DF} \qquad \text{or} \qquad \frac{AB}{DE} = \frac{AC}{DF}$$

Other aspects of similar triangles are:

1. Two triangles are similar if two angles of one triangle are congruent to two angles of another triangle.

2. Two right triangles are similar if an acute angle of one right triangle is congruent to an acute angle of another right triangle.

3. Two isosceles triangles are similar if they have congruent vertex angles.

4. Two triangles are similar if an angle of one triangle is congruent to an angle of another triangle and the sides included between the angles are proportional.

5. Two triangles are similar if their corresponding sides are proportional.

6. If a line intersects two sides of a triangle and is parallel to the third, then the triangles formed are similar.

7. If two triangles are similar, then their perimeters are proportional to any pair of corresponding sides.

8. If two triangles are similar, then their altitudes are proportional to any pair of corresponding sides.

9. If two triangles are similar, then their medians are proportional to any pair of corresponding sides.

10. A ray that bisects an angle of a triangle divides the opposite side into segments that are proportional to the lengths of the other two sides.

11. The lengths of the corresponding altitudes of two similar triangles have the same ratio as the lengths of any pair of corresponding sides.

12. If three parallel lines intersect two transversals, then the segments between each of the parallels are proportional.

13. The triangles formed by the altitude to the hypotenuse of a right triangle are similar to each other and to the original triangle.

14. In a right triangle, the altitude to the hypotenuse is the geometric mean between the segments it forms with the hypotenuse.

Examples

1. In the figure below, if $\triangle ABC \sim \triangle DEF$, find the lengths of sides x and y.

Since the triangles are similar, their corresponding sides are proportional. Hence,

$$\frac{AB}{DE} = \frac{AC}{DF} \quad \text{and} \quad \frac{AB}{DE} = \frac{BC}{EF}$$

Substituting in each equation and solving by cross multiplication, we have

$$\frac{4}{12} = \frac{5}{x} \qquad \frac{4}{12} = \frac{3}{y}$$

$$x = \frac{12 \bullet 5}{4} \quad \text{and} \quad y = \frac{12 \bullet 3}{4}$$

$$x = 15 \qquad y = 9$$

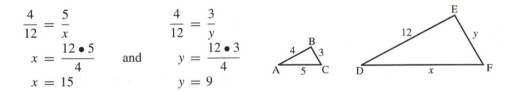

See **Cross Multiplication**.

2. In the figure below, if $\triangle AEB \sim \triangle DEF$, find the lengths of sides $\overline{DF}$ and $\overline{EF}$.

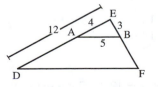

When one figure is inside another figure, it can be useful to split them into separate figures. If we pull $\triangle AEB$ out of $\triangle DEF$, the figures will look like below.

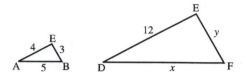

Letting $DF = x$ and $EF = y$, we see that the problem is the same as example 1 with

$$\frac{AE}{DE} = \frac{AB}{DF} \quad \text{and} \quad \frac{AE}{DE} = \frac{EB}{EF}$$

3. In the figure below, if $\overline{DE}$ is parallel to $\overline{AC}$, find the length of AD.

Since $\overline{DE}$ is parallel to $\overline{AC}$, we see from the other aspects of similar triangles, #6, that $\triangle DBE \sim \triangle ABC$. Let $AD = x$ and separate the figures.

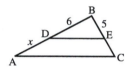

When the figures are separated, we see that $DB = 6$, $BE = 5$, $AB = x + 6$, and $BC = 15$. Since the triangles are similar, we know that their corresponding sides are proportional. Hence,

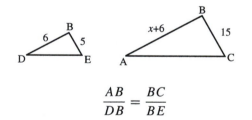

$$\frac{AB}{DB} = \frac{BC}{BE}$$

Substituting and solving, we can find the value of x. See **Cross Multiplication**.

$$\frac{x + 6}{6} = \frac{15}{5}$$
$$x + 6 = \frac{6 \bullet 15}{5}$$
$$x + 6 = 18 \qquad \text{See \textbf{Linear Equations, solutions of.}}$$
$$x = 18 - 6$$
$$x = 12$$

Since x is the length of $\overline{AD}$, $AD = 12$. ($\overline{AD}$ is a segment while AD is the measure of the segment.)

4. In the figure below, let $\triangle ABC$ be a right triangle and let $\overline{BD}$ be the altitude from B. Find the length of BD.

Since $\triangle ABC$ is a right triangle and $\overline{BD}$ is the altitude to the hypotenuse $\overline{AC}$, we see from the other aspects of similar triangles, #14, that BD is the geometric mean between AD and CD. Hence,

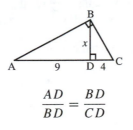

$$\frac{AD}{BD} = \frac{BD}{CD}$$

See **Geometric Mean**. Substituting and solving for x, we have

$$\frac{9}{x} = \frac{x}{4}$$
$$9 \bullet 4 = x^2$$
$$x^2 = 36 \qquad \text{Symmetric property.}$$
$$\sqrt{x^2} = \sqrt{36} \qquad \text{Use the principal root only, since } BD \text{ is distance (positive.)}$$
$$x = 6$$

Since $BD = x$, we have the length of $BD = 6$.

To solve the equation, see **Quadratic Equations, solutions of**.

➤ **Simple Interest**   See **Interest**.

➤ **Simplify Fractions**   See **Fractions, reducing**.

➤ **Simplify Radicals**   See **Radicals, simplifying (reducing)**.

➤ **Simplify Square Roots**   See **Radicals, simplifying (reducing)**.

➤ **Simplifying Like Terms**   See **Combining Like Terms**.

➤ **Simultaneous Solutions of Equations**   See **Systems of Linear Equations**.

➤ **Sine**   The **sine** is one of the six functions of trigonometry. Defined for a right triangle, it is a function of the **opposite** side to the **hypotenuse**. From angle A, $\sin A = \frac{a}{c}$. From angle B, $\sin B = \frac{b}{c}$.

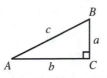

For more, see **Trigonometry, right triangle definition of**.

S

▶ **Skew Lines**   **Skew lines** are lines, not in the same plane, that do not intersect and are not parallel. The nonparallel flight paths of two airplanes continuing to fly at two different altitudes represent skew lines. Skew lines pass over one another but do not intersect.

▶ **Slant Height**   In a regular pyramid, the **slant height** is the length of the altitude of any one of its lateral faces. In a right circular cone, the slant height is the distance from any point on the circular base to the vertex. For an application, see **Pyramid** or **Cone (right circular).**

▶ **Slope**   If a nonvertical line passes through the points (x_1, y_1) and (x_2, y_2), then the slope of the line is defined to be the fraction

$$m = \frac{y_2 - y_1}{x_2 - x_1}$$

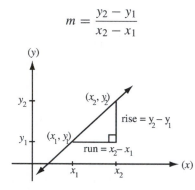

The change in the y-coordinates is called the rise and can be denoted by Δy (read "delta y"). The change in the x-coordinates is called the run and can be denoted by Δx. In reference to these terms, the slope is written

$$m = \frac{\text{change in } y\text{-coordinates}}{\text{change in } x\text{-coordinates}} = \frac{\text{rise}}{\text{run}} = \frac{\Delta y}{\Delta x}$$

The latter is the more familiar form seen in calculus.

Simply stated, slope deals with the steepness of a line, like the steepness of a hill. If the hill is less steep, the fraction of m is smaller. If the hill is more steep, the fraction of m is larger. Examples 1 and 2 show the change in the steepness.

A slope is **positive** if the line is increasing (going up as we move from left to right). See examples 1 and 2. A slope is **negative** if the line is decreasing (going down as we move from left to right). See example 3.

Examples 4 and 5 describe the slopes of horizontal and vertical lines. A **horizontal** line has a slope $m = 0$ while a **vertical** line has a slope $m =$ undefined.

The slopes of **parallel** lines are equal, $m_1 = m_2$. The slopes of **perpendicular** lines are negative reciprocals, $m_1 = \frac{-1}{m_2}$. See **Parallel Lines (slopes with)** and **Perpendicular Lines, slopes of.**

Examples

1. Given the points $A(1, 2)$ and $B(4, 8)$, find the slope of $\overline{AB}$ graphically (by using the graph) and analytically (by using the slope equation).

   To find the slope graphically, we need to plot the points, determine the rise and run, and then state the slope $m = \frac{\text{rise}}{\text{run}}$. From the graph, we can see that the change in the y-coordinates, the rise, is

$8 - 2 = 6$. Also, the change in the x-coordinates, the run, is $4 - 1 = 3$. Hence, the slope is

$$m = \frac{6}{3} = 2$$

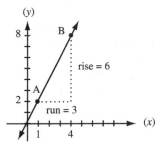

To find the slope analytically, we need to substitute the coordinates of A and B into the slope equation. Letting $(x_1, y_1) = (1, 2)$ and $(x_2, y_2) = (4, 8)$, we have

$$m = \frac{y_2 - y_1}{x_2 - x_1} = \frac{8 - 2}{4 - 1} = \frac{6}{3} = 2$$

Whether we choose the subscripts the way we did or whether we reverse the order makes no difference. Either way we will get the same slope. Suppose we chose $(x_2, y_2) = (1, 2)$ and $(x_1, y_1) = (4, 8)$, then

$$m = \frac{y_2 - y_1}{x_2 - x_1} = \frac{2 - 8}{1 - 4} = \frac{-6}{-3} = 2$$

2. Given the points $A(3, 2)$ and $B(7, 4)$, find the slope of $\overline{AB}$. From the graph, we can see that the slope is

$$m = \frac{\text{rise}}{\text{run}} = \frac{2}{4} = \frac{1}{2}$$

Or, we can find the slope using the slope equation. Letting $(x_1, y_1) = (3, 2)$ and $(x_2, y_2) = (7, 4)$, we have

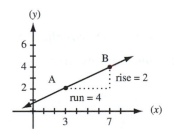

$$m = \frac{y_2 - y_1}{x_2 - x_1} = \frac{4 - 2}{7 - 3} = \frac{2}{4} = \frac{1}{2}$$

Comparing the graphs, as well as the slopes, of examples 1 and 2, we can see that the steepness of a line with $m = 2$ is more steep than a line with $m = \frac{1}{2}$. The smaller the slope, the less steepness there is in a line. The larger the slope, the more steepness there is in a line.

3. Given the points $A(-6, 4)$ and $B(-3, -2)$, find the slope of $\overline{AB}$.

We will first find the slope using the slope equation. We need to be careful with the negative signs. Letting $(x_1, y_1) = (-6, 4)$ and $(x_2, y_2) = (-3, -2)$, we have

$$m = \frac{y_2 - y_1}{x_2 - x_1} = \frac{-2 - 4}{-3 - (-6)} = \frac{-6}{-3 + 6} = \frac{-6}{3} = -2$$

See **Positive and Negative Numbers**.

Notice that since the line is decreasing (going down as we move from left to right) the sign of the slope is $-$. Decreasing lines have negative slopes. With this in mind, we can now find the slope by looking at the graph.

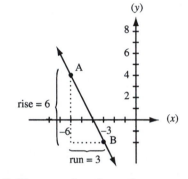

From the graph, we can see that the rise is 6 and the run is 3. These are lengths so they are positive. Distance is defined for positive values only. Hence, we have the slope

$$m = \frac{\text{rise}}{\text{run}} = -\frac{6}{3} = -2$$

Make the fraction negative because the line is decreasing.

4. Given the points $A(2, 3)$ and $B(6, 3)$, find the slope of $\overline{AB}$.

Letting $(x_1, y_1) = (2, 3)$, $(x_2, y_2) = (6, 3)$ and substituting in the slope equation, we have

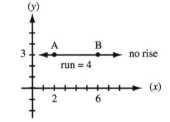

$$m = \frac{y_2 - y_1}{x_2 - x_1} = \frac{3 - 3}{6 - 2} = \frac{0}{4} = 0$$

The slope of a **horizontal line** is $m = 0$.

If we use the rise and run to calculate the slope, we have

$$m = \frac{\text{rise}}{\text{run}} = \frac{0}{4} = 0$$

5. Given the points $A(3, 2)$ and $B(3, 7)$, find the slope of $\overline{AB}$.

Letting $(x_1, y_1) = (3, 2)$, $(x_2, y_2) = (3, 7)$ and substituting in the slope equation, we have

$$m = \frac{y_2 - y_1}{x_2 - x_1} = \frac{7 - 2}{3 - 3} = \frac{5}{0} = \text{not defined}$$

The slope of a vertical line is not defined since division by zero is not defined.

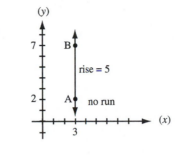

If we use the rise and run to calculate the slope, we have

$$m = \frac{\text{rise}}{\text{run}} = \frac{5}{0} = \text{not defined}$$

For more, see **Horizontal and Vertical Lines**. Also see **Linear Equations, graphs of** or **Linear Equation, finding the equation of a line**.

➤ **Slope Equation**   See **Slope**.

➤ **Slope-Intercept Form of an Equation of a Line**   An equation of a line that is written in the form

$$y = mx + b$$

where m is the slope and b is the y-intercept, is said to be in **slope-intercept form**. The other form of an equation of a line is **standard form**, written

$$Ax + By = C \qquad \text{or sometimes,} \qquad Ax + By + C = 0$$

See **Standard Form of an Equation of a Line**. To find an equation of a line by the slope-intercept method, see **Linear Equation, finding the equation of a line**, Slope-Intercept Method. To graph an equation of a line by the slope-intercept method, see **Linear Equations, graphs of**, Slope-Intercept Method.

To write an equation of a line in slope-intercept form, the equation must first be solved for y. The method used to solve the equation is that used to solve linear equations. See **Linear Equations, solutions of**. After the equation has been solved for y, the terms on the right-hand side of the equal sign must be simplified and placed in the proper order, the x-term first and the numerical term second. In this form, the coefficient of the x-term is the slope m of the linear equation while the numerical term b is the y-intercept.

Examples

1. Write the linear equation $8x - 4y = -12$ in slope-intercept form.

   To write the equation in slope-intercept form, we need to solve the equation for y and then simplify all terms into the form of $y = mx + b$:

$$8x - 4y = -12$$
$$-4y = -8x - 12 \qquad \text{Subtract } 8x \text{ from both sides.}$$
$$\frac{-4y}{-4} = \frac{-8x - 12}{-4} \qquad \text{Divide both sides by } -4.$$
$$y = \frac{-8x}{-4} - \frac{12}{-4} \qquad \text{See \textbf{Positive and Negative Numbers}.}$$
$$y = 2x + 3$$

Hence, we have the slope-intercept form of $8x - 4y = -12$, which is $y = 2x + 3$.

Comparing $y = 2x + 3$ with $y = mx + b$, we can determine the values of the slope and the y-intercept. The 2 is the slope $m = 2$ and the 3 is the y-intercept $b = 3$.

2. Write the linear equation $2y - (4x + 6) = -4(3x + 5)$ in slope-intercept form.

Solving for y and simplifying, we have

$$2y - (4x + 6) = -4(3x + 5)$$
$$2y - 4x - 6 = -12x - 20 \qquad \text{See \textbf{Distributive Property}.}$$
$$2y = -12x + 4x - 20 + 6 \qquad \text{Add } 4x \text{ and } 6 \text{ to both sides.}$$
$$2y = -8x - 14$$
$$\frac{2y}{2} = \frac{-8x - 14}{2}$$
$$y = \frac{-8x}{2} - \frac{14}{2}$$
$$y = -4x - 7$$

The equation $y = -4x - 7$ is the slope-intercept form of the given equation with a slope of $m = -4$ and y-intercept of $b = -7$.

3. Write the linear equation $Ax + By = C$ (standard form of a linear equation) in slope-intercept form.

$$Ax + By = C$$
$$By = -Ax + C$$
$$\frac{By}{B} = \frac{-Ax + C}{B}$$
$$y = \frac{-Ax}{B} + \frac{C}{B}$$
$$y = -\frac{A}{B}x + \frac{C}{B}$$

Comparing $y = -\frac{A}{B}x + \frac{C}{B}$ with $y = mx + b$, we see that the slope is $m = -\frac{A}{B}$ and the y-intercept is $b = \frac{C}{B}$. These equations can be useful when needing m and b and the equation is in standard form. See **Standard Form of an Equation of a Line**.

The equation in example 1 is written in standard form, $8x - 4y = -12$. Comparing this with $Ax + By = C$, we see that $A = 8$, $B = -4$, and $C = -12$. If we want the slope and the y-intercept of $8x - 4y = -12$, we need to substitute A, B, and C into $m = -\frac{A}{B}x$ and $b = \frac{C}{B}$. Therefore, the slope and y-intercept of $8x - 4y = -12$ are

$$m = -\frac{A}{B} = -\frac{8}{-4} = 2 \quad \text{and} \quad b = \frac{C}{B} = \frac{-12}{-4} = 3$$

➤ **Slope-Intercept Method of Finding an Equation of a Line**   See **Linear Equation, finding the equation of a line**, Slope-Intercept Method.

➤ **Slope-Intercept Method of Graphing a Line**   See **Linear Equations, graphs of**, Slope-Intercept Method.

➤ **Smaller and Bigger**   See the articles under **Inequalities**.

➤ **Smaller and Bigger Fractions**   See **Fractions, comparing**.

➤ **Solutions of Equations**   This article lists the categories of equations to be solved with an example of each type next to the heading. At the end of each heading is an article(s) of reference. Solutions of the equation will be found in that article.

1.   Absolute Value Equations: $-8 + |2x + 5| = 0$. See **Absolute Value Equations**.

2.   Complex Number Equations: $3z^2 + 24 = 0$ where $z = \pm 2i\sqrt{2}$. See **Complex Numbers**, example 6.

3.   Cubic Equations: $x^3 + 8 = 0$ or $4x^3 = 4x^2 + 80x$. See **Cubic Equations, solutions of**. Also see **Rational Roots Theorem**.

4.   Exponential Equations: $9^x = 81$ or $\left(\frac{1}{2}\right)^{1-x} = 16$. See **Exponential Equation, solution of**.

5.   Fractional Equations: fractional equations are rational equations. See category 12.

6.   Linear Equations: $n - 5 = -8$, $8n - 4 = 2n + 14$, or $-2(3n - 4) + 12 = -2n$. See **Linear Equations, solutions of**.

7.   Logarithmic Equations: $\log_7 x^3 - \log_7 x = 2$. See **Logarithmic Equations**.

8.   Matrix Equations: $\begin{bmatrix} 4 & -7 \\ 7 & -3 \end{bmatrix} \begin{bmatrix} x \\ y \end{bmatrix} = \begin{bmatrix} -13 \\ 5 \end{bmatrix}$. See **Matrix**.

9.   Polynomial Equations: for equations like $x^4 = x^2 + 12$ and $x^5 = 10x^3 - 9x$, see **Polynomial Equations**. For equations like $2x^3 - 5x^2 - 4x + 3 = 0$, see **Rational Roots Theorem**. Also see **Cubic Equations, solutions of**.

10.   Quadratic Equations: $x^2 - 25 = 0$ or $x^2 = -2x + 15$. See **Quadratic Equations, solutions of** or **Quadratic Formula**. Also, for $x^2 - 25 = 0$ or $x^4 - 16 - 0$, see **Root, nth**.

11.   Radical Equations: $\sqrt{2x - 6} - 6 = 0$ or $4 + \sqrt[3]{3x - 1} = 6$. See **Radicals, equations with**.

12.   Rational Equations: $\frac{x}{4} - \frac{9}{2} = \frac{-x}{5}$ or $\frac{1}{x} - \frac{5}{2x} = \frac{-3}{10}$ or $\frac{2}{x} + \frac{2}{x-3} = 1$. See **Rational Equations**.

13.   Simultaneous Solutions of Equations: see category 15.

14.   Square Root Equations: $\sqrt{2x - 6} - 6 = 0$ or $\sqrt{3x - 2} + \sqrt{2x - 2} - 1 = 0$. See **Radicals, equations with**.

15. Systems of Equations:

$$5x - 7y = -16$$
$$x + 4y = -13$$

or with three equations, see **Systems of Linear Equations, solutions of** or **Cramer's Rule.**

$$5x^2 - y^2 = 30$$
$$y^2 - 16 = 9x^2,$$

see **Systems of Nonlinear Equations, solutions of**. Also see **Matrix** or **Matrix, augmented (matrix) solution**.

16. Trigonometric Equations: $2 \sin x - 1 = 0$ or $1 - 3 \cos x = \sin^2 x$. See **Trigonometry, equations.**

➤ **Solution Set**   A **solution set** is a subset of the replacement set of a variable. It consists of the elements of the replacement set for which the variable is true or yields correct solutions.

Let the set of integers be the replacement set of the variable in the equation $n + 8 = 6$. Since the replacement set is the set from which the answer must be chosen, the solution set of the equation must be chosen from the set of integers. Solving the equation, we have

$$n + 8 = 6$$
$$n = 6 - 8$$
$$n = -2$$

Since -2 is an integer, the solution set of the equation is $\{-2\}$, read "the set consisting of the number -2," where $\{-2\}$ is a subset of the integers.

If the solution to the equation had been a fraction, the solution set of the equation would have been $\varnothing$ (the empty set). This is because a fraction is not an integer. A solution to an equation can only be accepted if it is an element of the replacement set, and a solution set can only be accepted if it is a subset of the replacement set.

Suppose, in our example, that the replacement set for the equation $n + 8 = 6$ had been the set of counting numbers. Even though we were able to solve the equation, the solution $n = -2$ is not a counting number, hence, the solution set would be $\varnothing$.

Examples

1. Determine the solution set of $x + 5 = 8$ if the replacement set is $\{1, 2, 3, 4, 5\}$.

$$x + 5 = 8$$
$$x = 8 - 5$$
$$x = 3$$

Since 3 is an element of the replacement set, the solution set is $\{3\}$.

2. Determine the solution set of $x + 8 = 10$, if the replacement set is $\{5, 6, 7, 8\}$.

$$x + 8 = 10$$
$$x = 10 - 8$$
$$x = 2$$

Since 2 is not an element of the replacement set, the solution set is $\varnothing$.

3. Write the solution set of $x > -2$ if the replacement set is the set of counting numbers.

Since the counting numbers begin with 1, any number less than 1 cannot be an element of the solution set. Hence, the solution set is $\{1, 2, 3, \ldots\}$.

➤ **Solving Equations**   See **Solutions of Equations**.

➤ **Solving Linear Inequalities**   See **Inequalities, linear solutions of**.

➤ **Solving Quadratic Inequalities**   See **Inequalities, quadratic solutions of**.

➤ **Solving Systems of Equations**   See **Systems of Linear Equations**.

➤ **Special Triangles**   See **Forty-Five Degree Triangle** or **Thirty-Sixty-Ninety Degree Triangle**.

➤ **Sphere**   A **sphere** is the set of all points that are a given distance from a given point. The given distance is called the **radius**, r, and the given point is called the **center** O. Similar to a circle, one of the diameters is $\overline{AB}$, $\overline{CD}$ is a **chord**, $\overleftrightarrow{CD}$ is a **secant line**, and the line t is a **tangent line**.

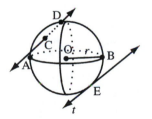

The volume of a sphere is given by the formula

$$V = \frac{4}{3}\pi r^3$$

The surface area of a sphere is given by the formula

$$A = 4\pi r^2$$

Example

1. Find the volume and surface area of a sphere with a diameter of 4 feet.

Since the radius is half of the diameter, the radius $r = 2$ feet. Substituting r into the formulas for volume and surface area, we have

Volume

$$V = \frac{4}{3}\pi r^3 = \frac{4}{3}\pi(2)^3 = \frac{4}{3}\pi \bullet 8 = \frac{32}{3}\pi \text{ ft}^3$$

Surface Area

$$A = 4\pi r^2 = 4\pi(2)^2 = 4\pi \bullet 4 = 16\pi \text{ ft}^2$$

➤ Square

A **square** is a rectangle with all sides congruent or equal in measure. See **Rectangle**. In addition, the angles of a square are right angles, both pairs of opposite sides are parallel, the diagonals are equal in measure, the diagonals are perpendicular, the diagonals bisect each other, the angles are bisected by the diagonals (each 45°), and two adjacent sides and their adjoining hypotenuse form a forty-five degree triangle.

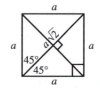

The area of a square is the product of two sides, $A = a \bullet a$ or $A = a^2$. The perimeter of a square is the sum of the lengths of the four sides, $P = a + a + a + a$ or $P = 4a$.

Example

---

1. Find the area, perimeter, and length of a diagonal of a square with one side of length 9 feet.

   Substituting into the equations for area and perimeter, we have

   $$A = a^2 = (9)^2 = 81 \text{ ft}^2$$
   $$P = 4a = 4(9) = 36 \text{ ft}$$
   $$\text{Diagonal} = a\sqrt{2} = 9\sqrt{2} \text{ ft} \qquad \text{See **Forty-Five Degree Triangle**.}$$

   We can also find the diagonal using the Pythagorean theorem. See **Pythagorean Theorem**.

   $$c^2 = a^2 + b^2$$
   $$c^2 = 9^2 + 9^2$$
   $$c^2 = 81 + 81$$
   $$c^2 = 162$$
   $$\sqrt{c^2} = \sqrt{162}$$
   $$c = \sqrt{162}$$
   $$c = \sqrt{81 \bullet 2} \qquad \text{See **Radicals, simplifying (reducing)**.}$$
   $$c = 9\sqrt{2}$$

➤ Square Root

The **square root** of a number is a number that when multiplied by itself will yield the given number. The square root of 9 is 3 because $3 \bullet 3 = 9$, and we write $\sqrt{9} = 3$.

Square roots are defined for positive values only. The square root of 9 is 3, not ± 3. The concept of $\pm$ (read "plus or minus") is developed from solutions of equations involving exponents and radicals. See **Root, nth**. For positive values of square roots, see **Principal nth Root**.

Examples

1. What is $\sqrt{25}$, $\sqrt{64}$, $-\sqrt{121}$, and $\sqrt{-4}$?

$$\sqrt{25} = 5 \qquad \text{because } 5 \bullet 5 = 25$$
$$\sqrt{64} = 8 \qquad \text{because } 8 \bullet 8 = 64$$
$$-\sqrt{121} = -11 \qquad \text{because } 11 \bullet 11 = 121$$
$$\sqrt{-4} = \text{ND} \qquad \text{"not defined," because no number when multiplied by itself will yield } -4.$$

Square roots are not defined for negative numbers inside the square root sign. To deal with square roots of negative numbers, see **Complex Numbers**.

➤ **Square Root Equations**   See **Radicals, equations with**.

➤ **Square Roots, operations with**   See the various articles under **Radical** and **Root**.

➤ **Square Roots, simplifying**   See **Radicals, simplifying (reducing)**.

➤ **Standard Equation**   For the **standard equation** of a circle, line (see **Linear Equation**), ellipse, hyperbola, or parabola, see the articles listed under each topic.

➤ **Standard Form of an Equation of a Line**   An equation of a line, which is written in the form $Ax + By = C$ (sometimes $Ax + By + C = 0$), is said to be in **standard form**, where A, B, and C are real numbers. If any of the terms A, B, and C are rational (fractions), the equation is usually rewritten, where the new A, B, and C are integers with a greatest common factor of 1 and the A term is positive. To write an equation in this form, we

1. Place the x- and y-terms on one side of the equal sign and all other terms (usually numerals) on the other side.

2. Multiply through the entire equation by the common denominator of all the terms. If there are any mixed fractions, change them to improper fractions first.

3. If the A term is negative, divide through the entire equation by -1.

The other form of an equation of a line is the **slope-intercept form**, written $y = mx + b$. See **Slope-Intercept Form of an Equation of a Line**.

Standard form can be a convenient form for graphing an equation of a line. To graph an equation of a line using standard form, see **Linear Equations, graphs of**, x-y Intercept Method.

Examples

1. Write $y = \frac{3}{4}x - \frac{2}{5}$ in standard form.

We need to place the term $\frac{3}{4}x$ on the left side of the equation, multiply through the equation by the common denominator for 4 and 5 (common denominator is 20), and then divide through by -1.

$$y = \frac{3}{4}x - \frac{2}{5}$$

$$-\frac{3}{4}x + y = -\frac{2}{5}$$ Subtract $\frac{3}{4}x$ from both sides.

$$20\left(-\frac{3}{4}x + y\right) = 20\left(-\frac{2}{5}\right)$$ Multiply by the common denominator.

$$-\frac{60}{4}x + 20y = -\frac{40}{5}$$ Also see **Distributive Property**.

$$-15x + 20y = -8$$

$$\frac{-15x}{-1} + \frac{20y}{-1} = \frac{-8}{-1}$$ Divide by -1.

$$15x - 20y = 8$$

2. Write the equations $x - 5 = 0$ and $\frac{-2}{3}y - 1 = 0$ in standard form.

The reason that each equation is missing a variable is because these are equations of a vertical and a horizontal line, respectively. See **Horizontal and Vertical Lines**. Regardless, we still apply the steps as we would for an equation with two variables:

$$x - 5 = 0 \qquad \frac{-2}{3}y - 1 = 0$$

$$x = 5 \qquad \frac{-2}{3}y = 1$$

$$3\left(\frac{-2}{3}y\right) = 3(1)$$

$$\frac{-6}{3}y = 3$$

$$-2y = 3$$

$$\frac{-2y}{-1} = \frac{3}{-1}$$

$$2y = -3$$

The standard form of the first equation is $x = 5$ and the second equation is $2y = -3$.

➤ **Standard Position of an Angle**   See **Angle, in trigonometry**.

➤ **Straight Angle**   In reference to trigonometry, the **straight angle** $\angle AOB$ is determined by the opposite rays $\overrightarrow{OA}$ and $\overrightarrow{OB}$ where the separation of the rays is $180°$ or π radians. For more, see **Angle, in trigonometry**.

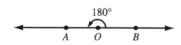

➤ **Subscripts**   Subscripts, usually whole numbers, are used to delineate one number from another. Instead of describing three numbers as x, y, and z, we refer to the same three numbers as a_0, a_1, and a_2 (read "a sub zero, a sub one, and a sub two").

If we let the three numbers be 4, 10, and 6, then using subscripts we can say that $a_0 = 4$, $a_1 = 10$, and $a_2 = 6$. The numbers do not necessarily have to be in order although the whole numbers of the subscripts do, unless otherwise specified.

➤ **Subset**   A **subset** is a set that is entirely contained in another set. If every element of set A is an element of set B, then set A is a subset of set B and we write $A \subset B$ (read "the set A is a subset of the set B").

Let $A = \{3, 4, 5, 6\}$ and $B = \{3, 4, 5, 6, 7, 8\}$. Since each element in A is an element of B, $A \subset B$. In this case, A is called a proper subset of B. Now consider the set B. Since each element in B is an element of B (is an element of its own set), B is a subset of itself. Such a subset is called an **improper subset**. Further, since $\varnothing \cup A = A$ for any set A, the empty set is a subset of any set. See **Union and Intersection.**

> ### Examples

1. Let $R = \{2, 5, 6\}$ and $S = \{1, 2, 3, 4, 5, 6, 7\}$. Is R a subset of S?

   Since each of the elements 2, 5, and 6 are elements of S, R is a subset of S and we write $R \subset S$.

2. Is $\{a, h, o, z\}$ a subset of {all letters of the alphabet}?

   Since each of the elements a, h, o, and z are letters of the alphabet, the set $\{a, h, o, z\}$ is a subset of {all letters of the alphabet}.

3. Is $P = \{$all prime numbers$\}$ a subset of $O = \{$all odd numbers$\}$?

   Since 2 is a prime number, 2 is an element of P. However, 2 is an even number and therefore not an element of O. Hence, P is not a subset of O and we write $P \not\subset O$.

4. Write all the subsets of $\{1, 2, 3\}$.

   Each of the elements is a subset and each pair is a subset. The set is a subset of itself and the empty set is a subset of every set.

   Hence, the subsets of $\{1, 2, 3\}$ are

   $$\{1\},\ \{2\},\ \{3\},\ \{1, 2\},\ \{2, 3\},\ \{1, 3\},\ \{1, 2, 3\},\ \text{and} \{\quad\}$$

   Since $\{1, 2\} = \{2, 1\}$, we do not consider equal sets twice. See **Set.**

➤ **Substitution Method for Systems of Equations**   See **Systems of Linear Equations** or **Systems of Nonlinear Equations, solutions of.**

➤ **Subtraction of Whole Numbers**   **Subtraction of whole numbers** is a process that determines the difference between two numbers. The larger number is called the **minuend** and the smaller number is called the **subtrahend**. The result, after the subtrahend has been subtracted from the minuend, is called the **difference** or **remainder**.

$$
\begin{array}{r}
\text{minued} \\
- \text{ subtrahend} \\
\hline
\text{difference or} \\
\text{remainder}
\end{array}
$$

S

When subtracting two numbers the units digits of each number must be aligned. Then each figure in the subtrahend is subtracted from the corresponding figure in its column in the minuend, proceeding from right to left. If the figure in the subtrahend is greater than the figure in the minuend, then we need to **borrow** 10 from the next highest digit of the minuend. We do this by decreasing the next highest digit by 1 and adding 10 to the corresponding figure of the minuend. Then we subtract.

To check the problem add the remainder and the subtrahend. The result should be the minuend.

Examples

1. Subtract 342 from 865. Check the answer.

We first line up the units digit of each number and then subtract corresponding figures, proceeding from right to left:

$$
\begin{array}{r}
865 \\
-\,342 \\
\hline
523
\end{array}
$$

Line up the 5 and 2.

Subtract 2 from 5, 4 from 6, and 3 from 8.

To check if the answer is correct, we add the remainder (answer) and the subtrahend to see if we get the minuend:

$$
\begin{array}{r}
523 \\
+\,342 \\
\hline
865
\end{array}
$$

Since 865 is the value of the minuend, the remainder (answer) is correct.

2. Subtract 62 from 358. Check the answer.

Since we cannot subtract 6 from 5 we will have to borrow from the next highest digit, the 3. We do this by decreasing the 3 by 1 and adding 10 to the 5:

$$
\begin{array}{r}
\overset{2\;\;1}{\cancel{3}58} \\
-\,62 \\
\hline
296
\end{array}
$$

Borrow 1 from the 3 and change the 5 to 15.

To check, we add 296 and 62 to see if we get 358:

$$
\begin{array}{r}
\overset{1}{296} \\
+\,62 \\
\hline
358
\end{array}
$$

Since we got 358 the answer, 296, is correct.

3. Take 1,751 from 4,031 and check the answer.

Since we cannot subtract 5 from 3, we need to borrow from the next highest digit, the 0. But 0 has no value from which to borrow, so we must borrow from its next highest digit, the 4. We decrease the 4 by 1 to get 3, add 10 to the 0 to get 10, decrease the 10 by 1 to get 9, and add 10 to the 3 to get 13. Then we can subtract:

$$
\begin{array}{r}
\overset{\scriptstyle 9}{} \\
\overset{\scriptstyle 3\ \ 1}{4{,}031} \\
-\ 1{,}751 \\
\hline
2{,}280
\end{array}
\qquad \text{Check:} \qquad
\begin{array}{r}
\overset{\scriptstyle 1\ 1}{} \\
2{,}280 \\
+\ 1{,}751 \\
\hline
4{,}031
\end{array}
$$

4. Find the difference between 7,000 and 6,459. Check the answer.

$$
\begin{array}{r}
\overset{\scriptstyle 9\,9}{} \\
\overset{\scriptstyle 6\ \ \ 1}{7{,}000} \\
-\ 6{,}459 \\
\hline
541
\end{array}
\qquad \text{Check:} \qquad
\begin{array}{r}
\overset{\scriptstyle 1\ 1}{} \\
{}_{1}\ 541 \\
+\ 6{,}459 \\
\hline
7{,}000
\end{array}
$$

5. Subtract 2,496,574 from 5,686,000. Check the answer.

$$
\begin{array}{r}
\overset{\scriptstyle 11\ \ 99}{} \\
\overset{\scriptstyle 575\ \ \ 1}{5{,}686{,}000} \\
-\ 2{,}496{,}574 \\
\hline
3{,}189{,}426
\end{array}
\qquad \text{Check:} \qquad
\begin{array}{r}
\overset{\scriptstyle 111\ 11}{} \\
3{,}189{,}426 \\
+\ 2{,}496{,}574 \\
\hline
5{,}686{,}000
\end{array}
$$

► **Subtraction Signs**   See **Positive and Negative Numbers**.

► **Sum and Difference Formulas**   The **sum and difference formulas** are one of several sets of identities used in trigonometry. See **Trigonometry, identities**. The sum and difference formulas for sine, cosine, and tangent are

$$
\sin(\alpha \pm \beta) = \sin\alpha \cos\beta \pm \cos\alpha \sin\beta
$$
$$
\cos(\alpha \pm \beta) = \cos\alpha \cos\beta \mp \sin\alpha \sin\beta
$$
$$
\tan(\alpha \pm \beta) = \frac{\tan\alpha \pm \tan\beta}{1 \mp \tan\alpha \tan\beta}
$$

If the formulas are written with the $\pm$ and $\mp$ signs, only three formulas need to be memorized instead of six formulas. When the top signs of the $\pm$ and $\mp$ signs are used, we get three of the six formulas. When the bottom signs of the $\pm$ and $\mp$ signs are used, we get the other three formulas. As an example:

Using Top Signs

$$
\tan(\alpha + \beta) = \frac{\tan\alpha + \tan\beta}{1 - \tan\alpha \tan\beta}
$$

Using Bottom Signs

$$
\tan(\alpha - \beta) = \frac{\tan\alpha - \tan\beta}{1 + \tan\alpha \tan\beta}
$$

S

Examples

1. Find the exact value of cos 75° using the sum and difference formula for cosine. Then calculate cos 75° to four decimal places.

   To find the exact value (not decimal approximation) of cos 75°, we need to use the thirty-sixty-ninety degree triangle and forty-five degree triangle. See the articles under those names.

   We observe that 75° = 30° + 45°. We will write 75° as the sum 30° + 45° because we know the exact values involving each from the 30-60-90 and 45 degree triangles. We will not write 75° as the sum 10° + 65°, or some similar sum, because we are not familiar with triangles involving those measures.

   Letting $\alpha = 30°$, $\beta = 45°$, and using the top signs in the formula $\cos(\alpha \pm \beta)$, we have

   $$\cos(\alpha + \beta) = \cos \alpha \cos \beta - \sin \alpha \sin \beta$$
   $$\cos(30° + 45°) = \cos 30° \cos 45° - \sin 30° \sin 45°$$

   See the articles on thirty-sixty-ninety or forty-five degree triangle. Also, see below. Rationalize $\frac{1}{\sqrt{2}}$ to get $\frac{\sqrt{2}}{2}$.

   $$= \left(\frac{\sqrt{3}}{2}\right)\left(\frac{1}{\sqrt{2}}\right) - \left(\frac{1}{2}\right)\left(\frac{1}{\sqrt{2}}\right)$$

   $$= \frac{\sqrt{3}}{2} \cdot \frac{\sqrt{2}}{2} - \frac{1}{2} \cdot \frac{\sqrt{2}}{2}$$

   $$= \frac{\sqrt{6}}{4} - \frac{\sqrt{2}}{4}$$

   See **Rationalizing the Denominator.**

   $$\cos 75° = \frac{\sqrt{6} - \sqrt{2}}{4}$$

   To calculate the four-decimal-place approximation of cos 75°, we have

   $$\cos 75° = \frac{\sqrt{6} - \sqrt{2}}{4} == \frac{2.44949 - 1.41421}{4} = \frac{1.03528}{4} = .2588$$

   To verify this, look up cos 75° in a table of values or use a calculator to determine cos 75° (put 75 in the calculator and press "cos").

   From the 30°-60°-90° triangle, we see that the sine from the 30° vertex is $\frac{1}{2}$ (opposite to hypotenuse) and the cosine from the 30° vertex is $\frac{\sqrt{3}}{2}$ (adjacent to hypotenuse). Hence, we have

   $$\sin 30° = \frac{1}{2} \quad \text{and} \quad \cos 30° = \frac{\sqrt{3}}{2}$$

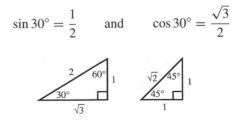

   From the 45° triangle, we see that the sine (opposite to hypotenuse) and cosine (adjacent to hypotenuse) are both the same, $\frac{1}{\sqrt{2}}$. If we rationalize this, we have

$$\frac{1}{\sqrt{2}} = \frac{1}{\sqrt{2}} \cdot \frac{\sqrt{2}}{\sqrt{2}} = \frac{\sqrt{2}}{\sqrt{4}} = \frac{\sqrt{2}}{2}$$

See **Rationalizing the denominator**. Hence,

$$\sin 45° = \frac{\sqrt{2}}{2} \qquad \text{and} \qquad \cos 45° = \frac{\sqrt{2}}{2}$$

2. Simplify $\sin 130° \cos 20° + \cos 130° \sin 20°$ and then evaluate.

Looking at the formulas for the sum and difference, we see that $\sin 130° \cos 20° + \cos 130° \sin 20°$ is of the form $\sin \alpha \cos \beta + \cos \alpha \sin \beta$ where $\alpha = 130°$ and $\beta = 20°$. Since $\sin \alpha \cos \beta + \cos \alpha \sin \beta$ is equal to $\sin(\alpha + \beta)$, we have

$$\sin 130° \cos 20° + \cos 130° \sin 20° = \sin(130° + 20°) = \sin 150° = \sin 30° = \frac{1}{2}$$

To evaluate $\sin 150°$, see **Trigonometry, reference angle**. To evaluate $\sin 30°$, see **Thirty-Sixty-Ninety Degree Triangle**.

Since the terminal side of $150°$ is in the second quadrant, the reference angle is $30°$. Hence, we need to consider a $30°$ triangle in the second quadrant where we see that the sine of the triangle is $\frac{1}{2}$ (opposite to hypotenuse). Therefore, $150° = \frac{1}{2}$.

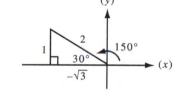

From the $30°$-$60°$-$90°$ triangle, we see that the sine of the $30°$ triangle is $\frac{1}{2}$ (opposite to hypotenuse). Hence, $\sin 30° = \frac{1}{2}$.

3. Derive the cofunction identity for $\sin(90° - \theta)$.

Letting $\alpha = 90°$, and $\beta = \theta$ and using the bottom signs in the formula of $\sin(\alpha \pm \beta)$, we have

$$\begin{aligned} \sin(\alpha - \beta) &= \sin \alpha \cos \beta - \cos \alpha \sin \beta \\ \sin(90° - \theta) &= \sin 90° \cos \theta - \cos 90° \sin \theta \\ &= (1) \cos \theta - (0) \sin \theta \qquad \text{See **Trigonometry, quadrant angles**.} \\ \sin(90° - \theta) &= \cos \theta \end{aligned}$$

To verify that $\sin(90° - \theta) = \cos \theta$, see **Cofunction Identities**.

S

4. Find the exact value of cot 165°. If you only need an example of the sum and difference formulas, find tan 165°.

We need to think of 165° in relation to the angles of the thirty-sixty-ninety degree triangle or the forty-five degree triangle. See the articles under those names. Thinking of these angles, we see that $165° = 120° + 45°$ where the reference angle of 120° is 60°. See **Trigonometry, reference angle**.

Since we do not have a formula for $\cot(\alpha \pm \beta)$, we will need to use the reciprocal identity $\cot \theta = \frac{1}{\tan \theta}$. See **Reciprocal Identities**. Our problem now becomes one of finding tan 165°. After that, we can determine cot 165° using the formula $\cot 165° = \frac{1}{\tan 165°}$.

Letting $\alpha = 120°$, $\beta = 45°$ and using the top signs in the formula $\tan(\alpha \pm \beta)$, we have

$$\tan(\alpha + \beta) = \frac{\tan \alpha + \tan \beta}{1 - \tan \alpha \tan \beta}$$

$$\tan(120° + 45°) = \frac{\tan 120° + \tan 45°}{1 - \tan 120° \tan 45°}$$
See the articles on thirty-sixty-ninety or forty-five degree triangles. Also see below.

$$= \frac{-\sqrt{3} + 1}{1 - (-\sqrt{3})(1)}$$

$$= \frac{1 - \sqrt{3}}{1 + \sqrt{3}}$$

$$= \frac{1 - \sqrt{3}}{1 + \sqrt{3}} \cdot \frac{1 - \sqrt{3}}{1 - \sqrt{3}}$$
See **Rationalizing the Denominator**.

$$= \frac{1 - 2\sqrt{3} + \sqrt{9}}{1 - \sqrt{9}} = \frac{1 - 2\sqrt{3} + 3}{1 - 3} = \frac{4 - 2\sqrt{3}}{-2}$$

$$= \frac{2(2 - \sqrt{3})}{-2} = \frac{2 - \sqrt{3}}{-1}$$

$$\tan 165° = -2 + \sqrt{3}$$

Now we can determine cot 165°. By substitution, we have

$$\cot 165° = \frac{1}{\tan 165°} = \frac{1}{-2 + \sqrt{3}}$$

$$= \frac{1}{-2 + \sqrt{3}} \cdot \frac{-2 - \sqrt{3}}{-2 - \sqrt{3}}$$
See **Rationalizing the Denominator**.

$$= \frac{-2 - \sqrt{3}}{4 - \sqrt{9}} = \frac{-2 - \sqrt{3}}{4 - 3} = \frac{-2 - \sqrt{3}}{1}$$

$$\cot 165° = -2 - \sqrt{3}$$

To avoid rationalizing the denominator twice, we could have used $\tan 165° = \frac{1 - \sqrt{3}}{1 + \sqrt{3}}$ before we rationalized. Then

$$\cot 165° = \frac{1}{\tan 165°} = \frac{1}{\frac{1 - \sqrt{3}}{1 + \sqrt{3}}}$$
Since $\frac{1}{\frac{a}{b}} = 1 \div \frac{a}{b} = 1 \cdot \frac{b}{a} = \frac{b}{a}$.

S

$$\cot 165° = \frac{1 + \sqrt{3}}{1 - \sqrt{3}} \qquad \text{See \textbf{Fractions, complex}.}$$

Then if we rationalize $\frac{1+\sqrt{3}}{1-\sqrt{3}}$ we will get $-2 - \sqrt{3}$.

In consideration of tan 120° and tan 45°, we need to look at the 30-60-90 and 45 degree triangles. Since the terminal side of 120° is in the second quadrant, the reference angle is 60°. Hence, we need to consider a 60° triangle in the second quadrant where we see that the tangent of the triangle is $\frac{\sqrt{3}}{-1} = -\sqrt{3}$ (opposite to adjacent). Therefore, tan 120° $= -\sqrt{3}$.

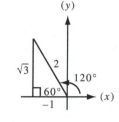

From the 45° triangle, we see that the tangent is $\frac{1}{1} = 1$. Hence, we have tan 45° = 1.

5. Determine the values of α and β so that 15°, 135°, 255°, and 285° can be written in either of the forms $\alpha + \beta$ or $\alpha - \beta$. Choose angles for α and β that have reference angles related to the 30-60-90, 45-degree triangles or the quadrant angles.

a.   15° = 45° − 30°   or   15° = 60° − 45°
b.   135° = 180° − 45°   or   135° = 90° + 45°   See **Trigonometry, quadrant angles**.
c.   255° = 210° + 45°   where the reference angle for 210° is 30°.
   or 255° = 300° − 45°   where the reference angle for 300° is 60°.
d.   285° = 330° − 45°   where the reference angle for 330° is 30°.
     285° = 315° − 30°   where the reference angle for 315° is 45°.

6. Prove: $\tan(360° - \theta) = -\tan \theta$.

Letting $\alpha = 360°$, $\beta = \theta$ and using the bottom sign in the formula for $\tan(\alpha \pm \beta)$, we have

$$\tan(\alpha - \beta) = \frac{\tan \alpha - \tan \beta}{1 + \tan \alpha \tan \beta}$$

$$\tan(360° - \theta) = \frac{\tan 360° - \tan \theta}{1 + \tan 360° \tan \theta} \qquad \text{See \textbf{Trigonometry, quadrant angles}.}$$

$$= \frac{0 - \tan \theta}{1 + (0) \tan \theta}$$

$$= \frac{-\tan \theta}{1}$$

$$\tan(360° - \theta) = -\tan \theta$$

S

➤ **Summation**   See **Sigma Notation**.

➤ **Sum of the Measures of the Angles of a Triangle**

In Euclidean geometry, the **sum of the measures of the angles of a triangle** is 180. Whether the triangle is acute, right, or obtuse the sum of the measures of any $\triangle ABC$ is

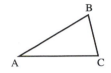

$$m\angle A + m\angle B + m\angle C = 180$$

Two special triangles involving an angle sum of 180 are the thirty-sixty-ninety degree triangle and the forty-five degree triangle. See the articles under those names. Also see **Measure of an Angle**.

In two of the non-Euclidean geometries, the sum of the measures of the angles of a triangle is either greater than 180 or less than 180. In Riemann's geometry, also called elliptic geometry, the sum is greater than 180. In Lobachevski's geometry, also called hyperbolic geometry, the sum is less than 180. The reason for the differences in the three geometries, Euclid's, Riemann's, and Lobachevski's is the parallel postulate. In each, a different postulate regarding parallel lines (or lack thereof) is used. As a result, each develops into a different geometry as well as a different angle sum.

In the following examples, the geometry considered is Euclid's.

Examples

1. In $\triangle ABC$ if the $m\angle A = 41$ and the $m\angle C = 32$, find the $m\angle B$.

    Since the sum of the measures of the angles of a triangle is 180, we can substitute the $m\angle A$ and the $m\angle C$ into the angle sum equation and then solve the equation for the $m\angle B$. See **Linear Equations, solutions of**.

    $$m\angle A + m\angle B + m\angle C = 180$$
    $$41 + m\angle B + 32 = 180$$
    $$m\angle B = 180 - 41 - 32$$
    $$m\angle B = 107$$

2. In right $\triangle ABC$ if the $m\angle C = 50$ find the $m\angle A$.

    Since $\triangle ABC$ is a right triangle, we know that the $m\angle B = 90$ (the center letter is the right angle). Hence, by substitution, we have

    $$m\angle A + m\angle B + m\angle C = 180$$
    $$m\angle A + 90 + 50 = 180$$
    $$m\angle A = 180 - 90 - 50$$
    $$m\angle A = 40$$

➤ **Sums**   See **Sigma Notation**.

➤ **Supplementary Angles**   Two angles whose measures have a sum of 180 are called **supplementary angles**. The definition is defined for two angles only, not three or more. Even though three angles may have a sum of 180, they are not supplementary.

Let $\angle 1$ and $\angle 2$ be **supplementary angles**. If $\angle 1$ and $\angle 2$ are supplementary angles then, by definition,

$$m\angle 1 + m\angle 2 = 180$$

The sum is 180 regardless if the angles are adjacent (share a common side). See **Adjacent Angles**. No matter what the orientation of the angles, if they are supplementary, then their measures will add up to 180.

Examples

1. If $\angle 1$ and $\angle 2$ are supplementary, find $m\angle 1$ if $m\angle 2 = 38$.

   To find $m\angle 1$, we need to substitute $m\angle 2 = 38$ into the formula for supplementary angles and then solve for $m\angle 1$:

   $$
   \begin{aligned}
   m\angle 1 + m\angle 2 &= 180 \\
   m\angle 1 + 38 &= 180 \\
   m\angle 1 &= 180 - 38 \qquad \text{See \textbf{Linear Equations, solutions of}.} \\
   m\angle 1 &= 142
   \end{aligned}
   $$

   The supplement of $38°$ is $142°$ because $38° + 142° = 180°$.

➤ **Symbols**   See **Math Symbols**.

➤ **Symmetric Property**   The **symmetric property** is one of the three axioms of equality. The other two are the **reflexive** and **transitive** properties. See the articles under these names.

   The symmetric property states that if a and b are real numbers, then the following holds: if $a = b$ then $b = a$. Also see **Property** or **Properties of the Real Number System**.

➤ **Symmetry**   Many graphs show a **symmetry** with respect to the x-axis, the y-axis, or the origin.

S

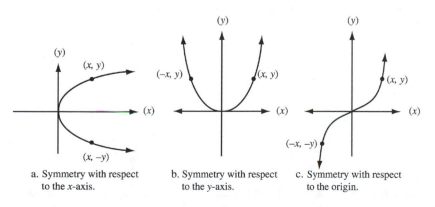

a. Symmetry with respect to the x-axis.

b. Symmetry with respect to the y-axis.

c. Symmetry with respect to the origin.

a. A graph is symmetric with respect to the x-axis if the x-axis divides the graph into two halves, each the mirror image of the other. In this case, any point (x, y) on one half of the graph is the image of the point $(x, -y)$ on the other half of the graph. To test the equation of the graph for symmetry with respect to the x-axis, we replace y with $-y$. If the equation remains equivalent (unchanged), the graph of the equation is symmetric with respect to the x-axis.

b. A graph is symmetric with respect to the y-axis if the y-axis divides the graph into two halves, each the mirror image of the other. In this case, any point (x, y) on one half of the graph is the image of the point $(-x, y)$ on the other half of the graph. To test the equation of the graph for symmetry with respect to the y-axis, we replace x with $-x$. If the equation remains equivalent, the graph of the equation is symmetric with respect to the y-axis.

c. A graph is symmetric with respect to the origin if the origin divides the graph into two halves, each the mirror image of the other. In this case, any point (x, y) on one half of the graph is the image of the point $(-x, -y)$ on the other half of the graph. To test the equation of the graph for symmetry with respect to the origin, we replace x with $-x$ and y with $-y$. If the equation remains equivalent, the graph of the equation is symmetric with respect to the origin.

Examples

1. Find the points symmetric to $(2, 4)$ and $(-3, -5)$ with respect to the x-axis, y-axis, and origin.

   To find the point symmetric with respect to the x-axis, we replace y with $-y$, with respect to the y-axis, we replace x with $-x$, and with respect to the origin, we replace x with $-x$ and y with $-y$.

Point	x-axis	y-axis	Origin
$(2, 4)$	$(2, -4)$	$(-2, 4)$	$(-2, -4)$
$(-3, -5)$	$(-3, 5)$	$(3, -5)$	$(3, 5)$

2. Test the equations $x^2 - y = 0$, $x - y^2 = 0$, $y = x^3$, and $x^2 + y^2 = 16$ for symmetry with respect to the x-axis, y-axis, and origin.

   To test symmetry with respect to the x-axis, we replace y with $-y$, with respect to the y-axis, we replace x with $-x$, and with respect to the origin, we replace x with $-x$ and y with $-y$.

	Equation	x-axis	y-axis	Origin
a.	$x^2 - y = 0$	$x^2 - (-y) = 0$ $x^2 + y = 0$, no	$(-x)^2 - y = 0$ $x^2 - y = 0$, yes	$(-x)^2 - (-y) = 0$ $x^2 + y = 0$, no
b.	$x - y^2 = 0$	$x - (-y)^2 = 0$ $x - y^2 = 0$, yes	$(-x) - y^2 = 0$ $-x - y^2 = 0$, no	$(-x) - (-y)^2 = 0$ $-x - y^2 = 0$, no
c.	$y = x^3$	$(-y) = x^3$ $-y = x^3$ $y = -x^3$, no	$y = (-x)^3$ $y = -x^3$, no	$(-y) = (-x)^3$ $-y = -x^3$ $y = x^3$, yes
d.	$x^2 + y^2 = 16$	$x^2 + (-y)^2 = 16$ $x^2 + y^2 = 16$, yes	$(-x)^2 + y^2 = 16$ $x^2 + y^2 = 16$, yes	$(-x)^2 + (-y)^2 = 16$ $x^2 + y^2 = 16$, yes

The equation $x^2 - y = 0$, which is the parabola $y = x^2$, is symmetric with respect to the y-axis. The equation $x - y^2 = 0$, which is the parabola $y = \sqrt{x}$, is symmetric with respect to the x-axis.

The equation $y = x^3$ is symmetric with respect to the origin. And the equation $x^2 + y^2 = 16$, which is a circle with a radius of 4 and its center at the origin, is symmetric to the x-axis, the y-axis, and the origin.

➤ **Synthetic Division**   **Synthetic division** is a method of dividing a polynomial $P(x)$ by a binomial $x - r$. It is much shorter than long division (see **Long Division of Polynomials**) and is very useful in synthetic substitution, factoring, and solving equations. See **Synthetic Substitution**, **Factor Theorem**, or **Rational Roots Theorem**.

Before dividing, the polynomial must first be placed in descending order. See **Ascending and Descending Order**. If a place value is missing, for example, $4x^3 + 2x - 3$ is missing the x^2 term, the position must be filled with a zero coefficient term. The polynomial in this case is then written $4x^3 + 0x^2 + 2x - 3$. This step is the same as with long division.

Next, we write down the coefficients of the polynomial only, including any zero coefficients. Now we can proceed with the process of synthetic division by $x - r$. We multiply the first coefficient by r and combine the product with the second coefficient. Then we multiply the result by r and combine that with the third coefficient. We continue the process until we have used up all of the coefficients. The resulting numbers are the coefficients of the quotient $Q(x)$, the first coefficient having a variable of one degree less than that of $P(x)$ and the last is the remainder. If the remainder is 0, then $x - r$ is said to divide $P(x)$ evenly or completely.

Examples

1. Find the quotient when $P(x) = 2x^3 - 9x^2 - 2x + 24$ is divided by $x - 4$.

   Since the divisor is $x - 4$, we begin the synthetic division with $r = 4$. If the divisor had been $x + 4$, we would have began with $r = -4$.

   Writing down the coefficients of $P(x)$ and applying the process of synthetic division, we have

   $$\begin{array}{r|rrrr} & 2 & -9 & -2 & 24 \\ 4 & & & & \end{array}$$
   Write the coefficients and the divisor in this format.

   $$\begin{array}{r|rrrr} & 2 & -9 & -2 & 24 \\ 4 & 2 & & & \end{array}$$
   Begin the process by bringing down the leading coefficient.

   $$\begin{array}{r|rrrr} & 2 & -9 & -2 & 24 \\ 4 & 2 & -1 & & \end{array}$$
   Multiply 2 by 4 and combine the product with -9 to get -1.

   $$\begin{array}{r|rrrr} & 2 & -9 & -2 & 24 \\ & & 8 & -4 & \\ \hline 4 & 2 & -1 & -6 & \end{array}$$
   Multiply -1 by 4 and combine the product with -2 to get -6.

   $$\begin{array}{r|rrrr} & 2 & -9 & -2 & 24 \\ & & 8 & -4 & -24 \\ \hline 4 & 2 & -1 & -6 & 0 \end{array}$$
   Multiply -6 by 4 and combine the product with 24 to get 0.

   The numbers 2, -1, and -6 are the coefficients of the quotient $Q(x)$ where the variable of the 2 is one degree less than that of $P(x)$. Hence, $Q(x) = 2x^2 - x - 6$. The remainder, 0, indicates that

S

$x - 4$ divides into $P(x)$ evenly. As a fraction, we have

$$\frac{2x^3 - 9x^2 - 2x + 25}{x - 4} = 2x^2 - x - 6$$

In reference to long division, we have

$$x - 4 \overline{)\,2x^3 - 9x^2 - 2x + 25} \quad \begin{array}{c} 2x^2 - x - 6 \quad \text{with a remainder of } 0 \end{array}$$

2. Find the quotient when $P(x) = x^4 - 6x^2 - 7$ is divided by $x + 3$.

Since $P(x)$ is missing the x^3 and x terms, we first need to reinsert them with 0 coefficients. This will hold their place value during the division. The new polynomial is $P(x) = x^4 + 0x^3 - 6x^2 + 0x - 7$.

Next, we need to rewrite the binomial $x + 3$ in the form $x - r$, i.e. $x - (-3)$. We will begin the process with $r = -3$. Hence, we have

$$\begin{array}{r|rrrrr} & 1 & 0 & -6 & 0 & -7 \\ & & -3 & 9 & -9 & 27 \\ \hline -3 & 1 & -3 & 3 & -9 & 20 \end{array}$$

Using the coefficients 1, -3, 3, and -9, we have the quotient $Q(x) = x^3 - 3x^2 + 3x - 9$ with a remainder of 20. If it is required to write the remainder as a fraction, $Q(x)$ is written $Q(x) = x^3 - 3x^2 + 3x - 9 + \frac{20}{x+3}$.

The steps applied during the process of the division are:

a. Bring down the 1.

b. Multiply 1 by -3 and combine their product (-3) with 0 to get -3.

c. Multiply -3 by -3 and combine their product (9) with -6 to get 3.

d. Multiply 3 by -3 and combine their product (-9) with 0 to get -9.

e. Multiply -9 by -3 and combine their product (27) with -7 to get the remainder 20.

3. Find the quotient when $P(x) = -3x^3 - 4x - 3$ is divided by $x + \frac{1}{2}$.

Inserting the 0 coefficient for the x^2 term and writing the binomial as $x - r = x - \left(-\frac{1}{2}\right)$, we have

$$\begin{array}{r|rrrr} & -3 & 0 & -4 & -3 \\ & & \frac{3}{2} & -\frac{3}{4} & \frac{19}{8} \\ \hline -\frac{1}{2} & -3 & \frac{3}{2} & -4\frac{3}{4} & -\frac{5}{8} \end{array}$$

With coefficients of -3, $\frac{2}{3}$, and $-4\frac{3}{4}$, the quotient is $Q(x) = -3x^2 + 1\frac{1}{2}x - 4\frac{3}{4}$ and the remainder is $-\frac{5}{8}$.

4. Find the quotient when $P(x) = 3x^4 - 6x^3 + 3x + 6$ is divided by $x - 2i$.

Inserting the 0 coefficient for the x^2 term and dividing with $r = 2i$, we have

$$
\begin{array}{c|ccccc}
 & 3 & -6 & 0 & 3 & 6 \\
 & & 6i & -12-12i & 24-24i & 48+54i \\
\hline
2i & 3 & -6+6i & -12-12i & 27-24i & 56+54i
\end{array}
$$

Hence, $Q(x) = 3x^3 - (6 - 6i)x^2 - (12 + 12i)x + (27 - 24i)$ with a remainder of $56 + 54i$.

➤ **Synthetic Substitution**   **Synthetic substitution** is an application of the remainder theorem. See **Remainder Theorem**. In general, if a polynomial $P(x)$ is divided by a binomial $x - r$, the remainder is $P(r)$. In other words, the remainder of the division is the same as the value obtained by substituting $x = r$ into the polynomial, the remainder $= P(r)$. See **Remainder Theorem**. Also see **Function Notation**.

Although long division can be used, synthetic division is the easiest form of division to use when applying synthetic substitution. See **Synthetic Division**.

Examples

1. Use synthetic substitution to find $P(4)$ if $P(x) = 2x^3 - 9x^2 - 2x + 38$.

Our goal is to find $P(4)$. To do this, we will divide $P(x)$ by 4 using synthetic division. When the division is completed, the remainder obtained will be the value of $P(4)$.

$$
\begin{array}{c|cccc}
 & 2 & -9 & -2 & 38 \\
 & & 8 & -4 & -24 \\
\hline
4 & 2 & -1 & -6 & 14
\end{array}
$$

Since the remainder is 14, $P(4) = 14$.

To verify our solution, we can find $P(4)$ by substituting $x = 4$ in $P(x)$:

$$
\begin{aligned}
P(x) &= 2x^3 - 9x^2 - 2x + 38 \\
P(4) &= 2(4)^3 - 9(4)^2 - 2(4) + 38 \\
&= 128 - 144 - 8 + 38 = 14
\end{aligned}
$$

Hence, $P(4) = 14$.

2. Use synthetic substitution to find $P(-5)$ if $P(x) = -4x^5 - 19x^4 + 20x - 10$.

Since $P(x)$ is missing the x^3 and x^2 terms, we first need to reinsert them with 0 coefficients. The polynomial we will use for division is $P(x) = -4x^5 - 19x^4 + 0x^3 + 0x^2 + 20x - 10$. Now we can divide and find $P(-5)$.

$$
\begin{array}{c|cccccc}
 & -4 & -19 & 0 & 0 & 20 & -10 \\
 & & 20 & -5 & 25 & -125 & 525 \\
\hline
-5 & -4 & 1 & -5 & 25 & -105 & 515
\end{array}
$$

Hence, $P(-5) = 515$.

We can verify the solution by evaluating $P(x)$ for $x = -5$:

$$P(x) = -4x^5 - 19x^4 + 20x - 10$$
$$P(-5) = -4(-5)^5 - 19(-5)^4 + 20(-5) - 10 = -4(-3125) - 19(625) - 100 - 10$$
$$= 12{,}500 - 11{,}875 - 110 = 515$$

► **Systems of Linear Equations**   A **system of linear equations**, in the plane, is a pair of linear equations in two variables. If one equation is the line $2x - 3y = 5$ and the other equation is the line $6x + 4y = 10$, then the two equations considered together are a system of linear equations.

If we look at the graph of a system of linear equations, one of the three situations can occur: the graphs of the lines intersect, the graphs of the lines are parallel, or the graphs of the lines graph one on top of the other (the lines are coincident). Each of these systems has a specific name.

If a system has a exactly one solution, it is called an **independent system**. If a system has at least one solution, it is called a **consistent system**. If a system has no solution, it is called an **inconsistent system**. And if a system has all solutions in common, it is called a **dependent system**. Therefore, in reference to these names, lines that intersect are called **consistent and independent**, lines that are coincident are called **consistent and dependent**, and lines that are parallel are called **inconsistent**. See **Systems of Linear Equations, solutions of**.

As can be seen from these definitions, there is some overlap and apparent redundancy in the descriptions of the systems. In general, lines either cross each other or they don't. If they do (intersecting and coincident lines), they are called consistent. If they don't (parallel lines), they are called inconsistent. Lines that cross each other are further divided into two other categories. If the lines cross each other in one point, as with intersecting lines, they are called independent. If they cross each other at every point, as with coincident lines, they are called dependent. This category is said to have infinitely many solutions. Hence, lines that cross each other at one point are **consistent and independent**. Coincident lines that cross each other at every point are **consistent and dependent**. And parallel lines are **inconsistent**.

Each of the three systems can be identified by observing the two equations. To do so, both of the equations must be in slope-intercept form ($y = mx + b$) or both of the equations must be in standard form ($Ax + By = C$). See **Slope-Intercept Form of an Equation of a Line** or **Standard Form of an Equation of a Line**.

Equations in Slope-Intercept Form

1. Consistent and independent (intersecting lines): the slopes of both equations are different.

2. Consistent and dependent (coincident lines): the slopes and intercepts of both equations are the same.

3. Inconsistent (parallel lines): the slopes are the same but the intercepts are different.

Equations in Standard Form

1. Consistent and independent (intersecting lines): either the A's are different, the B's are different, or both the A's and B's are different.

2. Consistent and dependent (coincident lines): the A's, B's, and C's of both equations are the same.

3. Inconsistent (parallel line): the A's and B's of both equations are the same but the C's are different.

Example

1. Identify each of the following systems as either consistent and independent, consistent and dependent, or inconsistent:

$$y - 4 = -x \qquad 2y + 6 = -4x \qquad 12x = 3(y + 3)$$
$$3 - 4x = -y + 2 \qquad -6x = 3y + 9 \qquad 2(y - 2) = 8x$$

To identify each system, we can either solve the equations for y and use the slope-intercept approach or write the equations in standard form and use the standard form approach. For clarity, we will do both.

Slope-Intercept Approach

In each of the problems, we need to solve each of the equations for y and then categorize their slopes or intercepts.

a. Solving each of the equations of the first system, we have

$$y - 4 = -x \qquad 3 - 4x = -y + 2$$
$$y = -x + 4 \qquad y = 4x + 2 - 3$$
$$y = 4x - 1$$

Since the slopes of both equations are different, $m = -1$ and $m = 4$, the system is consistent and independent. If graphed, the lines will cross each other at one point.

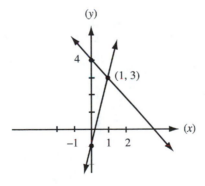

b. Solving each of the equations of the second system, we have

$$2y + 6 = -4x \qquad -6x = 3y + 9$$
$$2y = -4x - 6 \qquad -3y = 6x + 9$$
$$y = -2x - 3 \qquad y = -2x - 3$$

Since the slopes and the intercepts of both equations are the same, the system is consistent and dependent. If graphed, the lines will graph one on top of the other.

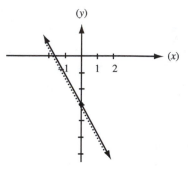

c. Solving each of the equations of the third system, we have

$$12x = 3(y + 3) \qquad 2(y - 2) = 8x$$
$$12x = 3y + 9 \qquad 2y - 4 = 8x$$
$$-3y = -12x + 9 \qquad 2y = 8x + 4$$
$$y = 4x - 3 \qquad y = 4x + 2$$

Since the slopes of both equations are the same, $m = 4$, the system is inconsistent. If graphed, the lines will be parallel.

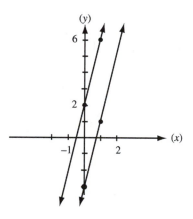

Standard Form Approach

In each of the problems, we need to write the equations in standard form and then categorize the equations according to the values of the A's, B's, and C's.

a. Writing each of the equations of the first system in standard form, we have

$$y - 4 = -x \qquad 3 - 4x = -y + 2$$
$$x + y = 4 \qquad -4x + y = 2 - 3$$
$$\qquad 4x - y = 1$$

Since both the A's ($A = 1$, $A = 4$) and the B's ($B = 1$ and $B = -1$) are different, the system is consistent and independent. If only the A's were different or only the B's were different, the system would still be consistent and independent.

b. Writing each of the equations of the second system in standard form, we have

$$2y + 6 = -4x \qquad\qquad -6x = 3y + 9$$
$$4x + 2y = -6 \qquad\qquad -6x - 3y = 9$$
$$2x + y = -3 \qquad\qquad\quad 6x + 3y = -9$$
$$\qquad\qquad\qquad\qquad\qquad 2x + y = -3$$

Since the A's, B's, and C's of both equations are the same, the system is consistent and dependent.

c. Writing each of the equations of the third system in standard form, we have

$$12x = 3(y + 3) \qquad\qquad 2(y - 2) = 8x$$
$$12x = 3y + 9 \qquad\qquad\quad 2y - 4 = 8x$$
$$12x - 3y = 9 \qquad\qquad -8x + 2y = 4$$
$$4x - y = 3 \qquad\qquad\quad 4x - y = -2 \qquad \text{Divide both sides of the equation by } -2.$$

Since the A's and B's of the equations are the same but the C's ($C = 3$ and $C = -2$) are different, the system is inconsistent.

➤ **Systems of Linear Equations, solutions by graphing**   Solving a system of linear equations with two variables by graphing means that we want to find the point of intersection of the two lines. To find the solution by algebraic methods, see **Systems of Linear Equations, solutions of**. Before beginning this article, the various methods of graphing must first be clear. See **Linear Equations, graphs of**.

To solve a system of linear equations by graphing, we first need to graph the lines. Then, by looking at the graph, we need to determine the point of intersection. Obviously, if the graph is not drawn well or if the point of intersection has difficult coordinates, the determined point may be slightly different from that of the actual, algebraic, solution.

Example

1. By graphing, find the solution to the following system of equations:

$$x + y = 5$$
$$x - 3y = -3$$

To graph the equations, we first need to decide which graphing method we will apply to each. Let's graph the first equation by the x-y intercept method and the second equation by the slope-intercept method. See **Linear Equations, graphs of**.

In the first equation, we need to find the x- and y-intercepts, respectively. If we let $y = 0$, we get the x-intercept, and if we let $x = 0$, we get the y-intercept. Letting $y = 0$, we have $x + 0 = 5$ or $x = 5$, the x-intercept. Letting $x = 0$, we have $0 + y = 5$ or $y = 5$, the y-intercept. Plotting $x = 5$ and $y = 5$, we have the graph of $x + y = 5$. See the graph.

In the second equation, we need to solve the equation for y, plot the y-intercept, use the rise and run to determine the second point, and then draw the graph of $x - 3y = -3$. See the graph.

$$x - 3y = -3$$
$$-3y = -x - 3$$
$$\frac{-3y}{-3} = \frac{-x}{-3}\frac{-3}{-3}$$
$$y = \frac{1}{3}x + 1$$

From the slope, $\frac{1}{3}$, we see that the rise is 1 and the run is 3.

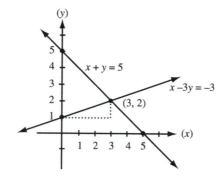

Since the graphs of the two lines intersect (cross) each other at the point $(3, 2)$, the point $(3, 2)$ is the solution of the system.

To verify this, we can substitute $(x, y) = (3, 2)$ into the original equations. If the substitutions remain balanced, the point is a solution of the system:

$$\begin{array}{ll} x + y = 5 & x - 3y = -3 \\ 3 + 2 = 5 & 3 - 3(2) = -3 \\ 5 = 5 & -3 = -3 \end{array}$$

Hence, $(3, 2)$ is a solution of the system.

► **Systems of Linear Equations, solutions of**   This article is divided into two sections, one that considers systems of linear equations with two variables (lines in the plane) and the other that considers systems of linear equations with three variables (lines in space). Each of the sections is further divided into two categories, solutions of linear systems by the **substitution method** and solutions of linear systems by the **linear combination** or **addition method**. For solutions of linear equations in two or three variables by determinants, see **Cramer's Rule**.

For solutions of linear equations in two or three variables by matrices, see **Matrix**. For solutions of linear equations in three variables by augmented matrices, see **Matrix, augmented (matrix) solution**. For solutions of linear equations in two variables by graphing, see **Systems of Linear Equations, solutions by graphing**.

A solution to a system with two variables is an ordered pair that satisfies both equations of the system. In reference to the graphs of the lines of the system, a solution to the system is the point where the two lines intersect. Hence, the ordered pair represents the coordinates of the point of intersection of the two lines. Similarly, a solution to a system with three variables is an ordered triple that satisfies each of the three equations of the system. The ordered triple represents the coordinates of the point of intersection of the three lines.

TWO VARIABLES

Substitution Method

1. Solve either of the equations for either of the variables.

2. Substitute the equation of step 1 into the other original equation and solve for the variable. This yields one coordinate of the solution.

3. Substitute the coordinate of step 2 into either original equation and solve for the variable. This yields the other coordinate of the solution.

4. Write the two coordinates as an ordered pair. This is the solution of the system. It represents the point of intersection of the two lines.

Linear Combination Method

a. Write both of the equations in standard form, $Ax + By = C$.

b. Multiply one, or both, of the equations (both sides of the equal sign) by a number(s) that will render either the x-terms or the y-terms the same but opposite in sign.

c. Combine the two equations and solve for the variable. This yields one coordinate of the solution.

d. To determine the other coordinate of the solution, we have two choices:

   i. Substitute the coordinate of step c into either original equation and solve for the variable. This yields the other coordinate of the solution. Finding the second variable this way utilizes the second half of the substitution method.

   ii. Or, apply step b all over again. This time, however, multiply by a number(s) that will render the other variable the same but opposite in sign. Combining these equations and solving for the variable will yield the other coordinate of the system.

e. Write the two coordinates as an ordered pair. This is the solution and point of intersection of the system.

THREE VARIABLES

Substitution Method

1. Solve one of the equations for one of the variables.

2. Substitute the equation in step 1 into each of the other two equations and simplify.

3. Solve the equations in step 2 by either the substitution method or the linear combination method. This will yield two of the three coordinates of the solution.

4. Substitute the variables in step 3 into any of the original equations and solve for the variable.

5. Write the three coordinates as an ordered triple. This is the point of intersection of the three lines.

Linear Combination Method

a. Write all three equations in standard form, $Ax + By + Cz = 0$.

b. Multiply any two of the equations by a number(s) that will render either the x-terms, the y-terms, or the z-terms the same but opposite in sign.

c. Combine the two equations and simplify into standard form.

d. Multiply two different equations by a number(s) that will render the same variable as in step b the same but opposite in sign.

e. Combine the two equations and simplify into standard form.

f. Solve the equations from steps c and e by the linear combination method, or the substitution method if you prefer. This will yield two of the three coordinates of the solution.

g. Substitute the variables in step f into any of the original equations and solve for the variable.

h. Write the three coordinates as an ordered triple. This is the point of intersection of the three lines.

Examples

1. Solve the following system by the substitution method:

$$5x - 7y = -16$$
$$x + 4y = 13$$

We need to solve either of the equations for one of the variables. If we solve the first equation for either x or y, we will get fractions. If possible, we need to avoid fractions so the solution will be easier. Looking at the second equation, we see that if we solve for x we will not have to deal with fractions. Sometimes avoiding fractions is not possible.

Solving the second equation for x, we have $x = -4y + 13$. We can now substitute $x = -4y + 13$ into $5x - 7y = -16$ and solve for y:

$$
\begin{aligned}
5x - 7y &= -16 \qquad \text{Solve } x + 4y = 13 \text{ for } x \text{ and substitute into } 5x - 7y = -16. \\
5(-4y + 13) - 7y &= -16 \\
-20y + 65 - 7y &= -16 \\
-27y &= -16 - 65 \\
-27y &= -81 \\
\frac{-27y}{-27} &= \frac{-81}{-27} \\
y &= 3
\end{aligned}
$$

Substituting $y = 3$ into either original equation, we can find x. Using $x + 4y = 13$, we have

$$x + 4y = 13$$
$$x + 4(3) = 13$$
$$x = 13 - 12$$
$$x = 1$$

The solution is an ordered pair (x, y) where $(x, y) = (1, 3)$.

To verify that $(1, 3)$ is correct, we can substitute the point into the original equations and see if the results are balanced. Substitution into one equation will suffice but both equations are complete:

$$
\begin{array}{ll}
5x - 7y = -16 & x + 4y = 13 \\
5(1) - 7(3) = -16 & 1 + 4(3) = 13 \\
5 - 21 = -16 & 13 = 13 \\
-16 = -16 &
\end{array}
$$

Hence, the point $(1, 3)$ is a solution of the system.

2. Solve the system in example 1 by the linear combination method.

Solving systems by the linear combination method uses a lot of mental observation. If we want to get the x's to cancel when we combine the equations, we need to multiply the second equation by -5. In this manner, the x's will be the same but opposite in sign ($5x$ and $-5x$). If we want to get the y's to cancel when we combine the equations, we need to multiply the first equations by 4 and the second equation by 7. In this manner, the y's will be the same but opposite in sign ($-28y$ and $28y$). Since multiplying one equation is simpler, let's cancel the x's.

Multiplying the second equation by -5, combining the equations, and solving for y, we have

$$
\begin{array}{l}
5x - 7y = -16 \\
\underline{x + 4y = 13}
\end{array}
$$

$$
\begin{array}{l}
5x - 7y = -16 \\
\underline{5(x + 4y) = -5(13)}
\end{array}
$$
Multiply both sides of the second equation by -5 so the x's can cancel.

$$
\begin{array}{l}
5x - 7y = -16 \\
5x - 20y = -65
\end{array}
$$

$$-27y = -81 \qquad \text{Combine the equations.}$$
$$\frac{-27y}{-27} = \frac{-81}{-27} \qquad \text{Divide by } -27.$$
$$y = 3$$

This is the y-coordinate of the ordered pair.

To find the x-coordinate of the ordered pair, we can either substitute $y = 3$ into either original equation, as we did in example 1, or apply the linear combination method to get the y's to cancel. Taking the latter approach, we have

S

$$5x - 7y = -16$$
$$\underline{x + 4y = 13}$$

$$4(5x - 7y) = 4(-16)$$ To get the y's to cancel we multiply by 4 and 7.
$$\underline{7(x + 4y) = 7(13)}$$

$$20x - 28y = -64$$
$$\underline{7x + 28y = 91}$$

$$27x + 0y = 27$$
$$27x = 27$$ Combine the equations and solve for x.
$$\frac{27x}{27} = \frac{27}{27}$$
$$x = 1$$

Hence, we have the ordered pair $(x, y) = (1, 3)$, which is the point of intersection of the two lines. For an example of finding the point of intersection by graphing, see **Systems of Linear Equations, solutions by graphing**.

3. Solve the following system by the substitution method and then by the linear combination method:

$$4x = 7y - 13$$
$$-3y - 5 = -7x$$

Substitution Method

We can solve either equation for either variable. Solving the first equation for x and substituting into the secod equation, we have

$$4x = 7y - 13$$ Solve for x.
$$\frac{4x}{4} = \frac{7y}{4} - \frac{13}{4}$$
$$x = \frac{7}{4}y - \frac{13}{4}$$

$$-3y - 5 = -7\left(\frac{7}{4}y - \frac{13}{4}\right)$$ Substitute x in the second equation and solve for y.

$$-3y - 5 = \frac{-49}{4}y + \frac{91}{4}$$

$$4(-3y - 5) = 4\left(\frac{-49}{4}y + \frac{91}{4}\right)$$ Clear fractions by multiplying by the common denominator.

$$-12y - 20 = \frac{4(-49)}{4}y + \frac{4(91)}{4}$$

$$-12y - 20 = -49y + 91$$
$$49y - 12y = 91 + 20$$
$$37y = 111$$
$$\frac{37y}{37} = \frac{111}{37}$$
$$y = 3$$ This is the y-coordinate.

To find the x-coordinate, we substitute $y = 3$ into either original equation. Substituting into the first equation, we have

$$4x = 7y - 13$$
$$4x = 7(3) - 13$$
$$4x = 8$$
$$x = 2$$

Hence, the point of intersection is the ordered pair $(x, y) = (2, 3)$.

Linear Combination Method

To solve by the linear combination method, we first need to write the equations in standard form and then determine which variable we want to cancel first.

$$\begin{array}{ll} 4x = 7y - 13 \\ 3x - 5 = -7x \end{array} \quad \rightarrow \quad \begin{array}{ll} 4x - 7y = -13 \\ 7x - 3y = 5 \end{array}$$

Neither the x's nor the y's are simple multiples of each other, as the x's were in example 1. As a result, it really doesn't matter which we eliminate first. If we want to eliminate the y's, we can multiply the first equation by 3 and the second equation by -7. This will yield $-21y$ and $+21y$. We could have multiplied the first equation by -3 and the second by 7. This would have worked as well. So would 6 and -14. We need products that will render the variables the same but opposite in sign.

$$\begin{array}{l} 4x - 7y = -13 \\ 7x - 3 = 5 \end{array} \quad \rightarrow \quad \begin{array}{l} 3(4x - 7y) = 3(-13) \\ -7(7x - 3y) = -7(5) \end{array} \quad \rightarrow \quad \begin{array}{l} 12x - 21y = -39 \\ -49x + 21y = -35 \end{array}$$

$$\begin{array}{l} -37x + 0y = -74 \\ -37x = -74 \\ \dfrac{-37x}{-37} = \dfrac{-74}{-37} \\ x = 2 \end{array}$$

To find y, we substitute $x = 2$ into either original equation or continue with the linear combination process. If we do the latter, we can eliminate the x's by multiplying the first equation by 7 and the second equation by -4:

$$\begin{array}{l} 4x - 7y = -13 \\ 7x - 3y = 5 \end{array} \quad \rightarrow \quad \begin{array}{l} 7(4x - 7y) = 7(-13) \\ -4(7x - 3y) = -4(5) \end{array} \quad \rightarrow \quad \begin{array}{l} 28x - 49y = -91 \\ -28x + 12y = -20 \end{array}$$

$$\begin{array}{l} -37y = -111 \\ \dfrac{-37y}{-37} = \dfrac{-111}{-37} \\ y = 3 \end{array}$$

Hence, the point of intersection of the two lines is $(x, y) = (2, 3)$.

4. Solve the following system by the substitution method:

$$2x + y - z = 3$$
$$x - 4y + 3z = 11$$
$$x + 2y + 2z = 13$$

We need to solve one of the equations for one of the variables and substitute it into the other two equations. Solving the second equation for x and substituting into the first and third equations, we have

$$x - 4y + 3z = 11$$
$$x = 4y - 3z + 11 \qquad \text{Solve second equation for } x.$$

First equation

$$2x + y - z = 3$$
$$2(4y - 3z + 11) + y - z = 3$$
$$8y - 6z + 22 + y - z = 3$$
$$9y - 7z = 3 - 22$$
$$9y - 7z = -19$$

Third equation

$$x + 2y + 2z = 13$$
$$(4y - 3z + 11) + 2y + 2z = 13$$
$$6y - z = 13 - 11$$
$$6y - z = 2$$

Now the problem has been reduced to solving a system of two equations. Solving by the substitution method, we have

$$6y - z = 2 \qquad \text{Solve for } z.$$
$$-z = -6y + 2$$
$$\frac{-z}{-1} = \frac{-6z}{-1} + \frac{2}{-1}$$
$$z = 6y - 2$$

Substitute z into $9y - 7z = -19$ and solve for y:

$$9y - 7z = -19$$
$$9y - 7(6y - 2) = -19$$
$$9y - 42y + 14 = -19$$
$$-33y = -19 - 14$$
$$-33y = -33$$
$$y = 1$$

Substitute $y = 1$ into $z = 6y - 2$ to find z:

$$z = 6y - 2$$
$$z = 6(1) - 2$$
$$z = 4$$

To find x, we substitute $y = 1$ and $z = 4$ into any of the original equations. Substituting into the second original equation, we have

$$x - 4y + 3z = 11$$
$$x - 4(1) + 3(4) = 11$$
$$x = 11 + 4 - 12$$
$$x = 3$$

Hence, the point of intersection of the three lines is the ordered triple $(x, y, z) = (3, 1, 4)$.

5. Solve the system in example 4 by the linear combination method.

Since the equations are already in standard form, we need to determine which of the variables we want to eliminate. The y's and the z's seem to be the best since they are already opposite in sign. If we choose one of them, we won't have to multiply by a negative. Let's choose the z's simply because they're at the end of the equations.

Eliminating the z's from the first and second equations, we have

$$\begin{aligned} 2x + y - z &= 3 \\ x - 4y + 3z &= 11 \end{aligned} \quad \rightarrow \quad \begin{aligned} 3(2x + y - z) &= 3(3) \\ x - 4y + 3z &= 11 \end{aligned} \quad \rightarrow \quad \begin{aligned} 6x + 3y - 3z &= 9 \\ x - 4y + 3z &= 11 \\ \hline 7x - y + 0z &= 20 \\ 7x - y &= 20 \end{aligned}$$

If we try to eliminate the z's from the second and third equations, we would have to multiply each by 2 and -3, or by -2 and 3, respectively. However, if we eliminate the z's from the first and third equations, we only have to multiply the first equation by 2:

$$\begin{aligned} 2x + y - z &= 3 \\ x + 2y + 2z &= 13 \end{aligned} \quad \rightarrow \quad \begin{aligned} 2(2x + y - z) &= 2(3) \\ x + 2y + 2z &= 13 \end{aligned} \quad \rightarrow \quad \begin{aligned} 4x + 2y - 2z &= 6 \\ x + 2y + 2z &= 13 \\ \hline 5x + 4y + 0z &= 19 \\ 5x + 4y &= 19 \end{aligned}$$

The problem has now been reduced to solving a system of two equations:

$$7x - y = 20$$
$$5x + 4y = 19$$

Eliminating the y's from these equations, we have

$$\begin{aligned} 7x - y &= 20 \\ 5x + 4y &= 19 \end{aligned} \quad \rightarrow \quad \begin{aligned} 4(7x - y) &= 4(20) \\ 5x + 4y &= 19 \end{aligned} \quad \rightarrow \quad \begin{aligned} 28x - 4y &= 80 \\ 5x + 4y &= 19 \\ \hline 33x + 0y &= 99 \\ 33x &= 99 \\ \frac{33x}{33} &= \frac{99}{33} \\ x &= 3 \end{aligned}$$

S

Substituting $x = 3$ into $7x - y = 20$ or $5x + 4y = 19$, we can find y:

$$5x + 4y = 19 \qquad \text{Substitute and solve for } y.$$
$$5(3) + 4y = 19$$
$$4y = 19 - 15$$
$$\frac{4y}{4} = \frac{4}{4}$$
$$y = 1$$

Finally, we need to substitute $x = 3$ and $y = 1$ into any of the original equations to find z. Substituting into the first equation, we have

$$2x + y - z = 3$$
$$2(3) + (1) - z = 3$$
$$-z = 3 - 6 - 1$$
$$\frac{-z}{-1} = \frac{-4}{-1}$$
$$z = 4$$

Hence, the solution to the system is the triple $(x, y, z) = (3, 1, 4)$.

To verify that $(3, 1, 4)$ is a correct solution, we can substitute it into any of the original equations. If the substitution remains balanced, the point is a solution. Using the first equation, we have

$$2x + y - z = 3$$
$$2(3) + (1) - 4 = 3$$
$$6 + 1 - 4 = 3$$
$$3 = 3$$

Therefore, the point is a solution to the system. For complete verification, we have to substitute the point into all three equations.

➤ **Systems of Nonlinear Equations, solutions of**   **Systems of nonlinear equations** can be solved either by the substitution method or by the combination method. The substitution method can be used if one of the equations can be solved for one of the variables or a similar term, i.e., solved for x^2 not x. The combination method can be used if the degree of the similar terms is the same. In this case, we need to multiply the equation(s) by a number(s) that will render the terms of a particular variable the same but opposite in sign. Combination of the equations will then yield a cancellation of that variable. For a review of these methods, see **Systems of Linear Equations, solutions of**.

Examples

1. Solve the following system by the substitution method and then by graphing:

$$x^2 + y^2 = 25$$
$$x + y = -7$$

If we solve the second equation for y (or for x) and substitute into the first equation, we will be able to find the x-values of the points of intersection. Then substituting each x-value into the second equation, we can find the corresponding y-value of each point:

$$x + y = -7 \qquad \text{Solve the second equation for } y.$$
$$y = -x - 7$$
$$x^2 + y^2 = 25 \qquad \text{Write the first equation.}$$
$$x^2 + (-x - 7)^2 = 25 \qquad \text{Substitute } y = -x - 7 \text{ into the first equation.}$$
$$x^2 + (-x - 7)(-x - 7) = 25 \qquad \text{Square the binomial.}$$
$$x^2 + x^2 + 14x + 49 = 25$$
$$2x^2 + 14x + 49 = 25$$
$$2x^2 + 14x + 24 = 0 \qquad \text{Set the equation equal to 0.}$$
$$x^2 + 7x + 12 = 0 \qquad \text{Divide through by 2.}$$
$$(x + 4)(x + 3) = 0 \qquad \text{Factor the trinomial.}$$

$$x + 4 = 0 \qquad \text{or} \qquad x + 3 = 0 \qquad \text{Solve each factor.}$$
$$x = -4 \qquad\qquad\qquad x = -3$$

Substituting each x-value into $x + y = -7$, we can determine their corresponding y-values:

For $x = -4$, we have

$$x + y = -7$$
$$-4 + y = -7$$
$$y = -7 + 4$$
$$y = -3$$

For $x = -3$, we have

$$x + y = -7$$
$$-3 + y = -7$$
$$y = -7 + 3$$
$$y = -4$$

Hence, the points of intersection are $(-4, -3)$ and $(-3, -4)$.

To verify the solutions, we need to substitute each into $x^2 + y^2 = 25$ and see if the substitutions remain balanced:

Substitution for $(-4, -3)$

$$x^2 + y^2 = 25$$
$$(-4)^2 + (-3)^2 = 25$$
$$16 + 9 = 25$$
$$25 = 25$$

Substitution for $(-3, -4)$

$$x^2 + y^2 = 25$$
$$(-3)^2 + (4)^2 = 25$$
$$9 + 16 = 25$$
$$25 = 25$$

Since each remains balanced, the points are solutions of the system.

To solve the system by graphing, we need to graph each equation and then determine the points of intersection by looking at the graph. The first equation is a circle with a center of $(0, 0)$ and a radius of $r = 5$. See **Circle**. The second equation is a line with an x-intercept of $x = -7$ and a y-intercept of $y = -7$. See **Linear Equations, graphs of**, x-y Intercept Method. Hence, we have the graphs of equations.

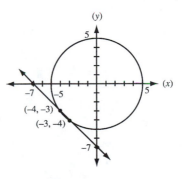

After graphing the circle and the line, it can be seen that their points of intersection are $(-4, -3)$ and $(-3, -4)$.

2. Find the solution of the following system:

$$5x^2 + y^2 = 30$$
$$y^2 - 16 = 9x^2$$

Since the variables are each of the same degree, we can use the combination method. To do this, we need each term of the first equation to match up with its corresponding term in the second equation. Then we need to determine which variable we would like to have cancel, multiply each equation by a number(s) that will get the variable to cancel, and then combine the equations. Matching corresponding terms, we have

$$\begin{aligned} 5x^2 + y^2 &= 30 \\ y^2 - 16 &= 9x^2 \end{aligned} \qquad \rightarrow \qquad \begin{aligned} 5x^2 + y^2 &= 30 \\ -9x^2 + y^2 &= 16 \end{aligned}$$

If we want to get the x^2 terms to cancel, we need to multiply the first equation by 9 and the second equation by 5. This will render the x^2 terms the same but opposite in sign, $+45x^2$ and $-45x^2$. If we want to get the y^2 terms to cancel we need only multiply one of the equations by -1, say the second equation. This will render the y^2 terms the same but opposite in sign, $+y^2$ and $-y^2$. Since canceling the y^2 terms is easier, we have

$$\begin{aligned} 5x^2 + y^2 &= 30 \\ -9x^2 + y^2 &= 16 \end{aligned} \quad \rightarrow \quad \begin{aligned} 5x^2 + y^2 &= 30 \\ -1\left(-9x^2 + y^2\right) &= -1(16) \end{aligned} \quad \rightarrow \quad \begin{aligned} 5x^2 + y^2 &= 30 \\ \underline{9x^2 - y^2 = -16} \\ 14x^2 + 0y^2 &= 14 \\ 14x^2 &= 14 \\ \frac{14x^2}{14} &= \frac{14}{14} \\ x^2 &= 1 \\ \sqrt{x^2} &= \pm\sqrt{1} \\ x &= \pm 1 \end{aligned}$$

These are the x-values of the points of intersection. To find the y-values, we need to substitute each x into one of the original equations and solve for y. Using the first equation, we have

$$\begin{array}{cc}
y\text{-values when } x = +1 & y\text{-values when } x = -1 \\
5x^2 + y^2 = 30 & 5x^2 + y^2 = 30 \\
5(1)^2 + y^2 = 30 & 5(-1)^2 + y^2 = 30 \\
y^2 = 30 - 5 & y^2 = 30 - 5 \\
y^2 = 25 & y^2 = 25 \\
\sqrt{y^2} = \pm\sqrt{25} & \sqrt{y^2} = \pm\sqrt{25} \\
y = \pm 5 & y = \pm 5
\end{array}$$

S

When $x = 1$, the y-values are $+5$ and -5. These yield the points of intersection $(1, 5)$ and $(1, -5)$. When $x = -1$, the y-values are $+5$ and -5. These yield two further points of intersection, $(-1, 5)$ and $(-1, -5)$. Hence, the solution to the system is the four points of intersection $(1, 5)$, $(1, -5)$, $(-1, 5)$, and $(-1, -5)$.

To verify the solutions, we need to substitute the points into the original equations. Substituting into one equation will suffice but that just proves the points satisfy that particular equation. To be complete, substitution should be made into both equations.

This system is conductive to a solution by the combination method because the variables of the corresponding terms are of the same degree. However, if we choose to apply the substitution method, we need to solve one of the equations for one of the variables, substitute into the other equation, solve for the remaining variable, and then calculate the values of the corresponding variable.

Solving the second equations for y, we have

$$
\begin{aligned}
y^2 - 16 &= 9x^2 \\
y^2 &= 9x^2 + 16 \qquad \text{Solve for } y. \text{ See \textbf{Root, } } n\textbf{th.} \\
\sqrt{y^2} &= \pm\sqrt{9x^2 + 16} \\
y &= \pm\sqrt{9x^2 + 16}
\end{aligned}
$$

Substituting into the first equation, we have

$$
\begin{aligned}
5x^2 + y^2 &= 30 \\
5x^2 + \left(\pm\sqrt{9x^2 + 16}\right)^2 &= 30 \qquad \text{Since the quantity is squared, the } \pm \text{ is positive, i.e.,} \\
5x^2 + 9x^2 + 16 &= 30 \qquad (\pm)^2 = (\pm)(\pm) = \left(\genfrac{}{}{0pt}{}{+}{+}\right) = +. \\
14x^2 &= 30 - 16 \\
14x^2 &= 14 \\
x^2 &= 1 \\
\sqrt{x^2} &= \pm\sqrt{1} \\
x &= \pm 1
\end{aligned}
$$

Hence, we have the x-values of the solution. To find the y-values, we need to substitute each x-value into one of the original equations, as we did earlier.

S

➤ **Table, in trigonometry**   See **Trigonometry Table, how to use**.

➤ **Tangent, in trigonometry**

The **tangent** is one of the six functions of trigonometry. Defined for a right triangle, it is a fraction of the **opposite** side to the **adjacent** side. From angle A, $\tan A = \frac{a}{b}$. From angle B, $\tan B = \frac{b}{a}$.

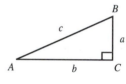

For more, see **Trigonometry, right triangle definition of**.

➤ **Tangent Line**

In relation to a curve, a **tangent line** is any line that intersects the curve in one point without passing through the curve. It touches the curve at one point.

In geometry, if we consider a circle, a tangent line is any line that contains a point of the circle but no interior points of circle. It touches the circle at one point. In the figure, we say that $\overleftrightarrow{PQ}$ (read "the line PQ") is tangent to the circle O at the point P.

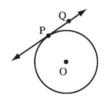

➤ **Term**   A **term** is a number, a variable, or the product of a number and a variable(s). The concept of a term is usually used to delineate one term from another.

In the expression $6x^3 - x^2 - 4xy + 5$, there are four terms: $6x^3$, $-x^2$, $-4xy$, and 5. Each term is separated by, yet includes, a positive ($+$) and a negative ($-$) sign. The term $6x^3$ is composed of the number $+6$, called the **numerical coefficient** or **coefficient**, and a variable with an exponent of 3. The term $-x^2$ is composed of the coefficient -1 and a variable with an exponent of 2. The -1 in this term is usually not written since $(-1)x^2$ is the same as $-x^2$. The term $-4xy$ is composed of the number -4 and the variables x and y. Finally, the last term 5, which has no variable, is referred to as a **constant** term. For more, see **Coefficient** or **Combining Like Terms**.

➤ **Terminal Side**   In trigonometry, the **terminal side** of an angle is a ray that is determined by rotating the initial side of the angle clockwise or counterclockwise. See **Angle, in trigonometry**.

➤ **Terminating Decimals**   See **Decimal Numbers, terminating**.

T

➤ **Terms**   See **Term**.

➤ **Theorem**   A **theorem** is a statement that can be proved to be logically sound by other theorems or postulates. See **Postulate** or **Proof in Geometry**. Synonymous with the term **proposition**, a theorem is used to develop even more complicated theorems. The end result is usually a theorem that is useful as a law or rule, i.e., the Pythagorean theorem, $a^2 + b^2 = c^2$.

➤ **Thirty-Sixty-Ninety Degree Triangle**   The **30-60-90 degree triangle** is one of two special triangles used in trigonometry. The other is the 45-45-90 degree triangle. See **Forty-Five Degree Triangle**.

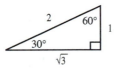

In a 30-60-90 degree triangle, the length of the shorter leg, the side opposite the 30° angle, is 1. The length of the longer leg, the side opposite the 60° angle, is $\sqrt{3}$. And the length of the hypotenuse, the side opposite the 90° angle, is 2.

To apply the 30-60-90 degree triangle, the concepts of cross multiplication and similar triangles must be clear. See the articles under those names. Also see **Trigonometry, reference angle**; **Direct Variation**; or **Proportion**.

Examples

1. Find the remaining sides of a 30-60-90 degree triangle if the length of the shorter leg is 4.

   What is required is to find the length of the longer leg and the length of the hypotenuse. We can do this by one of two methods. We can either compare the given triangle with the 30-60-90 degree triangle or find the sides using similar triangles.

   To find the sides by comparing the given triangle to the 30-60-90 degree triangle, we notice that the length of the shorter leg of the given triangle is 4 times that of the shorter leg of the 30-60-90 degree triangle.

   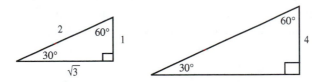

   Multiplying each of the sides of the 30-60-90 degree triangle by 4, we have the length of the shorter leg, $4(1) = 4$, the length of the longer leg, $4(\sqrt{3}) = 4\sqrt{3}$, and the length of the hypotenuse, $4(2) = 8$.

   To find the sides using similar triangles, we need to set up proportions between the given triangle and the 30-60-90 degree triangle.

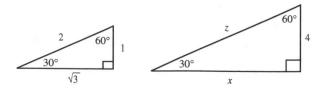

Letting x represent the length of the longer leg, we have the proportion $\frac{x}{\sqrt{3}} = \frac{4}{1}$. Letting z represent the length of the hypotenuse, we have the proportion $\frac{z}{2} = \frac{4}{1}$. Solving each proportion, we have the lengths of the sides:

$$\frac{x}{\sqrt{3}} = \frac{4}{1} \qquad \text{and} \qquad \frac{z}{2} = \frac{4}{1}$$

$$x = \frac{4\sqrt{3}}{1} \qquad\qquad\qquad z = \frac{4(2)}{1} \qquad \text{Cross multiply in each equation.}$$

$$x = 4\sqrt{3} \qquad\qquad\qquad z = 8$$

Hence, the length of the longer leg is $4\sqrt{3}$ and the length of the hypotenuse is 8.

2. Find the remaining sides of a 30-60-90 degree triangle if the length of the hypotenuse is 1.

To find the sides by comparing the given triangle to the 30-60-90 degree triangle, we notice that the length of the hypotenuse of the given triangle is $\frac{1}{2}$ that of the hypotenuse of the 30-60-90 degree triangle.

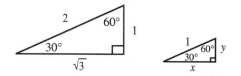

Multiplying each of the sides of the 30-60-90 degree triangle by $\frac{1}{2}$, we have the length of the hypotenuse, $\frac{1}{2}(2) = 1$, the length of the shorter leg $\frac{1}{2}(1) = \frac{1}{2}$, and the length of the longer leg, $\frac{1}{2}(\sqrt{3}) = \frac{\sqrt{3}}{2}$.

To find the sides using similar triangles, we need to set up proportions. Letting y represent the length of the shorter leg, we have the proportion $\frac{y}{1} = \frac{1}{2}$. Letting x represent the length of the longer leg, we have the proportion $\frac{x}{\sqrt{3}} = \frac{1}{2}$. Solving the proportions, we have

$$\frac{y}{1} = \frac{1}{2} \qquad \text{and} \qquad \frac{x}{\sqrt{3}} = \frac{1}{2}$$

$$y = \frac{1}{2} \qquad\qquad\qquad x = \frac{\sqrt{3}}{2} \qquad \text{Cross multiply.}$$

3. Find the remaining sides of a 30-60-90 degree triangle if the length of the longer leg is 4.

This type of problem is more difficult to solve by comparing triangles because the 4 is compared to $\sqrt{3}$. We need to think of a number that can multiply $\sqrt{3}$ to yield 4. The number is $\frac{4}{\sqrt{3}}$, since

$\frac{4}{\sqrt{3}}(\sqrt{3}) = 4$. Hence, the length of the shorter side is $\frac{4}{\sqrt{3}}(1) = \frac{4}{\sqrt{3}} = \frac{4\sqrt{3}}{3}$. See **Rationalizing the Denominator**. And the length of the hypotenuse is $\frac{4}{\sqrt{3}}(2) = \frac{8}{\sqrt{3}} = \frac{8\sqrt{3}}{3}$.

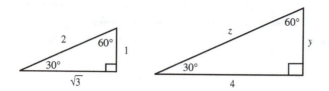

Usually a problem like this is solved by setting up proportions. Letting y represent the length of the shorter side, we have the proportion $\frac{y}{1} = \frac{4}{\sqrt{3}}$. Letting z represent the length of the hypotenuse, we have the proportion $\frac{z}{2} = \frac{4}{\sqrt{3}}$. Solving each, we have

$$\frac{y}{1} = \frac{4}{\sqrt{3}} \qquad \text{and} \qquad \frac{z}{2} = \frac{4}{\sqrt{3}}$$

$$y = \frac{4}{\sqrt{3}} \cdot \frac{\sqrt{3}}{\sqrt{3}} \qquad\qquad z = \frac{8}{\sqrt{3}} \cdot \frac{\sqrt{3}}{\sqrt{3}} \qquad \text{See \textbf{Rationalizing the Denominator}.}$$

$$y = \frac{4\sqrt{3}}{3} \qquad\qquad\qquad z = \frac{8\sqrt{3}}{3}$$

➤ **Transitive Property**   The **transitive property** is one of the three axioms of equality. The other two are the reflexive and symmetric properties. See articles under those names.

   The transitive property states that if a, b, and c are real numbers, then the following holds: if $a = b$ and $b = c$ then $a = c$. Also see **Property** or **Properties of the Real Number System**.

➤ **Translation or Reflection**   The **translation or reflection** of a graph moves a graph left or right, up or down, or turns the graph over or around. During a translation or reflection, the location of the graph changes but not its original size or shape. The concepts of translation or reflection allow for a much greater graphing versatility.

   Let $y = f(x)$ represent the graph of an original function. If t is a positive constant, then we have the following (see **Function Notation**):

1. $y = f(x) + t$ will translate the graph t units upward.
2. $y = f(x) - t$ will translate the graph t units downward.
3. $y = f(x + t)$ will translate the graph t units left.
4. $y = f(x - t)$ will translate the graph t units right.
5. $y = -f(x)$ will reflect the graph in the x-axis.
6. $y = f(-x)$ will reflect the graph in the y-axis.

The translations in 3 and 4 are opposite the signs in the parentheses.

   Along with the properties pertaining to translation or reflection, there are several graphs that are usually memorized. A list is provided below.

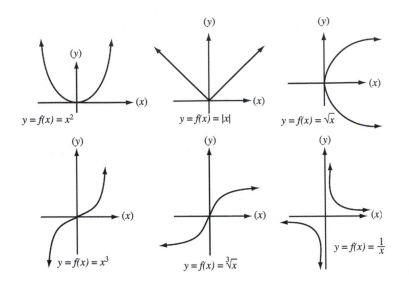

$y = f(x) = x^2$ $y = f(x) = |x|$ $y = f(x) = \sqrt{x}$

$y = f(x) = x^3$ $y = f(x) = \sqrt[3]{x}$ $y = f(x) = \frac{1}{x}$

Examples

1. Using the concepts of translation or reflection, graph the following:

$$y = x^2 + 1, \ y = x^2 - 1, \ y = (x + 1)^2, \ y = (x - 1)^2, \ y = -x^2, \ \text{and} \ y = (-x)^2$$

What we are asked to do is take the basic graph of $y = f(x) = x^2$ and move it through the four positions of translation and the two positions of reflection.

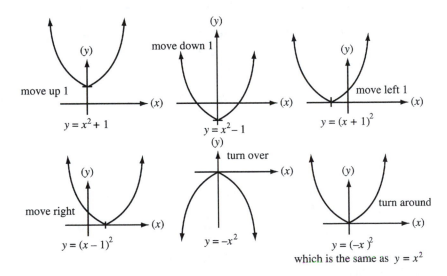

move up 1
$y = x^2 + 1$

move down 1
$y = x^2 - 1$

move left 1
$y = (x + 1)^2$

move right
$y = (x - 1)^2$

turn over
$y = -x^2$

turn around
$y = (-x)^2$
which is the same as $y = x^2$

For practice, it is a good idea to graph each of the six memorized graphs in each of the six positions of translation and reflection. By doing so, the equations of each graph will become more familiar and easier to recognize. Also, combinations of translation or reflection will become easier to handle, such as example 2.

2. Using the concepts of translation or reflection, graph $y = -|x - 3| + 2$.

In this equation, the following concepts are applied:

$$y = -f(x), \quad y = f(x - t), \quad \text{and } y = f(x) + t$$

The reflection $y = -f(x)$ is represented by the "$-$" in front of the absolute value sign. It will turn the graph over. The translation $y = f(x - t)$ is represented by the $x - 3$ inside the absolute value signs. It will move the graph right 3 units. The translation $y = f(x) + t$ is represented by the $+2$. It will move the graph up 2 units.

The graph that is being translated and reflected, actually reflected and then translated, is $y = f(x) = |x|$. The graph of the absolute value of x is to be turned over, moved right 3 units, and moved up 2 units.

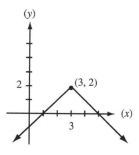

3. Using the concepts of translation and reflection, graph $y = \frac{1}{x-2} - 1$.

In this equation, we are applying $y = f(x - t)$ and $y = f(x) - t$. The graph of $y = f(x) = \frac{1}{x}$ is to be translated to the right 2 units and then translated down 1 unit.

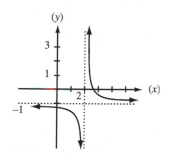

4. Using the concepts of translation or reflection, graph $y = \sqrt[3]{3 - x} - 2$.

Using the commutative property, we see that the given equation is equivalent to $y = \sqrt[3]{-x + 3} - 2$. In this form, we see that the following have been applied to $y = f(x) = \sqrt[3]{x}$: $y = f(-x)$, $y = f(x + t)$ and $y = f(x) - t$. The reflection $y = f(-x)$ is represented by the $-x$. It will turn the graph around, which in this case looks the same as turning the graph over. The translation

$y = f(x + t)$ is represented by the $-x + 3$. It will move the graph left 3 units. The translation $y = f(x) - t$ is represented by the -2. It will move the graph down 2 units.

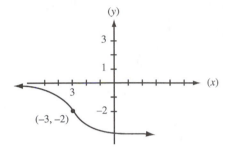

➤ **Transversal**   A **transversal** is a line that intersects (crosses) two or more lines in different points.

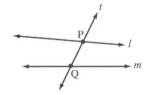

In the figure, the line t is a transversal. It intersects the lines l and m at the points P and Q. For more concerning the angles formed with a transversal, see **Interior and Exterior Angles (for lines)**. Also see **Parallel Lines (geometric)**.

➤ **Transverse Axis of an Ellipse**   See **Ellipse**.

➤ **Trapezoid**   A **trapezoid** is a quadrilateral with one pair (only) of parallel sides, called **bases**. The other two sides, called **legs**, are not parallel. If they were parallel, the quadrilateral would be a parallelogram, not a trapezoid. When the legs are congruent (equal in measure), the trapezoid is an **isosceles** trapezoid.

The area of a trapezoid is equal to the product of one-half the sum of the bases and the altitude.

$$A = \frac{1}{2}(b_1 + b_2)h$$

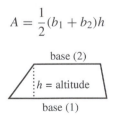

The perimeter of a trapezoid is found by adding the measures of the four sides. Other qualities of a trapezoid are:

1. The median (segment joining the midpoints of the legs) is parallel to the bases and equal to one-half the sum of their measures.

2. The base angles of an isosceles trapezoid are equal in measure, congruent.

3. The diagonals of an isosceles trapezoid are equal in measure, congruent.

Examples

1. Find the area of a trapezoid with a longer base of 28, shorter base of 14, and an altitude of 7. Letting $b_1 = 28, b_2 = 14, h = 7$ and substituting into the formula for the area of a trapezoid, we have

$$A = \frac{1}{2}(b_1 + b_2)h = \frac{1}{2}(28 + 14)7$$
$$= \frac{1}{2}(42)7 = 21 \bullet 7 = 147$$

The area of the trapezoid is 147.

2. If the area of a trapezoid is 147, the length of the longer base is 28, and the length of the shorter base is 14, find the length of the altitude.

This is just a reversal of example 1. Given the area and the bases, find the altitude. Letting $A = 147$, $b_1 = 28$, and $b_2 = 14$, we have

$$A = \frac{1}{2}(b_1 + b_2)h$$
$$147 = \frac{1}{2}(28 + 14)h \qquad \text{See \textbf{Linear Equations, solutions of}.}$$
$$147 = \frac{1}{2}(42)h$$
$$147 = 21h$$
$$\frac{147}{21} = \frac{21}{21}h$$
$$7 = h$$
$$h = 7$$

Hence, the length of the altitude is 7.

► **Triangle**   A **triangle** is the figure formed by the union of the three segments determined by three noncollinear points. If all the sides of a triangle are of unequal measure, the triangle is called a **scalene** triangle. If two of the sides of a triangle are of equal measure (congruent), the triangle is called an **isosceles** triangle. If all three sides of a triangle are of equal measure (congruent), the triangle is called an **equilateral** triangle.

Triangles are also categorized by their angles. If all of the angles of a triangle are acute (measure less than 90), the triangle is called an **acute** triangle. If one of the angles of a triangle is obtuse (measure greater than 90), the triangle is called an **obtuse** triangle. If one of the angles of a triangle is a right angle (measure equal to 90), the triangle is called a **right triangle**. And, if all of the angles of a triangle are equal in measure (congruent), the triangle is called an **equiangular** triangle. Since the angle sum of a triangle is 180°, the measure of each angle of an equiangular triangle is 60°.

T

Some of the qualities of a triangle are:

1. The sum of the angles of a triangle is $180°$.

2. If two of the angles of one triangle are congruent to the corresponding angles of another triangle, then the third angles are also congruent.

3. If three angles of one triangle are congruent to three angles of another triangle, the triangles are similar.

4. The angles of an equilateral or equiangular triangle have a measure of 60.

5. In a right triangle, the acute angles are complementary.

6. The base angles of an isosceles triangle have equal measure (are congruent).

7. In an isosceles triangle, the segment joining the midpoints of the legs is parallel to the third side and equal to half its measure.

8. The measure of any exterior angle is equal to the sum of the measures of its remote interior angles. See **Remote Interior Angles** or **Exterior Angles (for a triangle)**.

The **area** of a triangle is equal to $\frac{1}{2}$ the product of a side, called the base, and its altitude. Letting b equal the base and h equal the altitude, we have $A = \frac{1}{2}bh$.

altitude = h
base = b

The **perimeter** of a triangle is found by adding the measures of the three sides.

Example

1. Find the area of $\triangle ABC$ if $AC = 12$ and the altitude to $\overline{AC}$ is $h = 5$.

   Substituting $h = 5$ and $b = AC = 12$ into the area formula, we have

   $$A = \frac{1}{2}bh = \frac{1}{2}(12)(5)$$
   $$= \frac{1}{2}(60) = \frac{60}{2} = 30$$

B
h = 5
A   base = 12   C

   The area of the triangle is 30.

➤ **Triangle Inequality**   The **triangle inequality** is one of the properties of absolute values. See **Properties of the Real Number System**. It states that if a and b are real numbers, then

$$|a + b| \leqslant |a| + |b|$$

➤ **Triangle Sum**   See **Sum of the Measures of the Angles of a Triangle**.

➤ **Trigonometric Form of a Complex Number**   See **Complex Number, trigonometric form**.

➤ **Trigonometry**   Developed in antiquity, **trigonometry** was originally derived for measuring the sides of triangles in relation to their distances and their interrelated angles. Directly used in surveying, there was also a great application of trigonometry to navigation and astronomy.

    Throughout the centuries as an understanding of the various sciences increased, and especially by the time of Newton and Leibniz with the development of calculus, trigonometry became imbedded in the modern viewpoint of mathematics. The necessity of trigonometry increased as applications to phenomena of a periodic or repetitive nature became apparent. Concepts pertaining to rotation and vibration were included in the domain of trigonometry. Among these are pendulums, vibration of strings, sound waves, light waves, orbits of the planets or atomic particles, and, in general, the study of mechanical vibrations, electromagnetic theory, and thermodynamics.

➤ **Trigonometry, application to a non-right triangle**   See **Law of Cosines** or **Law of Sines**. Also see **Trigonometry, area formula**.

➤ **Trigonometry, application to a right triangle**   See **Trigonometry, right triangle definition of**.

➤ **Trigonometry, area formula**   The **area formula in trigonometry** basically has two purposes. One, of course, is to find the area of a triangle. The other, and more important, is to derive the law of sines. The law of sines in conjunction with the law of cosines allows for solutions of non-right triangles. See **Law of Sines** or **Law of Cosines**.

    To derive the area formula, we use the equation for the area of a triangle, $A = \frac{1}{2}bh$ (see **Triangle**), and the sine function (see **Trigonometry, right triangle definition of**). In $\triangle ABC$ with an altitude of h, we know from geometry that its area K is

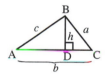

$$K = \frac{1}{2}bh$$

From trigonometry, we know that in $\triangle ABD$ the $\sin A = \frac{h}{c}$. Solving this equation for h and substituting in $K = \frac{1}{2}bh$, we have

$$\sin A = \frac{h}{c} \qquad\qquad K = \frac{1}{2}bh$$

$$c \sin A = h \qquad\qquad K = \frac{1}{2}b(c \sin A) \qquad \text{Substitute for } h.$$

$$h = c \sin A \qquad\qquad = \frac{1}{2}bc \sin A$$

Notice that the formula $K = \frac{1}{2}bc \sin A$ gives the area in terms of the sides b and c and the angle A. It does not involve the altitude h. In a similar fashion, we can derive the area formula from each of the other two vertices,

$$K = \frac{1}{2}ac \sin B \qquad \text{and} \qquad K = \frac{1}{2}ab \sin C$$

T

In general, the area of a triangle is equal to one half the product of the lengths of any two sides and the sine of their included angle.

Examples

1. Find the area of $\triangle ABC$ to the nearest tenth if $a = 8$, $c = 14$, and $m\angle B = 110$.

   From $\angle B$, the area formula is $K = \frac{1}{2}ac \sin B$. Substituting a, c, and $m\angle B$ into the equation, we have

$$K = \frac{1}{2}ac \sin B$$
$$= \frac{1}{2}(8)(14) \sin 110°$$
$$= 56(\sin 70°) \qquad \text{Use supplementary angles}$$
$$= 56(.9397) \qquad \text{for angles greater than 90°.}$$
$$= 52.6$$

To simplify $\sin 110° = \sin 70° = .9397$; see **Trigonometry, reference angle**. Also see **Trigonometry Table, how to use**.

2. In $\triangle ABC$, find the measure of $\angle A$ to the nearest degree if $b = 14$, $c = 6$, and the area of the triangle is 36.

   From $\angle A$, the area formula is $K = \frac{1}{2}bc \sin A$. Substituting b, c, and the area $K = 36$ into the formula, we have

$$K = \frac{1}{2}bc \sin A$$
$$36 = \frac{1}{2}(14)(6) \sin A$$
$$\frac{36}{42} = \frac{42 \sin A}{42}$$
$$.8571 = \sin A$$
$$m\angle A = \arcsin .8571 \qquad \text{See \textbf{Trigonometry, inverse (arc) functions}}$$
$$\qquad\qquad\qquad \text{or \textbf{Trigonometry Table, how to use}}.$$
$$m\angle A = 59°$$

For more accuracy of the angle, use interpolation. See **Interpolation (linear)**.

➤ **Trigonometry, converting from degrees to radians**   See **Converting Degrees to Radians**.

➤ **Trigonometry, converting from radians to degrees**   See **Converting Radians to Degrees**.

➤ **Trigonometry, equations**   A **trigonometric equation** is an equation that has at least one of the trigonometric functions in it. Not to be confused with an identity, which can be solved for all values of the variable, an equation in trigonometry is solved for particular values of variable. These are the only values that will satisfy the equation.

To solve a trigonometric equation, we need to do two things:

1. First, using identities and the techniques of algebra, we need to isolate the trigonometric function.

2. Next, considering the value the function is equal to, we need to determine what value(s) of the variable will yield the value of the function. This is done by observation or by applying the inverse of the given function.

It is the second step that causes most confusion. Equations in trigonometry do not solve as simply as those in algebra. When a variable has been isolated in algebra, the problem is finished. When the function has been isolated in trigonometry, we are not finished. A whole new problem now has to be considered. We need to consider what value of the variable will yield the value of the function. Or, we need to apply the inverse of the function. A familiarity with the following articles will help: **Thirty-Sixty-Ninety Degree Triangle**; **Forty-Five Degree Triangle**; **Trigonometry, signs of the six functions**; **Trigonometry Table, how to use**; and **Coterminal Angles**. Also see **Trigonometry, identities**.

To isolate the trigonometric functions, here are some things to try:

1. Solve for the trigonometric function. See example 1.

2. Use reciprocal or quotient identities. See example 2.

3. Collect similar terms and then solve for the trigonometric function. See example 3.

4. Factor by FOIL or factor common monomials and then solve each factor. See examples 4 and 5.

5. Substitute using identities. See examples 5, 6, and 8.

6. Take the square root of both sides of the equation. See example 7.

7. Square both sides of the equation. See example 8.

Examples

1. Solve $2 \sin x - 1 = 0$ for x.

We first need to isolate the $\sin x$:

$$2 \sin x - 1 = 0$$
$$2 \sin x = 1$$
$$\frac{2 \sin x}{2} = \frac{1}{2}$$
$$\sin x = \frac{1}{2}$$

Now we need to determine what value of x will yield a $\frac{1}{2}$. Notice that the sign of $\frac{1}{2}$ is positive (+). Since $\frac{1}{2}$ is the sine of 30° and the sine is + in the 1st and 2nd quadrants, we have (see **Thirty-Sixty-**

Ninety Degree Triangle or **Trigonometry, signs of the six functions)**

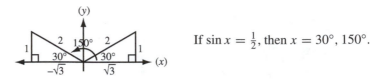

If $\sin x = \frac{1}{2}$, then $x = 30°, 150°$.

The values $x = 30°$ and $x = 150°$ are the two values for $0° \leqslant x \leqslant 360°$ (x between and including $0°$ and $360°$) that have a sine of $+\frac{1}{2}$. Therefore, if we go around the circle once, these are the answers. If we go around the circle again, whether in the positive or negative directions, all coterminal angles are also solutions. Each coterminal angle will have a sine of $+\frac{1}{2}$. Hence, if we are not limited, i.e., $0° \leqslant x \leqslant 360°$, we need to consider all coterminal angles as part of our answer.

To include coterminal angles in our solution, we need to consider the sequences of coterminal solutions. In the positive direction, the angles coterminal with $30°$ are $30° + 360° = 390°$, $30° + 720° = 750°$, and so on. In the negative direction, the angles coterminal with $30°$ are $30° - 360° = -330°$, $30° - 720° = -690°$, and so on. Therefore, the sequence of coterminal angles for $30°$ is

$$\cdots, -690°, -330°, 30°, 390°, 750°, \cdots$$

The expression that will generate this sequence is $30° + n \bullet 360°$ where n is an integer. Hence, the first half of the solution for $\sin x = \frac{1}{2}$ for the $30°$ angle is $x = 30° + n \bullet 360°$. To verify the solution, let n equal $-2, -1, 0, 1$, and 2 and substitute n into $x = 30° + n \bullet 360°$. This will generate the sequence stated above. With a little practice, generating sequences will get easier.

In the positive direction, the angles coterminal with $150°$ are $150° + 360° = 510°$, $150° + 720° = 870°$, and so on. In the negative direction, the angles coterminal with $150°$ are $150° - 360° = -210°$, $150° - 720° = -570°$, and so on. Therefore, the sequence of coterminal angles for $150°$ is

$$\cdots, -570°, -210°, 150°, 510°, 870°, \cdots$$

The expression that will generate this sequence is $150° + n \bullet 360°$ where n is an integer. Hence, the solution for $\sin x = \frac{1}{2}$ for the $150°$ angle is $x = 150° + n \bullet 360°$.

Combined, the general solution for $2\sin x - 1 = 0$ is

$$x = 30° + n \bullet 360° \qquad \text{and} \qquad x = 150° + n \bullet 360°$$

If the solution is desired in radians, we have

$$x = \frac{\pi}{6} + 2\pi n \qquad \text{and} \qquad x = \frac{5\pi}{6} + 2\pi n$$
$$\text{or} \quad = \frac{\pi}{6} + 2n\pi \qquad \text{or} \quad = \frac{5\pi}{6} + 2n\pi$$

where $\frac{\pi}{6} = 30°$, $\frac{5\pi}{6} = 150°$, and $2\pi = 360°$. See **Converting Degrees to Radians**.

If it is required to solve the problem by applying the inverse of the sine, we need to take the arcsin of both sides of $\sin x = \frac{1}{2}$. See **Trigonometry, inverse (arc) functions**.

$$\sin x = \frac{1}{2}$$

$$\text{arcsin}(\sin x) = \text{arcsin}\frac{1}{2}$$

$x = 30°$ The arcsin and sin cancel each other. Also, look up $\frac{1}{2} = .5$ in the chart or use $\sin^{-1} .5$ in a calculator.

In either case only the 1st quadrant answers will be determined. The rest of the problem has to be solved as we did before.

2. Solve $\sin x - \cos x = 0$ for x.

There is no simple identity for $\sin x$ or $\cos x$ that can simplify the equation. However, if we set $\sin x = \cos x$ and then divide by the $\cos x$, we can get $\tan x$.

$$\sin x - \cos x = 0$$
$$\sin x = \cos x$$
$$\frac{\sin x}{\cos x} = \frac{\cos x}{\cos x}$$
$$\tan x = 1 \qquad \text{See \textbf{Reciprocal Identities} or \textbf{Quotient Identities}.}$$

Of course, one could say "how was I supposed to know how to do this?" The answer is, "experience." Now that you've seen the process, remember it. Keep it in your memory of things to do when solving equations in trigonometry.

To finish the problem, we need to consider when the tangent is $+1$. The tangent is $+1$ in the 1st and 3rd quadrants.

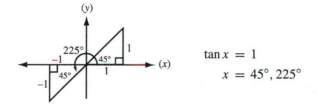

$$\tan x = 1$$
$$x = 45°, 225°$$

If we limit x to $0 \leqslant x \leqslant 360$ (which is $0 \leqslant x \leqslant 2\pi$ in radians), the solutions to the equation are $x = 45°$ and $x = 225°$. If we want to give a general solution to the equation, we need to derive sequences for the positive and negative coterminal angles. First, subtracting $n \bullet 360$ from $45°$ and then adding $n \bullet 360$ to $45°$, we get the sequence

$$\cdots, -675°, -315°, 45°, 405°, 765°, \cdots$$

which is generated by $45° + n \bullet 360°$. Therefore, one solution is $45° + n \bullet 360°$. Repeating the process with $225°$, we get the sequence

$$\cdots, -495°, -135°, 225°, 585°, 945°, \cdots$$

which is generated by $225° + n \bullet 360°$. Therefore, the other solution is $x = 225° + n \bullet 360°$. It would be sufficient to say that the solutions are

$$x = 45° + n \bullet 360° \qquad \text{and} \qquad x = 225° + n \bullet 360°$$

However, this is one of those situations where the two sequences can be put together. Combining the sequences, we have

$$\cdots, -675°, -495°, -315°, -135°, 45°, 225°, 405°, 585°, 765°, 945°, \cdots$$

We notice that each of the terms of the sequence are $180°$ apart. This sequence can be generated by $45° + n \bullet 180°$. Therefore, the combined solution is $x = 45° + n \bullet 180°$. All of the answers are represented by this one solution. Hence, the solution to $\sin x - \cos x = 0$ is

$$x = 45° + n \bullet 180°$$
$$\text{or} \qquad x = \frac{\pi}{4} + n\pi$$

The two answers can be combined because they are separated by $180°$, $225° - 45° = 180°$. The answer in example 1 could not because they are separated by $150° - 30° = 120°$.

3. Find all solutions of $\cos x - \sqrt{3} = -\cos x$.

Solving for $\cos x$, we have

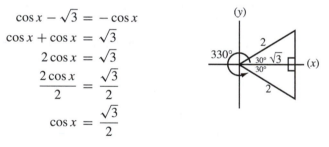

$$\cos x - \sqrt{3} = -\cos x$$
$$\cos x + \cos x = \sqrt{3}$$
$$2 \cos x = \sqrt{3}$$
$$\frac{2 \cos x}{2} = \frac{\sqrt{3}}{2}$$
$$\cos x = \frac{\sqrt{3}}{2}$$

The values of x that yield a cosine of $+\frac{\sqrt{3}}{2}$ are $x = 30°$ and $x = 330°$:

$$\text{If } \cos x = \frac{+\sqrt{3}}{2}$$
$$\text{then } x = 30°, 330°$$

The angles that are coterminal with $30°$ and $330°$ are $30° + n \bullet 360°$ and $330° + n \bullet 360°$. Hence, the general solutions are $x = 30° + n \bullet 360°$ and $x = 330° + n \bullet 360°$. In radians, the solutions are $x = \frac{\pi}{6} + 2n\pi$ and $x = \frac{11\pi}{6} + 2n\pi$.

4. Solve $2 \sin^2 x = \sin x + 1$ for $0° \leqslant x < 360°$.

Since we are limited with $0° \leqslant x < 360°$, we want to find values of x in this interval, including $0°$ but not $360°$. Noticing the $\sin^2 x$, we might think of starting by substituting the Pythagorean identity

$\sin^2 x = 1 - \cos^2 x$. But this will take us nowhere. What is more obvious is that the equation has three terms. If we set the equation equal to 0, we can see if the terms will factor by FOIL. If so, we can use the product theorem of zero to solve each factor. See **Product Theorem of Zero**.

$$2 \sin^2 x = \sin x + 1$$

$$2 \sin^2 x - \sin x - 1 = 0 \qquad \text{Set equal to 0.}$$

$$(2 \sin x + 1)(2 \sin x - 1) = 0 \qquad \text{Factor by FOIL.}$$

$2 \sin x + 1 = 0 \qquad$ or $\qquad \sin x - 1 = 0 \qquad$ Solve each factor.

$2 \sin x = -1 \qquad\qquad\qquad \sin x = 1$

$\sin x = -\dfrac{1}{2} \qquad\qquad\qquad x = 90° \qquad$ See **Trigonometry, quadrant angles**.

$x = 210°, 330° \qquad\qquad$ or $x = \dfrac{\pi}{2}$ in radians

or $x = \dfrac{7\pi}{6}, \dfrac{11\pi}{6}$ in radians

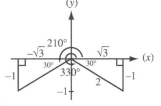

5. Solve $1 - 3 \cos x = \sin^2 x$ for $0° \leqslant x < 360°$.

Our first thought might be to set the equation equal to 0 and try factoring it. But that doesn't work. Since $\sin^2 x = 1 - \cos^2 x$ by the Pythagorean identity, we can try substituting for $\sin^2 x$:

$$1 - 3 \cos x = \sin^2 x$$

$$1 - 3 \cos x = 1 - \cos^2 x \qquad \text{Substitute for } \sin^2 x.$$

$$\cos^2 x - 3 \cos x = 0 \qquad \text{Set equal to 0.}$$

$$\cos x (\cos x - 3) = 0 \qquad \text{Factor the monomial.}$$

$\cos x = 0 \qquad$ or $\qquad \cos x - 3 = 0 \qquad$ Solve each factor. See **Product**

$x = 90°, 270° \qquad\qquad\qquad \cos x = 3 \qquad$ **Theorem of Zero**.

or $x = \dfrac{\pi}{2}, \dfrac{3\pi}{2} \qquad\qquad$ no solutions

The cosine is 0 at two locations, 90° and 270°. Therefore, we have two solutions, $x = 90°$ and $x = 270°$. See **Trigonometry, quadrant angles**. Since the cosine is never larger than 1, $\cos x = 3$ makes no sense. There is no solution for $\cos x = 3$. We do not need to consider the general solutions since x was limited with $0° \leqslant x < 360°$.

6. Solve $\cos 2x = -\sin x$ for $0° \leqslant x < 360°$.

To solve an equation like this, we first need to simplify the double angle, $\cos 2x$, to a single angle, $\cos 2x = 1 - 2 \sin^2 x$. See **Double-Angle Formula**. Of the three choices for the double angle, we use $\cos 2x = 1 - 2\sin^2 x$ because it will make the original equation only have sine terms in it. That should render the equation easier to deal with. If we had chosen either of the other two double-angle formulas, we would have had to further use the Pythagorean identity:

$$\cos 2x = -\sin x$$
$$1 - 2\sin^2 x = -\sin x \qquad \text{Substitute for } \cos 2x.$$
$$-2\sin^2 x + \sin x + 1 = 0 \qquad \text{Set equal to 0.}$$
$$\frac{-2\sin^2 x}{-1} + \frac{\sin x}{-1} + \frac{1}{-1} = \frac{0}{-1}$$
$$2\sin^2 x - \sin x - 1 = 0$$

Now the equation can be factored and each factor solved. The rest of the problem from here is the same as the solution in example 4.

7. Solve $4\sin^2 x - 3 = 0$ for $0° \leqslant x \leqslant 360°$.

We can isolate the $\sin x$ by applying the same methods we would for solving the equation $4A^2 - 3 = 0$. See **Quadratic Equations, solutions of**, example 3.

$$4\sin^2 x - 3 = 0$$
$$4\sin^2 x = 3$$
$$\sin^2 x = \frac{3}{4}$$
$$\sqrt{\sin^2 x} = \pm\sqrt{\frac{3}{4}}$$
$$\sin x = \pm\frac{\sqrt{3}}{2}$$

$$\sin x = +\frac{\sqrt{3}}{2} \qquad \text{or} \qquad \sin x = -\frac{\sqrt{3}}{2}$$
$$x = 60°, 120° \qquad\qquad\qquad x = 240°, 300°$$
$$\text{or} \quad x = \frac{\pi}{3}, \frac{2\pi}{3} \qquad \text{or} \qquad x = \frac{4\pi}{3}, \frac{5\pi}{3}$$

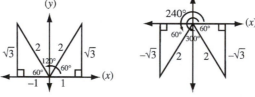

8. Solve $\cos x = \sin x - 1$ for $0° \leqslant x < 360°$.

Dividing through by either $\sin x$ or $\cos x$ does not help. Also, there are no simple identities for $\sin x$ or $\cos x$. So we're stuck. However, a method that is sometimes used is to square both sides of the equation. Put this in your memory of things to do.

$$\cos x = \sin x - 1$$
$$(\cos x)^2 = (\sin x - 1)^2$$
$$\cos^2 x = (\sin x - 1)(\sin x - 1)$$
$$\cos^2 x = \sin^2 x - 2 \sin x + 1$$

Now we change the $\cos^2 x$ to an expression involving sin only by using $\cos^2 x = 1 - \sin^2 x$. Then we can solve the equation.

$$1 - \sin^2 x = \sin^2 x - 2 \sin x + 1$$
$$-2 \sin^2 x + 2 \sin x = 0$$
$$\frac{-2 \sin^2 x}{-2} + \frac{2 \sin x}{-2} = \frac{0}{-2}$$
$$\sin^2 x - \sin x = 0$$
$$\sin x (\sin x - 1) = 0$$

$$\sin x = 0 \qquad \text{or} \qquad \sin x - 1 = 0$$
$$x = 0°, 180° \qquad \qquad \sin x = 1$$
$$\text{or} \qquad x = 0, \pi \qquad \qquad x = 90°$$
$$\text{or} \qquad x = \frac{\pi}{2}$$

Since the original equation was squared, we might have introduced extraneous solutions. Therefore, we need to check each of the answers.

a. $\cos 0° = \sin 0° - 1$
    $1 = 0 - 1$
    $1 = -1$ Therefore, 0° is not a solution.

b. $\cos 180° = \sin 180° - 1$
    $-1 = 0 - 1$
    $-1 = -1$ Therefore, 180° is a solution.

c. $\cos 90° = \sin 90° - 1$
    $0 = 1 - 1$
    $0 = 0$ Therefore, 90° is a solution.

Hence, we see that $x = 180°$ and $x = 90°$ are solutions to the equation.

For practice, instead of squaring both sides of $\cos x = \sin x - 1$, try squaring both sides of $\cos x + 1 = \sin x$ or $1 = \sin x - \cos x$. The eventual answers will be the same.

➤ **Trigonometry, evaluating functions of any angle**   See **Trigonometry, reference angle.**

➤ **Trigonometry, even and odd functions (functions or identities of negative angles)**   Even and odd functions in trigonometry are defined by $f(-\theta) = f(\theta)$ if the function is even and $f(-\theta) = -f(\theta)$ if the function is odd. See **Even and Odd Functions**.

Even Functions   **Odd Functions**

$\cos(-\theta) = \cos\theta$ $\sin(-\theta) = -\sin\theta$

$\sec(-\theta) = \sec\theta$ $\csc(-\theta) = -\csc\theta$

$\tan(-\theta) = -\tan\theta$

$\cot(-\theta) = -\cot\theta$

If it is preferred to write the functions in order, we have

$\sin(-\theta) = -\sin\theta$, odd $\csc(-\theta) = -\csc\theta$, odd

$\cos(-\theta) = \cos\theta$, even $\sec(-\theta) = \sec\theta$, even

$\tan(-\theta) = -\tan\theta$, odd $\cot(-\theta) = -\cot\theta$, odd

To verify the equations, consider the points $P(-a, b)$ and $Q(-a, -b)$ where θ and $-\theta$ are angles with their initial sides on the positive x-axis and their terminal sides from P and Q to the origin, respectively.

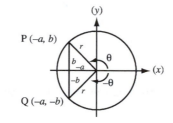

$$\sin(-\theta) = \frac{-b}{r} = -\frac{b}{r} = -\sin\theta, \qquad \text{since } \sin\theta = \frac{b}{r}$$

$$\cos(-\theta) = \frac{-a}{r} = \cos\theta, \qquad \text{since } \cos\theta = \frac{-a}{r}$$

$$\tan(-\theta) = \frac{-b}{-a} = -\frac{b}{-a} = -\tan\theta, \qquad \text{since } \tan\theta = \frac{b}{-a}$$

Similarly, we can find the values of $\csc(-\theta)$, $\sec(-\theta)$, and $\cot(-\theta)$. Another way of finding the remaining three functions is to apply the reciprocal identities.

Examples

1. Find the exact value of $\sin(-60°)$.

   Using $\sin(-\theta) = -\sin\theta$, we have

   $$\sin(-60°) = -\sin 60°$$

   $$= \frac{\sqrt{3}}{2} \qquad \text{Since } \sin 60° = \frac{\sqrt{3}}{2}. \text{ See \textbf{Thirty-Sixty-Ninety Degree Triangle}.}$$

2. Find the exact value of $\cos(-45°)$.

Using $\cos(-\theta) = \cos\theta$, we have

$$\cos(-45°) = \cos 45° = \frac{\sqrt{2}}{2} \qquad \text{See \textbf{Forty-Five Degree Triangle}.}$$

3. Find the exact value of $\csc(-330°)$.

Using $\csc(-\theta) = -\csc\theta$, we have

$$\csc(-330°) = -\csc(-330°) = -(-\csc 30°)$$

Since $\csc 330° = -\csc 30°$. See **Trigonometry, reference angle**. Also see **Trigonometry, signs of the six functions**.

$$= \csc 30° = \frac{2}{1} = 2$$

4. Find the exact value of $\sec(-210°)$.

Using $\sec(-\theta) = \sec\theta$, we have

$$\sec(-210°) = \sec 210° = -\sec 30°$$

See **Trigonometry, reference angle**.

$$= -\frac{2}{\sqrt{3}} = -\frac{2}{\sqrt{3}} \cdot \frac{\sqrt{3}}{\sqrt{3}}$$

See **Rationalizing the Denominator**.

$$= \frac{-2\sqrt{3}}{3}$$

► **Trigonometry, functions**   In this article, the six functions of trigonometry are defined in relation to a unit circle. If you are interested in the six functions defined in relation to opposite, adjacent, and hypotenuse, if you are looking for an application of trigonometry, or if you are looking for a beginning of trigonometry, how it works, and what it can do, see **Trigonometry, right triangle definition of**.

On a unit circle (a circle with a radius of 1) if we let $P(x, y)$ be a point on the circle corresponding to a real number t, then we define the following where x and y are functions of t:

$$\sin t = y \qquad\qquad \csc t = \frac{1}{y}, \qquad \text{with } y \neq 0$$

$$\cos t = x \qquad\qquad \sec t = \frac{1}{x}, \qquad \text{with } x \neq 0$$

$$\tan t = \frac{y}{x}, \qquad \text{with } x \neq 0 \qquad \cot t = \frac{x}{y}, \qquad \text{with } y \neq 0$$

Since the radius of the circle is 1, the real number t is determined by the equation

$$t = r\theta = (1)\theta = \theta$$

where θ is a central angle in standard position and is measured in radians (see **Arc Length (in geometry)**). In other words, since $t = \theta$, the length of the arc subtended by θ is the arc length t. If $\theta = \frac{\pi}{3}$, then the length of the arc subtended by θ is $t = \frac{\pi}{3}$. In conjunction, when t is determined the point $P(x, y)$ is determined.

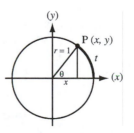

T

1. On a unit circle, evaluate the six trigonometric functions that correspond with the real number $\frac{\pi}{3}$.

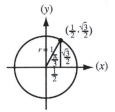

From the 30-60-90 degree triangle (see **Thirty-Sixty-Ninety Degree Triangle**, example 2), we know that $\frac{\pi}{3} = \frac{180°}{3} = 60°$ corresponds with the point $\left(\frac{1}{2}, \frac{\sqrt{3}}{2}\right)$ on the unit circle. This follows from the fact that a given 30-60-90 degree triangle with a hypotenuse of 1 is $\frac{1}{2}$ that of the hypotenuse of the 30-60-90 degree triangle.

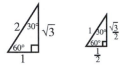

Therefore, we need to multiply the sides of the 30-60-90 degree triangle by $\frac{1}{2}$. Hence, the hypotenuse is $\frac{1}{2}(2) = 1$, the length of the shorter leg is $\frac{1}{2}(1) = \frac{1}{2}$, and the length of the longer leg is $\frac{1}{2}(\sqrt{3}) = \frac{\sqrt{3}}{2}$. The shorter and longer legs are the coordinates of the point $\left(\frac{1}{2}, \frac{\sqrt{3}}{2}\right)$. Since $\frac{\pi}{3}$ corresponds with $(x, y) = \left(\frac{1}{2}, \frac{\sqrt{3}}{2}\right)$, we have

$$\sin\frac{\pi}{3} = y = \frac{\sqrt{3}}{2} \qquad \csc\frac{\pi}{3} = \frac{1}{y} = \frac{1}{\frac{\sqrt{3}}{2}} = \frac{2}{\sqrt{3}} = \frac{2\sqrt{3}}{3}$$

See **Rationalizing the Denominator**.

$$\cos\frac{\pi}{3} = x = \frac{1}{2} \qquad \sec\frac{\pi}{3} = \frac{1}{x} = \frac{1}{\frac{1}{2}} = 2$$

$$\tan\frac{\pi}{3} = \frac{y}{x} = \frac{\frac{\sqrt{3}}{2}}{\frac{1}{2}} = \sqrt{3} \qquad \cot\frac{\pi}{3} = \frac{x}{y} = \frac{\frac{1}{2}}{\frac{\sqrt{3}}{2}} = \frac{1}{\sqrt{3}} = \frac{\sqrt{3}}{3}$$

Notice that the functions in the second column are reciprocals of those in the first column, $\frac{1}{y}$ with y, $\frac{1}{x}$ with x, and $\frac{x}{y}$ with $\frac{y}{x}$. As a result, after the values of the first column are determined, we can calculate their reciprocals to determine the values of the second column:

$$\csc\frac{\pi}{3} = \frac{2}{\sqrt{3}} = \frac{2\sqrt{3}}{3} \qquad \frac{2}{\sqrt{3}} \text{ is the reciprocal of } \frac{\sqrt{3}}{2}$$

$$\sec\frac{\pi}{3} = 2 \qquad 2 \text{ is the reciprocal of } \frac{1}{2}$$

$$\cot\frac{\pi}{3} = \frac{1}{\sqrt{3}} = \frac{\sqrt{3}}{3} \qquad \frac{1}{\sqrt{3}} \text{ is the reciprocals of } \sqrt{3}$$

T

To simplify the $\tan\frac{\pi}{3}$ where $\tan\frac{\pi}{3} = \frac{\frac{\sqrt{3}}{2}}{\frac{1}{2}} = \sqrt{3}$, see **Fractions, complex**. In general,

$$\frac{\frac{a}{b}}{\frac{c}{d}} = \frac{a}{b} \div \frac{c}{d} = \frac{a}{b} \bullet \frac{d}{c} = \frac{ad}{bc}$$

Hence, we have

$$\frac{\frac{\sqrt{3}}{2}}{\frac{1}{2}} = \frac{\sqrt{3}}{2} \bullet \frac{2}{1} = \sqrt{3}$$

To simplify the $\csc\frac{\pi}{3}$ where $\csc\frac{\pi}{3} = \frac{2}{\sqrt{3}}$, see **Rationalizing the Denominator**.

$$\frac{2}{\sqrt{3}} = \frac{2}{\sqrt{3}} \bullet \frac{\sqrt{3}}{\sqrt{3}} = \frac{2\sqrt{3}}{\sqrt{9}} = \frac{2\sqrt{3}}{3}$$

To convert $\frac{\pi}{3} = \frac{180°}{3} = 60°$, see **Converting Radians to Degrees**.

2. On a unit circle, evaluate the six trigonometric functions that correspond with the real number $-\frac{3\pi}{4}$.

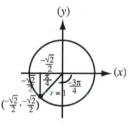

From the 45-degree triangle (see **Forty-Five Degree Triangle**), we know that $-\frac{3\pi}{4} = \frac{-3(180°)}{4} = -135°$ corresponds with the point $\left(\frac{-\sqrt{2}}{2}, \frac{-\sqrt{2}}{2}\right)$ on the unit circle (also see **Trigonometry, reference angle** or **Angle, in trigonometry**). This follows from the fact that a given 45-degree triangle with a hypotenuse of 1 is $\frac{1}{\sqrt{2}} = \frac{\sqrt{2}}{2}$ that of the hypotenuse of the 45-degree triangle.

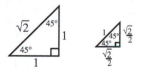

This means that we need to multiply $\sqrt{2}$ by $\frac{\sqrt{2}}{2}$ to get 1. Therefore, if we multiply all of the sides by $\frac{\sqrt{2}}{2}$, we have the length of the hypotenuse $\frac{\sqrt{2}}{2}(\sqrt{2}) = \frac{\sqrt{4}}{2} = \frac{2}{2} = 1$ and the length of the legs (both are the same) $\frac{\sqrt{2}}{2}(1) = \frac{\sqrt{2}}{2}$. The legs determine the coordinates of the point associated with a 45-degree triangle in the 3rd quadrant, $\left(\frac{-\sqrt{2}}{2}, \frac{-\sqrt{2}}{2}\right)$.

Since $\frac{-3\pi}{4}$ corresponds with $(x, y) = \left(-\frac{\sqrt{2}}{2}, -\frac{\sqrt{2}}{2}\right)$, we have

$$\sin\frac{-3\pi}{4} = y = \frac{-\sqrt{2}}{2} \qquad \csc\frac{-3\pi}{4} = -\frac{2}{\sqrt{2}} = -\sqrt{2}, \text{ reciprocal of the sin}$$

$$\cos\frac{-3\pi}{4} = x = \frac{-\sqrt{2}}{2} \qquad \sec\frac{-3\pi}{4} = \frac{-2}{\sqrt{2}} = -\sqrt{2}, \text{ reciprocal of the cos}$$

$$\tan\frac{-3\pi}{4} = \frac{y}{x} = \frac{\frac{-\sqrt{2}}{2}}{\frac{-\sqrt{2}}{2}} = 1 \qquad \cot\frac{-3\pi}{4} = 1, \text{ reciprocal of the tan}$$

3. On a unit circle, evaluate the six trigonometric functions that correspond with 0, $\frac{\pi}{2}$, π, $\frac{3\pi}{2}$, and 2π $(0°, 90°, 180°, 270°, \text{ and } 360°)$.

Since 0 corresponds with $(x, y) = (1, 0)$, we have

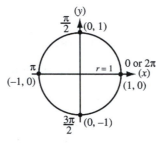

$$\sin 0 = y = 0 \qquad \csc 0 = \frac{1}{y} = \frac{1}{0} = \text{not defined (can't divide by 0)}$$

$$\cos 0 = x = 1 \qquad \sec 0 = \frac{1}{x} = \frac{1}{1} = 1$$

$$\tan 0 = \frac{y}{x} = \frac{0}{1} = 0 \qquad \cot 0 = \frac{x}{y} = \frac{1}{0} = \text{(not defined)}$$

Since $\frac{\pi}{2}$ corresponds with $(x, y) = (0, 1)$, we have

$$\sin\frac{\pi}{2} = 1 \qquad \csc\frac{\pi}{2} = \frac{1}{1} = 1$$

$$\cos\frac{\pi}{2} = 0 \qquad \sec\frac{\pi}{2} = \frac{1}{0} = \text{ND}$$

$$\tan\frac{\pi}{2} = \frac{1}{0} = \text{ND} \qquad \cot\frac{\pi}{2} = \frac{0}{1} = 0$$

Since π corresponds with $(x, y) = (-1, 0)$, we have

$$\sin\pi = 0 \qquad \csc\pi = \frac{1}{0} = \text{ND}$$

$$\cos\pi = -1 \qquad \sec\pi = \frac{1}{-1} = -1$$

$$\tan\pi = \frac{0}{-1} = 0 \qquad \cot\pi = \frac{-1}{0} = \text{ND}$$

Since $\frac{3\pi}{2}$ corresponds with $(x, y) = (0, -1)$, we have

$$\sin\frac{3\pi}{2} = -1 \qquad\qquad \csc\frac{3\pi}{2} = \frac{1}{-1} = -1$$

$$\cos\frac{3\pi}{2} = 0 \qquad\qquad \sec\frac{3\pi}{2} = \frac{1}{0} = \text{ND}$$

$$\tan\frac{3\pi}{2} = \frac{-1}{0} = \text{ND} \qquad\qquad \cot\frac{3\pi}{2} = \frac{0}{-1} = 0$$

Since 2π corresponds with $(x, y) = (1, 0)$, the functions for 0 and 2π are the same.

For purposes of evaluating and graphing, it is important to remember the six functions of the quadrant angles 0, $\frac{\pi}{2}$, π, $\frac{3\pi}{2}$, and 2π (the same as 0). If it is preferred, use $0°$, $90°$, $180°$, $270°$, and $360°$ instead of radians. One way of remembering them is to memorize the values on two circles and then generate the values of four other circles.

Memorize these circles.

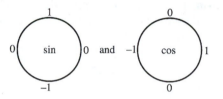

The cos circle is the same as the sin circle, just rotated $90°$ to the right.

From these circles, we know that $\sin 0 = 0$, $\sin\frac{\pi}{2} = 1$, etc. Also $\cos 0 = 1$, $\cos\frac{\pi}{2} = 0$, etc.

Using these circles, we can generate the circle for tangent. To do this, we need to apply the quotient identity $\tan\theta = \frac{\sin\theta}{\cos\theta}$. Thinking of the sin circle placed above the cos circle, we can divide their values.

generates the tan circle

Since cosecant, secant, and cotangent are reciprocals of sine, cosine, and tangent, we can reciprocate each of their values to generate the remaining three circles. We will use the idea that the reciprocal of $0 = \frac{0}{1}$ is $\frac{1}{0} = \text{ND}$. Also the reciprocal of $\text{ND} = \frac{1}{0}$ is $\frac{0}{1} = 0$.

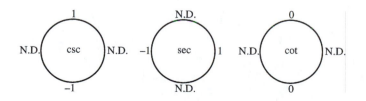

➤ **Trigonometry, functions of negative angles (Functions or identities of negative angle).** See **Trigonometry, even and odd functions**.

➤ **Trigonometry, graphs of the six functions** There are basically two methods used to **graph the six functions of trigonometry**: graphing by charting, or graphing by using the amplitude, period, or asymptotes. To graph by charting, we choose several x-values, calculate their corresponding y-values, plot the points (x, y), and draw the graph. This can be a tedious as well as time-consuming process. As a result, a shorter method is desired, the method of using the amplitude, period, or asymptotes.

The method that uses the amplitude, period, or asymptotes is developed from observing the x-intercept, vertices, asymptotes, period, and amplitude from the charting method. To graph by this method, we determine the amplitude (where applicable) and mark its values on the y-axis (A and $-A$), calculate the period, mark the period on the x-axis, mark the remaining five key values $\left(0, \frac{P}{4}, \frac{P}{2}, \frac{3P}{4},\ \text{as well}\ P\right)$, plot the five key points or key points and asymptotes of the trigonometric function (see memorized graphs below), then draw the graph.

Among the many facts that become apparent by graphing with the charting method is the fact that all graphs of each function are similar. All sine graphs are similar. All cosine graphs are similar, and so on. What delineates one graph from another is their x-intercepts, vertices, amplitude (where applicable), and period. Otherwise the appearance of the graphs of each function is similar. A sketch of the graphs of the six functions is given below and should be memorized.

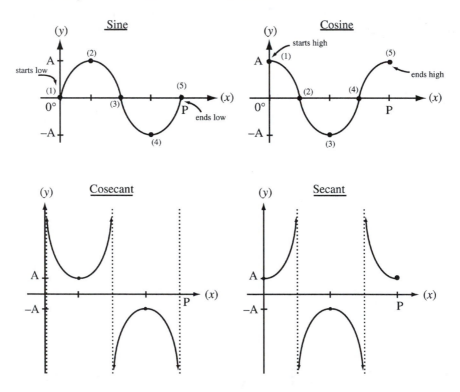

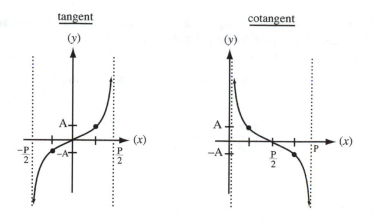

To determine the amplitude, period, or asymptotes of these graphs, see example 2 first, then go to individual examples.

To see how to divide the period into its five key values (for $\sin x$, $\cos x$, $\csc x$, and $\sec x$), see examples 3 and 4. The examples include degrees as well as radians.

For a clearer understanding of how to mark values on the x-axis, see the end of example 5. This gives a comparison between degrees, radians, and the values on the x-axis.

Examples

1. Graph $y = \sin x$ by the charting method.

To graph $y = \sin x$ by the charting method, we need to choose several x-values, calculate their corresponding y-values, plot the points (x, y), and then draw the graph. A previous knowledge of quadrant angles, reference angles, the thirty-sixty-ninety degree triangle, and the forty-five degree triangle will help when deciding what values of x to choose as well as calculating each value of y (see the articles under these names). Whether x is chosen in degrees or radians is a matter of preference. We will use both. Choosing values between $0°$ and $360°$ or between 0 radians and 2π radians, we have

$x°$	x^R	$y = \sin x$
$0°$	0	$\sin 0° = 0$
$30°$	$\frac{\pi}{6}$	$\sin 30° = \frac{1}{2} = .5$
$45°$	$\frac{\pi}{4}$	$\sin 45° = \frac{\sqrt{2}}{2} \approx .7$
$60°$	$\frac{\pi}{3}$	$\sin 60° = \frac{\sqrt{3}}{2} \approx .86$
$90°$	$\frac{\pi}{2}$	$\sin 90° = 1$
$135°$	$\frac{3\pi}{4}$	$\sin 135° = \frac{\sqrt{2}}{2} \approx .7$
$180°$	π	$\sin 180° = 0$
$270°$	$\frac{3\pi}{2}$	$\sin 270° = -1$
$360°$	2π	$\sin 360° = 0$

The graph represents one period (cycle) of the sine curve. It has an amplitude of 1 and a period of 360° or 2π radians. If the graph continued to the left of 0° or to the right of 360°, it would begin to repeat the cycle. A period shows one cycle only. See **Amplitude**, **Period**, or **Periodic Function**.

2. Use the charting method to graph $y = \sin x$, $y = 2\sin x$, $y = \sin 2x$, and $y = \sin\frac{1}{2}x$ on the same set of axes.

x	$y = \sin x$	$y = 2\sin x$	$y = \sin 2x$	$y = \sin\frac{1}{2}x$
0°	$\sin 0° = 0$	$2\sin 0° = 0$	$\sin 2(0°) = \sin 0° = 0$	$\sin\frac{1}{2}(0°) = \sin 0° = 0$
45°			$\sin 2(45°) = \sin 90° = 1$	
90°	$\sin 90° = 1$	$2\sin 90° = 2$	$\sin 2(90°) = \sin 180° = 0$	
135°			$\sin 2(135°) = \sin 270° = -1$	
180°	$\sin 180° = 0$	$2\sin 180° = 0$	$\sin 2(180°) = \sin 360° = 0$	$\sin\frac{1}{2}(180°) = \sin 90° = 1$
270°	$\sin 270° = -1$	$2\sin 270° = -2$		
360°	$\sin 360° = 0$	$2\sin 360° = 0$		$\sin\frac{1}{2}(360°) = \sin 180° = 0$
$\vdots$				
540°				$\sin\frac{1}{2}(540°) = \sin 270° = -1$
720°				$\sin\frac{1}{2}(720°) = \sin 360° = 0$

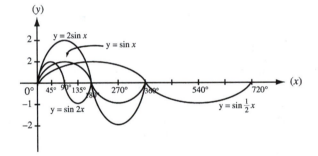

It is from comparative graphing like this that an easier method of graphing can be derived, the method of using the amplitude, period, or asymptotes.

Consider A and B in the equation $y = A\sin Bx$. When $A = 1$ and $B = 1$, we have the basic sine curve $y = \sin x$ which has an amplitude of 1 and a period of 360° or 2π radians. Notice that when $A = 2$, the graph of $y = 2\sin x$ increases in amplitude. As a result, we know that the A determines the amplitude of the graph.

Now consider the graphs of $y = \sin 2x$ and $y = \sin\frac{1}{2}x$ where $B = 2$ and $B = \frac{1}{2}$, respectively. Comparing these to $y = \sin x$, we see that $y = \sin 2x$ has a period of 180° or π radians and $y = \sin\frac{1}{2}x$ has a period of 720° or 4π radians. The graph of $y = \sin 2x$ has a period that is half that of $y = \sin x$ while the graph of $y = \sin\frac{1}{2}x$ has a period that is twice that of $y = \sin x$. The period of each is represented by the equation $P = \frac{360°}{B}$ or $P = \frac{2\pi}{B}$ in radians. Therefore, we realize that B can be used to determine the period of the graph.

Lastly, we notice that changing the values of A and B only, changes the amplitude or period of the graph; the basic shape always remains the same. This is the key aspect in developing our new method. Since the shape is always basically the same, all we have to do is determine the amplitude and period, then we can sketch in the sine graph.

By comparative graphing of the remaining five functions, it can be shown that this method works for them as well. The only deviation is the period of the tangent and cotangent graphs; they have a period of $P = \frac{180°}{B}$ or $P = \frac{\pi}{B}$ in radians.

In general, to graph a trigonometric equation by using the amplitude, period, or asymptotes, we do the following:

a. For sine and cosine,

    1) Determine the amplitude and mark off A and $-A$ on the y-axis.

    2) Use B to determine the period $P = \frac{360°}{B}$ or $P = \frac{2\pi}{B}$ in radians.

    3) Mark off P on the x-axis, then half of that $\left(\frac{1}{2}P\right)$, half of that again $\left(\frac{1}{2}P\right)$, and then recognizing the pattern mark the value of the fourth x-value (there are five altogether). For example, if $P = 180°$ then half of that is $90°$, half again is $45°$, and then recognizing the pattern $0°$, $45°$, $90°$, ?, we see the fourth x-value is $135°$.

    4) Sketch in the graph of the sine or the cosine.

b. For cosecant or secant,

    1) Determine the amplitude and mark off A and $-A$ on the y-axis.

    2) Use B to determine the period $P = \frac{360°}{B}$ or $P = \frac{2\pi}{B}$ in radians.

    3) Using the period, mark the five key values on the x-axis as described in 3) above.

    4) Draw in the asymptotes for the given function.

    5) Sketch in the graph of the cosecant or the secant.

c. For tangent or cotangent,

    1) Use B to determine the period $P = \frac{360°}{B}$ or $P = \frac{2\pi}{B}$ in radians.

    2) For tangent, mark $\frac{1}{2}P$ and $-\frac{1}{2}P$ on the x-axis. For cotangent, mark P and $\frac{1}{2}P$ on the x-axis.

    3) For tangent and cotangent, further mark the midpoints of each interval in 3) above and plot the points determined by the midpoints and A and $-A$.

    4) Draw in the asymptotes for the given function.

    5) Sketch in the graph of the tangent or the cotangent.

3. Using amplitude and period, sketch the graph of $y = 3\cos 2x$.

Comparing $y = 3\cos 2x$ to $y = A\cos Bx$, we see that the amplitude is $A = 3$ and the period is $P = \frac{360°}{B} = \frac{360°}{2} = 180°$ or $P = \frac{2\pi}{B} = \frac{2\pi}{2} = \pi$ in radians. Marking 3 and -3 on the y-axis and marking the period and the remaining five values on the x-axis, we have

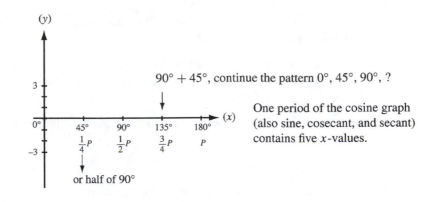

$90° + 45°$, continue the pattern $0°$, $45°$, $90°$, ?

One period of the cosine graph (also sine, cosecant, and secant) contains five x-values.

or half of $90°$

Next, from the figures of previously memorized graphs, we plot the five key points and sketch in the graph of the cosine. It starts high and ends high.

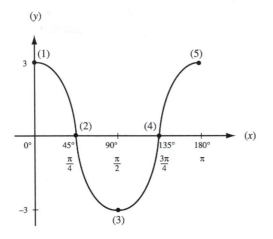

Plot the five key points of the cosine graph first, then sketch in the graph.

An easy way to recognize the pattern for radians is to divide the period (π) by 2 and then divide that result ($\frac{\pi}{2}$) by 2 to get $\frac{\pi}{4}$.

We then have the pattern or sequence

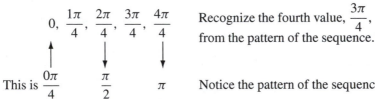

$$0, \frac{1\pi}{4}, \frac{2\pi}{4}, \frac{3\pi}{4}, \frac{4\pi}{4}$$

Recognize the fourth value, $\frac{3\pi}{4}$, from the pattern of the sequence.

This is $\frac{0\pi}{4}$ $\frac{\pi}{2}$ π Notice the pattern of the sequence with $0 = \frac{0\pi}{4}$.

Then we reduce the fractions to get the sequence

$$0, \frac{\pi}{4}, \frac{\pi}{2}, \frac{3\pi}{4}, \pi$$

4. Using amplitude, period, and asymptotes, sketch the graphs of $y = 2\csc 3x$ and $y = 2\sec 3x$.

The amplitude of each graph is the same, $A = 2$. The period of each graph is also the same, $P = \frac{360°}{3} = 120°$ or $P = \frac{2\pi}{3}$ in radians.

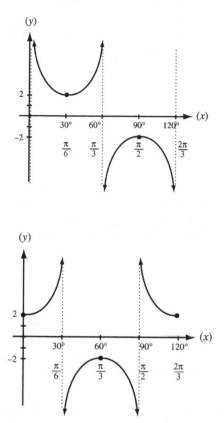

Plot the amplitude (maximum and minimum points) and the asymptotes at $0°$, $60°$, and $120°$. Then sketch the cosecant graph.

Plot the amplitude (maximum and minimum points) and the asymptotes at $30°$ and $90°$. Then sketch the secant graph.

To determine the sequence for the five x-values in radians, we divide $\frac{2\pi}{3}$ by 2 to get $\frac{2\pi}{6} = \frac{\pi}{3}$. Then we divide $\frac{\pi}{3}$ by 2 to get $\frac{\pi}{6}$. This gives us the sequence

$$0, \frac{1\pi}{6}, \frac{2\pi}{6}, \frac{3\pi}{6}, \frac{4\pi}{6} \qquad \text{or} \qquad 0, \frac{\pi}{6}, \frac{\pi}{3}, \frac{\pi}{2}, \frac{2\pi}{3}$$

5. Using the period and asymptotes, sketch the graph of $y = 2\tan 3x$ and $y = 2\cot 3x$.

The period for tangent and cotangent is different from the other four functions, $P = \frac{180°}{B}$ or $P = \frac{\pi}{B}$ in radians. With $B = 3$, we have

$$P = \frac{180°}{3} = 60° \qquad \text{or} \qquad P = \frac{\pi}{3}$$

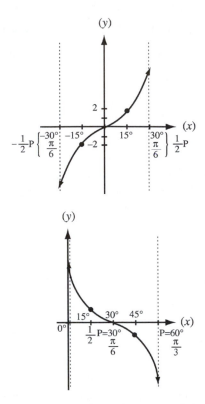

Mark $\frac{1}{2}P = \frac{1}{2}(60°) = 30°$ and $-\frac{1}{2}P = -\frac{1}{2}(60°) = -30°$ on the x-axis. With midpoints of $-15°$ and $15°$, respectively, we plot the points corresponding with -2 and 2 (from $A = 2$). Then we draw the asymptotes and sketch the tangent graph.

Mark $P = 60°$ and $\frac{1}{2}P = \frac{1}{2}(60°) = 30°$ on the x-axis. With midpoints of $15°$ and $45°$, respectively, we plot the points corresponding with -2 and 2 (from $A = 2$). Then we draw the asymptotes and sketch the cotangent graph.

In each graph, the midpoints and their corresponding y-values are only used for better graphing accuracy. However, it is not necessary to include them. Sketching the graphs using the period and asymptotes is sufficient. The A-value only indicates the amount of bend that occurs in each graph.

Notice that in each of the graphs we have graphed so far the separation of the x-values of one graph is not necessarily the same as the separation of the x-values of the other graphs. Since the periods of some of the graphs are smaller, a consistent set of numbers on the x-axis would have left us with graphs that were too compact to work with. So we stretched them out, so to speak, by changing the lengths of the intervals of the x-axis. The intervals

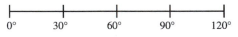

are the same as the intervals

We do this as a matter of convenience. Otherwise, we might have no space to work in. If the two intervals shown above were drawn on the same axis, we would have

Working with a graph that had a period of $120°$ could be difficult with this interval. So, we stretch them out.

If accuracy is needed, however, we need to compare degrees with radians and then consider the location of the radians on the number line. Before continuing it is necessary to be familiar with the following articles: **Converting Degrees to Radians** or **Converting Radians to Degrees**.

Since we can convert each degree to its corresponding radian measure, we need only concern ourselves with the location of the radians on the x-axis (number line). We do this by using a decimal approximation of π. In hundredths, $\pi \approx 3.14$ (pi is approximately equal to 3.14). Since 180° is the same as π radians, they each are located (approximately) at 3.14 on the number line.

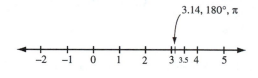

Similarly, other angles such as 30° or $\frac{\pi}{6} \approx \frac{3.14}{6} \approx .52$, 90° or $\frac{\pi}{2} \approx \frac{3.14}{2} \approx 1.57$, 360° or $2\pi \approx 2(3.14) \approx 6.28$, and $-135°$ or $-\frac{3\pi}{4} \approx \frac{-3(3.14)}{4} \approx -2.36$ can be located on the number line.

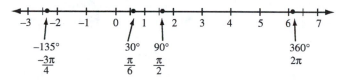

6. Graph $y = \cos 2x$ on the interval $-90° \leqslant x \leqslant 270°$.

So far, all of our graphs, except the tangent graph, have been graphed on the interval from 0° to P. The tangent was graphed on the interval from $-\frac{P}{2}$ to $\frac{P}{2}$. In all cases, only one period of the graph was drawn. If it is required to go beyond the boundaries of one period, as in this example, an expanded approach must be taken.

Two approaches are generally taken when graphing beyond one period of a function: graph only what is required, in this case the graph for $-90° \leqslant x \leqslant 270°$, or graph complete periods beyond what is required and then erase the unneeded portions.

To graph $y = \cos 2x$ on the interval $-90° \leqslant x \leqslant 270°$, we first need to plot the key points of one period of the graph. Then we need to plot the key points beyond the period, in either direction, until we get to the endpoints of the required interval. With an amplitude of $A = 1$ and a period of $P = \frac{360°}{2} = 180°$ or $P = \frac{2\pi}{2} = \pi$ in radians, we have

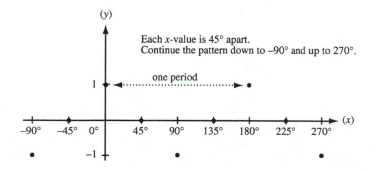

The x-values in radians are $-\frac{\pi}{2}$, $\frac{-\pi}{4}$, 0, $\frac{\pi}{4}$, $\frac{\pi}{2}$, $\frac{3\pi}{4}$, π, $\frac{5\pi}{4}$, and $\frac{3\pi}{2}$.

After the key points have been plotted, we sketch in the graph for the cosine.

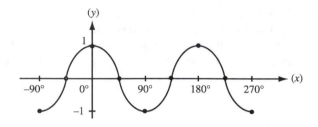

For more, see the article on **Period** or **Periodic Function**.

7. Graph $y = 2\sin\left(x - \frac{\pi}{2}\right)$ and $y = 2\sec\left(x + \frac{\pi}{4}\right)$.

Adding or subtracting a number to the angle of a trigonometric function will translate (move) the graph of the function either left or right. If the number is added to the angle, the graph is translated left. If the number is subtracted from the angle, the graph is translated right. See **Translation or Reflection**.

Graphs of equations such as $y = 2\sin\left(x - \frac{\pi}{2}\right)$ and $y = 2\sec\left(x + \frac{\pi}{4}\right)$ involve the concept of **phase shift**. The graphs have the same shape as $y = 2\sin x$ and $y = 2\sec x$ except they have been shifted (moved over) $\frac{\pi}{2} = 90°$ to the right or $\frac{\pi}{4} = 45°$ to the left, respectively. The graphs of these equations can be seen in the article **Phase Shift**.

8. Graph $y = \sin x + 2$ for $0° \leqslant x \leqslant 360°$.

Adding or subtracting a number to a trigonometric function will translate (move) the graph of the function either up or down, respectively. See **Translation or Reflection**. In this case, the graph of $y = \sin x$ has been translated up 2 units. To graph the equation, we place a dotted line through 2 on the y-axis and then graph $y = \sin x$ on the dotted line instead of the x-axis. (We do the same with all of the trigonometric functions.) With $A = 1$ and $P = \frac{360°}{1} = 360°$, we have

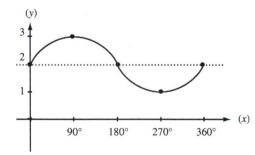

Plot the key points on the dotted line, not the x-axis.

9. Graph $y = -\sin x$, $y = \sin(-x)$, and $y = -\sin(-x)$ on the interval $0° \leqslant x \leqslant 360°$.

Any of the trigonometric functions that have been made negative have graphs that are reflected in the x-axis (flipped over). See **Translation or Reflection**. To graph an equation like $y = -\sin x$, where the function is negative, we reflect the graph of $y = \sin x$ in the x-axis (flip it over).

Any of the trigonometric functions that have angles that have been made negative have graphs that follow the concepts of even and odd functions. See **Trigonometry, even and odd functions (functions or identities of negative angles)**. Since $y = \sin(-x) = -\sin x$, we graph $y = -\sin x$ as stated above. In a similar fashion, we can graph $y = \csc(-x) = -\csc x$, $y = \tan(-x) = -\tan x$ and $y = \cot(-x) = -\cot x$. The graphs of $y = \cos(-x) = \cos x$ and $y = \sec(-x) = \sec x$ remain the same. They don't reflect. Since $\sin(-x) = -\sin x$, the graphs of $y = -\sin x$ and $y = \sin(-x)$ are the same.

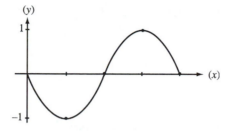

The graph of $y = -\sin(-x)$ is the same as $y = \sin x$. This follows logically since each of the negative signs in $y = -\sin(-x)$ will flip the graph. We can see this from algebraic manipulation as well:

$$y = -\sin(-x) = -\big(\sin(-x)\big) = -(-\sin x) \qquad \text{Since } \sin(-x) = -\sin x.$$
$$= \sin x$$

➤ **Trigonometry, identities**   An identity is a statement that holds true for any value in the domain of its variable. (See **Domain and Range**). It is an equation but is dealt with in a different fashion than the usual types of equations. Whereas an equation is usually solved for its variable, an identity is used to simplify expressions, determine values, aid in the solutions of equations, or determine other identities. The following are examples of algebraic identities: $3x - 6 = 3(x - 2)$, $x^2 - 25 = (x + 5)(x - 5)$, and $\frac{1}{2x} = \frac{x^2}{2x^3}$ where $x \neq 0$.

As an example of an application of an identity, suppose we wanted to solve the equation $3(x - 2) = 18$. We could not solve the equation unless there was some way of dealing with the expression $3(x - 2)$. Since we have the identity $3x - 6 = 3(x - 2)$ (distributive property), we can replace $3(x - 2)$ with $3x - 6$. This gives us the next line in the solution, $3x - 6 = 18$. Then the solution can be completed from here.

In this article, a list of the identities is given. For further information on each type of identity, see the article by that name. At the end of the identity list, some hints on proving identities are given, followed by a selection of proofs.

Reciprocal Identities

$$\sin\theta = \frac{1}{\csc\theta} \qquad \csc\theta = \frac{1}{\sin\theta}$$
$$\cos\theta = \frac{1}{\sec\theta} \qquad \sec\theta = \frac{1}{\cos\theta}$$
$$\tan\theta = \frac{1}{\cot\theta} \qquad \cot\theta = \frac{1}{\tan\theta}$$

T

Quotient (Tangent and Cotangent) Identities

$$\tan\theta = \frac{\sin\theta}{\cos\theta} \qquad \cot\theta = \frac{\cos\theta}{\sin\theta}$$

Pythagorean Identities

$$\sin^2\theta + \cos^2\theta = 1 \qquad \sin^2\theta \text{ means } (\sin\theta)^2$$
$$1 + \tan^2\theta = \sec^2\theta$$
$$1 + \cot^2\theta = \csc^2\theta$$

Cofunction Identities

(If preferred, let $90° = \frac{\pi}{2}$)

$$\sin(90° - \theta) = \cos\theta \qquad \csc(90° - \theta) = \sec\theta$$
$$\cos(90° - \theta) = \sin\theta \qquad \sec(90° - \theta) = \csc\theta$$
$$\tan(90° - \theta) = \cot\theta \qquad \cot(90° - \theta) = \tan\theta$$

Even and Odd Identities or Negative Angle Identities

$$\sin(-\theta) = -\sin\theta \qquad \csc(-\theta) = -\csc\theta$$
$$\cos(-\theta) = \cos\theta \qquad \sec(-\theta) = \sec\theta$$
$$\tan(-\theta) = -\tan\theta \qquad \cot(-\theta) = -\cot\theta$$

Sum and Difference Identities (Formulas)

(Also known as Addition Formulas)

$$\sin(\theta \pm \phi) = \sin\theta \cos\phi \pm \cos\theta \sin\phi$$
$$\cos(\theta \pm \phi) = \cos\theta \cos\phi \mp \sin\theta \sin\phi$$
$$\tan(\theta \pm \phi) = \frac{\tan\theta \pm \tan\phi}{1 \mp \tan\theta \tan\phi}$$

There are six identities in the list, depending on the choice of top or bottom signs for $+$ or $-$.

Double-Angle Identities (Formula)

$$\sin 2\theta = 2\sin\theta \cos\theta$$
$$\cos 2\theta = \cos^2\theta - \sin^2\theta$$
$$\text{or} \qquad = 2\cos^2\theta - 1$$
$$\text{or} \qquad = 1 - 2\sin^2\theta$$
$$\tan 2\theta = \frac{2\tan\theta}{1 - \tan^2\theta}$$

Half-Angle Identities (Formulas)

$$\sin\frac{\theta}{2} = \pm\sqrt{\frac{1-\cos\theta}{2}}$$

$$\cos\frac{\theta}{2} = \pm\sqrt{\frac{1+\cos\theta}{2}}$$

$$\tan\frac{\theta}{2} = \pm\sqrt{\frac{1-\cos\theta}{1+\cos\theta}} \quad \text{or} \quad = \frac{1-\cos\theta}{\sin\theta} \quad \text{or} \quad = \frac{\sin\theta}{1+\cos\theta}$$

The $\pm$ signs are chosen as either $+$ or $-$ depending on the quadrant in which the angle $\frac{\theta}{2}$ lies. See **Trigonometry, signs of the six functions**.

Product to Sum or Sum to Product Identities (Formulas)

Product to Sum

$$\sin\theta \sin\phi = \frac{1}{2}\left[\cos(\theta-\phi) - \cos(\theta+\phi)\right]$$

$$\cos\theta \cos\phi = \frac{1}{2}\left[\cos(\theta-\phi) + \cos(\theta+\phi)\right]$$

$$\sin\theta \cos\phi = \frac{1}{2}\left[\sin(\theta+\phi) + \sin(\theta-\phi)\right]$$

$$\cos\theta \sin\phi = \frac{1}{2}\left[\sin(\theta+\phi) - \sin(\theta-\phi)\right]$$

Sum to Product

$$\sin\theta + \sin\phi = 2\sin\left(\frac{\theta+\phi}{2}\right)\cos\left(\frac{\theta-\phi}{2}\right)$$

$$\sin\theta - \sin\phi = 2\cos\left(\frac{\theta+\phi}{2}\right)\sin\left(\frac{\theta-\phi}{2}\right)$$

$$\cos\theta + \cos\phi = 2\cos\left(\frac{\theta+\phi}{2}\right)\cos\left(\frac{\theta-\phi}{2}\right)$$

$$\cos\theta - \cos\phi = -2\sin\left(\frac{\theta+\phi}{2}\right)\sin\left(\frac{\theta-\phi}{2}\right)$$

Proving identities teaches a student how to simplify trigonometric expressions. Many areas of trigonometry rely on this talent, especially simplifying trigonometric expressions, calculating trigonometric values, and solving trigonometric equations.

To "prove an identity" means that we want to show that either the left side of the identity can be transformed into the right side (exactly the same) or the right side of the identity can be transformed into the left side. Work should be done on one side or the other only, not both. Our goal is to learn from solving the identity, not just to get the identity done. If a student is allowed to work on both sides, the identity is more easily simplified but the lesson for future work is sacrificed.

Although proving identities is one of those things that gets better with practice, there are a few helpful hints.

1. Usually work with the more complicated side.

2. Quite often it is useful to change all of the terms to sin and cos, usually using the reciprocal or the quotient identities.

3. If the identity can be written with fractions (quite often involving sin and cos), combine them:

$$\frac{a}{b} + \frac{c}{d} = \frac{ad + bc}{bd} \qquad \text{or} \qquad 1 + \frac{a}{b} = \frac{b + a}{b} \qquad \text{or} \qquad a + \frac{b}{c} = \frac{ac + b}{c}$$

4. Sometimes separating a fraction can help:

$$\frac{a + b}{c} = \frac{a}{c} + \frac{b}{c} \qquad \text{or} \qquad \frac{a + b}{b} = \frac{a}{b} + \frac{b}{b} = \frac{a}{b} + 1$$

5. If the identity has complex fractions, simplify them. Sometimes this will lead back to hint 2:

$$\frac{\frac{a}{b}}{\frac{c}{d}} = \frac{a}{b} \bullet \frac{d}{c} = \frac{ad}{bc}$$

6. If the denominator of a fraction is in the form $a + b$ or $a - b$, multiply both the denominator and numerator by either $a - b$ or $a + b$, respectively:

$$\frac{c}{a + b} = \frac{c}{a + b} \bullet \frac{a - b}{a - b} = \frac{c(a - b)}{a^2 - b^2}$$

$$\frac{d}{a - b} = \frac{d}{a - b} \bullet \frac{a + b}{a + b} = \frac{d(a + b)}{a^2 - b^2}$$

Quite often this will leave the identity with some form of a Pythagorean identity.

7. When there are terms with squares in them (an exponent of 2), think of making substitutions from the Pythagorean identities.

8. Factoring sometimes helps:

$$ab + ac = a(b + c)$$
$$a^2 - b^2 = (a + b)(a - b) \qquad \text{i.e.} \qquad 1 - \sin^2 x = (1 + \sin x)(1 - \sin x)$$

T

Examples

1. Prove: $\cos\theta \tan\theta = \sin\theta$.

$$\cos\theta \tan\theta = \sin\theta$$
$$\cos\theta \frac{\sin\theta}{\cos\theta} = \sin\theta \qquad \text{Quotient identity.}$$
$$\sin\theta = \sin\theta$$

2. Prove: $2\cos^2\theta - 1 = 1 - 2\sin^2\theta$.

$$2\cos^2\theta - 1 = 1 - 2\sin^2\theta$$
$$2(1 - \sin^2\theta) = 1 - 2\sin^2\theta \qquad \text{Pythagorean identity, } \cos^2\theta = 1 - \sin^2\theta.$$
$$2 - 2\sin^2\theta - 1 = 1 - 2\sin^2\theta$$
$$1 - 2\sin^2\theta = 1 - 2\sin^2\theta$$

We could have worked with the other side just as easily, substituting $\sin^2\theta = 1 - \cos^2\theta$.

3. Prove: $\frac{\sin^2\theta + \cos^2\theta}{\sin\theta} = \csc\theta$.

$$\frac{\sin^2\theta + \cos^2\theta}{\sin\theta} = \csc\theta$$
$$\frac{1}{\sin\theta} = \csc\theta \qquad \text{Pythagorean identity, } 1 = \sin^2\theta + \cos^2\theta.$$
$$\csc\theta = \csc\theta \qquad \text{Reciprocal identity.}$$

The original problem could have been given to us as $\sin\theta + \cos\theta\cot\theta = \csc\theta$. Do you see that if the original problem is separated and simplified that this is the result?

$$\frac{\sin^2\theta}{\sin\theta} + \frac{\cos^2\theta}{\sin\theta} = \csc\theta$$
$$\sin\theta + \frac{\cos\theta\cos\theta}{\sin\theta} = \csc\theta$$
$$\sin\theta + \cos\theta \bullet \frac{\cos\theta}{\sin\theta} = \csc\theta$$
$$\sin\theta + \cos\theta\cot\theta = \csc\theta \qquad \text{Since the quotient identity gives } \frac{\cos\theta}{\sin\theta} = \cot\theta.$$

We show this to illustrate the purpose of the third hint. If we reverse the steps, the result of the third hint can be achieved. See example 4.

4. Prove: $\sin\theta + \cos\theta\cot\theta = \csc\theta$.

To prove this identity, we can use hints 2 and 3. We will change the left side of the identity to sin and cos and then add the fractions:

$$\sin\theta + \cos\theta\cot\theta = \csc\theta$$
$$\sin\theta + \cos\theta\frac{\cos\theta}{\sin\theta} = \csc\theta \qquad \text{Quotient identity.}$$
$$\sin\theta + \frac{\cos^2\theta}{\sin\theta} = \csc\theta \qquad \text{Multiply, } \cos\theta\cos\theta = (\cos\theta)^2 = \cos^2\theta.$$
$$\frac{\sin^2\theta + \cos^2\theta}{\sin\theta} = \csc\theta \qquad \text{Use hint 3, } a + \frac{b}{c} = \frac{ac+b}{c}.$$

The rest of the problem is like example 3.

5. Prove: $\frac{1+\cos\theta}{\cos\theta} = 1 + \sec\theta$.

By separating on the left:

$$\frac{1+\cos\theta}{\cos\theta} = 1 + \sec\theta$$

$$\frac{1}{\cos\theta} + \frac{\cos\theta}{\cos\theta} = 1 + \sec\theta$$

$$\sec\theta + 1 = 1 + \sec\theta$$

$$1 + \sec\theta = 1 + \sec\theta$$

Use the reciprocal identity $\frac{1}{\cos\theta} = \sec\theta$

By reciprocal identity on the right:

$$\frac{1+\cos\theta}{\cos\theta} = 1 + \sec\theta$$

$$\frac{1+\cos\theta}{\cos\theta} = 1 + \frac{1}{\cos\theta}$$

$$\frac{1+\cos\theta}{\cos\theta} = \frac{\cos\theta + 1}{\cos\theta}$$

$$\frac{1+\cos\theta}{\cos\theta} = \frac{1+\cos\theta}{\cos\theta}$$

To add $1 + \frac{1}{\cos\theta}$, use hint 2.

6. Prove: $(1 - \sin\theta)(1 + \csc\theta) = \cot\theta\cos\theta$.

$$(1 - \sin\theta)(1 + \csc\theta) = \cot\theta\cos\theta$$

$$(1 - \sin\theta)\left(1 + \frac{1}{\sin\theta}\right) = \cot\theta\cos\theta \qquad \text{Reciprocal identity.}$$

$$(1 - \sin\theta)\left(\frac{\sin\theta + 1}{\sin\theta}\right) = \cot\theta\cos\theta \qquad \text{Hint 3.}$$

$$(1 - \sin\theta)\left(\frac{1 + \sin\theta}{\sin\theta}\right) = \cot\theta\cos\theta \qquad \text{Commutative property.}$$

$$\frac{1 - \sin^2\theta}{\sin\theta} = \cot\theta\cos\theta \qquad \text{FOIL.}$$

$$\frac{\cos^2\theta}{\sin\theta} = \cot\theta\cos\theta \qquad \text{Pythagorean identity.}$$

$$\frac{\cos\theta}{\sin\theta}\cos\theta = \cot\theta\cos\theta \qquad (\cos\theta)^2 = \cos\theta\cos\theta$$

$$\cot\theta\cos\theta = \cot\theta\cos\theta \qquad \text{Quotient identity.}$$

The product $(1 - \sin\theta)(1 + \sin\theta)$ is $1 + \sin\theta - \sin\theta - (\sin\theta)^2$, which simplifies to $1 - \sin^2\theta$. See **FOIL**.

7. Prove: $\sin^2\theta = \frac{\sec^2\theta - 1}{\sec^2\theta}$.

By observation, three approaches are apparent: use the second Pythagorean identity, separate the fraction, or write in terms of sin and cos.

Pythagorean identity:

$$\sin^2\theta = \frac{\sec^2\theta - 1}{\sec^2\theta}$$

$$= \frac{\tan^2\theta}{\sec^2\theta} \qquad \text{Pythagorean identity.}$$

$$= \frac{\frac{\sin^2 \theta}{\cos^2 \theta}}{\frac{1}{\cos^2 \theta}}$$

Quotient identity.
Reciprocal identity.

$$= \frac{\sin^2 \theta}{\cos^2 \theta} \bullet \frac{\cos^2 \theta}{1}$$

Hint 5.

$$\sin^2 \theta = \sin^2 \theta$$

Separate the fraction:

$$\sin^2 \theta = \frac{\sec^2 \theta - 1}{\sec^2 \theta}$$

$$= \frac{\sec^2 \theta}{\sec^2 \theta} - \frac{1}{\sec^2 \theta}$$

$$= 1 - \cos^2 \theta$$

Reciprocal identity, $\frac{1}{\sec^2 \theta} = \cos^2 \theta$.

$$\sin^2 \theta = \sin^2 \theta$$

Pythagorean identity.

We would think to use the Pythagorean identity because there are squared terms in the problem. From the second Pythagorean identity, we have $\tan^2 \theta = \sec^2 \theta - 1$. Making this substitution and changing the terms to sin and cos gives us a solution.

We would think to separate the fraction because the problem is in the form $\frac{a-1}{a}$ which separates into $\frac{a}{a} - \frac{1}{a} = 1 - \frac{1}{a}$. This, along with the reciprocal identity, gives us a solution.

Lastly, we might have thought to change each term to sin or cos. This also works.

Write in terms of sin and cos:

$$\sin^2 \theta = \frac{\sec^2 \theta - 1}{\sec^2 \theta}$$

$$= \frac{\frac{1}{\cos^2 \theta} - 1}{\frac{1}{\cos^2 \theta}}$$

Reciprocal identity.

$$= \frac{1 - \cos^2 \theta}{\cos^2 \theta} \bullet \frac{\cos^2 \theta}{1}$$

Hints 3 and 5.

$$= 1 - \cos^2 \theta$$

$$\sin^2 \theta = \sin^2 \theta$$

Pythagorean identity.

8. Prove: $\frac{1}{1-\sin \theta} + \frac{1}{1+\sin \theta} = 2 \sec^2 \theta$.

We would think to add the fractions since the common denominator (a product of the denominators) is something we should recognize: $(1 - \sin \theta)(1 + \sin \theta) = 1 - \sin^2 \theta = \cos^2 \theta$ from the first Pythagorean identity. Applying this with hint 3, we have

$$\frac{1}{1-\sin\theta}+\frac{1}{1+\sin\theta}=2\sec^2\theta$$

$$\frac{1+\sin\theta+1-\sin\theta}{1-\sin^2\theta}=2\sec^2\theta \qquad \text{Hint 3 and FOIL.}$$

$$\frac{2}{\cos^2\theta}=2\sec^2\theta \qquad \text{Pythagorean identity.}$$

$$2\bullet\frac{1}{\cos^2\theta}=2\sec^2\theta$$

$$2\sec^2\theta=2\sec^2\theta \qquad \text{Reciprocal identity.}$$

9. Prove: $\frac{\sec\theta}{1-\cos\theta}=\frac{\sec\theta+1}{\sin^2\theta}$.

Looking at the two denominators, we see that if the first denominator were $1-\cos^2\theta$, then it would be the same as the second denominator ($1-\cos^2\theta=\sin^2\theta$ by the first Pythagorean identity). We can make this happen by multiplying $1-\cos\theta$ by $1+\cos\theta$ and counterbalancing this by multiplying $\sec\theta$ by $1+\cos\theta$. See hint 6.

$$\frac{\sec\theta}{1-\cos\theta}=\frac{\sec\theta+1}{\sin^2\theta}$$

$$\frac{\sec\theta(1+\cos\theta)}{(1-\cos\theta)(1+\cos\theta)}=\frac{\sec\theta+1}{\sin^2\theta} \qquad \text{Multiply the denominator and the numerator}$$
$$\text{by } 1+\cos\theta. \text{ See hint 6.}$$

$$\frac{\sec\theta+\sec\theta\cos\theta}{1-\cos^2\theta}=\frac{\sec\theta+1}{\sin^2\theta}$$

$$\frac{\sec\theta+1}{\sin^2\theta}=\frac{\sec\theta+1}{\sin^2\theta} \qquad \text{From } \frac{1}{\sec\theta}=\cos\theta \text{ we have } 1=\sec\theta\cos\theta.$$

Another approach is to change $\sin^2\theta$ to $1-\cos^2\theta$ and then factor it. See hint 8.

$$\frac{\sec\theta}{1-\cos\theta}=\frac{\sec\theta+1}{\sin^2\theta}$$

$$\frac{\sec\theta}{1-\cos\theta}=\frac{\frac{1}{\cos\theta}+1}{1-\cos^2\theta} \qquad$$
Reciprocal identity, $\sec\theta=\frac{1}{\cos\theta}$.
Pythagorean identity, $\sin^2\theta=1-\cos^2\theta$.
Add $\frac{1}{\cos\theta}+1$; see hint 3.

$$\frac{\sec\theta}{1-\cos\theta}=\frac{\frac{1+\cos\theta}{\cos\theta}}{\frac{(1+\cos\theta)(1-\cos\theta)}{1}} \qquad \text{Factor; see hint 8.}$$

$$\frac{\sec\theta}{1-\cos\theta}=\frac{1+\cos\theta}{\cos\theta}\bullet\frac{1}{(1+\cos\theta)(1-\cos\theta)}$$

$$\frac{\sec\theta}{1-\cos\theta}=\frac{1}{\cos\theta}\bullet\frac{1}{1-\cos\theta}$$

$$\frac{\sec\theta}{1-\cos\theta}=\sec\theta\frac{1}{1-\cos\theta} \qquad \text{Reciprocal identity.}$$

$$\frac{\sec\theta}{1-\cos\theta}=\frac{\sec\theta}{1-\cos\theta}$$

We write $(1+\cos\theta)(1-\cos\theta)$ over 1 to indicate the complex fraction. The intention is to multiply by the reciprocal.

➤ **Trigonometry, inverse (arc) functions**   An **inverse** or **arc function in trigonometry** is an angle. It is an angle that will yield the value of a given function. If the sine function has a value of $.5 = \frac{1}{2}$, the angle that will yield a sine of $\frac{1}{2}$ is 30°. See **Thirty-Sixty-Ninety Degree Triangle**. The angle that yields a sine of $\frac{1}{2}$ is called the inverse sine or arcsine, we write

$$30° = \arcsin\frac{1}{2} \quad \text{or} \quad 30° = \sin^{-1}\frac{1}{2}$$

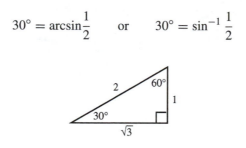

Do not confuse the -1 with a negative exponent. It implies an inverse only.

As another example, consider the cosine of 30°. In one direction, we say $\cos 30° = \frac{\sqrt{3}}{2}$. In the reverse or inverse direction, we say $\arccos\frac{\sqrt{3}}{2} = 30°$ or $\cos^{-1}\frac{\sqrt{3}}{2} = 30°$. Dividing $\sqrt{3}$ by 2 where $\frac{\sqrt{3}}{2} = .866$, we also have $\arccos .866 = 30°$ or $\cos^{-1} .866 = 30°$.

This can be verified by looking up .866 in a trigonometry chart. The .866 of the cos column corresponds with 30°. Also, enter .866 in a calculator and press the "INV" or "2nd" button and then the "$\cos^{-1}$" button. The result is 30°. The cosine of 30° is .866 and the inverse cosine or arccos of .866 is 30°. For more along these lines, see **Trigonometry Table, how to use**.

Formally, we have the definitions of all six inverse or arc trigonometric functions.

Inverse Function	Domain	Range
$y = \arcsin x$ if and only if $\sin y = x$	$-1 \leqslant x \leqslant 1$	$-\frac{\pi}{2} \leqslant y \leqslant \frac{\pi}{2}$
$y = \arccos x$ if and only if $\cos y = x$	$-1 \leqslant x \leqslant 1$	$0 \leqslant y \leqslant \pi$
$y = \arctan x$ if and only if $\tan y = x$	$-\infty < x < \infty$	$-\frac{\pi}{2} < y < \frac{\pi}{2}$
$y = \text{arccsc}\, x$ if and only if $\csc y = x$	$-\infty < x \leqslant -1$ and $1 \leqslant x < \infty$	$-\frac{\pi}{2} \leqslant y < 0$ and $0 < y \leqslant \frac{\pi}{2}$
$y = \text{arcsec}\, x$ if and only if $\sec y = x$	$-\infty < x \leqslant -1$ and $1 \leqslant x < \infty$	$0 \leqslant y < \frac{\pi}{2}$ and $\frac{\pi}{2} < y \leqslant \pi$
$y = \text{arccot}\, x$ if and only if $\cot y = x$	$-\infty < x < \infty$	$0 < y < \pi$

If preferred, the domain and range for the inverse cosecant and secant can be written in alternate form:

$$y = \text{arccsc}\, x \quad \text{if and only if } \csc y = x \quad |x| \geqslant 1 \quad -\frac{\pi}{2} \leqslant y \leqslant \frac{\pi}{2}, \ y \neq 0$$

$$y = \text{arcsec}\, x \quad \text{if and only if } \sec y = x \quad |x| \geqslant 1 \quad 0 \leqslant y \leqslant \pi, \ y \neq \frac{\pi}{2}$$

Before continuing, a familiarity with trigonometry of a right triangle is paramount. See **Trigonometry, right triangle definition of**.

T

Graphs of the Six Inverse or Arc Trigonometric Functions

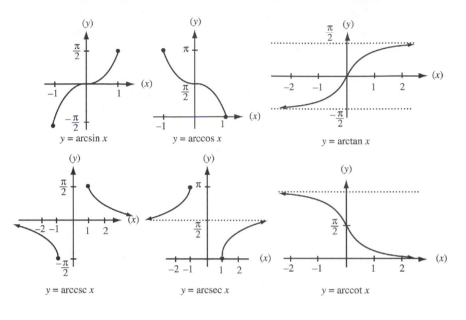

$y = \arcsin x$ $y = \arccos x$ $y = \arctan x$

$y = \text{arccsc } x$ $y = \text{arcsec } x$ $y = \text{arccot } x$

Examples

1. Evaluate $\arcsin \frac{1}{2}$.

What is required is to find an angle θ such that $\sin \theta = \frac{1}{2}$, or $\theta = \arcsin \frac{1}{2}$. Although the sine is a positive $\frac{1}{2}$ in the 1st and 2nd quadrants (see **Trigonometry, signs of the six functions**), the definition for arcsin, $-\frac{\pi}{2} \leqslant \theta \leqslant \frac{\pi}{2}$, excludes the 2nd quadrant. Hence, the angle in the 1st quadrant that yields a sine of $\frac{1}{2}$ is $\theta = 30°$. Therefore, $\theta = \arcsin \frac{1}{2} = 30°$ or $\frac{\pi}{6}$ in radians.

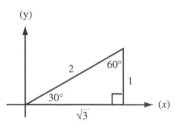

2. Evaluate $\arccos \frac{-\sqrt{3}}{2}$.

We want to find an angle θ that will yield a cosine of $\frac{-\sqrt{3}}{2}$, or $\theta = \arccos \frac{-\sqrt{3}}{2}$. Although the cosine is $\frac{-\sqrt{3}}{2}$ in the 2nd and 3rd quadrants, the definition for arccosine, $0 \leqslant \theta \leqslant \pi$, excludes the 3rd quadrant. Hence, the angle in the 2nd quadrant that yields a cosine of $\frac{-\sqrt{3}}{2}$ is $\theta = 150°$ or $\frac{5\pi}{6}$ in radians. Therefore, $\theta = \arccos \frac{-\sqrt{3}}{3} = 150°$ or $\frac{5\pi}{6}$ in radians.

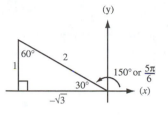

3. Evaluate arctan 0.

   It is required to find an angle θ that has a tangent of 0, or $\theta = \arctan 0$. Although the tangent is 0 for $\theta = 0°$ and $\theta = 180°$ (see **Trigonometry, quadrant angles**), the definition for arctangent, $-\frac{\pi}{2} \leqslant \theta \leqslant \frac{\pi}{2}$, excludes 180°. Hence, the angle that yields a tangent of 0 is $\theta = 0°$, or $\theta = \arctan 0 = 0°$.

4. Evaluate $\cos\left(\arcsin\frac{\sqrt{3}}{2}\right)$, which is the same as $\cos\left(\sin^{-1}\frac{\sqrt{3}}{2}\right)$.

   Similar to the first three examples, we first want to find an angle θ where $\theta = \arcsin\frac{\sqrt{3}}{2}$, then we want to find $\cos\theta$. Since $\frac{\sqrt{3}}{2}$ is positive, the definition for arcsine, $-\frac{\pi}{2} \leqslant \theta \leqslant \frac{\pi}{2}$, has us consider a triangle in the 1st quadrant where $\sin 60° = \frac{\sqrt{3}}{2}$. Therefore, $\theta = \arcsin\frac{\sqrt{3}}{2} = 60°$ or $\frac{\pi}{3}$ in radians.

   Next, we need to evaluate $\cos\theta$ or $\cos 60°$. From the same triangle, we see that $\cos 60° = \frac{1}{2}$. Hence, $\cos\left(\arcsin\frac{\sqrt{3}}{2}\right) = \cos\left(\sin^{-1}\frac{\sqrt{3}}{2}\right) = \cos 60° = \frac{1}{2}$.

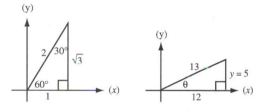

5. Evaluate $\cot\left(\arccos\frac{12}{13}\right)$.

   Since arccosine is defined for $0 \leqslant \theta \leqslant \pi$, the angle $\theta = \arccos\frac{12}{13}$ is a 1st quadrant angle. We do not need to know the actual value of θ. We are concerned only in the $\cot\theta$. Therefore, we see that $\cot\theta = \frac{12}{y}$.

   To complete the problem, we need to determine the value of y. By the Pythagorean theorem (see **Pythagorean Theorem**), we have

   $$x^2 + y^2 = r^2$$
   $$12^2 + y^2 = 13^2$$
   $$y^2 = 169 - 144$$
   $$y^2 = 25$$
   $$\sqrt{y^2} = \sqrt{25}$$
   $$y = 5$$

   Hence, we have $\cot\left(\arccos\frac{12}{13}\right) = \cot\theta = \frac{12}{5}$.

6. Graph $y = \arcsin(x + 1)$.

The graph of $y = \arcsin(x+1)$ is similar to the graph of $y = \arcsin x$, only translated 1 unit left. See **Translation or Reflection**. For a familiarity with graphing techniques, see **Trigonometry, graphs of the six functions**.

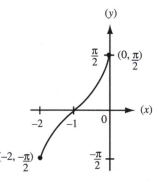

7. Graph $y = 2\arccos\frac{1}{3}x$.

In the equation $y = 2\arccos\frac{1}{3}x$, the 2 and the $\frac{1}{3}$ have an effect on the range and the domain, respectively. We can see this by observing some values of a chart.

x	$y = 2\arccos\frac{1}{3}x$
-3	$2\arccos\frac{1}{3}(-3) = 2\arccos(-1) = 2(180°) = 360°$
1	$2\arccos\frac{1}{3}(0) = 2\arccos 0 = 2(90°) = 180°$
3	$2\arccos\frac{1}{3}(3) = 2\arccos 1 = 2(0°) = 0°$

Since arccosine is defined for a domain between -1 and 1 inclusive, $\arccos\frac{1}{3}x$ is defined for $-1 \leqslant \frac{1}{3}x \leqslant 1$. Multiplying through the inequality by 3 gives us $-3 \leqslant x \leqslant 3$, the domain of $y = 2\arccos\frac{1}{3}x$. Notice that $\frac{1}{3}(-3) = -1$ and $\frac{1}{3}(3) = 1$. Any x that would give us a value less than -1 or greater than 1 cannot be part of the domain. Hence, we choose x-values between -3 and 3 inclusive.

Next, notice that the 2 of $y = 2\arccos\frac{1}{3}x$ multiplies each of the arccosine values. This increases the range from 180° to 360°. Hence, the y-values range between 0° and 360° inclusive. Similarly, the x- and y-values of the remaining inverse trigonometric fiunctions can be determined.

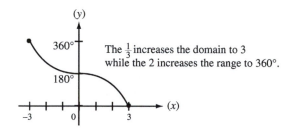

The $\frac{1}{3}$ increases the domain to 3 while the 2 increases the range to 360°.

➤ **Trigonometry, law of cosines**   See **Law of Cosines**.

➤ **Trigonometry, law of sines**   See **Law of Sines**.

➤ **Trigonometry, number line in degrees and radians**   See **Trigonometry, graphs of the six functions**, end of example 5.

➤ **Trigonometry of a Right Triangle**   See **Trigonometry, right triangle definition of**.

➤ **Trigonometry, quadrant angles**   A **quadrant angle** is an angle whose initial side lies on the positive x-axis and whose terminal side lies on an axis. The angles $0°$, $90°$, $180°$, $270°$, and $360°$ are quadrant angles. All of the angles that are coterminal with these angles are also quadrant angles.

Of particular interest is the value of the six trigonometric functions at each of the quadrant angles. There are generally three approaches used to derive these values: application of the unit circle to the definitions of the trigonometric functions, application of a general circle of radius r to the right triangle definition of the six functions, or the application of the limit concept to the six trigonometric functions.

The derivation of the values of the quadrant angles of the six functions by an application of the unit circle is completely covered in the article **Trigonometry, functions**, example 3. The article contains some helpful hints for memorizing the values. The other two approaches are covered here.

The second approach applies the right triangle definitions of the six functions to the key values around a circle of radius r, $(r, 0)$ for $0°$, $(0, r)$ for $90°$, $(-r, 0)$ for $180°$, $(0, -r)$ for $270°$, and $(r, 0)$ for $360°$. As a matter of review, we will write the six definitions of the function here. For more, see **Trigonometry, right triangle definition of**.

$$\sin \theta = \frac{y}{r} \qquad \csc \theta = \frac{r}{y}$$

$$\cos \theta = \frac{x}{r} \qquad \sec \theta = \frac{r}{x}$$

$$\tan \theta = \frac{y}{x} \qquad \cot \theta = \frac{x}{y}$$

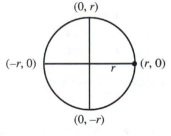

To calculate the value of each function for each quadrant angle, we need to substitute the x- or y-value of each key value into the function. The abbreviation ND is used for division by zero, not defined.

Trigonometric Quadrant Values

For $\sin\theta = \dfrac{y}{r}$:

$$\sin 0° = \frac{0}{r} = 0$$

$$\sin 90° = \frac{r}{r} = 1$$

$$\sin 180° = \frac{0}{r} = 0$$

$$\sin 270° = \frac{-r}{r} = -1$$

$$\sin 360° = \frac{0}{r} = 0$$

For $\cos\theta = \dfrac{x}{r}$:

$$\cos 0° = \frac{r}{r} = 1$$

$$\cos 90° = \frac{0}{r} = 0$$

$$\cos 180° = \frac{-r}{r} = -1$$

$$\cos 270° = \frac{0}{r} = 0$$

$$\cos 360° = \frac{r}{r} = 1$$

For $\tan\theta = \dfrac{y}{x}$:

$$\tan 0° = \frac{0}{r} = 0$$

$$\tan 90° = \frac{r}{0} = \text{ND}$$

$$\tan 180° = \frac{0}{-r} = 0$$

$$\tan 270° = \frac{-r}{0} = \text{ND}$$

$$\tan 360° = \frac{0}{r} = 0$$

For $\csc\theta = \dfrac{r}{y}$:

$$\csc 0° = \frac{r}{0} = \text{ND}$$

$$\csc 90° = \frac{r}{r} = 1$$

$$\csc 180° = \frac{r}{0} = \text{ND}$$

$$\csc 270° = \frac{r}{-r} = -1$$

$$\csc 360° = \frac{r}{0} = \text{ND}$$

For $\sec\theta = \dfrac{r}{x}$:

$$\sec 0° = \frac{r}{r} = 1$$

$$\sec 90° = \frac{r}{0} = \text{ND}$$

$$\sec 180° = \frac{r}{-r} = -1$$

$$\sec 270° = \frac{r}{0} = \text{ND}$$

$$\sec 360° = \frac{r}{r} = 1$$

For $\cot\theta = \dfrac{x}{y}$:

$$\cot 0° = \frac{r}{0} = \text{ND}$$

$$\cot 90° = \frac{0}{r} = 0$$

$$\cot 180° = \frac{-r}{0} = \text{ND}$$

$$\cot 270° = \frac{0}{-r} = 0$$

$$\cot 360° = \frac{r}{0} = \text{ND}$$

One way of remembering the values for $\sin\theta$ and $\cos\theta$ is to memorize the values on two circles. From these circles, we know that $\sin 0° = 0$, $\sin 90° = 1$, etc. Also $\cos 0° = 1$, $\cos 90° = 0$, etc. Instead of memorizing six circles, the circles for $\sin\theta$ and $\cos\theta$ can be used to generate the other four. See **Trigonometry, functions**, example 3.

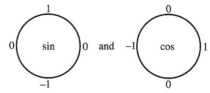

The last approach of deriving the quadrant values of the six trigonometric functions uses the concept of a limit. As the angle θ approaches each of the quadrant angles, we need to observe the limiting values of the fractions of each trigonometric function.

Consider the $\sin\theta = \dfrac{y}{r}$. As θ approaches $0°$ in the first quadrant, the value of y gets smaller. When $\theta = 0°$, we say that $y = 0$ and, hence, $\sin 0° = \dfrac{y}{r} = \dfrac{0}{r} = 0$.

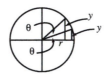

In contrast, as θ approaches $90°$ in the first quadrant, the value of y gets larger. When $\theta = 90°$, we say that y becomes the length of the radius, r, and, hence, $\sin 90° = \frac{y}{r} = \frac{r}{r} = 1$.

As θ passes $90°$ and we move into the 2nd quadrant, the value of y gets smaller again. When $\theta = 180°$, we say that $y = 0$ and, hence, $\sin 180° = \frac{y}{r} = \frac{0}{r} = 0$.

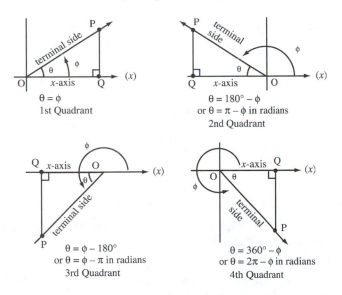

Continuing this process in all of the quadrants will produce all of the quadrant values for the $\sin \theta$. Similarly, all of the quadrant values for the remaining five functions can be determined.

▶ **Trigonometry, reference angle**   Let ϕ be an angle in standard position (see **Angle, in trigonometry**). Then the **reference angle** of ϕ is the angle θ formed by the terminal side of ϕ and the x-axis. Further, let P be any point on the terminal side of θ and let Q be the point on the x-axis such that $\overline{PQ}$ is perpendicular to the x-axis. Then the triangle formed by P, the origin O, and Q is called a **reference triangle**, where $\angle POQ$ is the **reference angle**.

If ϕ is a positive angle

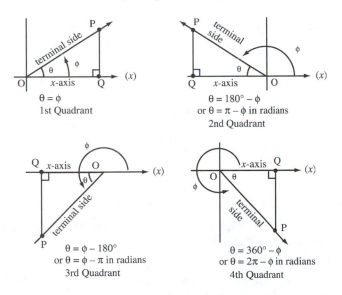

If φ is a negative angle

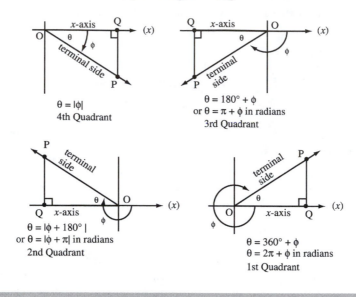

Examples

1. State the reference angle of 30°, 150°, 210°, and 330°.

   These angles are in the 1st, 2nd, 3rd, and 4th quadrants, respectively.

   a. In the 1st quadrant, the reference angle θ is $\theta = \phi$, therefore, $\theta = 30°$.

   b. In the 2nd quadrant, the reference angle θ is $\theta = 180° - \phi$, therefore, $\theta = 180° - 150° = 30°$.

   c. In the 3rd quadrant, the reference angle θ is $\theta = \phi - 180°$, therefore, $\theta = 210° - 180° = 30°$.

   d. In the 4th quadrant, the reference angle θ is $\theta = 360° - \phi$, therefore, $\theta = 360° - 330° = 30°$.

2. Draw the reference angle for 225°, −240°, and −30°.

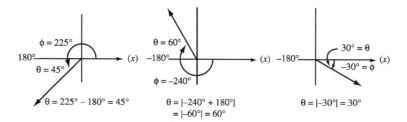

3. Write cos 150° in terms of its reference angle, then evaluate its exact value.

   To write cos 150° in terms of its reference angle, we need to do two things: determine the reference angle of 150° and determine the + or − sign of the cosine in the 2nd quadrant (see **Trigonometry, signs of the six functions**).

   Since 150° is in the 2nd quadrant, its reference angle is $\theta = 180° - 150° = 30°$. Also, from the chart for the signs of the six functions, we know that the cosine is negative in the 2nd quadrant.

Therefore, $\cos 150° = -\cos 30°$. It is negative because it is in the 2nd quadrant and $30°$ because this is the value of the reference angle.

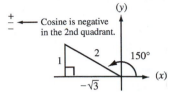

To evaluate the exact value of $\cos 150°$, we need to draw the reference triangle for $150°$, label the values of its sides, and then determine the cosine. Since the reference angle is $30°$, the reference triangle is a 30-60-90 degree triangle in the 2nd quadrant (see **Thirty-Sixty-Ninety Degree Triangle**). Therefore, $\cos 150° = \frac{\text{adjacent}}{\text{hypotenuse}} = \frac{-\sqrt{3}}{2}$ (see **Trigonometry, right triangle definition of**).

4. Write $\sin(-120°)$ in terms of its reference angle, then evaluate its exact value.

Since $-120°$ is a 3rd quadrant angle with a reference angle of $\theta = 180° - 120° = 60°$, the reference triangle is a sixty degree triangle. Also, from the chart for the signs of the six functions, we know that the sine is negative in the 3rd quadrant. Therefore, $\sin(-120°) = -\sin 60°$ and the exact value is $\sin(-120°) = \frac{\text{opposite}}{\text{hypotenuse}} = \frac{-\sqrt{3}}{2}$.

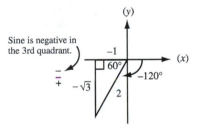

5. In radians, draw the reference angles for $\frac{5\pi}{4}$, $\frac{-4\pi}{3}$, and $-\frac{\pi}{6}$, respectively. For a comparison with degrees, see example 2.

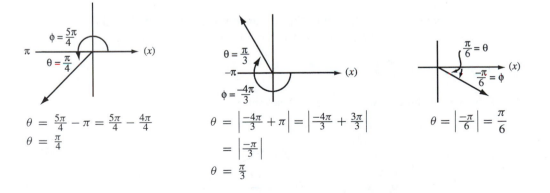

$\theta = \frac{5\pi}{4} - \pi = \frac{5\pi}{4} - \frac{4\pi}{4}$
$\theta = \frac{\pi}{4}$

$\theta = \left|\frac{-4\pi}{3} + \pi\right| = \left|\frac{-4\pi}{3} + \frac{3\pi}{3}\right|$
$= \left|\frac{-\pi}{3}\right|$
$\theta = \frac{\pi}{3}$

$\theta = \left|\frac{-\pi}{6}\right| = \frac{\pi}{6}$

$$\theta = \frac{5\pi}{4} - \pi \qquad \theta = \left| -\frac{4\pi}{3} + \pi \right| \qquad \theta = \left| -\frac{\pi}{6} \right| = \frac{\pi}{6}$$

$$= \frac{5\pi}{4} - \frac{4\pi}{4} \qquad \theta = \left| -\frac{4\pi}{3} + \frac{3\pi}{3} \right|$$

$$= \frac{\pi}{4} \qquad \theta = \left| -\frac{\pi}{3} \right|$$

$$\theta = \frac{\pi}{3}$$

▶ **Trigonometry, reference triangle**   See **Trigonometry, reference angle**.

▶ **Trigonometry, right triangle definition of**

There are two ways of defining the six trigonometric functions. One way is strictly through a functional approach; see **Trigonometry, functions**. The other way is through the **right triangle definition of trigonometry**. This approach defines the six functions in relation to a right triangle and then places the triangle into any of the four quadrants as a reference triangle (see **Trigonometry, reference angle**). In this manner, values for any degree or radian can be calculated.

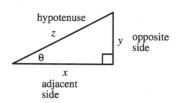

Let θ be one of the acute angles of a right triangle as shown in the figure. From the point of view of θ, the side y is opposite θ, the side x is adjacent to θ, and the side z is the hypotenuse. Sometimes confusion whether x or z is the adjacent side can occur. This can be avoided by remembering that the hypotenuse is the longest side of the triangle and is directly opposite the ninety degree angle.

After the sides of a right triangle have been identified in relation to one of its acute angles, there are six fractions (ratios) that can be considered. Each fraction is defined (given a name) as follows:

$$\sin\theta = \frac{\text{opp}}{\text{hyp}} = \frac{y}{z} \qquad \csc\theta = \frac{\text{hyp}}{\text{opp}} = \frac{z}{y}$$

$$\cos\theta = \frac{\text{adj}}{\text{hyp}} = \frac{x}{z} \qquad \sec\theta = \frac{\text{hyp}}{\text{adj}} = \frac{z}{x}$$

$$\tan\theta = \frac{\text{opp}}{\text{adj}} = \frac{y}{x} \qquad \cot\theta = \frac{\text{adj}}{\text{opp}} = \frac{x}{y}$$

It must be remembered that each of the six functions is a fraction. The $\sin\theta$ is a fraction. It is a fraction of the y-side with the z-side, and so on for the remaining functions.

Examples

1. Using $\triangle ABC$, find the sine, cosine, and tangent from $\angle A$ and from $\angle C$.

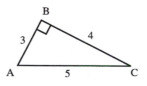

From the point of view of $\angle A$, the opposite side is 4, the adjacent side is 3, and the hypotenuse is 5. From the point of view of $\angle C$, the opposite side is 3, the adjacent side is 4, and the hypotenuse is the same, 5.

From $\angle A$:   From $\angle C$:

$$\sin A = \frac{\text{opp}}{\text{hyp}} = \frac{4}{5} \qquad \sin C = \frac{\text{opp}}{\text{hyp}} = \frac{3}{5}$$

$$\cos A = \frac{\text{adj}}{\text{hyp}} = \frac{3}{5} \qquad \cos C = \frac{\text{adj}}{\text{hyp}} = \frac{4}{5}$$

$$\tan A = \frac{\text{opp}}{\text{adj}} = \frac{4}{3} \qquad \tan C = \frac{\text{opp}}{\text{adj}} = \frac{3}{4}$$

2. Find the exact values of the six trigonometric functions for an angle of $150°$.

The reference triangle of a $150°$ angle is a triangle in the 2nd quadrant with a reference angle of $30°$; see **Trigonometry, reference angle**. To find the exact values of the six trigonometric functions means that we want to use the values of the sides of the reference triangle to determine the ratios. We are not allowed to use a calculator or the trigonometric table.

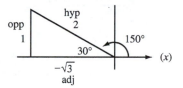

From the values of the sides of a 30-60-90 degree triangle (see **Thirty-Sixty-Ninety Degree Triangle**), we see that the opposite side is 1, the adjacent side is $-\sqrt{3}$, and the hypotenuse is 2. Applying these values to the definitions of the six trigonometric functions, we have

$$\sin 150° = \frac{\text{opp}}{\text{hyp}} = \frac{1}{2} \qquad\qquad \csc 150° = \frac{2}{1} = 2$$

$$\cos 150° = \frac{\text{adj}}{\text{hyp}} = -\frac{\sqrt{3}}{2} \qquad\qquad \sec 150° = \frac{2}{-\sqrt{3}} = \frac{-2\sqrt{3}}{3}$$

$$\tan 150° = \frac{\text{opp}}{\text{adj}} = \frac{1}{-\sqrt{3}} = \frac{-\sqrt{3}}{3} \qquad\qquad \cot 150° = \frac{-\sqrt{3}}{1} = -\sqrt{3}$$

To simplify $\frac{1}{-\sqrt{3}}$ or $\frac{2}{-\sqrt{3}}$, see **Rationalizing the Denominator**.

$$\frac{1}{-\sqrt{3}} = \frac{1}{-\sqrt{3}} \cdot \frac{\sqrt{3}}{\sqrt{3}} = \frac{\sqrt{3}}{-\sqrt{9}} = -\frac{\sqrt{3}}{3}$$

$$\frac{2}{-\sqrt{3}} = \frac{2}{-\sqrt{3}} \cdot \frac{\sqrt{3}}{\sqrt{3}} = \frac{2\sqrt{3}}{-\sqrt{9}} = \frac{-2\sqrt{3}}{3}$$

3. Find the exact value of $\cos\left(-\frac{7\pi}{4}\right)$.

Since $-\frac{7\pi}{4}$ is in the 1st quadrant, the reference angle is

$$\theta = 2\pi + \frac{-7\pi}{4} = \frac{8\pi}{4} - \frac{7\pi}{4} = \frac{\pi}{4} \qquad \text{See \textbf{Trigonometry, reference angle}.}$$

If it isn't clear that $\frac{-7\pi}{4}$ is in the 1st quadrant, change it to degrees first:

$$-\frac{7\pi}{4} = -\frac{7(180°)}{4} = -315°$$

The reference angle for $-315°$ is $45°$ or $\frac{\pi}{4}$ in radians.

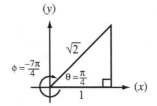

From the values of the sides of a 45-degree triangle (see **Forty-Five Degree Triangle**), we see that the adjacent side is 1 and the hypotenuse is $\sqrt{2}$. Therefore, we have

$$\cos\left(-\frac{7\pi}{4}\right) = \frac{\text{adj}}{\text{hyp}} = \frac{1}{\sqrt{2}} = \frac{\sqrt{2}}{2}$$

To simplify, see **Rationalizing the Denominator**.

$$\frac{1}{\sqrt{2}} = \frac{1}{\sqrt{2}} \bullet \frac{\sqrt{2}}{\sqrt{2}} = \frac{\sqrt{2}}{\sqrt{4}} = \frac{\sqrt{2}}{2}$$

4. Given the point $(-3, -4)$, find the exact values of the six trigonometric functions.

Since $(x, y) = (-3, -4)$, two sides of the triangle are known. To determine the third side, we can apply the Pythagorean theorem. See **Pythagorean Theorem**.

$$r^2 = x^2 + y^2$$
$$r^2 = (-3)^2 + (-4)^2$$
$$r^2 = 9 + 16$$
$$r^2 = 25$$
$$\sqrt{r^2} = \sqrt{25} \qquad \text{Use the positive square root since distance is defined as a positive value.}$$
$$r = 5$$

With $x = -3$, $y = -4$, and $r = 5$, we can draw the triangle associated with the point $(-3, -4)$ where the reference angle is θ. Hence, we have the values of the six trigonometric functions:

$$\sin \theta = \frac{-4}{5} \qquad \csc \theta = \frac{5}{-4} = -\frac{5}{4}$$

$$\cos \theta = \frac{-3}{5} \qquad \sec \theta = \frac{5}{-3} = -\frac{5}{3}$$

$$\tan \theta = \frac{-4}{-3} = \frac{4}{3} \qquad \cot \theta = \frac{-3}{-4} = \frac{3}{4}$$

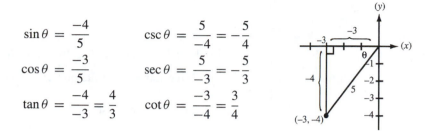

5. Solve right $\triangle ABC$ where $m\angle A = 42°$, $AB = 12$, and $\angle C$ is a right angle. Calculate the sides to the nearest tenth.

To **solve a triangle** means that we want to find the remaining sides and angles. In other words, we want to find $m\angle B$ and the lengths of a and b.

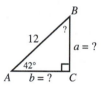

Since $m\angle C = 90°$ (see **Right Angle**), we know that $m\angle A + m\angle B = 90°$ (see **Sum of the Measures of the Angles of a Triangle**). Therefore, we have the measure of $\angle B$:

$$m\angle A + m\angle B = 90°$$
$$42° + m\angle B = 90°$$
$$m\angle B = 90° - 42°$$
$$m\angle B = 48°$$

To find the lengths of a and b, we need to determine which of the six trigonometric functions, we are going to use. Since there are three parts to each of the six equations, we need to know two of the three parts. Then we can solve for the third.

This means that if we know the measure of an angle, we need to make a fraction between a side we want to know and a side we already know. Whether these sides are the opposite, the adjacent, or the hypotenuse determines which function we use.

From the point of view of angle A, which we know is $42°$, we can make a fraction involving the opposite side a (which we don't know) with the hypotenuse, 12. Since the fraction $\frac{\text{opp}}{\text{hyp}}$ is associated with the sine function, we know which equation to use (or cosecant equation if we used $\frac{\text{opp}}{\text{hyp}}$).

Substituting into the equation for the sine and solving, we have

$$\sin A = \frac{\text{opp}}{\text{hyp}}$$

$$\sin 42° = \frac{a}{12}$$

$$.6691 = \frac{a}{12} \qquad \text{Look up } 42° \text{ in the chart or use a calculator.}$$

$$12(.6691) = a \qquad \text{See } \textbf{Trigonometry Table, how to use.}$$

$$a = 8.0292$$

$$a = 8.0 \qquad \text{Rounded to the nearest tenth.}$$

Now that we know two sides of the right triangle, we could use the Pythagorean theorem to find the third side or use one of the remaining trigonometric functions. We'll do the latter.

From the point of view of angle A, we can either make a fraction involving the adjacent side b with the hypotenuse, 12, or the opposite side a, 8.0. Since the opposite side is an approximation, we will have more accuracy if we use the hypotenuse. Hence, we see that the fraction $\frac{\text{adj}}{\text{hyp}}$ is associated with the cosine and we have (or secant equation if we used $\frac{\text{hyp}}{\text{adj}}$)

$$\cos A = \frac{\text{adj}}{\text{hyp}}$$

$$\cos 42° = \frac{b}{12}$$

$$.7431 = \frac{b}{12}$$

$$12(.7431) = b$$

$$b = 8.9172$$

$$b = 8.9 \qquad \text{Rounded to the nearest tenth.}$$

Therefore, the triangle has been solved with side $a = 8.0$, side $b = 8.9$, and $m\angle B = 48°$.

6. Solve right $\triangle ABC$ if $a = 2$, $b = 7$, and $\angle C$ is a right angle. Calculate the side to the nearest tenth, and the angles to the nearest tenth and nearest minute.

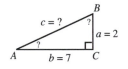

To solve the triangle means that we want to find side c, $m\angle A$, and $m\angle B$. When two sides of a triangle are known, the third side can be found by the Pythagorean theorem (see **Pythagorean Theorem**).

$$c^2 = a^2 + b^2$$

$$c^2 = 2^2 + 7^2$$

$$c^2 = 4 + 49$$

$$c^2 = 53$$
$$\sqrt{c^2} = \sqrt{53}$$
$$c \approx 7.3$$

To determine the angles, we need to decide which of the six functions we are going to use. Since side c is an approximation, we need to use a and b. From the point of view of $\angle A$, we see that side a is the opposite side and side b is the adjacent side. The fraction involving these two sides is $\frac{\text{opp}}{\text{adj}}$, and, hence, we know we are going to use the tangent equation (or cotangent equation if we used the fraction $\frac{\text{adj}}{\text{opp}}$)

$$\tan A = \frac{\text{opp}}{\text{adj}}$$
$$\tan A = \frac{2}{7}$$
$$\tan A = .2857$$
$$m\angle A = \arctan .2857 \qquad \text{See \textbf{Trigonometry Table, how to use}.}$$
$$m\angle A = 15.9° \text{ or } 15°57' \qquad \text{See \textbf{Degree-Minutes-Seconds}.}$$

Depending on the type of trigonometry table that is used to find arctan .2857, interpolation may have to be used. See **Interpolation (linear)**. If a calculator is used, i.e., $\tan^{-1}.2857$, then round the answer to the nearest tenth.

To convert from decimal degrees to degrees and minutes, we need to use the decimal 15.944°. Multiplying by 60, we have $(.944)(60) = 56.64$ or $57'$. See **Degree-Minutes-Seconds**.

From the point of view of $\angle B$, we see that side a is the adjacent side and side b is the opposite side. The fraction involving these two sides is $\frac{\text{opp}}{\text{adj}}$, which corresponds with tangent equation (or cotangent equation if we used the fraction $\frac{\text{adj}}{\text{opp}}$)

$$\tan B = \frac{\text{opp}}{\text{adj}}$$
$$\tan B = \frac{7}{2}$$
$$\tan B = 3.5$$
$$m\angle B = \arctan 3.5$$
$$m\angle B = 74.1° \text{ or } 74°3'$$

Of course, an easier way of finding $\angle B$ after $\angle A$ is determined is to subtract from 90°:

$$m\angle A + m\angle B = 90°$$
$$15.9° + m\angle B = 90°$$
$$m\angle B = 90° - 15.9°$$
$$m\angle B = 74.1°$$

For a review of $90° - 15°57' = 74°3'$, see **Degree-Minutes-Seconds**.

➤ **Trigonometry, signs of the six functions**   Consider a 30-60-90 degree triangle in the 1st and 2nd quadrants where the reference angle is 30°.

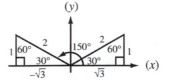

Notice that the distinguishing characteristic between each of the triangles is the $\sqrt{3}$. The fact that the triangle in the 2nd quadrant has a $-\sqrt{3}$ is what makes it a 2nd quadrant triangle.

Now consider the cosine of the two triangles (see **Trigonometry, right triangle definition of**). The cosine of the triangle in the 1st quadrant is $\frac{\sqrt{3}}{2}$ and the cosine of the triangle in the 2nd quadrant is $\frac{-\sqrt{3}}{2}$, or $\cos 30° = \frac{\sqrt{3}}{2}$ and $\cos 150° = \frac{-\sqrt{3}}{2}$. The numerical value of each is the same. They differ only in reference to their positive ($+$) and negative ($-$) signs.

In fact, this is the case for all six functions in each of the four quadrants. They differ with one another, from one quadrant to the next, only in reference to their $+$ or $-$ signs. It is these signs that are referred to as the **signs of the six functions in trigonometry**.

To determine the $+$ or $-$ signs of the six functions in each of the quadrants, we can utilize the figure below. The sides of each triangle are shown only in reference to their $+$ or $-$ signs and the reference angle θ. The hypotenuse of each triangle is $+$ because the Pythagorean theorem always renders the hypotenuse positive.

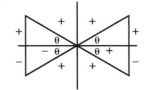

The sign of each function is calculated using the right triangle definition of the function; see **Trigonometry, right triangle definition of**. We only need to calculate the sine, cosine, and tangent in each quadrant since their reciprocals (cosecant, secant, and cotangent) have the same signs.

1st Quadrant

$$\sin \theta = \frac{\text{opp}}{\text{hyp}} = \frac{+}{+} = +$$

$$\cos \theta = \frac{\text{adj}}{\text{hyp}} = \frac{+}{+} = +$$

$$\tan \theta = \frac{\text{opp}}{\text{adj}} = \frac{+}{+} = +$$

2nd Quadrant

$$\sin \theta = \frac{+}{+} = +$$

$$\cos \theta = \frac{-}{+} = -$$

$$\tan \theta = \frac{+}{-} = -$$

3rd Quadrant

$$\sin \theta = \frac{-}{+} = -$$

$$\cos \theta = \frac{-}{+} = -$$

$$\tan \theta = \frac{-}{-} = +$$

4th Quadrant

$$\sin \theta = \frac{-}{+} = -$$

$$\cos \theta = \frac{+}{+} = +$$

$$\tan \theta = \frac{-}{+} = -$$

T

An easy way to remember the signs is to draw the four quadrants with the signs in them:

$$
\begin{array}{c|c}
\begin{matrix} + \\ - \\ - \end{matrix} & \begin{matrix} + \\ + \\ + \end{matrix} \\
\hline
\begin{matrix} - \\ - \\ + \end{matrix} & \begin{matrix} + \\ + \\ - \end{matrix}
\end{array}
$$

Examples

1. Write cos 135° and sin 330° in terms of their reference angles.

   a. Since 135° is a 2nd quadrant angle, its sign is −. Therefore, cos 135° = − cos 45°.

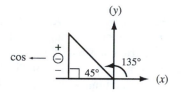

   b. Since 330° is a 4th quadrant angle, its sign is −. Therefore, sin 330° = − sin 60°.

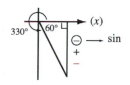

   To determine the reference angles of 45° and 60°, respectively, see **Trigonometry, reference angle**.

2. Name the quadrant determined in each situation:

   a. $\sin \theta > 0$ and $\cos \theta < 0$

   b. $\cos \theta > 0$ and $\tan \theta < 0$

   c. $\cot \theta < 0$ and $\sec \theta > 0$

   a. To say $\sin \theta > 0$ means that the sine is positive. The sine is positive in two quadrants, I and II. To say $\cos \theta < 0$ means that the cosine is negative. The cosine is negative in two quadrants, II and III. Hence, the sine is positive and the cosine is negative at the same time in quadrant II.

   b. The cosine is positive in quadrants I and IV while the tangent is negative in quadrants II and IV. The cosine is positive and the tangent is negative at the same time in quadrant IV.

c. The cotangent is negative when the tangent is negative, in quadrants II and IV. The secant is positive when the cosine is positive, in quadrants I and IV. The cotangent is negative and the secant is positive at the same time in quadrant IV.

➤ **Trigonometry Table, how to use**   A trigonometry table is a list of approximate values of the trigonometric functions for positive acute angles. The table at the end of the book gives the angles in degrees and minutes as well as radians. In the first column, the angles are listed from $0°00'$ to $45°00'$ in intervals of 10 minutes (the symbol $'$ means minutes). In the last column, the angles are listed from bottom to top as $45°00'$ to $90°00'$. The second column and second to last column give the listings in radians, from .0000 to .7854 and from .7854 to 1.5708, respectively. See **Converting Degrees to Radians**.

$$45° = \frac{\pi}{4} = \frac{3.1416}{4} = .7854 \text{ radians}$$

$$90° = \frac{\pi}{2} = \frac{3.1416}{2} = 1.5708 \text{ radians}$$

To find the value of any of the six trigonometric functions from $0°$ to $90°$, we read down the first column or up the last column until we locate the angle under consideration, then from the top of the page we match the column of the function with the row of the angle and read the value at their intersection. We do this for angles from $0°$ to $45°$. For angles from $45°$ to $90°$, we find the function at the bottom of the page and match its column with the row of the angle and read the value at their intersection.

This can be more easily seen by example. First, let's find sin $14°$. From $14°00'$, we match the column of the sin with the row of $14°00'$ and read the value at their intersection, .2419. Hence, sin $14° = .2419$.

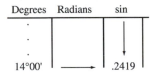

We can verify this by using a calculator. Enter 14 in the calculator and press sin. The value should be .2419219, which we round off to .2419. Make sure that the calculator is in degree mode.

Next, let's find cos $76°$. Since $76°$ is in the last column, we read up from the bottom of the page. From $76°00'$, we match the column of the cos (from the bottom of the page) with the row of $76°00'$ and read the value at their intersection, .2419. Hence, cos $76° = .2419$. The most common mistake, of course, is reading from the top of the page instead of the bottom.

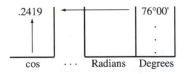

To determine the value of a trigonometric function whose angle is between consecutive entries in the table, the process of linear interpolation must be applied. See **Interpolation (linear)**. Through interpolation, values beyond those given in the table can be determined. During interpolation, remember to consider that as θ varies from $0°$ to $90°$, the values of the sin, tan, and sec increase while those of the cos, csc, and cot decrease.

Examples

1. Use the trigonometric table to evaluate $\cos 23°20'$, $\tan .7447$, and $\cot 59°40'$.

   a. Reading down the degree column (since $23°20'$ is less than $45°00'$), we locate $23°20'$. From $23°20'$, we match the column of the cos with the row of $23°20'$ and read the value at their intersection, .9182. Hence, $\cos 23°20' = .9182$.

   b. Reading down the radian column (since .7447 is greater than .7854), we locate .7447. From .7447, we match the column of the tan with the row of .7447 and read the value at their intersection, .9217. Hence, $\tan .7447 = .9217$.

   c. Since $59°40'$ is greater than $45°00'$, we read up the last column until we locate $59°40'$. Remember to read above $59°$ to $40'$ and not below. The one below is for $58°40'$. From $59°40'$, we match the column of the cot (from the bottom of the page) with the row of $59°40'$ and read the value at their intersection, .5851. Hence, $\cot 59°40' = .5851$. The most common mistake is reading from the top of the page instead of the bottom.

2. Find $\cos 231°50'$.

   Since the chart only goes to $90°00'$, we need to consider the reference angle of $231°50'$ (see **Trigonometry, reference angle**).

$$231°50' - 180°00' = 51°50'$$

In particular, we are concerned with the fact that $\cos 231°50' = -\cos 51°50'$. It is negative $(-)$ because $231°50'$ is a 3rd quadrant angle and the cosine is negative in the 3rd quadrant (see **Trigonometry, signs of the six functions**).

To find $-\cos 51°50'$, we need to match the column of the cos (from the bottom of the page) with the row of $51°50'$ and read the value at their intersection, .6180. Hence, we have

$$\cos 231°50' = -\cos 51°50' = -.6180$$

T

➤ **Trinomial**   A **trinomial** is a polynomial with three terms. The following are examples of trinomials: $3x^2 + 5x + 2$, $8y^3 - 5y^3 + 6y$, $8a^4 - 3a^2 - 2a$, and $-x^3y^2 + 12xy - 5y^7$. Each of the terms of a trinomial is separated by a positive $(+)$ or negative $(-)$ sign. It is this fact that determines a trinomial and not the amount of variables in each term.

For more, see **Polynomial**. To factor a trinomial, see **Factoring Trinomials**. Also see **Perfect Square Trinomial**.

▶ **Trinomial Equations**   See **Solutions of Equations**.

▶ **Trinomial Factoring**   See **Factoring Trinomials**.

▶ **Truth Table**   A **truth table** is a table of truth or falsity set up for statements of conjunction, disjunction, or other statements in logic. See **Conjunction** or **Disjunction**.

The conjunction $p \wedge q$ is a true statement when both p and q are true statements. Otherwise, it is false. The disjunction $p \vee q$ is a false statement when both p and q are false statements. Otherwise, it is true. Hence, we have the truth table for conjunction and disjunction.

p	q	$p \wedge q$	$p \vee q$
T	T	T	T
T	F	F	T
F	T	F	T
F	F	F	F

In the first two columns, all the possible combinations of truth or falsity are represented. Either p and q are both true, p is true when q is false, p is false when q is true, or both p and q are false. The only way $p \wedge q$ can be true is when both p and q are true, the first row. There are three possibilities for $p \vee q$ to be true. By definition, $p \vee q$ means p is true, q is true, or both p and q are true. The first part of the definition allows for one or the other of p or q to be false, yet the statement is true, and the last part allows both to be true.

When dealing with truth tables, it is important to keep the concept of "logical equivalence" in mind. If two statements have exactly the same entries in each row of truth tables, then they are logically equivalent. In this manner, $p \rightarrow q$ and $q \rightarrow p$, or $p \leftrightarrow q$. In other words, equivalent statements are biconditional statements. See **Biconditional (equivalent) Statement.**

Examples

1. Show whether the statements $\sim(p \wedge q)$ and $\sim p \vee \sim q$ are logically equivalent or not.

If $\sim(p \wedge q)$ and $\sim p \vee \sim q$ are logically equivalent, then the entries in each of the rows of their respective truth tables must be the same. Hence, we first need to derive a truth table for each of the statements.

To derive the truth table for $\sim(p \wedge q)$, we first need to write the table for $p \wedge q$ and then negate each of these entries. See **Negation**.

p	q	$p \wedge q$	$\sim(p \vee q)$
T	T	T	F
T	F	F	T
F	T	F	T
F	F	F	F

We write the truth table for the conjunction $p \wedge q$ and then write the negation of each entry.

To derive the truth table for $\sim p \vee \sim q$, we first need to write the entries for $\sim p$ and $\sim q$, and then write the disjunction of those, $\sim p \vee \sim q$. See **Disjunction**.

T

p	q	$\sim p$	$\sim q$	$\sim p \vee \sim q$
T	T	F	F	F
T	F	F	T	T
F	T	T	F	T
F	F	T	T	T

We write the disjunction for the entries in columns 3 and 4 according to the definition of disjunction for $p \vee q$.

Since all of the entries in each of the columns for $\sim(p \wedge q)$ and $\sim p \vee \sim q$ are the same, the statements are logically equivalent.

When deriving the disjunction for $\sim p$ and $\sim q$, a comparison between two tables occurs. In the 1st row of the table for $\sim p$ and $\sim q$, we see the entries F and F. These two are then compared to the 4th row of the table for a disjunction, F and F yields an F. Hence, $\sim p \vee \sim q$ is false. In the 2nd row of the table for $\sim p$ and $\sim q$, we see the entries F and T. These two are then compared to the 3rd row of the table for a disjunction, F and T yield a T. Hence, $\sim p \vee \sim q$, is true, and so on.

2. Write the truth tables for the converse, inverse, and contrapositive.

To derive the truth tables for the converse, inverse, and contrapositive, we need to compare their entries with those for a conditional statement. Or, more simply, just remember that when the hypothesis is true and the conclusion is false, the conditional statement is false. Otherwise, it is true.

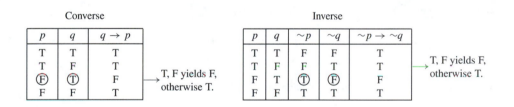

Converse

p	q	$q \to p$
T	T	T
T	F	T
(F)	(T)	F
F	F	T

T, F yields F, otherwise T.

Inverse

p	q	$\sim p$	$\sim q$	$\sim p \to \sim q$
T	T	F	F	T
T	F	F	T	T
F	T	(T)	(F)	F
F	F	T	T	T

T, F yields F, otherwise T.

Contrapositive

p	q	$\sim p$	$\sim q$	$\sim q \to \sim p$
T	T	F	F	T
T	F	(F)	(T)	F
F	T	T	F	T
F	F	T	T	T

T, F yields F, otherwise T.

➤ **Turning Point**   See **Vertex**.

➤ **Two-Column Deductive Proof**   See **Proof in Geometry**.

➤ **Two Equations in Two Unknowns**   See **Systems of Linear Equations, solutions of**.

➤ **Two-Point Equation**   See **Linear Equation, finding the equation of a line**, Two-Point Method.

➤ **Two Variables, functions of**   See **Function (two variables)**.

Uu

➤ **Union and Intersection**   The **union** of two sets A and B is defined as the set of elements that are in A, or in B, or in both, and we write $A \cup B$ (read A union B or the union of A and B). The **intersection** of two sets A and B is defined as the set of elements that belong to both A and B (at the same time), and we write $A \cap B$ (read A intersection B or the intersection of A and B). If the sets A and B have no elements in common, they are said to be **disjoint**. For an example, see **Disjoint**. In addition to union and intersection, see **Complement of a Set**.

Examples

1. Given $A = \{1, 2, 3, 4\}$, $B = \{3, 4, 5\}$, and $C = \{5, 6, 7\}$, find $A \cup B$, $A \cap B$, $A \cup C$, and $A \cap C$.

   a. $A \cup B = \{1, 2, 3, 4, 5\}$; do not include the duplicates.

   b. $A \cap B = \{3, 4\}$; include only what is common to both.

   c. $A \cup C = \{1, 2, 3, 4, 5, 6, 7\}$.

   d. $A \cap C = \varnothing$, the empty set, since they have nothing in common. The empty set may also be written as $\{\ \}$.

2. Given $P = \{0, 1, 3, 6, 7\}$, $Q = \{3, 5, 7\}$, and $R\{1, 2, 4, 6\}$, find $(P \cup Q) \cap R$ and $(P \cap R) \cup (Q \cap R)$.

   a. To find $(P \cup Q) \cap R$, we first need to determine $P \cup Q$, then intersect the result with R:

   $$P \cup Q = \{0, 1, 3, 5, 6, 7\}$$
   $$(P \cup Q) \cap R = \{0, 1, 3, 5, 6, 7\} \cap \{1, 2, 4, 6\} = \{1, 6\}$$

   b. To find $(P \cap R) \cup (Q \cap R)$, we first need to determine $P \cap R$ and $Q \cap R$, then we can unite their results. Since $P \cap R = \{1, 6\}$ and $Q \cap R = \varnothing$, we have

   $$(P \cap R) \cup (Q \cap R) = \{1, 6\} \cup \varnothing = \{1, 6\}$$

➤ **Unit Circle**   A **unit circle** is a circle with a radius of 1 unit and is centered at the origin $(0, 0)$. If P is any point on the circle, then the distance from the origin to P is 1 unit, the length of the radius.

U

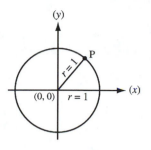

➤ **Universal Set**   The **universal set** or **universe** is a set that contains all of the other sets in the discussion as subsets. It is denoted by U. For an application, see **Venn Diagram**. The domain is also considered as a universal set. See **Domain and Range**.

➤ **Unlike Terms**   **Unlike terms** are terms that are not "like" or "similar." Whereas the expression $6a + 2a - 3a$ is composed of three like terms, the expression $8a + 3b - 5$ is composed of three unlike terms. For more on like terms, see **Combining Like Terms**.

U

Vv

➤ **Variable**   Any letter that can be replaced by a number or another letter is called a **variable**. The expression $6y + 2$ contains the variable y. The value of $6y + 2$ depends on the value of y. When y is determined, so is $6y + 2$. If we let $y = 3$, then $6y + 2 = 6(3) + 2 = 20$. If $y = 5$, then $6y + 2 = 6(5) + 2 = 32$. If $y = a + 3$, then $6y + 2 = 6(a + 3) + 2 = 6a + 18 + 2 = 6a + 20$, and a new variable has been introduced.

➤ **Variable, dependent**   In the equation $y = 3x + 4$, y is called the **dependent variable** since it is determined by x. Other examples are $y = 4x^2 - 11$ and $y = 5\sqrt{x} + 2$. The x is called the independent variable.

➤ **Variable, independent**   In the equation $y = 3x + 4$, x is called the **independent variable**. It is independent of y. In contrast, y is called the dependent variable. Its value depends on the value of x. When calculating values of the equation, x is chosen independently and the value of y is then determined. It depends on the value of x.

➤ **Variation, combined**   See **Combined Variation**.

➤ **Variation, direct**   See **Direct Variation**.

➤ **Variation, inverse**   See **Inverse Variation**.

➤ **Variation, joint**   See **Joint Variation**.

➤ **Vector**   Let $\overrightarrow{PQ}$ be a directed line, then the set of all directed line segments in the plane of $\overrightarrow{PQ}$ that are equivalent to $\overrightarrow{PQ}$ is called a **vector v**, where $\mathbf{v} = \overrightarrow{PQ}$. Vectors are denoted by lowercase letters written in boldface, i.e. $\mathbf{u}$, $\mathbf{v}$, or $\mathbf{w}$.

A directed line segment $\overrightarrow{PQ}$ is a segment with an initial point P, a terminal point Q, and is determined by a given magnitude and direction. The length of $\overrightarrow{PQ}$ is denoted $\|PQ\|$.

The motivation for a directed line segment is the desire in physics, as well as in other sciences, to define quantities such as force and velocity that involve both magnitude and direction. These quantities cannot be determined by a single value alone.

If two directed line segments have the same magnitude (length) and direction (usually an angle), they are said to be equivalent. It is in the sense of equivalence that we define a vector to be the set of all directed line segments that have the same length (magnitude) and angle (direction).

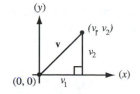

If the initial point of a directed line segment is placed at the origin of a coordinate system, we say that it is representative of the set of equivalent directed line segments and is called a vector **v** in **standard position**. If we let the terminal point of **v** be denoted by the point (v_1, v_2), then the coordinate v_1 is called the **x-component** of **v** and the coordinate v_2 is called the **y-component** of **v**.

In general, if we consider any two points $P(p_1, p_2)$ where $Q(q_1, q_2)$, where $\mathbf{v} = \overrightarrow{PQ}$, then the x-component of **v** is $v_1 = q_1 - p_1$ and the y-component of **v** is $v_2 = q_2 - p_2$. The magnitude of **v** is

$$\|\mathbf{v}\| = \sqrt{(q_1 - p_1)^2 + (q_2 - p_2)^2} = \sqrt{v_1^2 + v_2^2}$$

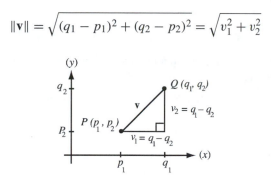

There are two basic operations that are defined for vectors: scalar multiplication and vector addition (or subtraction). Scalar multiplication is the product of a vector **v** and a scalar k where k is a real number, and we write $k\mathbf{v}$. If k is positive, then the direction of $k\mathbf{v}$ is the same as the direction of **v**. If k is negative, then the direction of $k\mathbf{v}$ is opposite the direction of **v**.

Vector addition can be defined either geometrically or using components. To add two vectors geometrically, we place the initial point of one vector (**v**) on the terminal point of a second vector (**u**), keeping the length and direction the same. The resulting vector $\mathbf{u} + \mathbf{v}$ has an initial point that is coincident with the initial point of **u** and a terminal point that is coincident with the terminal point of **v**. Notice that $\mathbf{u} + \mathbf{v}$ is the diagonal of a parallelogram. Hence, this form of vector addition is called the **parallelogram law**.

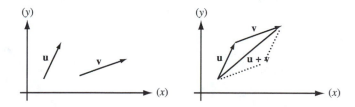

The subtraction of two vectors is defined by addition, $\mathbf{u} - \mathbf{v} = \mathbf{u} + (-\mathbf{v})$. By scalar multiplication, we know that $-\mathbf{v}$ is a vector that has a direction opposite that of **v**. Placing the initial point of $-\mathbf{v}$ on the terminal point of **u**, we have the resulting vector $\mathbf{u} + (-\mathbf{v}) = \mathbf{u} - \mathbf{v}$, where the initial point of $\mathbf{u} - \mathbf{v}$

V

coincides with the initial point of **u** and the terminal point of **u** − **v** coincides with the terminal point of −**v**. In addition of vectors, the resulting vector is called the **resultant**.

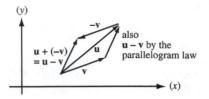

To add two vectors **u** and **v** by using components, we consider the components of both vectors.

Let u_1 and u_2 be the x-component and y-component of **u**, respectively. Further, let v_1 and v_2 be the x-component and y-component of **v**, respectively. Then the x-component of **u** + **v** is $u_1 + v_1$ and the y-component of **u** + **v** is $u_2 + v_2$.

Examples

1. Find the x-component, y-component, and length of the vector **v** with an initial point $P(2, -1)$ and a terminal point $Q(-1, 4)$.

   If we let $P(2, -1) = (p_1, p_2)$ and $Q(-1, 4) = (q_1, q_2)$, then the components of **v** are

   $$v_1 = q_1 - p_1 = -1 - 2 = -3$$
   $$v_2 = q_2 - p_2 = 4 - (-1) = 5$$

   The length of **v** is

   $$\|v\| = \sqrt{(q_1 - p_1)^2 + (q_2 - p_2)^2}$$
   $$= \sqrt{(-3)^2 + (5)^2}$$
   $$= \sqrt{9 + 25}$$
   $$= \sqrt{34}$$

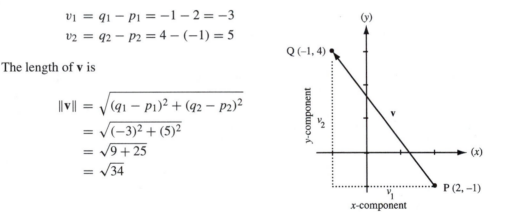

2. Write the vectors **u** and **v** in terms of the vectors **a** and **b** and then write the vectors **a** and **b** in terms of the vectors **u** and **v**.

   Since the initial point of **u** coincides with the initial point of **a** and the terminal point of **u** coincides with the terminal point of **b**, we have

   $$u = a + b$$

   Similarly, we have

   $$v = 2a + b$$

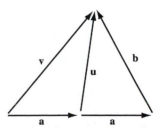

To write the vectors **a** and **b** in terms of the vectors **u** and **v** we need to solve the equations $\mathbf{u} = \mathbf{a} + \mathbf{b}$ and $\mathbf{v} = 2\mathbf{a} + \mathbf{b}$ simultaneously (see **Systems of Linear Equations**). Solving the first equation for **a** and then for **b**, and then substituting into the second equation, we have

$$\begin{aligned}
\mathbf{u} &= \mathbf{a} + \mathbf{b} & \mathbf{u} &= \mathbf{a} + \mathbf{b} \\
\mathbf{u} - \mathbf{b} &= \mathbf{a} & \mathbf{u} - \mathbf{a} &= \mathbf{b} \\
& & & \\
\mathbf{v} &= 2\mathbf{a} + \mathbf{b} & \mathbf{v} &= 2\mathbf{a} + \mathbf{b} \\
\mathbf{v} &= 2(\mathbf{u} - \mathbf{b}) + \mathbf{b} & \mathbf{v} &= 2\mathbf{a} + (\mathbf{u} - \mathbf{a}) \\
\mathbf{v} &= 2\mathbf{u} - 2\mathbf{b} + \mathbf{b} & \mathbf{v} &= \mathbf{a} + \mathbf{u} \\
\mathbf{b} &= 2\mathbf{u} - \mathbf{v} & \mathbf{a} &= -\mathbf{u} + \mathbf{v}
\end{aligned}$$

3. Determine the x-component and y-component of the vector with a magnitude of 12 and a direction of 135°.

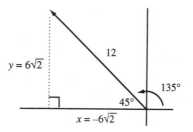

Since 135° is a second-quadrant angle, its reference angle is 45° (see **Trigonometry, reference angle**). Comparing the 45-degree triangle with a hypotenuse of 12 to the 45-degree triangle with a hypotenuse of $\sqrt{2}$, we can derive the lengths of the legs of the reference triangle, $x = -6\sqrt{2}$ and $y = 6\sqrt{2}$. See **Forty-Five Degree Triangle**. Hence, we have the x-component, $y = -6\sqrt{2}$, and the y-component, $y = 6\sqrt{2}$.

Rationalize:

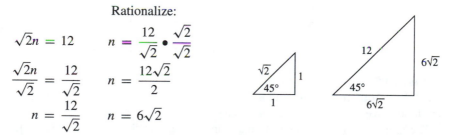

$$\sqrt{2}n = 12 \qquad n = \frac{12}{\sqrt{2}} \cdot \frac{\sqrt{2}}{\sqrt{2}}$$

$$\frac{\sqrt{2}n}{\sqrt{2}} = \frac{12}{\sqrt{2}} \qquad n = \frac{12\sqrt{2}}{2}$$

$$n = \frac{12}{\sqrt{2}} \qquad n = 6\sqrt{2}$$

There are two other ways of determining the sides of the reference triangle. One involves the Pythagorean theorem and the other involves trigonometry.

Using the Pythagorean Theorem (see **Pythagorean theorem**), we have

$$\begin{aligned}
x^2 + y^2 &= z^2 \\
x^2 + x^2 &= 12^2 \qquad y = x \text{ in a 45-degree triangle.} \\
2x^2 &= 144 \\
x^2 &= 72
\end{aligned}$$

$$\sqrt{x^2} = \sqrt{72}$$
$$x = \sqrt{36 \bullet 2}$$
$$x = 6\sqrt{2}$$

Since x is negative in the second quadrant and y is positive, we have $x = -6\sqrt{2}$ and $y = 6\sqrt{2}$.

Using trigonometry to determine the sides of the reference triangle will give us a decimal approximation. This approach can be used when the reference angle is not familiar, i.e., 37°. We don't know the sides of a 37-degree triangle.

x-component	y-component
$\cos 45° = \dfrac{x}{12}$	$\sin 45° = \dfrac{y}{12}$
$12(.7071) = x$	$12(.7071) = y$
$x = 8.4852$	$y = 8.4852$

Since x is negative in the second quadrant and y is positive, we have $x = -8.4852$ and $y = 8.4852$. See **Trigonometry, right triangle definition of**.

4. If **u** and **v** are two vectors in standard position with terminal points of $(1, 2)$ and $(3, 1)$, respectively, find the x-component and the y-component of the resultants of $\mathbf{u} + \mathbf{v}$ and $\mathbf{u} - \mathbf{v}$.

If we let $(u_1, u_2) = (1, 2)$ and $(v_1, v_2) = (3, 1)$, then the x-component of $\mathbf{u}+\mathbf{v}$ is $u_1+v_1 = 1+3 = 4$ and the y-component of $\mathbf{u} + \mathbf{v}$ is $u_2 + v_2 = 2 + 1 = 3$. Similarly, the x-component of $\mathbf{u} - \mathbf{v}$ is $u_1 - v_1 = 1 - 3 = -2$ and the y-component of $\mathbf{u} - \mathbf{v}$ is $u_2 - v_2 = 2 - 1 = -1$.

➤ **Venn Diagram**   A **Venn Diagram** is a drawing that aids in the comprehension of problems related to sets. The sets are closed curves that are contained within a greater set called the universal set or universe, denoted by U. The universe contains all the sets in the discussion as subsets.

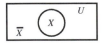

If we let $U = \{1, 2, 3, 4, 5, 6, 7\}$ and $X = \{3, 4, 5\}$, then the region of U, not lying inside X, is called the **complement** of U and is denoted by $\overline{X}$ (read X-bar), where $\overline{X}\{1, 2, 6, 7\}$. The complement is not the universal set. It is a set that contains all of the elements of the universe that are not in X.

The greatest attribute of a Venn diagram is its use of shading to represent the union, intersection, and complement of sets (see **Union and Intersection**). In the following, let $U = \{1, 2, 3, 4, 5, 6, 7\}$, $X = \{3, 4, 5\}$, $Y = \{\text{even numbers in } U\} = \{2, 4, 6\}$, and $Z = \{\text{odd numbers in } U\} = \{1, 3, 5, 7\}$.

Union: $Y \cup Z = \{\text{numbers in } Y \text{ or } Z, \text{ or both}\}$

$Y \cup Z = \{1, 2, 3, 4, 5, 6, 7\}$

Intersection: $Y \cup Z = \{$numbers in X and $Z\}$

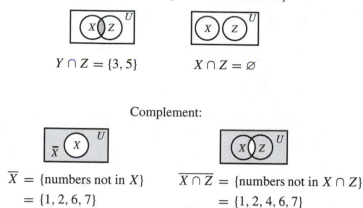

$Y \cap Z = \{3, 5\}$ $X \cap Z = \emptyset$

Complement:

$\overline{X} = \{$numbers not in $X\}$ $\overline{X \cap Z} = \{$numbers not in $X \cap Z\}$
$\quad = \{1, 2, 6, 7\}$ $\quad = \{1, 2, 4, 6, 7\}$

Examples

1. If $B \subset A$ (read B is a subset of A, see **Subset**), use a Venn diagram to demonstrate $A \cup B$, $A \cap B$, and $(A \cap B) \cup \overline{A}$.

$$A \cup B \qquad A \cap B \qquad (A \cap B) \cup \overline{A}$$

2. If A, B, and C are mutually overlapping sets, find $(A \cup B) \cup C$, $(A \cap B) \cap C$, $\overline{A \cap C}$, and $\overline{A} \cap B$.

$(A \cup B) \cup C \qquad (A \cap B) \cap C$

$\overline{A \cap C} \qquad \overline{A} \cap B$

V

➤ **Vertex**   For the vertex of a parabola, see **Parabola** or **Quadratic Equations, graphs of**. For the vertex of an ellipse, see **Ellipse**.

An angle is defined as the union of two noncolinear rays with the same endpoint. Their common endpoint is called a **vertex**. In the figure, the point B is the vertex of $\angle ABC$. In this sense, the term vertex is used to denote the vertex angle of an isosceles triangle, the vertex of a cone, or the vertices of a triangle, polyhedron, rectangular solids, and the like.

➤ **Vertical and Horizontal Lines**   See **Horizontal and Vertical Lines**.

➤ **Vertical Angle**   When two lines intersect, the nonadjacent angles formed are called **vertical angles**. In the figure, $\angle 1$ and $\angle 2$ form one pair of vertical angles; $\angle 3$ and $\angle 4$ form another.

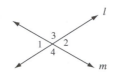

It can be shown, through a proof, that vertical angles are congruent. See **Proof in Geometry**. This follows from the fact that $\angle 1$ and $\angle 2$ are supplementary to the same angle, $\angle 3$ (or $\angle 4$). If they are supplementary to the same angle, then they are congruent to each other, $\angle 1 \cong \angle 2$. Similarly, $\angle 3 \cong \angle 4$.

➤ **Vertical Axis**   In the coordinate plane, the y-axis is called the **vertical axis**. See **Coordinate Plane**.

➤ **Vertical Line Test**   A function is a relation in which each of the elements of the domain is paired with exactly one of the elements of the range. To assure that a relation is a function, we can apply the vertical line test to its graph. If a vertical line can be drawn intersecting the graph in more than one point, then the graph is not a function. For examples involving graphs or for more on functions, see **Function**.

➤ **Volume**   **Volume** represents the amount of cubic units that can be placed in a solid. If the figure is measured in inches, then the unit of volume is measured in cubic inches. If the figure is measured in feet, then the unit of volume is measured in cubic feet, and so on.

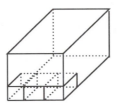

Listed below are some of the more commonly used formulas for determining volume.

V

Rectangular Solid:

$$V = lwh \qquad \text{where } l = \text{length, } w = \text{width, and } h = \text{height}$$

Cube:

$$V = s^3 \qquad \text{where } s = \text{length of a side}$$

Volume of a Right Circular Cylinder:

$$V = \pi r^2 h \qquad \text{where } r = \text{radius and } h = \text{height}$$

Volume of a Right Circular Cone:

$$V = \frac{1}{3}\pi r^2 h \qquad \text{where } r = \text{radius and } h = \text{height}$$

Volume of a Sphere:

$$V = \frac{4}{3}\pi r^3 \qquad \text{where } r = \text{radius}$$

Volume of a Right Triangular Prism:

$$V = Bh \qquad \text{where } B = \text{area of the triangular base and } h = \text{height}$$

Volume of a Pyramid:

$$V = \frac{1}{3}Bh \qquad \text{where } B = \text{area of the base and } h = \text{height}$$

Example

1. Find the volume of a rectangular solid with a length of 8 feet, width of 5 feet, and height of 3 feet.

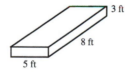

From the formula for the volume of a rectangular solid, we have

$$V = lwh = (8)(5)(3) = 120 \text{ ft}^3$$

V

➤ **Whole Numbers**   **Whole numbers** are a subset of the real numbers. They begin with 0 and include the counting numbers.

$$\text{Whole numbers} = \{0, 1, 2, 3, 4, \ldots\}$$

Whole numbers do not include negative numbers, fractions, or irrational numbers. For more, see **Real Number System**.

➤ **X-Axis**   See **Coordinate Plane**.

➤ **X-Intercept**   The **x-intercept** is a point where the graph of an equation crosses the x-axis. Its coordinates are $(x, 0)$. For an example, see **Linear Equations, graphs of**, x-y intercept method.

➤ **X-Y Intercept Method of Graphing**   See **Linear Equations, graphs of**.

➤ **Y-Axis**   See **Coordinate Plane**.

➤ **Y-Intercept**   The **y-intercept** is a point where the graph of an equation crosses the y-axis. Its coordinates are $(0, y)$. For an example, see **Linear Equations, graphs of**, x-y intercept method.

➤ **Zero as an Exponent**   See **Exponents, rules of**, special case.

➤ **Zero Exponents**   See **Exponents, rules of**.

➤ **Zero Factorial**   See **Factorial**.

➤ **Zeros of a Function or Polynomial**   The solution or solutions of an equation are called the **zeros of the function**. For an application, see **Product Theorem of Zero**.

Appendix A

Natural Logarithmic Tables

	0.00	0.01	0.02	0.03	0.04	0.05	0.06	0.07	0.08	0.09
1.0	0.0000	0.0100	0.0198	0.0296	0.0392	0.0488	0.0583	0.0677	0.0770	0.0862
1.1	0.0953	0.1044	0.1133	0.1222	0.1310	0.1398	0.1484	0.1570	0.1655	0.1740
1.2	0.1823	0.1906	0.1989	0.2070	0.2151	0.2231	0.2311	0.2390	0.2469	0.2546
1.3	0.2624	0.2700	0.2776	0.2852	0.2927	0.3001	0.3075	0.3148	0.3221	0.3293
1.4	0.3365	0.3436	0.3507	0.3577	0.3646	0.3716	0.3784	0.3853	0.3920	0.3988
1.5	0.4055	0.4121	0.4187	0.4253	0.4318	0.4383	0.4447	0.4511	0.4574	0.4637
1.6	0.4700	0.4762	0.4824	0.4886	0.4947	0.5008	0.5068	0.5128	0.5188	0.5247
1.7	0.5306	0.5365	0.5423	0.5481	0.5539	0.5596	0.5653	0.5710	0.5766	0.5822
1.8	0.5878	0.5933	0.5988	0.6043	0.6098	0.6152	0.6206	0.6259	0.6313	0.6366
1.9	0.6419	0.6471	0.6523	0.6575	0.6627	0.6678	0.6729	0.6780	0.6831	0.6881
2.0	0.6931	0.6981	0.7031	0.7080	0.7129	0.7178	0.7227	0.7275	0.7324	0.7372
2.1	0.7419	0.7457	0.7514	0.7561	0.7608	0.7655	0.7701	0.7747	0.7793	0.7839
2.2	0.7885	0.7930	0.7975	0.8020	0.8065	0.8109	0.8154	0.8198	0.8242	0.8286
2.3	0.8329	0.8372	0.8416	0.8459	0.8502	0.8544	0.8587	0.8629	0.8671	0.8713
2.4	0.8755	0.8796	0.8838	0.8879	0.8920	0.8961	0.9002	0.9042	0.9083	0.9123
2.5	0.9163	0.9203	0.9243	0.9282	0.9322	0.9361	0.9400	0.9439	0.9478	0.9517
2.6	0.9555	0.9594	0.9632	0.9670	0.9708	0.9746	0.9783	0.9821	0.9858	0.9895
2.7	0.9933	0.9969	1.0006	1.0043	1.0080	1.0116	1.0152	1.0188	1.0225	1.0260
2.8	1.0296	1.0332	1.0367	1.0403	1.0438	1.0473	1.0508	1.0543	1.0578	1.0613
2.9	1.0647	1.0682	1.0716	1.0750	1.0784	1.0818	1.0852	1.0886	1.0919	1.0953
3.0	1.0986	1.1019	1.1053	1.1086	1.1119	1.1151	1.1184	1.1217	1.1249	1.1282
3.1	1.1314	1.1346	1.1378	1.1410	1.1442	1.1474	1.1506	1.1537	1.1569	1.1600
3.2	1.1632	1.1663	1.1694	1.1725	1.1756	1.1787	1.1817	1.1848	1.1878	1.1909
3.3	1.1939	1.1969	1.2000	1.2030	1.2060	1.2090	1.2119	1.2149	1.2179	1.2208
3.4	1.2238	1.2267	1.2296	1.2326	1.2355	1.2384	1.2413	1.2442	1.2470	1.2499
3.5	1.2528	1.2556	1.2585	1.2613	1.2641	1.2669	1.2698	1.2726	1.2754	1.2782
3.6	1.2809	1.2837	1.2865	1.2892	1.2920	1.2947	1.2975	1.3002	1.3029	1.3056
3.7	1.3083	1.3110	1.3137	1.3164	1.3191	1.3218	1.3244	1.3271	1.3297	1.3324
3.8	1.3350	1.3376	1.3203	1.3429	1.3455	1.3481	1.3507	1.3533	1.3558	1.3584
3.9	1.3610	1.3635	1.3661	1.3686	1.3712	1.3737	1.3762	1.3788	1.3813	1.3838
4.0	1.3863	1.3888	1.3913	1.3938	1.3962	1.3987	1.4012	1.4036	1.4061	1.4085
4.1	1.4110	1.4134	1.4159	1.4183	1.4207	1.4231	1.4255	1.4279	1.4303	1.4327
4.2	1.4351	1.4375	1.4398	1.4422	1.4446	1.4469	1.4493	1.4516	1.4540	1.4563
4.3	1.4586	1.4609	1.4633	1.4656	1.4679	1.4702	1.4725	1.4748	1.4770	1.4793
4.4	1.4816	1.4839	1.4861	1.4884	1.4907	1.4929	1.4951	1.4974	1.4996	1.5019
4.5	1.5041	1.5063	1.5085	1.5107	1.5129	1.5151	1.5173	1.5195	1.5217	1.5239
4.6	1.5261	1.5282	1.5304	1.5326	1.5347	1.5369	1.5390	1.5412	1.5433	1.5454
4.7	1.5476	1.5497	1.5518	1.5539	1.5560	1.5581	1.5602	1.5623	1.5644	1.5665
4.8	1.5686	1.5707	1.5728	1.5748	1.5769	1.5790	1.5810	1.5831	1.5851	1.5872
4.9	1.5892	1.5913	1.5933	1.5953	1.5974	1.5994	1.6014	1.6034	1.6054	1.6074
5.0	1.6094	1.6114	1.6134	1.6154	1.6174	1.6194	1.6214	1.6233	1.6253	1.6273
5.1	1.6292	1.6312	1.6332	1.6351	1.6371	1.6390	1.6409	1.6429	1.6448	1.6467
5.2	1.6487	1.6506	1.6525	1.6544	1.6563	1.6582	1.6601	1.6620	1.6639	1.6658
5.3	1.6677	1.6696	1.6715	1.6734	1.6752	1.6771	1.6790	1.6808	1.6827	1.6845
5.4	1.6864	1.6882	1.6901	1.6919	1.6938	1.6956	1.6974	1.6993	1.7011	1.7029

Natural Logarithmic Tables (Continued)

	0.00	0.01	0.02	0.03	0.04	0.05	0.06	0.07	0.08	0.09
5.5	1.7047	1.7066	1.7084	1.7102	1.7120	1.7138	1.7156	1.7174	1.7192	1.7210
5.6	1.7228	1.7246	1.7263	1.7281	1.7299	1.7317	1.7334	1.7352	1.7370	1.7387
5.7	1.7405	1.7422	1.7440	1.7457	1.7475	1.7492	1.7509	1.7527	1.7544	1.7561
5.8	1.7579	1.7596	1.7613	1.7630	1.7647	1.7664	1.7681	1.7699	1.7716	1.7733
5.9	1.7750	1.7766	1.7783	1.7800	1.7817	1.7834	1.7851	1.7867	1.7884	1.7901
6.0	1.7918	1.7934	1.7951	1.7967	1.7984	1.8001	1.8017	1.8034	1.8050	1.8066
6.1	1.8083	1.8099	1.8116	1.8132	1.8148	1.8165	1.8181	1.8197	1.8213	1.8229
6.2	1.8245	1.8262	1.8278	1.8294	1.8310	1.8326	1.8342	1.8358	1.8374	1.8390
6.3	1.8405	1.8421	1.8437	1.8453	1.8469	1.8485	1.8500	1.8516	1.8532	1.8547
6.4	1.8563	1.8579	1.8594	1.8610	1.8625	1.8641	1.8656	1.8672	1.8687	1.8703
6.5	1.8718	1.8733	1.8749	1.8764	1.8779	1.8795	1.8810	1.8825	1.8840	1.8856
6.6	1.8871	1.8886	1.8901	1.8916	1.8931	1.8946	1.8961	1.8976	1.8991	1.9006
6.7	1.9021	1.9036	1.9051	1.9066	1.9081	1.9095	1.9110	1.9125	1.9140	1.9155
6.8	1.9169	1.9184	1.9199	1.9213	1.9228	1.9242	1.9257	1.9272	1.9286	1.9301
6.9	1.9315	1.9330	1.9344	1.9359	1.9373	1.9387	1.9402	1.9416	1.9430	1.9445
7.0	1.9459	1.9473	1.9488	1.9502	1.9516	1.9530	1.9544	1.9559	1.9573	1.9587
7.1	1.9601	1.9615	1.9629	1.9643	1.9657	1.9671	1.9685	1.9699	1.9713	1.9727
7.2	1.9741	1.9755	1.9769	1.9782	1.9796	1.9810	1.9824	1.9838	1.9851	1.9865
7.3	1.9879	1.9892	1.9906	1.9920	1.9933	1.9947	1.9961	1.9974	1.9988	2.0001
7.4	2.0015	2.0028	2.0042	2.0055	2.0069	2.0082	2.0096	2.0109	2.0122	2.0136
7.5	2.0149	2.0162	2.0176	2.0189	2.0202	2.0215	2.0229	2.0242	2.0255	2.0268
7.6	2.0281	2.0295	2.0308	2.0321	2.0334	2.0347	2.0360	2.0373	2.0386	2.0399
7.7	2.0412	2.0425	2.0438	2.0451	2.0464	2.0477	2.0490	2.0503	2.0516	2.0528
7.8	2.0541	2.0554	2.0567	2.0580	2.0592	2.0605	2.0618	2.0631	2.0643	2.0656
7.9	2.0669	2.0681	2.0694	2.0707	2.0719	2.0732	2.0744	2.0757	2.0769	2.0782
8.0	2.0794	2.0807	2.0819	2.0832	2.0844	2.0857	2.0869	2.0882	2.0894	2.0906
8.1	2.0919	2.0931	2.0943	2.0956	2.0968	2.0980	2.0992	2.1005	2.1017	2.1029
8.2	2.1041	2.1054	2.1066	2.1078	2.1090	2.1102	2.1114	2.1126	2.1138	2.1150
8.3	2.1163	2.1175	2.1187	2.1199	2.1211	2.1223	2.1235	2.1247	2.1258	2.1270
8.4	2.1282	2.1294	2.1306	2.1318	2.1330	2.1342	2.1353	2.1365	2.1377	2.1389
8.5	2.1401	2.1412	2.1424	2.1436	2.1448	2.1459	2.1471	2.1483	2.1494	2.1506
8.6	2.1518	2.1529	2.1541	2.1552	2.1564	2.1576	2.1587	2.1599	2.1610	2.1622
8.7	2.1633	2.1645	2.1656	2.1668	2.1679	2.1691	2.1702	2.1713	2.1725	2.1736
8.8	2.1748	2.1759	2.1770	2.1782	2.1793	2.1804	2.1815	2.1827	2.1838	2.1849
8.9	2.1861	2.1872	2.1883	2.1894	2.1905	2.1917	2.1928	2.1939	2.1950	2.1961
9.0	2.1972	2.1983	2.1994	2.2006	2.2017	2.2028	2.2039	2.2050	2.2061	2.2072
9.1	2.2083	2.2094	2.2105	2.2116	2.2127	2.2138	2.2148	2.2159	2.2170	2.2181
9.2	2.2192	2.2203	2.2214	2.2225	2.2235	2.2246	2.2257	2.2268	2.2279	2.2289
9.3	2.2300	2.2311	2.2322	2.2332	2.2343	2.2354	2.2364	2.2375	2.2386	2.2396
9.4	2.2407	2.2418	2.2428	2.2439	2.2450	2.2460	2.2471	2.2481	2.2492	2.2502
9.5	2.2513	2.2523	2.2534	2.2544	2.2555	2.2565	2.2576	2.2586	2.2597	2.2607
9.6	2.2618	2.2628	2.2638	2.2649	2.2659	2.2670	2.2680	2.2690	2.2701	2.2711
9.7	2.2721	2.2732	2.2742	2.2752	2.2762	2.2773	2.2783	2.2793	2.2803	2.2814
9.8	2.2824	2.2834	2.2844	2.2854	2.2865	2.2875	2.2885	2.2895	2.2905	2.2915
9.9	2.2925	2.2935	2.2946	2.2956	2.2966	2.2976	2.2986	2.2996	2.3006	2.3016

Appendix B

Common Logarithmic Tables

	0.00	0.01	0.02	0.03	0.04	0.05	0.06	0.07	0.08	0.09
1.0	0.0000	0.0043	0.0086	0.0128	0.0170	0.0212	0.0253	0.0294	0.0334	0.0374
1.1	0.0414	0.0453	0.0492	0.0531	0.0569	0.0607	0.0645	0.0682	0.0719	0.0755
1.2	0.0792	0.0828	0.0864	0.0899	0.0934	0.0969	0.1004	0.1038	0.1072	0.1106
1.3	0.1139	0.1173	0.1206	0.1239	0.1271	0.1303	0.1335	0.1367	0.1399	0.1430
1.4	0.1461	0.1492	0.1523	0.1553	0.1584	0.1614	0.1644	0.1673	0.1703	0.1732
1.5	0.1761	0.1790	0.1818	0.1847	0.1875	0.1903	0.1931	0.1959	0.1987	0.2014
1.6	0.2041	0.2068	0.2095	0.2122	0.2148	0.2175	0.2201	0.2227	0.2253	0.2279
1.7	0.2304	0.2330	0.2355	0.2380	0.2405	0.2430	0.2455	0.2480	0.2504	0.2529
1.8	0.2553	0.2577	0.2601	0.2625	0.2648	0.2672	0.2695	0.2718	0.2742	0.2765
1.9	0.2788	0.2810	0.2833	0.2856	0.2878	0.2900	0.2923	0.2945	0.2967	0.2989
2.0	0.3010	0.3032	0.3054	0.3075	0.3096	0.3118	0.3139	0.3160	0.3181	0.3201
2.1	0.3222	0.3243	0.3263	0.3284	0.3304	0.3324	0.3345	0.3365	0.3385	0.3404
2.2	0.3424	0.3444	0.3464	0.3483	0.3502	0.3522	0.3541	0.3560	0.3579	0.3598
2.3	0.3617	0.3636	0.3655	0.3674	0.3692	0.3711	0.3729	0.3747	0.3766	0.3784
2.4	0.3802	0.3820	0.3838	0.3856	0.3874	0.3892	0.3909	0.3927	0.3945	0.3962
2.5	0.3979	0.3997	0.4014	0.4031	0.4048	0.4065	0.4082	0.4099	0.4116	0.4133
2.6	0.4150	0.4166	0.4183	0.4200	0.4216	0.4232	0.4249	0.4265	0.4281	0.4298
2.7	0.4314	0.4330	0.4346	0.4362	0.4378	0.4393	0.4409	0.4425	0.4440	0.4456
2.8	0.4472	0.4487	0.4502	0.4518	0.4533	0.4548	0.4564	0.4579	0.4594	0.4609
2.9	0.4624	0.4639	0.4654	0.4669	0.4683	0.4698	0.4713	0.4728	0.4742	0.4757
3.0	0.4771	0.4786	0.4800	0.4814	0.4829	0.4843	0.4857	0.4871	0.4886	0.4900
3.1	0.4914	0.4928	0.4942	0.4955	0.4969	0.4983	0.4997	0.5011	0.5024	0.5038
3.2	0.5052	0.5065	0.5079	0.5092	0.5105	0.5119	0.5132	0.5145	0.5159	0.5172
3.3	0.5185	0.5198	0.5211	0.5224	0.5237	0.5250	0.5263	0.5276	0.5289	0.5302
3.4	0.5315	0.5328	0.5340	0.5353	0.5366	0.5378	0.5391	0.5403	0.5416	0.5428
3.5	0.5441	0.5453	0.5465	0.5478	0.5490	0.5502	0.5514	0.5527	0.5539	0.5551
3.6	0.5563	0.5575	0.5587	0.5599	0.5611	0.5623	0.5635	0.5647	0.5658	0.5670
3.7	0.5682	0.5694	0.5705	0.5717	0.5729	0.5740	0.5752	0.5763	0.5775	0.5786
3.8	0.5798	0.5809	0.5821	0.5832	0.5843	0.5855	0.5866	0.5877	0.5888	0.5899
3.9	0.5911	0.5922	0.5933	0.5944	0.5955	0.5966	0.5977	0.5988	0.5999	0.6010
4.0	0.6021	0.6031	0.6042	0.6053	0.6064	0.6075	0.6085	0.6096	0.6107	0.6117
4.1	0.6128	0.6138	0.6149	0.6160	0.6170	0.6180	0.6191	0.6201	0.6212	0.6222
4.2	0.6232	0.6243	0.6253	0.6263	0.6274	0.6284	0.6294	0.6304	0.6314	0.6325
4.3	0.6335	0.6345	0.6355	0.6365	0.6375	0.6385	0.6395	0.6405	0.6415	0.6425
4.4	0.6435	0.6444	0.6454	0.6464	0.6474	0.6484	0.6493	0.6503	0.6513	0.6522
4.5	0.6532	0.6542	0.6551	0.6561	0.6571	0.6580	0.6590	0.6599	0.6609	0.6618
4.6	0.6628	0.6637	0.6646	0.6656	0.6665	0.6675	0.6684	0.6693	0.6702	0.6712
4.7	0.6721	0.6730	0.6739	0.6749	0.6758	0.6767	0.6776	0.6785	0.6794	0.6803
4.8	0.6812	0.6821	0.6830	0.6839	0.6848	0.6857	0.6866	0.6875	0.6884	0.6893
4.9	0.6902	0.6911	0.6920	0.6928	0.6937	0.6946	0.6955	0.6964	0.6972	0.6981
5.0	0.6990	0.6998	0.7007	0.7016	0.7024	0.7033	0.7042	0.7050	0.7059	0.7067
5.1	0.7076	0.7084	0.7093	0.7101	0.7110	0.7118	0.7126	0.7135	0.7143	0.7152
5.2	0.7160	0.7168	0.7177	0.7185	0.7193	0.7202	0.7210	0.7218	0.7226	0.7235
5.3	0.7243	0.7251	0.7259	0.7267	0.7275	0.7284	0.7292	0.7300	0.7308	0.7316
5.4	0.7324	0.7332	0.7340	0.7348	0.7356	0.7364	0.7372	0.7380	0.7388	0.7396

Common Logarithmic Tables (Continued)

	0.00	0.01	0.02	0.03	0.04	0.05	0.06	0.07	0.08	0.09
5.5	0.7404	0.7412	0.7419	0.7427	0.7435	0.7443	0.7451	0.7459	0.7466	0.7474
5.6	0.7482	0.7490	0.7497	0.7505	0.7513	0.7520	0.7528	0.7536	0.7543	0.7551
5.7	0.7559	0.7566	0.7574	0.7582	0.7589	0.7597	0.7604	0.7612	0.7619	0.7627
5.8	0.7634	0.7642	0.7649	0.7657	0.7664	0.7672	0.7679	0.7686	0.7694	0.7701
5.9	0.7709	0.7716	0.7723	0.7731	0.7738	0.7745	0.7752	0.7760	0.7767	0.7774
6.0	0.7782	0.7789	0.7796	0.7803	0.7810	0.7818	0.7825	0.7832	0.7839	0.7846
6.1	0.7853	0.7860	0.7868	0.7875	0.7882	0.7889	0.7896	0.7903	0.7910	0.7917
6.2	0.7924	0.7931	0.7938	0.7945	0.7952	0.7959	0.7966	0.7973	0.7980	0.7987
6.3	0.7993	0.8000	0.8007	0.8014	0.8021	0.8028	0.8035	0.8041	0.8048	0.8055
6.4	0.8062	0.8069	0.8075	0.8082	0.8089	0.8096	0.8102	0.8109	0.8116	0.8122
6.5	0.8129	0.8136	0.8142	0.8149	0.8156	0.8162	0.8169	0.8176	0.8182	0.8189
6.6	0.8195	0.8202	0.8209	0.8215	0.8222	0.8228	0.8235	0.8241	0.8248	0.8254
6.7	0.8261	0.8267	0.8274	0.8280	0.8287	0.8293	0.8299	0.8306	0.8312	0.8319
6.8	0.8325	0.8331	0.8338	0.8344	0.8351	0.8357	0.8363	0.8370	0.8376	0.8382
6.9	0.8388	0.8395	0.8401	0.8407	0.8414	0.8420	0.8426	0.8432	0.8439	0.8445
7.0	0.8451	0.8457	0.8463	0.8470	0.8476	0.8482	0.8488	0.8494	0.8500	0.8506
7.1	0.8513	0.8519	0.8525	0.8531	0.8537	0.8543	0.8549	0.8555	0.8561	0.8567
7.2	0.8573	0.8579	0.8585	0.8591	0.8597	0.8603	0.8609	0.8615	0.8621	0.8627
7.3	0.8633	0.8639	0.8645	0.8651	0.8657	0.8663	0.8669	0.8675	0.8681	0.8686
7.4	0.8692	0.8698	0.8704	0.8710	0.8716	0.8722	0.8727	0.8733	0.8739	0.8745
7.5	0.8751	0.8756	0.8762	0.8768	0.8774	0.8779	0.8785	0.8791	0.8797	0.8802
7.6	0.8808	0.8814	0.8820	0.8825	0.8831	0.8837	0.8842	0.8848	0.8854	0.8859
7.7	0.8863	0.8871	0.8876	0.8882	0.8887	0.8893	0.8899	0.8904	0.8910	0.8915
7.8	0.8921	0.8927	0.8932	0.8938	0.8943	0.8949	0.8954	0.8960	0.8965	0.8971
7.9	0.8976	0.8982	0.8987	0.8993	0.8998	0.9004	0.9009	0.9015	0.9020	0.9025
8.0	0.9031	0.9036	0.9042	0.9047	0.9053	0.9058	0.9063	0.9069	0.9074	0.9079
8.1	0.9085	0.9090	0.9096	0.9101	0.9106	0.9112	0.9117	0.9122	0.9128	0.9133
8.2	0.9138	0.9143	0.9149	0.9154	0.9159	0.9165	0.9170	0.9175	0.9180	0.9186
8.3	0.9191	0.9196	0.9201	0.9206	0.9212	0.9217	0.9222	0.9227	0.9232	0.9238
8.4	0.9243	0.9248	0.9253	0.9258	0.9263	0.9269	0.9274	0.9279	0.9284	0.9289
8.5	0.9294	0.9299	0.9304	0.9309	0.9315	0.9320	0.9325	0.9330	0.9335	0.9340
8.6	0.9345	0.9350	0.9355	0.9360	0.9365	0.9370	0.9375	0.9380	0.9385	0.9390
8.7	0.9395	0.9400	0.9405	0.9410	0.9415	0.9420	0.9425	0.9430	0.9435	0.9440
8.8	0.9445	0.9450	0.9455	0.9460	0.9465	0.9469	0.9474	0.9479	0.9484	0.9489
8.9	0.9494	0.9499	0.9504	0.9509	0.9513	0.9518	0.9523	0.9528	0.9533	0.9538
9.0	0.9542	0.9547	0.9552	0.9557	0.9562	0.9566	0.9571	0.9576	0.9581	0.9586
9.1	0.9590	0.9595	0.9600	0.9605	0.9609	0.9614	0.9619	0.9624	0.9628	0.9633
9.2	0.9638	0.9643	0.9647	0.9652	0.9657	0.9661	0.9666	0.9671	0.9675	0.9680
9.3	0.9685	0.9689	0.9694	0.9699	0.9703	0.9708	0.9713	0.9717	0.9722	0.9727
9.4	0.9731	0.9736	0.9741	0.9745	0.9750	0.9754	0.9759	0.9764	0.9768	0.9773
9.5	0.9777	0.9782	0.9786	0.9791	0.9795	0.9800	0.9805	0.9809	0.9814	0.9818
9.6	0.9823	0.9827	0.9832	0.9836	0.9841	0.9845	0.9850	0.9854	0.9859	0.9863
9.7	0.9868	0.9872	0.9877	0.9881	0.9886	0.9890	0.9894	0.9899	0.9903	0.9908
9.8	0.9912	0.9917	0.9921	0.9926	0.9930	0.9934	0.9939	0.9943	0.9948	0.9952
9.9	0.9956	0.9961	0.9965	0.9969	0.9974	0.9978	0.9983	0.9987	0.9991	0.9996

Table of Values of the Trigonometric Functions

θ Deg.	θ Rad.	sin θ	cos θ	tan θ	cot θ	sec θ	csc θ		
0°00′	.0000	.0000	1.0000	.0000		1.000		1.5708	90°00′
10′	.0029	.0029	1.0000	.0029	343.77	1.000	343.8	1.5679	50′
20′	.0058	.0058	1.0000	.0058	171.89	1.000	171.9	1.5650	40′
30′	.0087	.0087	1.0000	.0087	114.59	1.000	114.6	1.5621	30′
40′	.0116	.0116	.9999	.0116	85.940	1.000	85.95	1.5592	20′
50′	.0145	.0245	.9999	.0145	68.750	1.000	68.76	1.5563	10′
1°00′	.0175	.0175	.9998	.0175	57.290	1.000	57.30	1.5533	89°00′
10′	.0204	.0204	.9998	.0204	49.104	1.000	49.11	1.5504	50′
20′	.0233	.0233	.9997	.0233	42.964	1.000	42.98	1.5475	40′
30′	.0262	.0262	.9997	.0262	38.188	1.000	38.20	1.5446	30′
40′	.0291	.0291	.9996	.0291	34.368	1.000	34.38	1.5417	20′
50′	.0320	.0320	.9995	.0320	31.242	1.001	31.26	1.5388	10′
2°00′	.0349	.0349	.9994	.0349	28.636	1.001	28.65	1.5359	88°00′
10′	.0378	.0378	.9993	.0378	26.432	1.001	26.45	1.5330	50′
20′	.0407	.0407	.9992	.0407	24.542	1.001	24.56	1.5301	40′
30′	.0436	.0436	.9990	.0437	22.904	1.001	22.93	1.5272	30′
40′	.0465	.0465	.9989	.0466	21.470	1.001	21.49	1.5243	20′
50′	.0495	.0494	.9988	.0495	20.206	1.001	20.23	1.5213	10′
3°00′	.0524	.0523	.9986	.0524	19.081	1.001	19.11	1.5184	87°00′
10′	.0553	.0552	.9985	.0553	18.075	1.002	18.10	1.5155	50′
20′	.0582	.0581	.9983	.0582	17.169	1.002	17.20	1.5126	40′
30′	.0611	.0610	.9981	.0612	16.350	1.002	16.38	1.5097	30′
40′	.0640	.0640	.9980	.0641	15.605	1.002	15.64	1.5068	20′
50′	.0669	.0669	.9978	.0670	14.924	1.002	14.96	1.5039	10′
4°00′	.0698	.0698	.9976	.0699	14.301	1.002	14.34	1.5010	86°00′
10′	.0727	.0727	.9974	.0729	13.727	1.003	13.76	1.4981	50′
20′	.0756	.0756	.9971	.0758	13.197	1.003	13.23	1.4952	40′
30′	.0785	.0785	.9969	.0787	12.706	1.003	12.75	1.4923	30′
40′	.0814	.0814	.9967	.0816	12.251	1.003	12.29	1.4893	20′
50′	.0844	.0843	.9964	.0846	11.826	1.004	11.87	1.4864	10′
5°00′	.0873	.0872	.9962	.0875	11.430	1.004	11.47	1.4835	85°00′
10′	.0902	.0901	.9959	.0904	11.059	1.004	11.10	1.4806	50′
20′	.0931	.0929	.9957	.0934	10.712	1.004	10.76	1.4777	40′
30′	.0960	.0958	.9954	.0963	10.385	1.005	10.43	1.4748	30′
40′	.0989	.0987	.9951	.0992	10.078	1.005	10.13	1.4719	20′
50′	.1018	.1016	.9948	.1022	9.7882	1.005	9.839	1.4690	10′
6°00′	.1047	.1045	.9945	.1051	9.5144	1.006	9.567	1.4661	84°00′
10′	.1076	.1074	.9942	.1080	9.2553	1.006	9.309	1.4632	50′
20′	.1105	.1103	.9939	.1110	9.0098	1.006	9.065	1.4603	40′
30′	.1134	.1132	.9936	.1139	8.7769	1.006	8.834	1.4573	30′
40′	.1164	.1161	.9932	.1169	8.5555	1.007	8.614	1.4544	20′
50′	.1193	.1190	.9929	.1198	8.3450	1.007	8.405	1.4515	10′
7°00′	.1222	.1219	.9925	.1228	8.1443	1.008	8.206	1.4486	83°00′
10′	.1251	.1248	.9922	.1257	7.9530	1.008	8.016	1.4457	50′
20′	.1280	.1276	.9918	.1287	7.7704	1.008	7.834	1.4428	40′
30′	.1309	.1305	.9914	.1317	7.5958	1.009	7.661	1.4399	30′
40′	.1338	.1334	.9911	.1346	7.4287	1.009	7.496	1.4370	20′
50′	.1367	.1363	.9907	.1376	7.2687	1.009	7.337	1.4341	10′
8°00′	.1396	.1392	.9903	.1405	7.1154	1.010	7.185	1.4312	82°00′
10′	.1425	.1421	.9899	.1435	6.9682	1.010	7.040	1.4283	50′
20′	.1454	.1449	.9894	.1465	6.8269	1.011	6.900	1.4254	40′
30′	.1484	.1478	.9890	.1495	6.6912	1.011	6.765	1.4224	30′
40′	.1513	.1507	.9886	.1524	6.5606	1.012	6.636	1.4195	20′
50′	.1542	.1536	.9881	.1554	6.4348	1.012	6.512	1.4166	10′
9°00′	.1571	.1564	.9877	.1584	6.3138	1.012	6.392	1.4137	81°00′
	cos θ	sin θ	cot θ	tan θ	csc θ	sec θ	θ Rad.	θ Deg.	

Table of Values of the Trigonometric Functions

θ Deg.	θ Rad.	$\sin\theta$	$\cos\theta$	$\tan\theta$	$\cot\theta$	$\sec\theta$	$\csc\theta$		
9°00′	.1571	.1564	.9877	.1584	6.3138	1.012	6.392	1.4137	81°00′
10′	.1600	.1593	.9872	.1614	6.1970	1.013	6.277	1.4108	50′
20′	.1629	.1622	.9868	.1644	6.0844	1.013	6.166	1.4079	40′
30′	.1658	.1650	.9863	.1673	5.9758	1.014	6.059	1.4050	30′
40′	.1687	.1679	.9858	.1703	5.8708	1.014	5.955	1.4021	20′
50′	.1716	.1708	.9853	.1733	5.7694	1.015	5.855	1.3992	10′
10°00′	.1745	.1736	.9848	.1763	5.6713	1.015	5.759	1.3963	80°00′
10′	.1774	.1765	.9843	.1793	5.5764	1.016	5.665	1.3934	50′
20′	.1804	.1794	.9838	.1823	5.4845	1.016	5.575	1.3904	40′
30′	.1833	.1822	.9833	.1853	5.3955	1.017	5.487	1.3875	30′
40′	.1862	.1851	.9827	.1883	5.3093	1.018	5.403	1.3846	20′
50′	.1891	.1880	.9822	.1914	5.2257	1.018	5.320	1.3817	10′
11°00′	.1920	.1908	.9816	.1944	5.1446	1.019	5.241	1.3788	79°00′
10′	.1949	.1937	.9811	.1974	3.0658	1.019	5.164	1.3759	50′
20′	.1978	.1965	.9805	.2004	4.9894	1.020.	5.089	1.3730	40′
30′	.2007	.1994	.9799	.2035	4.9152	1.020	5.016	1.3701	30′
40′	.2036	.2022	.9793	.2065	4.8430	1.021	4.945	1.3672	20′
50′	.2065	.2051	.9787	.2095	4.7729	1.022	4.876	1.3643	10′
12°00′	.2094	.2079	.9781	.2126	4.7046	1.022	4.810	1.3614	78°00′
10′	.2123	.2108	.9775	.2156	4.6382	1.023	4.745	1.3584	50′
20′	.2153	.2136	.9769	.2186	4.5736	1.024	4.682	1.3555	40′
30′	.2182	.2164	.9763	.2217	4.5107	1.024	4.620	1.3526	30′
40′	.2211	.2193	.9757	.2247	4.4494	1.025	4.560	1.3497	20′
50′	.2240	.2221	.9750	.2278	4.3897	1.026	4.502	1.3468	10′
13°00′	.2269	.2250	.9744	.2309	4.3315	1.026	4.445	1.3439	77°00′
10′	.2298	.2278	.9737	.2339	4.2747	1.027	4.390	1.3410	50′
20′	.2327	.2306	.9730	.2370	4.2193	1.028	4.336	1.3381	40′
30′	.2356	.2334	.9724	.2401	4.1653	1.028	4.284	1.3352	30′
40′	.2385	.2363	.9717	.2432	4.1126	1.029	4.232	1.3323	20′
50′	.2414	.2391	.9710	.2462	4.0611	1.030	4.182	1.3294	10′
14°00′	.2443	.2419	.9703	.2493	4.0108	1.031	4.134	1.3265	76°00′
10′	.2473	.2447	.9696	.2524	3.9617	1.031	4.086	1.3235	50′
20′	.2502	.2476	.9689	.2555	3.9136	1.032	4.039	1.3206	40′
30′	.2531	.2504	.9681	.2586	3.8667	1.033	3.994	1.3177	30′
40′	.2560	.2532	.9674	.2617	3.8208	1.034	3.950	1.3148	20′
50′	.2589	.2560	.9667	.2648	3.7760	1.034	3.906	1.3119	10′
15°00′	.2618	.2588	.9659	.2679	3.7321	1.035	3.864	1.3090	75°00′
10′	.2647	.2616	.9652	.2711	3.6891	1.036	3.822	1.3061	50′
20′	.2676	.2644	.9644	.2742	3.6470	1.037	3.782	1.3032	40′
30′	.2705	.2672	.9636	.2773	3.6059	1.038	3.742	1.3003	30′
40′	.2734	.2700	.9628	.2805	3.5656	1.039	3.703	1.2974	20′
50′	.2763	.2728	.9621	.2836	3.5261	1.039	3.665	1.2945	10′
16°00′	.2793	.2756	.9613	.2867	3.4874	1.040	3.628	1.2915	74°00′
10′	.2822	.2784	.9605	.2899	3.4495	1.041	3.592	1.2886	50′
20′	.2851	.2812	.9596	.2931	3.4124	1.042	3.556	1.2857	40′
30′	.2880	.2840	.9588	.2962	3.3759	1.043	3.521	1.2828	30′
40′	.2909	.2868	.9580	.2994	3.3402	1.044	3.487	1.2799	20′
50′	.2938	.2896	.9572	.3026	3.3052	1.045	3.453	1.2770	10′
17°00′	.2967	.2924	.9563	.3057	3.2709	1.046	3.420	1.2741	73°00′
10′	.2996	.2952	.9555	.3089	3.2371	1.047	3.388	1.2712	50′
20′	.3025	.2979	.9546	.3121	3.2041	1.048	3.356	1.2683	40′
30′	.3054	.3007	.9537	.3153	3.1716	1.049	3.326	1.2654	30′
40′	.3083	.3035	.9528	.3185	3.1397	1.049	3.295	1.2625	20′
50′	.3113	.3062	.9520	.3217	3.1084	1.050	3.265	1.2595	10′
18°00′	.3142	.3090	.9511	.3249	3.0777	1.051	3.236	1.2566	72°00′
		$\cos\theta$	$\sin\theta$	$\cot\theta$	$\tan\theta$	$\csc\theta$	$\sec\theta$	θ Rad.	θ Deg.

Table of Values of the Trigonometric Functions

θ Deg.	θ Rad.	sin θ	cos θ	tan θ	cot θ	sec θ	csc θ		
18°00′	.3142	.3090	.9511	.3249	3.0777	1.051	3.236	1.2566	72°00′
10′	.3171	.3118	.9502	.3281	3.0475	1.052	3.207	1.2537	50′
20′	.3200	.3145	.9492	.3314	3.0178	1.053	3.179	1.2508	40′
30′	.3229	.3173	.9483	.3346	2.9887	1.054	3.152	1.2479	30′
40′	.3258	.3201	.9474	.3378	2.9600	1.056	3.124	1.2450	20′
50′	.3287	.3228	.9465	.3411	2.9319	1.057	3.098	1.2421	10′
19°00′	.3316	.3256	.9455	.3443	2.9042	1.058	3.072	1.2392	71°00′
10′	.3345	.3283	.9446	.3476	2.8770	1.059	3.046	1.2363	50′
20′	.3374	.3311	.9436	.3508	2.8502	1.060	3.021	1.2334	40′
30′	.3403	.3338	.9426	.3541	2.8239	1.061	2.996	1.2305	30′
40′	.3432	.3365	.9417	.3574	2.7980	1.062	2.971	1.2275	20′
50′	.3462	.3393	.9407	.3607	2.7725	1.063	2.947	1.2246	10′
20°00′	.3491	.3420	.9397	.3640	2.7475	1.064	2.924	1.2217	70°00′
10′	.3520	.3448	.9387	.3673	2.7228	1.065	2.901	1.2188	50′
20′	.3549	.3475	.9377	.3706	2.6985	1.066	2.878	1.2159	40′
30′	.3578	.3502	.9367	.3739	2.6746	1.068	2.855	1.2130	30′
40′	.3607	.3529	.9356	.3772	2.62511	1.069	2.833	1.2101	20′
50′	.3636	.3557	.9346	.3805	2.6279	1.070	2.812	1.2072	10′
21°00′	.3665	.3584	.9336	.3839	2.6051	1.071	2.790	1.2043	69°00′
10′	.3694	.3611	.9325	.3872	2.5826	1.072	2.769	1.2014	50′
20′	.3723	.3638	.9315	.3906	2.5605	1.074	2.749	1.1985	40′
30′	.3752	.3665	.9304	.3939	2.5386	1.075	2.729	1.1956	30′
40′	.3782	.3692	.9293	.3973	2.5172	1.076	2.709	1.1926	20′
50′	.3811	.3719	.9283	.4006	2.4960	1.077	2.689	1.1897	10′
22°00′	.3840	.3746	.9272	.4040	2.4751	1.079	2.669	1.1868	68°00′
10′	.3869	.3773	.9261	.4074	2.4545	1.080	2.650	1.1839	50′
20′	.3898	.3800	.9250	.4108	2.4342	1.081	2.632	1.1810	40′
30′	.3927	.3827	.9239	.4142	2.4142	1.082	2.613	1.1781	30′
40′	.3956	.3854	.9228	.4176	2.3945	1.084	2.595	1.1752	20′
50′	.3985	.3881	.9216	.4210	2.3750	1.085	2.577	1.1723	10′
23°00′	.4014	.3907	.9205	.4245	2.3559	1.086	2.559	1.1694	67°00′
10′	.4043	.3934	.9194	.4279	2.3369	1.088	2.542	1.1665	50′
20′	.4072	.3961	.9182	.4314	2.3183	1.089	2.525	1.1636	40′
30′	.4102	.3987	.9171	.4348	2.2998	1.090	2.508	1.1606	30′
40′	.4131	.4014	.9159	.4383	2.2817	1.092	2.491	1.1577	20′
50′	.4160	.4041	.9147	.4417	2.2637	1.093	2.475	1.1548	10′
24°00′	.4189	.4067	.9135	.4452	2.2460	1.095	2.459	1.1519	66°00′
10′	.4218	.4094	.9124	.4487	2.2286	1.096	2.443	1.1490	50′
20′	.4247	.4120	.9112	.4522	2.2113	1.097	2.427	1.1461	40′
30′	.4276	.4147	.9100	.4557	2.1943	1.099	3.411	1.1432	30′
40′	.4305	.4173	.9088	.4592	2.1775	1.100	2.396	1.1403	20′
50′	.4334	.4200	.9075	.4628	2.1609	1.102	2.381	1.1374	10′
25°00′	.4363	.4226	.9063	.4663	2.1445	1.103	2.366	1.1345	65°00′
10′	.4392	.4253	.9051	.4699	2.1283	1.105	2.352	1.1316	50′
20′	.4422	.4279	.9038	.4734	2.1123	1.106	2.337	1.1286	40′
30′	.4451	.4305	.9026	.4770	2.0965	1.108	2.323	1.1257	30′
40′	.4480	.4331	.9013	.4806	2.0809	1.109	2.309	1.1228	20′
50′	.4509	.4358	.9001	.4841	2.0655	1.111	2.295	1.1199	10′
26°00′	.4538	.4384	.8988	.4877	2.0503	1.113	2.281	1.1170	64°00′
10′	.4567	.4410	.8975	.4913	2.0353	1.114	2.268	1.1141	50′
20′	.4596	.4436	.8962	.4250	2.0204	1.116	2.254	1.1112	40′
30′	.4625	.4462	.8949	.4986	2.0057	1.117	2.241	1.1083	30′
40′	.4654	.4488	.8936	.5022	1.9912	1.119	2.228	1.1054	20′
50′	.4683	.4514	.8923	.5059	1.9768	1.121	2.215	1.1025	10′
27°00′	.4712	.4540	.8910	.5095	1.9626	1.122	2.203	1.0996	63°00′
		cos θ	sin θ	cot θ	tan θ	csc θ	sec θ	θ Rad.	θ Deg.

Table of Values of the Trigonometric Functions

θ Deg.	θ Rad.	sin θ	cos θ	tan θ	cot θ	sec θ	csc θ		
27°00′	.4712	.4540	.8910	.5095	1.9626	1.122	2.203	1.0996	63°00′
10′	.4741	.4566	.8897	.5132	1.9486	1.124	2.190	1.0966	50′
20′	.4771	.4592	.8884	.5169	1.9347	1.126	2.178	1.0937	40′
30′	.4800	.4617	.8870	.5206	1.9210	1.127	2.166	1.0908	30′
40′	.4829	.4643	.8857	.5243	1.9074	1.129	2.154	1.0879	20′
50′	.4858	.4669	.8843	.5280	1.8940	1.131	2.142	1.0850	10′
28°00′	.4887	.4695	.8829	.5317	1.8807	1.133	2.130	1.0821	62°00′
10′	.4916	.4720	.8816	.5354	1.8676	1.134	2.118	1.0792	50′
20′	.4945	.4746	.8802	.5392	1.8546	1.136	2.107	1.0763	40′
30′	.4974	.4772	.8788	.5430	1.8418	1.138	2.096	1.0734	30′
40′	.5003	.4797	.8774	.5467	1.8291	1.140	2.085	1.0705	20′
50′	.5032	.4823	.8760	.5505	1.8165	1.142	2.074	1.0676	10′
29°00′	.5061	.4848	.8746	.5543	1.8040	1.143	2.063	1.0647	61°00′
10′	.5091	.4874	.8732	.5581	1.7917	1.145	2.052	1.0617	50′
20′	.5120	.4899	.8718	.5619	1.7796	1.147	2.041	1.0588	40′
30′	.5149	.4924	.8704	.5658	1.7675	1.149	2.031	1.0559	30′
40′	.5178	.4950	.8689	.5696	1.7556	1.151	2.020	1.0530	20′
50′	.5207	.4975	.8675	.5735	1.7437	1.153	2.010	1.0501	10′
30°00′	.5236	.5000	.8660	.5774	1.7321	1.155	2.000	1.0472	60°00′
10′	.5265	.5025	.8646	.5812	1.7205	1.157	1.990	1.0443	50′
20′	.5294	.5050	.8631	.5851	1.7090	1.159	1.980	1.0414	40′
30′	.5323	.5075	.8616	.5890	1.6977	1.161	1.970	1.0385	30′
40′	.5352	.5100	.8601	.5930	1.6864	1.163	1.961	1.0356	20′
50′	.5381	.5125	.8587	.5969	1.6753	1.165	1.951	1.0327	10′
31°00′	.5411	.5150	.8572	.6009	1.6643	1.167	1.942	1.0297	59°00′
10′	.5440	.5175	.8557	.6048	1.6534	1.169	1.932	1.0268	50′
20′	.5469	.5200	.8542	.6088	1.6426	1.171	1.923	1.0239	40′
30′	.5498	.5225	.8526	.6128	1.6319	1.173	1.914	1.0210	30′
40′	.5527	.5250	.8511	.6168	1.6212	1.175	1.905	1.0181	20′
50′	.5556	.5275	.8496	.6208	1.6107	1.177	1.896	1.0152	10′
32°00′	.5585	.5299	.8480	.6249	1.6003	1.179	1.887	1.0123	58°00′
10′	.5614	.5324	.8465	.6289	1.5900	1.181	1.878	1.0094	50′
20′	.5643	.5348	.8450	.6330	1.5798	1.184	1.870	1.0065	40′
30′	.5672	.5373	.8434	.6371	1.5697	1.186	1.861	1.0036	30′
40′	.5701	.5398	.8418	.6412	1.5597	1.188	1.853	1.0007	20′
50′	.5730	.5422	.8403	.6453	1.5497	1.190	1.844	.9977	10′
33°00′	.5760	.5446	.8387	.6494	1.5399	1.192	1.836	.9948	57°00′
10′	.5789	.5471	.8371	.6536	1.5301	1.195	1.828	.9919	50′
20′	.5818	.5495	.8355	.6577	1.5204	1.197	1.820	.9890	40′
30′	.5847	.5519	.8339	.6619	1.5108	1.199	1.812	.9861	30′
40′	.5876	.5544	.8323	.6661	1.5013	1.202	1.804	.9832	20′
50′	.5905	.5568	.8307	.6703	1.4919	1.204	1.796	.9803	10′
34°00′	.5934	.5592	.8290	.6745	1.4826	1.206	1.788	.9774	56°00′
10′	.5963	.5616	.8274	.6787	1.4733	1.209	1.781	.9745	50′
20′	.5992	.5640	.8258	.6830	1.4641	1.211	1.773	.9716	40′
30′	.6021	.5664	.8241	.6873	1.4550	1.213	1.766	.9687	30′
40′	.6050	.5688	.8225	.6916	1.4460	1.216	1.758	.9657	20′
50′	.6080	.5712	.8208	.6959	1.4370	1.218	1.751	.9628	10′
35°00′	.6109	.5736	.8192	.7002	1.4281	1.221	1.743	.9599	55°00′
10′	.6138	.5760	.8175	.7046	1.4193	1.223	1.736	.9570	50′
20′	.6167	.5783	.8158	.7089	1.4106	1.226	1.729	.9541	40′
30′	.6196	.5807	.8141	.7133	1.4019	1.228	1.722	.9512	30′
40′	.6225	.5831	.8124	.7177	1.3934	1.231	1.715	.9483	20′
50′	.6254	.5854	.8107	.7221	1.3848	1.233	1.708	.9454	10′
36°00′	.6283	.5878	.8090	.7265	1.3764	1.236	1.701	.9425	54°00′
		cos θ	sin θ	cot θ	tan θ	csc θ	sec θ	θ Rad.	θ Deg.

Table of Values of the Trigonometric Functions

θ Deg.	θ Rad.	sin θ	cos θ	tan θ	cot θ	sec θ	csc θ		
36°00′	.6283	.5878	.8090	.7265	1.3764	1.236	1.701	.9425	54°00′
10′	.6312	.5901	.8073	.7310	1.3680	1.239	1.695	.9396	50′
20′	.6341	.5925	.8056	.7355	1.3597	1.241	1.688	.9367	40′
30′	.6370	.5948	.8039	.7400	1.3514	1.244	1.681	.9338	30′
40′	.6400	.5972	.8021	.7445	1.3432	1.247	1.675	.9308	20′
50′	.6429	.5995	.8004	.7490	1.3351	1.249	1.668	.9279	10′
37°00′	.6458	.6018	.7986	.7536	1.3270	1.252	1.662	.9250	53°00′
10′	.6487	.6041	.7969	.7581	1.3190	1.255	1.655	.9221	50′
20′	.6516	.6065	.7951	.7627	1.3111	1.258	1.649	.9192	40′
30′	.6545	.6088	.7934	.7673	1.3032	1.260	1.643	.9163	30′
40′	.6574	.6111	.7916	.7720	1.2954	1.263	1.636	.9134	20′
50′	.6603	.6134	.7898	.7766	1.2876	1.266	1.630	.9105	10′
38°00′	.6632	.6157	.7880	.7813	1.2799	1.269	1.624	.9076	52°00′
10′	.6661	.6180	.7862	.7860	1.2723	1.272	1.618	.9047	50′
20′	.6690	.6202	.7844	.7907	1.2647	1.275	1.612	.9018	40′
30′	.6720	.6225	.7826	.7954	1.2572	1.278	1.606	.8988	30′
40′	.6749	.6248	.7808	.8002	1.2497	1.281	1.601	.8959	20′
50′	.6778	.6271	.7790	.8050	1.2423	1.284	1.595	.8930	10′
39°00′	.6807	.6293	.7771	.8089	1.2349	1.287	1.589	.8901	51°00′
10′	.6836	.6316	.7753	.8146	1.2276	1.290	1.583	.8872	50′
20′	.6865	.6338	.7735	.8195	1.2203	1.293	1.578	.8843	40′
30′	.6894	.6361	.7716	.8243	1.2131	1.296	1.572	.8814	30′
40′	.6923	.6383	.7698	.8292	1.2059	1.299	1.567	.8785	20′
50′	.6952	.6406	.7679	.8342	1.1988	1.302	1.561	.8756	10′
40°00′	.6981	.6428	.7660	.8391	1.1918	1.305	1.556	.8727	50°00′
10′	.7010	.6450	.7642	.8441	1.1847	1.309	1.550	.8698	50′
20′	.7039	.6472	.7623	.8491	1.1778	1.312	1.545	.8668	40′
30′	.7069	.6494	.7604	.8541	1.1708	1.315	1.540	.8639	30′
40′	.7098	.6517	.7585	.8591	1.1640	1.318	1.535	.8610	20′
50′	.7127	.6539	.7566	.8642	1.1571	1.322	1.529	.8581	10′
41°00′	.7156	.6561	.7547	.8693	1.1504	1.325	1.524	.8552	49°00′
10′	.7185	.6583	.7528	.8744	1.1436	1.328	1.519	.8523	50′
20′	.7214	.6604	.7509	.8796	1.1369	1.332	1.514	.8494	40′
30′	.7243	.6626	.7490	.8847	1.1303	1.335	1.509	.8465	30′
40′	.7272	.6648	.7470	.8899	1.1237	1.339	1.504	.8436	20′
50′	.7301	.6670	.7451	.8952	1.1171	1.342	1.499	.8407	10′
42°00′	.7330	.6691	.7431	.9004	1.1106	1.346	1.494	.8378	48°00′
10′	.7359	.6713	.7412	.9057	1.1041	1.349	1.490	.8348	50′
20′	.7389	.6734	.7392	.9110	1.0977	1.353	1.485	.8319	40′
30′	.7418	.6756	.7373	.9163	1.0913	1.356	1.480	.8290	30′
40′	.7447	.6777	.7353	.9217	1.0850	1.360	1.476	.8261	20′
50′	.7476	.6799	.7333	.9271	1.0786	1.364	1.471	.8232	10′
43°00′	.7505	.6820	.7314	.9325	1.0724	1.367	1.466	.8203	47°00′
10′	.7534	.6841	.7294	.9380	1.0661	1.371	1.462	.8174	50′
20′	.7563	.6862	.7274	.9435	1.0599	1.375	1.457	.8145	40′
30′	.7592	.6884	.7254	.9490	1.0538	1.379	1.453	.8116	30′
40′	.7621	.6905	.7234	.9545	1.0477	1.382	1.448	.8087	20′
50′	.7650	.6926	.7214	.9601	1.0416	1.386	1.444	.8058	10′
44°00′	.7679	.6947	.7193	.9657	1.0355	1.390	1.440	.8029	46°00′
10′	.7709	.6967	.7173	.9713	1.0295	1.394	1.435	.7999	50′
20′	.7738	.6988	.7153	.9770	1.0235	1.398	1.431	.7970	40′
30′	.7767	.7009	.7133	.9827	1.0176	1.402	1.427	.7941	30′
40′	.7796	.7030	.7112	.9884	1.0117	1.406	1.423	.7912	20′
50′	.7825	.7050	.7092	.9942	1.0058	1.410	1.418	.7883	10′
45°00′	.7854	.7071	.7071	1.0000	1.0000	1.414	1.414	.7854	45°00′
		cos θ	sin θ	cot θ	tan θ	csc θ	sec θ	θ Rad.	θ Deg.

Appendix D

Table of Conversions

To convert from one unit to another, divide both sides of the conversion equation by the conversion factor (the number other than one). For example, to find 1 m = ? ft, divide both sides of the equation 1 ft = .3048 m by the conversion factor .3048,

$$1 \text{ ft} = .3048 \text{ m}$$
$$\frac{1 \text{ ft}}{.3048} = \frac{.3048 \text{ m}}{.3048}$$
$$3.28 \text{ ft} = 1 \text{m}$$

Looking through the table for length, you can see that 1 m = 3.28 ft.

To change from a given number of units of one denomination, to another denomination we do the following:

1. To change from a smaller unit to a larger unit, we divide by the conversion factor.

2. To change from a larger unit to a smaller unit, we multiply by the conversion factor.

Example 1:

48 in = ? ft [smaller (in) to larger (ft)]
Since 12 in = 1 ft, we divide by 12.
48 in = 48 ÷ 12 = 4 ft

Length:

1 in = 25.4 mm = 2.54 cm = .0254 m
1 ft = 12 in = .3048 m
1 yd = 3 ft = .914 m
1 mi = 5280 ft = 1760 yd =1.609 km
1 cm = 10 mm = .3937 in
1 m = 100 cm = 39.37 in = 3.28 ft = 1.09 yd
1 km = 1000 m = 3,281 ft = .621 mi

Volume:

$1 \text{ cm}^3 = .061 \text{ in}^3$
$1 \text{ m}^3 = 35.3145 \text{ ft}^3 = 1.308 \text{ yd}^3$
$1 \text{ km}^3 = .239 \text{ mi}^3$
$1 \text{ L}^3 = .0353 \text{ ft}^3 = 1000 \text{ cm}^3$
$1 \text{ in}^3 = 16.3872 \text{ cm}^3$
$1 \text{ ft}^3 = .0283 \text{ m}^3 = 28.316 \text{ L}^3$
$1 \text{ yd}^3 = .7646 \text{ m}^3$
$1 \text{ mi}^3 = 4.165 \text{ km}^3$

Example 2:

4 ft = ? in [larger (ft) to smaller (in)]
Since 12 in = 1 ft, we multiply by 12.
4 ft = 4 × 12 = 48 in

Area:

$1 \text{ cm}^2 = .155 \text{ in}^2$
$1 \text{ m}^2 = 10.7639 \text{ ft}^2 = 1.196 \text{ yd}^2$
$1 \text{ km}^2 = .386 \text{ mi}^2$
$1 \text{ in}^2 = 6.4516 \text{ cm}^2$
$1 \text{ ft}^2 = .0929 \text{ m}^2$
$1 \text{ yd}^2 = .8361 \text{ m}^2$
$1 \text{ mi}^2 = 2.589 \text{ km}^2$
$1 \text{ acre} = 4{,}047 \text{ m}^2 = 43{,}562 \text{ ft}^2$

Liquid Measure:

1 pt = 16 oz
1 qt = 2 pt = .9463 L
1 gal = 4 qt
1 L = 1.0567 qt

Dry Measure:

1 qt = 2 pt = 1.1012 L
1 ipk = 8 qt
1 bu = 4 pk
1 L = .9081 qt

Measure of Weight:

1 oz = 28.3495 g
1 lb = 16 oz = 453.6 g
1 short ton = 2,000 lb
1 long ton = 2,240 lb
1 g = .0353 oz
1 kg = 2.20 lb

Capacity:

1 g = 1000 mg
1 kg = 1000 g

Index

About the Author

Chris Kornegay has taught high school for 25 years at Albuquerque High School in Albuquerque, New Mexico. During that time, he has been named Teacher of the Year by students and teachers and honored by the Chamber of Commerce and other organizations in Albuquerque. He is the author of *Strong's Math Dictionary & Solution Guide*. He also served with distinction in the Vietnam War and is the recipient of two Purple Hearts as well as other medals.